Synthetic Biology

Tools for Engineering Biological Systems

A subject collection from *Cold Spring Harbor Perspectives in Biology*

OTHER SUBJECT COLLECTIONS FROM *COLD SPRING HARBOR PERSPECTIVES IN BIOLOGY*

Cilia
Microbial Evolution
Learning and Memory
DNA Recombination
Neurogenesis
Size Control in Biology: From Organelles to Organisms
Mitosis
Glia
Innate Immunity and Inflammation
The Genetics and Biology of Sexual Conflict
The Origin and Evolution of Eukaryotes
Endocytosis
Mitochondria
Signaling by Receptor Tyrosine Kinases

SUBJECT COLLECTIONS FROM *COLD SPRING HARBOR PERSPECTIVES IN MEDICINE*

Tissue Engineering and Regenerative Medicine
Chromatin Deregulation in Cancer
Malaria: Biology in the Era of Eradication
Antibiotics and Antibiotic Resistance
The p53 Protein: From Cell Regulation to Cancer
Aging: The Longevity Dividend
Epilepsy: The Biology of a Spectrum Disorder
Molecular Approaches to Reproductive and Newborn Medicine
The Hepatitis B and Delta Viruses
Intellectual Property in Molecular Medicine
Retinal Disorders: Genetic Approaches to Diagnosis and Treatment
The Biology of Heart Disease
Human Fungal Pathogens
Tuberculosis
The Skin and Its Diseases
MYC *and the Pathway to Cancer*
Bacterial Pathogenesis
Transplantation
Cystic Fibrosis: A Trilogy of Biochemistry, Physiology, and Therapy

Synthetic Biology

Tools for Engineering Biological Systems

A subject collection from *Cold Spring Harbor Perspectives in Biology*

EDITED BY

Daniel G. Gibson
J. Craig Venter Institute

Clyde A. Hutchison III
J. Craig Venter Institute

Hamilton O. Smith
J. Craig Venter Institute

J. Craig Venter
J. Craig Venter Institute

COLD SPRING HARBOR LABORATORY PRESS
Cold Spring Harbor, New York • www.cshlpress.org

Synthetic Biology: Tools for Engineering Biological Systems
A subject collection from *Cold Spring Harbor Perspectives in Biology*
Articles online at www.cshperspectives.org

Executive Editor	Richard Sever
Managing Editor	Maria Smit
Senior Project Manager	Barbara Acosta
Permissions Administrator	Carol Brown
Production Editor	Diane Schubach
Cover Design	Denise Weiss
Publisher	John Inglis

Front cover artwork: Synthetic *Mycoplasma* bacteria. Colored scanning electron micrograph of *M. mycoides* JCVI-syn1.0 cells. This was the first self-replicating bacterium controlled by a synthetic genome. It was developed by scientists at the J. Craig Venter Institute and announced in May 2010. To produce the cells, a genome based on the *M. mycoides* genome was synthesized and inserted into a *M. capricolum* shell that had had its genetic material removed. Magnification: 38,000× when printed at 10 cm across. (Image credit: Thomas Deerinck and Mark Ellisman, National Center for Microscopy and Imaging Research [NCMIR], and John Glass and J. Craig Venter, J. Craig Venter Institute.)

Library of Congress Cataloging-in-Publication Data

Names: Gibson, Daniel G., editor. | Hutchison, Clyde A., III, editor. | Smith, Hamilton O. (Hamilton Othanel), 1931- editor. | Venter, J. Craig, editor.
Title: Synthetic biology : tools for engineering biological systems / edited by Daniel G. Gibson, Clyde A. Hutchison III, Hamilton O. Smith, J. Craig Venter, J. Craig Venter Institute.
Description: Cold Spring Harbor, New York : Cold Spring Harbor LaboratoryPress, [2016] | Series: A subject collection from Cold Spring Harbor perspectives in biology | Includes bibliographical references and index.
Identifiers: LCCN 2016026977 | ISBN 9781621821182 (hardcover : alk. paper)

Subjects: LCSH: Synthetic biology--Industrial applications. | Genetic engineering--Technique. | Transgenic organisms.
Classification: LCC TP248.6 .S98 2016 | DDC 660.6/5--dc23
LC record available at https://lccn.loc.gov/2016026977

All World Wide Web addresses are accurate to the best of our knowledge at the time of printing.

For a complete catalog of all Cold Spring Harbor Laboratory Press publications, visit our website at www.cshlpress.org.

Contents

APPLICATIONS TO HIGHER ORGANISMS

Preface

THIS VOLUME PRESENTS A WIDE RANGE of perspectives on synthetic biology, presented by leaders in the field. But what is "synthetic biology"? This is a rapidly evolving field of research, whose definition is addressed by the authors of several chapters in this book. Terminology is important and useful, but the name of a new field is largely a linguistic and philosophical issue, not a biological one. The name of a new area of research is not a driver of innovation but rather a retrospective attempt to describe the emergence of a new research community. Nonetheless, the editors cannot resist the opportunity to present a perspective on "synthetic biology."

With the advent of methods for the chemical synthesis of DNA, synthetic DNA gradually became a more powerful part of the toolkit of molecular biology. As DNA synthesis methods improved, it became possible to synthesize whole genomes of small viruses, and eventually complete bacterial genomes, as well as some eukaryotic chromosomes. Presumably, the complete synthesis of a whole eukaryotic genome is now not far off. This line of work, often called "synthetic genomics," may be simply defined and includes the design, synthesis, and testing of complete genomes and chromosomes. "Synthetic biology" includes "synthetic genomics" but is not restricted to completely synthetic genetic elements. In practice it has become a more up-to-date term for "genetic engineering." In current usage "synthetic biology" involves the design and assembly of biological systems for either practical purposes or basic research. Synthetic DNA, recently developed genome-editing methods, computational design tools, and other recent innovations are hallmarks of the field. An engineering perspective is often quite explicit, complete with design–build–test cycles and, all too frequently, milestones.

Another, more pragmatic, definition of synthetic biology is "those things presented at synthetic biology conferences." This volume includes a diverse set of 15 such chapters that we grouped into five topics as follows.

1. *Design, synthesis, and manipulation of genomes:* Three chapters deal with chemical synthesis of DNA, minimal cellular genomes, and enzymatic methods for genome editing.

2. *Modifying the central dogma:* Two chapters discuss synthetic alternatives to the conventional Watson–Crick DNA structure and altering the genetic code to expand the repertoire of amino acids encoded.

3. *Experimental and computational approaches to the design and improvement of biological processes:* Three chapters cover the use of cell-free systems to develop genetic circuits, the engineering of modified cellular organization, and automation of the design process in synthetic biology.

4. *Applications to small molecule production:* Two chapters describe integrated approaches to metabolic engineering and engineering microbial systems for the production of natural products.

5. *Applications to higher organisms:* This is the longest section of the book. It consists of five chapters, covering engineering gene circuits for mammalian cells; applying synthetic biology methods to engineer organs for transplantation; engineering the morphogenesis of tissues and organs; the use of synthetic biology to engineer plants; and the use of synthetic biology methods to engineer bacteriophage for fighting bacterial infections in humans.

As you can see, this book covers a broad range of topics. These include method development, the application of synthetic biology in basic biological research, and also practical applications. These applications include the development of microbial factories for production of high-value molecular products. Medical applications based on the engineering of cells and tissues and of microbes that interact with humans are also discussed.

Synthetic biology is still a young field and will likely be a hot topic for some years. But, because of its great power, it may eventually be destined to permeate all areas of biology and to therefore disappear as a distinct discipline.

DANIEL G. GIBSON
CLYDE A. HUTCHISON III
HAMILTON O. SMITH
J. CRAIG VENTER

Synthetic DNA Synthesis and Assembly: Putting the Synthetic in Synthetic Biology

Randall A. Hughes[1] and Andrew D. Ellington[2]

[1]Applied Research Laboratories, The University of Texas at Austin, Austin, Texas 78758
[2]Department of Molecular Biosciences, The University of Texas at Austin, Austin, Texas 78712
Corresponding authors: randall.a.hughes@gmail.com

The chemical synthesis of DNA oligonucleotides and their assembly into synthons, genes, circuits, and even entire genomes by gene synthesis methods has become an enabling technology for modern molecular biology and enables the design, build, test, learn, and repeat cycle underpinning innovations in synthetic biology. In this perspective, we briefly review the techniques and technologies that enable the synthesis of DNA oligonucleotides and their assembly into larger DNA constructs with a focus on recent advancements that have sought to reduce synthesis cost and increase sequence fidelity. The development of lower-cost methods to produce high-quality synthetic DNA will allow for the exploration of larger biological hypotheses by lowering the cost of use and help to close the DNA read–write cost gap.

DNA neither cares nor knows. DNA just is. And we dance to its music.

—Richard Dawkins
River Out of Eden: A Darwinian Life

It has been said that the 20th century was the "century of the atom" in which discoveries on the physical and chemical properties of the elements led to breakthroughs such as atomic energy (and weaponry), medical diagnostics, computers, and the microchip to name just a few. These advances had a dramatic effect on our way of life and helped shape the promise and possibilities of science and technology. In the early part of the 21st century, we are witnessing what could very likely become known as the "century of DNA." As the score to life's intricate symphony, DNA provides the blueprint for biological function. Advances over the last few decades in both reading (sequencing) and writing (synthesis) DNA sequences have made marked changes in our ability to understand and engineer biological systems. These advancements have led to the development of groundbreaking technologies for the design, assembly, and manipulation of DNA encoded genes, materials, circuits, and metabolic pathways, which are allowing for an ever greater manipulation of biological systems and even entire organisms.

Thanks to next-generation sequencing (NGS) technologies capable of generating an estimated 15 petabases of sequence data per year worldwide (Schatz and Phillippy 2012), the current megagenomics era has led to the swelling of biological sequence repositories with DNA sequences isolated from every organism and environment imaginable. Associated

Cite this article as *Cold Spring Harb Perspect Biol* doi: 10.1101/cshperspect.a023812

improvements in bioinformatics techniques and software allow researchers to obtain, analyze, and manipulate these DNA sequences in ways easier than ever. The ever-increasing availability of biological sequence information from all branches of the "tree of life" has deepened our understanding of biological systems and the interrelated nature of organisms at the genetic level. NGS technologies have led to a deeper understanding of human diseases, the microbiome, and the genetic diversity of organisms in our environment. This sequence boom is also allowing for the expansion of scientific disciplines such as metabolic engineering and synthetic biology as researchers seek to use novel sequences in the manipulation of biological systems for anthropocentric means. In addition, this wealth of information is leading to the development of a variety of diagnostics and therapeutics, which will contribute to the long-term improvement of human health.

This biological sequence data bonanza has been aided by a stream of innovations in instrumentation and techniques for generating sequencing data with high fidelity, increased throughput, and decreased cost, which has contributed to making NGS a go-to technology for many applications in biology. The ability to generate large DNA sequence data sets has allowed researchers to create and test large scoped biological hypotheses using NGS technologies that were not possible before their development. Beyond analyzing biological systems at the DNA sequence level, the need to construct DNA from designed sequences has driven the parallel development of methods and instrumentation to produce synthetic DNA at scale to enable the testing of engineered biological components. DNA reading by NGS when combined with modern DNA synthesis technologies form the two foundational technologies driving synthetic biology efforts and will eventually instill the predictability and reliability to engineered biological systems that chemical engineering has brought to chemical systems. This being said, our ability to sequence DNA is currently better than our ability to synthesize DNA de novo, although new technologies are helping to close the gap.

Synthetic biology is emerging as an important discipline with the potential to impact a number of academic and industrial applications including the creation of novel therapeutics, materials, biosensing, and manufacturing capabilities. Although our current understanding of biological systems is vast, it is still far from complete. Adapting exogenous DNA sequences and the functional components they encode that have naturally evolved within a network of interrelated sequences remains a principle challenge of engineering biology. Currently, the engineering of biological systems requires a heavy dose of empirical trial and error to evaluate novel enzymes, expression systems, and pathways for the desired function. Typically, this process is accomplished by first designing a desired synthetic biological circuit or pathway using computer-aided design tools (Fig. 1). Next, the resulting construct DNA is divided into smaller overlapping pieces (typically 200–1500-bp segments) that are easier to synthesize. These DNA components are then synthesized from a set of overlapping single-stranded oligonucleotides in-house (or by commercial vendors). The resulting overlapping synthons are then assembled into larger pieces of DNA and cloned into an expression vector and the sequence of the resulting construct is then verified. The sequence-verified constructs are then transformed into a cell and assayed for function. Depending on the results, changes to the construct design can be made and further iterations of the test cycle repeated. This design, build, test, learn, and repeat process has become the backbone of synthetic biology (Fig. 1), which in turn has put a premium on automated processes and methods that can shorten the development cycle and increase throughput.

One of the attributes that makes biological systems attractive from an engineering perspective is the fact that biological functions are encoded to a large part in DNA. Therefore, a gross simplification of biological engineering can be reduced to the design, production, and testing of DNA sequences. As researchers seek to engineer biological systems with novel DNA sequences, the need for custom synthetic DNA sequences has grown. This is particularly true

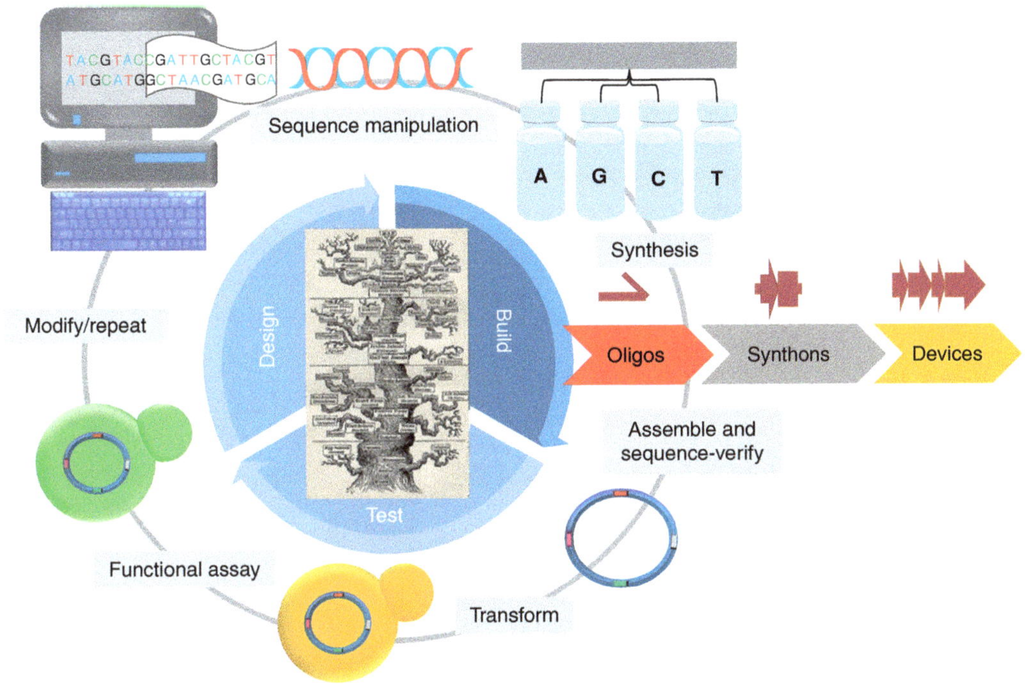

Figure 1. The synthetic biology test cycle. (From *top*, clockwise) Synthetic DNA constructs are designed and manipulated using computer-aided design software. The designed DNA is then divided into synthesizable pieces (synthons) up to 1–1.5 kbp. The synthons are then broken up into overlapping single-stranded oligonucleotide sequences and chemically synthesized. The oligonucleotides are then assembled together into the designed synthons using gene synthesis techniques. If necessary, multiple synthons can be assembled together into larger DNA assemblies or devices. The assembled DNAs are then typically cloned into an expression vector and sequence-verified. Once verified, the synthetic constructs are transformed into a cell and the function of the synthetic construct is assayed. Depending on the results the constructs can then be modified or refined and the test cycle is repeated until a DNA construct is obtained that produces the desired function.

when the sequences to be engineered are derived from metagenomic sequences in which no organism may be available from which to isolate the DNA via other methods. The synthesis of synthetic DNA is often referred to generically as "gene synthesis," which specifically is the synthesis of gene-length pieces of DNA (250–2000 bp) directly from single-stranded synthetic DNA oligonucleotides. Unfortunately, whereas the cost of sequencing has decreased precipitously over time, the cost of gene synthesis and oligonucleotide synthesis in general has not kept pace, although technological innovation and market forces are progressively lowering the cost of synthetic DNA. The cost of gene synthesis is typically directly tied to the cost of oligonucleotide synthesis from which the genes are made. The cost of oligonucleotide synthesis has not decreased appreciably in more than a decade, generally ranging from \$0.05 to \$0.17 per base depending on the synthesis scale, the length of the oligonucleotide, and the supplier (Kosuri and Church 2014). Traditionally, this cost floor has been carried through to the production of gene synthesis products. These synthetic genes currently range in cost from \$0.10 to \$0.30 per bp (\$100–\$300 for a 1-kb gene) from traditional commercial suppliers, although companies exploiting newer lower-cost synthesis methods are starting to bring the cost down.

Lowering the cost of gene synthesis would enable the generation of larger data sets by making the cost of gene construction less expensive, meaning that more constructs could be generated for the same cost investment. This would allow researchers the ability to sample a greater amount of the design landscape. With a greater understanding of what does and does not work in a given construct design, a set of general design rules could be created that would improve the success of the design process and eventually shorten the design–build–test–learn process. Because the major cost of gene synthesis is the reagents that are needed for oligonucleotide synthesis, approaches that reduce reagent consumption, improve the robustness and accuracy of the gene assembly process, and enable increased throughput have become valuable tools for advancing the usage of synthetic DNA by allowing for lower cost of use. In this perspective, a brief overview of the methods and technologies that have contributed to the production of synthetic DNA via gene synthesis methods will be highlighted as well as some of the challenges that have had to be overcome to reduce the cost, increase the throughput, and ensure the fidelity of synthetic DNA.

OLIGONUCLEOTIDE SYNTHESIS

Modern de novo gene synthesis techniques can trace their beginnings to the mid-1960s and Gobind Khorana and colleagues' efforts to decipher the genetic code and to reconstitute biological function synthetically (Scheuerbrandt et al. 1963; Khorana et al. 1968). These early efforts focused on chemically synthesizing and ligating together 17 oligonucleotides, which encoded the gene for a 77-nucleotide alanine tRNA using synthetic chemistries still in their infancy in an effort that took more than 5 years to complete (Agarwal et al. 1970; Khorana et al. 1972). Within 7 years of this work, the first reported synthetic gene encoding the 14-amino-acid hormone, somatostatin, was expressed in recombinant *Escherichia coli* (Itakura et al. 1977). Subsequent improvements to oligonucleotide synthesis chemistries and techniques in the early 1980s (reviewed in Caruthers 2011; Roy and Caruthers 2013) led to the development of solid-phase phosphoramidite chemistries whose robustness and fidelity allowed for automated methods to be developed to enable scalable oligonucleotide synthesis that is used to this day for the commercial synthesis of oligonucleotides. Synthetic oligonucleotides for gene synthesis applications are generally synthesized using variations of the phosphoramidite chemistry methods either on traditional column-based synthesizers or microarray-based synthesizers that will be discussed in the following sections.

COLUMN-BASED OLIGONUCLEOTIDE SYNTHESIS

The commonly used phosphoramidite synthesis chemistry consists of a four-step chain elongation cycle that adds one base per cycle onto a growing oligonucleotide chain attached to a solid support matrix (Fig. 2). In the first step, a dimethoxytrityl (DMT)-protected nucleoside phosphoramidite that is attached to a solid support (usually contained within a synthesis column) is deprotected by the addition of trichloroacetic acid. This activates the support-attached phosphoramidite for chain elongation with the next phosphoramidite monomer. In the second step, the next base in the sequence is added in the form of a DMT-protected phosphoramidite and is coupled to the 5′-hydroxyl group of the previous nucleoside phosphoramidite in the sequence forming a phosphite triester. Third, any unreacted 5′-hydroxyl groups are capped by acylation to render any unextended sequences inert in subsequent rounds of the chain elongation cycle and thus reducing deletion errors in the finished oligonucleotide sequences. In the fourth step, the phosphite triester linkage between the monomers is converted to a phosphate linkage via oxidation with an iodine solution to produce a cyanoethyl-protected phosphate backbone. The synthesis cycle then repeats for the next base in the sequence via the removal of the 5′-terminal DMT protecting group. After the desired sequence has been synthesized from the 3′ to 5′, the oligonucleotide is chemically cleaved from the solid synthesis support and the protecting groups on the bases and the backbone are removed. This process is highly amenable to automation and forms the basis for oligonucleotide synthesizers, which can synthesize 96–

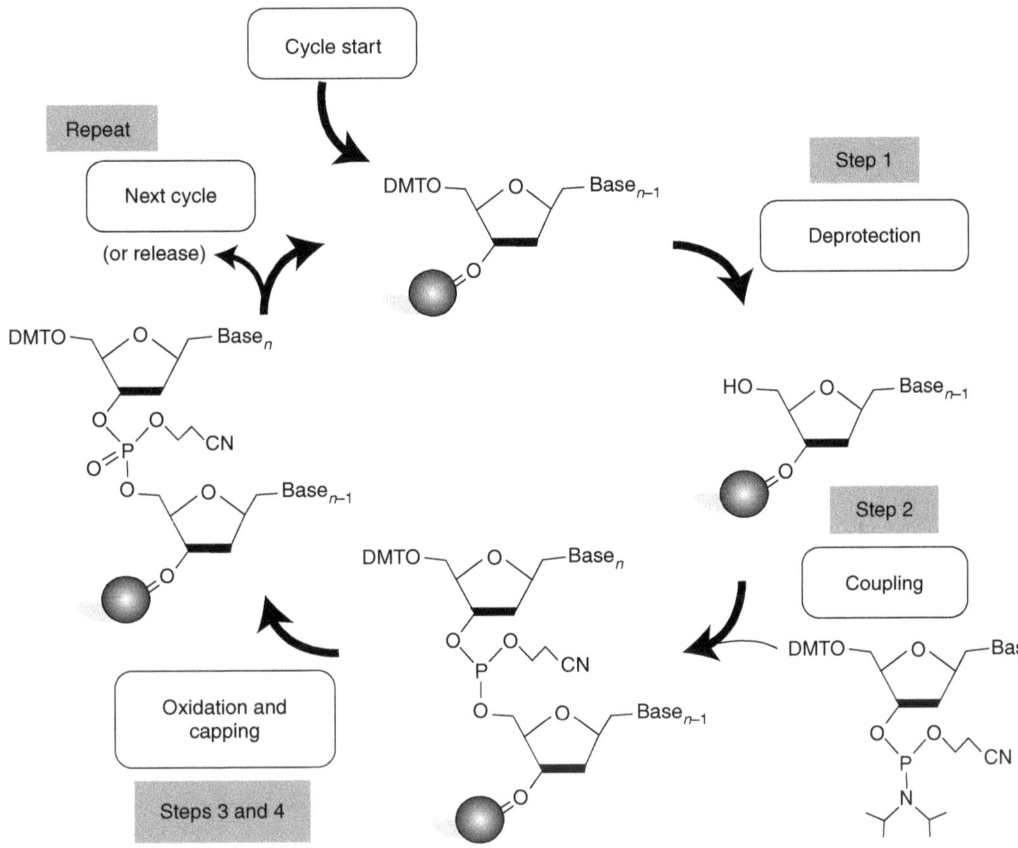

Figure 2. Phosphoramidite-based synthesis of oligonucleotides. This synthesis process is the most commonly used for the synthesis of DNA oligonucleotides for gene synthesis.

1536 distinct oligonucleotides simultaneously (Fig. 3A) (Rayner et al. 1998; Cheng et al. 2002). These synthesizers typically synthesize oligonucleotides at scales ranging from 10 to 1000 nmol with a cost ranging between \$0.05 and \$0.17 per base. These oligonucleotides can generally be synthesized up to ~100 nt with error rates at or below 0.5%. Column-based coupling yields are generally quite efficient (typically >99% per coupling); however, the yield of full-length oligonucleotides typically decrease with increasing oligonucleotide length. Yields of full-length oligonucleotides can be affected by spurious depurination of the synthesized oligomer, especially at adenosines during the acid deprotection steps of the synthesis cycle and becomes especially problematic for longer oligonucleotide sequences (Efcavitch and Heiner 1985; Septak 1996; LeProust et al. 2010). These spurious abasic sites can reduce the yields of full-length oligonucleotides by promoting the cleavage of the oligonucleotide phosphate backbone during the removal of the remaining nucleobase and backbone protecting groups following the final synthesis cycle. An additional reduction of synthesis quality is caused by the introduction of synthesis-related errors into the synthesized sequences, primarily in the form of single-base deletions. The main source of this type of error is the incomplete removal of the DMT protecting group or the combined inefficiencies of the coupling and capping steps during the synthesis cycle. Although complete removal of synthesis-related errors will likely be impossible because of less than 100% reaction efficiencies in a step-wise

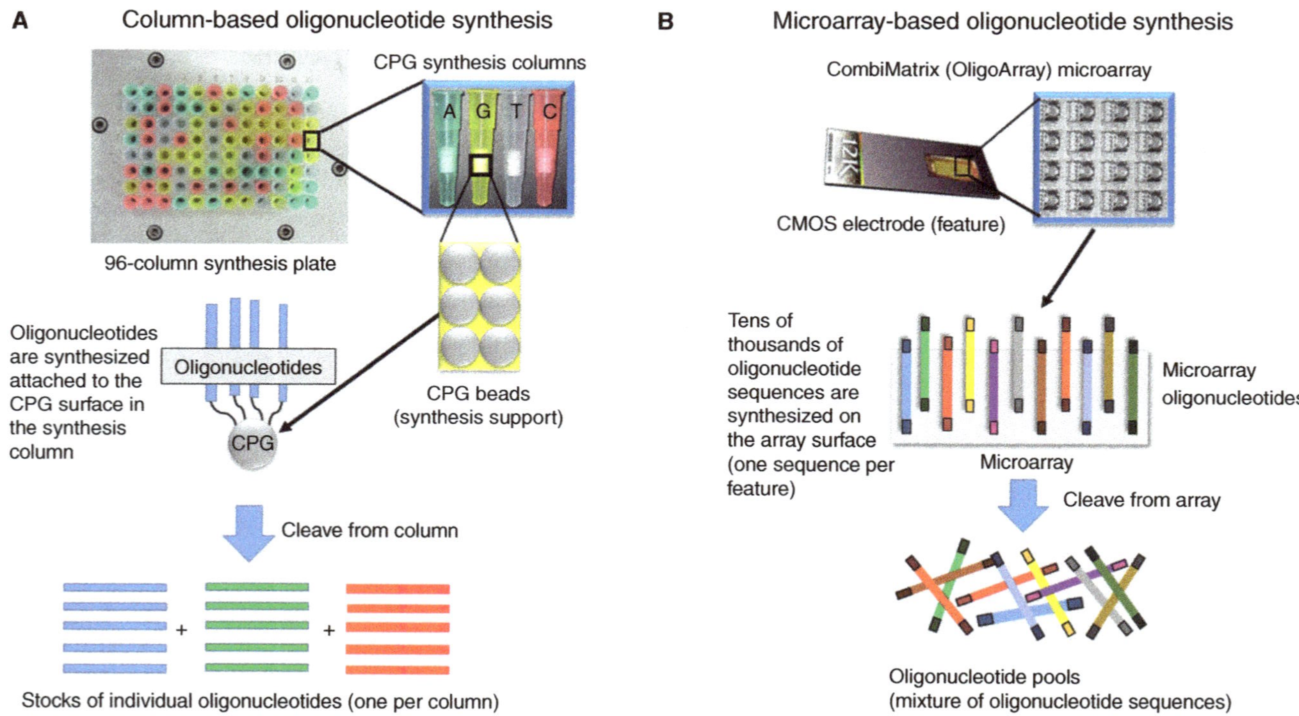

Figure 3. Methods for solid-phase synthesis of oligonucleotides. (*A*) Column-based oligonucleotide synthesis. This is the traditional method of synthesizing DNA using solid-phase phosphoramidite chemistry. The synthesis of a unique oligonucleotide sequence is done on the surface of controlled-porosity glass beads (CPG) contained with a synthesis column. During the synthesis process, the reagents flow through the column and across the packed CPG matrix and the oligonucleotide "grows" off from the bead surface. Only one sequence can be synthesized per column, although high-throughput synthesizers exist that can synthesize on multiple columns at once. A 96-column synthesis plate is shown as an example. (*B*) Microarray-based oligonucleotide synthesis. In this method, microarray chips containing tens of thousands of distinct features synthesize unique oligonucleotide sequences at once with one unique oligonucleotide sequence synthesized per chip feature. On standard arrays, there are no physical barriers between features, so following cleavage of the synthesized oligonucleotides from the chip surface the end product is a pool of sequences containing every oligonucleotide synthesized on the array. Subsequent processing steps are required to "fish" the desired oligonucleotide sequences out of the synthesis pool for subsequent gene synthesis. The OligoArray CMOS microarray chip is shown as an example, although a handful of different array formats exist for oligonucleotide synthesis.

Cite this article as *Cold Spring Harb Perspect Biol* doi: 10.1101/cshperspect.a023812

multiple reaction synthesis, recent improvements in the synthesis process and the chemistries used will likely lead to gains in the length and quality of oligonucleotides synthesized in the not-too-distant future (LeProust et al. 2010). For gene synthesis applications, because every oligonucleotide synthesized is made on individual synthesis columns the oligonucleotides necessary to assemble a gene must be added together into an assembly pool postsynthesis, which inherently reduces the oligonucleotide synthesis pool complexity because only the oligonucleotides needed for a given synthesis are included in the assembly mixture.

MICROARRAY-BASED OLIGONUCLEOTIDE SYNTHESIS

An alternative to traditional column-based oligonucleotide synthesis has emerged from the use of microarray oligonucleotide synthesis (Fig. 3B). These technologies, which were originally developed for producing oligonucleotide microarrays for diagnostic applications, have emerged as a promising low-cost alternative to column-based oligonucleotide synthesis. Early versions of these synthesizers synthesized the oligonucleotides attached to a microchip surface using a modification of the phosphoramidite synthesis process. Affymetrix was one of the early pioneers in this field and developed light-activated chemistries to control the spatially separated synthesis of oligonucleotides on the chip surface using standard mask-based photolithography techniques to selectively deprotect special photolabile nucleoside phosphoramidites (Fodor et al. 1991; Pease et al. 1994). Companies such as NimbleGen and LC Sciences further simplified the light-directed synthesis procedures by eliminating the photolithography masks by using programmable micromirror devices to precisely control the light-based chemistries (Singh-Gasson et al. 1999; Gao et al. 2001). Alternatively, ink-jet printing-based technologies developed by Agilent and semiconductor-based chip technologies by CombiMatrix (CustomArray) allow for the use of standard phosphoramidites and reagents without the need for expensive micromirror controllers or photomasking. In particular, these two synthesis technologies currently provide the best combination of cost, oligonucleotide synthesis length, and accuracy, which has led to their use in several recent gene synthesis applications (Cleary et al. 2004; Warner et al. 2010; Patwardhan et al. 2012). Using microarray-based synthesizers, oligonucleotides are typically synthesized at femtomolar synthesis scales, that is, typically two to four orders of magnitude lower than that used in column-based synthesis. This lower synthesis scale leads to a concomitant decrease in the quantity of reagents needed for the synthesis. The reduction of reagent use in the synthesis process leads to dramatically lower costs. In addition to the reduced synthesis scale, chip-based oligonucleotide synthesizers offer levels of multiplexing not possible on traditional column-based synthesizers that allows for a greater number of unique oligonucleotide sequences to be synthesized in a given synthesis run. With possible feature densities in the tens of thousands per chip and the capability for some synthesizers to synthesize multiple chips simultaneously, the ability of microarray-based synthesis methods far outstrip the capacity of even the largest column-based instruments. The combined reduced-synthesis scale and massive multiplexing capability means that oligonucleotide synthesis costs from array-based platforms can range from \$0.00001 to \$0.0001 per nucleotide depending on the platform, oligonucleotide length and synthesis scale. When compared with the \$0.05–\$0.10 cost per base attainable for column-synthesized oligonucleotides, the appeal of chip-synthesized oligonucleotides for gene synthesis applications becomes obvious.

Although array-based platforms offer superior synthesis capabilities in terms of multiplexing and cost there are some challenges with using these platforms for making oligonucleotides for gene synthesis applications. One problem is that planar array synthesized oligonucleotides tend to be relatively low quality, which leads to more synthesis-related errors when compared with column-synthesized oligonucleotides (Kosuri and Church 2014; Wan et al. 2014). Despite this, recent improvements in chip design and refinements to the synthesis process have led to an increase in the quality of array-synthesized oligonucleotides. One such source of reduced synthesis quality is the depurination of oligonucleotide sequences on chips, which can occur spontaneously because of prolonged exposure of the nascent oligonucleotides to the deprotecting reagents during the synthesis cycle. Optimizations to the reaction conditions and the flow of reagents to the chip during the synthesis cycle has been shown to effectively improve the quality of the oligonucleotides synthesized on chips by reducing depurination and increasing the overall synthesis yields, which in turn enables the synthesis of oligonucleotides up to 200 nts in length (LeProust et al. 2010). Another source of reduced synthesis quality with chip produced oligonucleotides is because of "edge effects," which is caused by misalignment of the reagent droplets with the selected chip feature, or inaccurate selective deprotection of the proper synthesis features caused by improper reagent sequestration or light beam drifts in synthesizers using light-activated synthesis chemistries. Edge effects reduce oligo quality by reducing the selective control of the synthesis reactions in adjacent oligonucleotides, which can lead to substitution or deletion errors in oligonucleotides synthesized on adjacent features. Work to reduce edge effects during the synthesis process has shown that refinements to the chip design can improve the fidelity of array-synthesized oligonucleotides to rival column-synthesized oligonucleotides (Saaem et al. 2010). Continued improvement in array design and refinements to synthesis reagents and processes used will lead to routine robust synthesis of high-quality long oligonucleotides from arrays, which will further their use as the go to source for low-cost oligonucleotides for gene synthesis applications.

GENE SYNTHESIS FROM OLIGONUCLEOTIDES

Because DNA is a polymer made up of four different nucleotide monomers, gene synthesis and DNA assembly methods are in effect a form of hierarchical polymer synthesis. For synthetic DNA, individual phosphoramidite monomers are combined together to create individual oligonucleotides 60–100 nt in length. To facilitate the assembly of a synthetic double-stranded DNA (synthon) from single-stranded oligonucleotides, adjacent oligonucleotides are designed to contain overlapping sequences between the oligonucleotides encoded on the opposing strands of the DNA duplex and are assembled together during the gene synthesis process to make double-stranded synthons from 200 to 2000 bp in length. Sequence-verified synthons can then be assembled together by numerous methods to make larger synthetic DNA constructs encoding entire metabolic pathways or genomes.

Early oligonucleotide assembly methods relied on the ligation of adjacent oligonucleotides via the use of a DNA ligase (Agarwal et al. 1970; Sekiya et al. 1976, 1979). Initially, the ligation of adjacent oligonucleotides was done sequentially, but as assembly methods and the quality of the synthetic oligonucleotides improved and the use of thermostable ligases were used to improve the stringency of the assembly reaction the one-pot assembly of DNA synthons via ligation became possible (Au et al. 1998; Xiong et al. 2008b). Shortly after the discovery of the polymerase chain reaction (PCR) in the 1980s ligation-free methods for using synthetic single-stranded oligonucleotides were developed to make completely synthetic pieces of DNA without the need for a naturally derived DNA template (Dillon and Rosen 1990; Jayaraman et al. 1991; Stemmer et al. 1995). Since this time, dozens of different methods have been created to assemble double-stranded DNA from single-stranded oligonucleotides via PCR-like methods. A summary of many of these methods has been reviewed in the literature (Xiong et al. 2008a; Czar et al. 2009; Hughes et al. 2011; Ma et al. 2012) and will not be reviewed extensively here. All of these methods use single-stranded synthetic oligonucleotides with complementary overlapping sequences between adjacent oligonucleotides to assemble double-stranded DNA synthons using a thermostable DNA polymerase and PCR. The only differences between the myriad of PCR-based DNA assembly methods is in how the substituent oligonucleotides are designed to be assembled together and the reaction conditions under which they are assembled, the end product of all of these methods is the same, a double-stranded DNA (dsDNA) synthon. Of these assembly methods, variations of the polymerase cycle assembly (PCA) method is by far the most commonly used to assemble DNA synthons from oligonucleotides (Fig. 4) (Stemmer et al. 1995). Using PCA, a dsDNA sequence is divided into oligonucleotide sequences (typically 60–80 nt), which encode both strands of the DNA duplex with overlaps between adjacent oligonucleotides that range from 15 to 25 nt in length. Typically, adjacent oligonucleotides are designed with gaps between the forward and the reverse overlapping regions of the assembly oligonucleotides to reduce the amount of oligonucleotide synthesis required to synthesize a given sequence. Once designed and synthesized, the substituent oligonucleotides for an assembly are pooled together in equimolar concentrations and cycled in a one-pot "assembly" reaction in which adjacent oligonucleotides are randomly extended in a nonexponential manner by a DNA polymerase to produce a mixture of oligonucleotide extension products of various lengths. This mixture is then used as a template to seed a second PCR reaction, in which the desired "full-length" product is amplified from the assembly mixture in the presence of an excess of the outermost assembly primers. In addition to ligation and PCA-based assembly methods, Gibson and colleagues, as part of their pioneering work in synthetic genomics, have developed both in vitro (Gibson et al. 2009; Gibson 2011a), and in vivo (Gibson 2009,

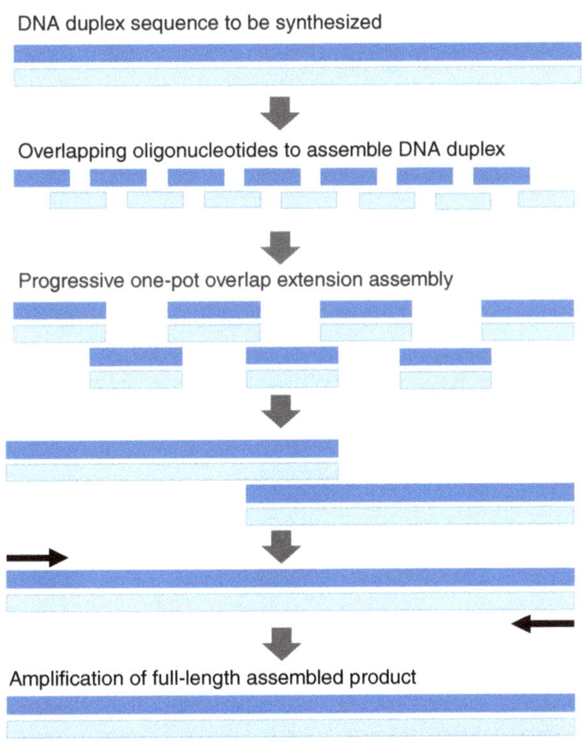

Figure 4. Polymerase chain assembly. Overlapping oligonucleotides encoding a DNA duplex are assembled together via progressive overlap extension assembly in a one-pot reaction. Following assembly the full-length assembled synthon is amplified out of the assembly mixture by polymerase chain reaction (PCR) with the outermost primers.

2011b, 2012) one-pot, single-step methods for simultaneously assembling DNA synthons from oligonucleotides and cloning them directly into plasmids. In some variation, these three assembly methods are used in most commercial and academic gene synthesis applications reported to date.

GENE SYNTHESIS FROM ARRAY-DERIVED OLIGONUCLEOTIDE POOLS

Over the last decade, the desire for cheaper sources of synthetic DNA has fueled an interest in array-synthesized DNA as a way to produce oligonucleotides inexpensively. Beyond early demonstrations of gene synthesis from array produced oligonucleotides (Richmond et al. 2004; Tian et al. 2004; Zhou et al. 2004), recent innovations have made substantial improvements to the quality, efficiency, and scalability of microarray-based oligonucleotide synthesis and gene assembly. To use microarray-synthesized DNA for gene synthesis, several hurdles have had to be overcome to make this source of DNA an attractive alternative to oligonucleotides synthesized by conventional column-based approaches. Some of the properties that make array-synthesized oligonucleotides attractive as a low-cost source of DNA for gene synthesis present challenges for actually using oligonucleotides synthesized in this way for gene synthesis. The reduced synthesis scale of array-produced oligonucleotides is attractive from a cost perspective because of the concomitant reduction in reagent consumption needed to synthesize a given DNA sequence. However, the amount of DNA produced (fmol) is generally much less than what is needed (pmol) to reliably assemble them into larger synthons. Similarly, the massive multiplexing capabilities of array-based oligonucleotide synthesis means that tens of thousands of unique oligonucleotide sequences can be synthesized simultaneously on the array surface. However, the sequence complexity of the oligonucleotide pools produced on the array can lead to problems with assembling them into larger constructs due to the interference between oligonucleotides that contain even modest sequence homology. Last, as mentioned previously, array-synthesized oligonucleotides are in general lower quality as compared with their column-synthesized counterparts, which manifests itself as sequence errors in the synthetic DNA assembled from them. So, in general, the use of array-synthesized oligonucleotides requires the use of some sort of error reduction strategy to reduce the work required to obtain sequence-verified constructs.

To overcome these limitations, a couple of different methods have been developed to increase the effective oligonucleotide pool concentration to ensure reliable assembly and reproducibility (Tian et al. 2004; Kosuri et al. 2010; Quan et al. 2011). Tian et al. (2004) addressed the concentration issue by including a common priming sequence followed by a nicking endonuclease recognition site on the 3′-end of every oligonucleotide synthesized on the array (Fig. 5A). The array-synthesized oligonucleotides serve as the template strand for a primer extension reaction using the common priming sequence incorporated into the array-attached oligonucleotide pool. Following primer extension, a nicking endonuclease included in the reaction mixture removes the common primer sequences to release a single-stranded DNA oligonucleotide that is complementary to each oligonucleotide synthesized on the array. This process amplifies the array-synthesized oligonucleotide pool to increase the concentration of the oligonucleotides to levels suitable for gene synthesis. In this method, because the array-synthesized oligonucleotides remain attached to the array, spatial control of oligonucleotide assembly can occur on chip. To exploit this feature and to improve the multiplexing capabilities of chip-based oligonucleotide synthesis, Quan et al. (2011) used a custom ink-jet synthesizer that synthesized oligonucleotides on specialized chips. Within these specialized chips, micro-compartmentalized assembly wells can be designed into the synthesis array with the oligonucleotides necessary to assemble a unique synthon, synthesized, amplified, and assembled within individual microwells (Fig. 5A). Using such a method effectively reduces the sequence complexity of the localized oligonucleotide

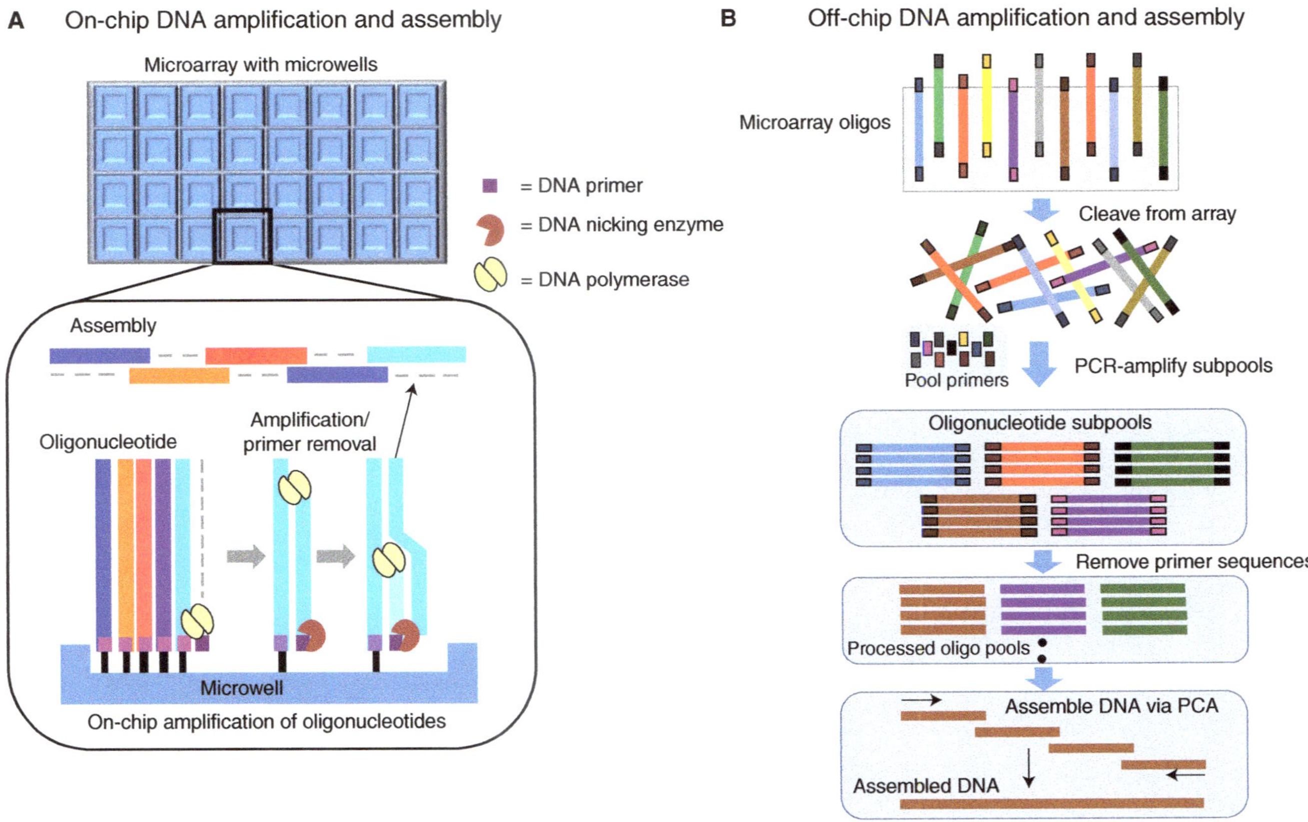

Figure 5. Gene synthesis from microarray-synthesized oligonucleotides. (*A*) Synthons can be assembled using on-chip synthesis and assembly by including a single priming site into the 3′-end of every oligonucleotide synthesized on the microarray. The oligonucleotides can then be amplified within microwells designed into the array by incubating with a common primer and a DNA polymerase. The primer sequence is removed from the assembly oligonucleotides using a nicking endonuclease, freeing the oligonucleotides to be assembled together via polymerase chain assembly within the same well. (*B*) Synthons can also be synthesized off-chip by first cleaving the oligonucleotide pools from the array. The cleaved synthesis oligonucleotide pool can be amplified into subpools by PCR using priming sites incorporated into both ends of the synthesized oligonucleotide. The amplified subpools can then be further subdivided into assembly oligonucleotide pools by additional unique priming sites included in the oligonucleotide flanking sequences. Following segregation by amplification the priming sequences are removed from the assembly oligonucleotides by restriction enzyme digestion. The processed oligonucleotides can then be assembled together by polymerase cycle assembly.

pool, which in turn increases the robustness of assembly for gene synthesis while also allowing for synthesis multiplexing that can occur in each of the many microwells designed into the chip.

Another method first introduced by Kosuri et al. (2010) solves the oligonucleotide concentration and pool complexity problems by including "bar-coded" priming sequences into each of the oligonucleotides synthesized on the array (Fig. 5B). In this method, the oligonucleotides required to synthesize a desired synthon is flanked on both ends by sets of predefined primer sequences. The outermost priming sequences allow for amplification of the array-synthesized oligonucleotide pool, which has been cleaved from the chip surface following synthesis. This step allows for one set of unique primers to amplify the entire oligonucleotide pool such that a portion of the pool can be used for subsequent assembly or archival purposes. Inboard of the pool amplification primers is a set of flanking primer-binding sequences that allow for selective amplification of assembly subpools from the synthesis pool. Each unique synthon would be given a unique set of flanking subpool primer sequences such that the oligonucleotides necessary to assemble any given synthon would be selectively amplified from the synthesis pool for subsequent assembly. Following amplification of the subpool oligonucleotides, the priming sequences are removed by restriction digestion of the amplified dsDNA oligonucleotides using a unique enzyme recognition sequence flanking the assembly oligonucleotide sequence. This oligonucleotide design scheme simultaneously solves the oligonucleotide concentration and pool complexity problems associated with array-derived oligonucleotides and does not require specialized chips or array synthesizers. This method is also highly amenable to automation as each subpool can be assembled in unique wells in a multiwell plate format using PCA assembly techniques.

ERROR CORRECTION AND SEQUENCE VALIDATION

One of the chief challenges in the synthesis and assembly of synthetic DNA from unpurified oligonucleotides is dealing with sequence errors introduced into the sequences by synthesis by-products and the assembly process itself. Following assembly, a synthetic DNA can be thought of as a population of sequences containing a mixture of correct and incorrect sequences. Traditionally, the synthetic DNA would be cloned into a vector and individual clones isolated and sequenced by Sanger sequencing. However, the probability of any given sequence containing an error increases with the length of DNA, which in turn means that more clones must be sequenced to obtain a completely correct clone. This procedure is time-consuming and the setup work required adds significantly to the cost of the DNA being produced. To reduce the number of clones necessary to obtain a correct sequence, some postsynthesis method must be applied to the assembled DNA to sieve correct sequences from incorrect ones. As a result, methods have been developed that approximate the error-correcting process that cells use to maintain sequence fidelity during DNA replication. In general, most methods for removing DNA sequence errors from synthetic DNA start with the formation of a DNA heteroduplex. This is done by heating up the synthetic DNA to denature and disassociate the strands followed by cooling the sample to reanneal the strands together. For any given position within a DNA sequence, most of the duplexes in the synthetic mixture will contain the correct base at that position with errors sprinkled throughout the population. Because sequence errors occur randomly within an assembled DNA sequence this denaturation and reannealing process leads to the formation of heteroduplexed DNA at positions that contain errors. Positions that contain mutations within these heteroduplexes can be acted on by proteins, which specifically recognize sequence mutations in DNA. One such group of methods relies on the sequence mismatch recognition capabilities of the MutS protein to specifically bind to sequence mismatches in synthetic DNA duplexes. Selective binding of MutS to error-containing DNA can be used to sieve error-free sequences from those that contain errors (Carr et al. 2004; Binkowski et al. 2005; Wan et al. 2014). These methods usually

 Cite this article as *Cold Spring Harb Perspect Biol* doi: 10.1101/cshperspect.a023812

immobilize MutS to a solid matrix material and then purify column-bound (error-containing) DNA sequences from unbound material (error-reduced). Error-containing heteroduplex DNA can also be sieved using enzymes that recognize and cut the DNA duplex at the site of the base mismatch (Young and Dong 2004; Fuhrmann et al. 2005; Kosuri et al. 2010; Saaem et al. 2012; Dormitzer et al. 2013). The use of endonuclease enzymes (or enzyme cocktails), which recognize and cleave DNA heteroduplexes at the sites of mismatches, has been shown to be highly effective at reducing synthesis-related errors in synthetic genes allowing for time and material savings such that in some cases the treated genes can be used directly in functional assays without cloning and sequence verification (Dormitzer et al. 2013). These methods are a relatively straightforward and cost-effective approach for sieving error-containing sequences from synthetic DNA and can reduce the observed error rate from 1/100 bases to 1/10,000 bases (Dormitzer et al. 2013; Kosuri and Church 2014).

An alternative to enzyme-mediated error correction techniques is the direct sequencing of oligonucleotides and assembled synthons using NextGen sequencing techniques. Although more expensive on a per-sequence basis than enzyme-based error correction techniques, NextGen sequencing techniques offers tremendous multiplexing capabilities in which thousands of sequence reads can be taken on multiple oligonucleotides or synthons simultaneously. In one such application, Matzas et al. (2010) used Roche 454 sequencing combined with a bead-picking robot to selectively remove oligonucleotide sequences (attached to beads) from the sequencing array that showed perfect sequences that were then used for gene synthesis. Kim et al. (2012) used random sequence tags to mark individual synthetic sequences in a 454 sequencing reaction and used these tags to selectively amplify correct sequences from the sequenced pool. In a similar approach, Schwartz et al. (2012) used Illumina sequencing barcodes to "dial-out" correct sequences via selective amplification of the sequence-verified bar-coded sequences. Recently, they have refined this approach to improve the multiplexing capacity of microarray-synthesized DNA to pairwise assemble 2271 131–250 bp synthons from a single oligonucleotide pool and used bar-coded primers to selectively isolate sequence-verified individuals (Klein et al. 2016). All of these methods point to ways in which NextGen technologies can be used to improve the quality of synthetic DNA. In addition, because sequence-based approaches evaluate collections of individual molecules of DNA, they are suitable for the sequence verification of synthetic DNA, which may contain regions of sequence degeneracy such as libraries for directed evolution. Currently, each of these methods are limited by the sequence read-length capabilities of the NextGen instrument used and by sequencing-related errors. However, as NextGen instrumentation and techniques continue to improve, these limitations will become less significant and allow for the accurate verification of longer pieces of synthetic DNA.

MAKE IT BIGGER: SYNTHESIS OF LONGER DNA ASSEMBLIES

Recent advancements in oligonucleotide synthesis and techniques for gene synthesis have largely focused on reducing costs, increasing throughput, and the quality of synthetic DNA. Concomitant advances in assembling DNA into longer and longer pieces have led to methods to construct large enzyme complexes (Kodumal et al. 2004), entire metabolic pathways (Temme et al. 2012), and even entire genomes (Smith et al. 2003; Gibson et al. 2008a; Hutchison et al. 2016) synthetically. Methods to assemble synthons have been developed to assemble synthons and larger DNA assemblies in both in vitro (Gibson et al. 2010) and in vivo (Gibson et al. 2008b). Although restriction enzyme-based cloning techniques have been the de rigueur choice for manipulating DNA constructs for a couple of decades and were the basis of early BioBrick and similar assembly methods (Shetty et al. 2008; Lee et al. 2011), the need to simplify the cloning/assembly process while reducing the limitations on sequence design has led to the development of scar-less restriction-enzyme-free cloning and assembly techniques. Seam-less assembly and cloning methods available to the modern DNA jockey include Gibson assembly (Gibson et al. 2009; Gibson 2011a), Golden Gate assembly (Engler et al. 2008), sequence and ligation-independent cloning (SLIC) (Li and Elledge 2007), ligation cycling reaction (de Kok et al. 2014), paper-clip assembly (Trubitsyna et al. 2014), yeast assembly (Gibson et al. 2008b; Gibson 2011b), and circular polymerase extension cloning (CPEC) (Quan and Tian 2009), to name just a few. Which DNA assembly technique to use is largely a matter of choice, and multiple approaches are often applied in parallel. Of these methods, Gibson assembly is probably the most commonly used to assemble multiple pieces of DNA together into larger constructs. This method uses a one-pot isothermal technique, which uses an enzyme mixture containing a thermostable DNA polymerase, DNA ligase, and exonuclease to chew-back, anneal and repair adjacent overlapping DNA sequences to assemble the desired construct. Recent innovations in the design of unique overlapping sequences to direct the assembly process has further expanded the usage of the Gibson assembly method for combinatorial assembly of large DNA sequences (Guye et al. 2013; Torella et al. 2014).

One of the hallmarks of synthetic biology is the application of rational design principles to the design and assembly of biological components; however, because it is often difficult to know a priori how well a given DNA construct will work once introduced into a cell, it is often necessary to try several versions of the construct to find which one will work best. Therefore, a greater emphasis on the modular design of DNA parts enables the assembly of a greater variety of potential constructs through mix-and-match combinatorial assembly of DNA components. In addition to simplifying the overall assembly process, modular design and assembly of DNA components makes automation of the process possible, which can reduce the time, labor, and cost of making and testing multiple constructs. Most of the aforementioned assembly methods can be used for the assembly in vitro, and Gibson assembly has been applied to the assembly of DNA segments multiple kilobases in length (Gibson et al. 2009). In another such example, Gibson assembly was used to assemble the 16.3-kb mouse mitochondrial genome directly from 60 mer oligonucleotides (Gibson et al. 2010). Efficient assembly of even larger synthetic DNA segments can also be performed in vivo by using the homologous recombination capabilities of the yeast *Saccharomyces cerevisiae.* In an example of the exceptional ability of yeast to assemble exogenous DNA into larger assemblies from overlapping synthons or subassemblies, researchers at the J. Craig Venter Institute have successfully used yeast to assemble multiple 0.5–1 Mbp bacterial genomes (Gibson et al. 2008a; Hutchison et al. 2016) and even assembled synthons directly from overlapping oligonucleotides (Gibson 2009, 2011b, 2012). Each of the aforementioned assembly techniques could be automated to further increase the throughput for constructing larger synthetic DNAs and enable the exploration and testing of large biological hypotheses.

CONCLUDING REMARKS

The development of automatable robust chemistries for chemical DNA synthesis over the last 40 years has contributed to the advancement of our understanding of biology and has laid the groundwork for the predictable engineering of biological systems. Synthetic DNA is central to the development of methods to engineer biology and when combined with the massive amounts of sequence data being generated by NGS efforts will contribute to the advancement of synthetic biology toward applications heretofore unimaginable. To date, there have been a handful of moonshot demonstrations such as the complete synthesis of an entire yeast chromosome (Annaluru et al. 2014), an entire bacterial genome (Gibson et al. 2008a), and the subsequent synthesis of a minimal bacterial genome (Hutchison et al. 2016), which illustrate the use of synthetic DNA and the capabilities of existing gene synthesis methods to accomplish large-scale synthetic biology efforts. These examples, when combined with numerous projects in which synthetic DNA has been used to evaluate functional biological components (Salis et al. 2009; Callura et al. 2012; Brophy and Voigt

2016), create synthetic vaccines (Dormitzer et al. 2013), and construct synthetic genetic circuits (Salis et al. 2009; Sowa et al. 2015; Nielsen et al. 2016; Rubens et al. 2016) and when applied to wholesale genomic editing (Wang et al. 2009; Doudna and Charpentier 2014), point to a future where the nuances of biological function can in part be understood by the design, synthesis, and assay of interchangeable synthetic components akin to synthetic drug development.

To enable larger-scale engineering efforts and to democratize the use of synthetic biology techniques and design principles, recent innovations have sought to lower the cost of synthesizing DNA oligonucleotides and their subsequent assembly into synthetic constructs ready for use. The use of microarray/microfluidic oligonucleotide synthesis techniques, when combined with automated gene synthesis protocols and improvements in synthetic DNA quality, have made great strides toward this goal. In response to market pressures, commercialization of these technologies and the continued development of other efficient DNA synthesis techniques will further enhance the availability of increasingly inexpensive synthetic DNA.

Although the synthesis of relatively small synthons (<1 kb) can often be straightforward and accomplished at the undergraduate level the assembly of larger pieces of synthetic DNA can still pose several technical challenges to even the most experienced DNA jockey. Problems created by a high degree of secondary structure in the target sequence, as well as repetitive DNA sequences, can pose challenges to their assembly. Sequences that are unknowingly toxic to the host cells used for intermediate manipulations of the DNA (e.g., cloning for sequence verification) or assay of synthetic DNA constructs can also pose problems for the successful assembly and use of some synthetic DNA constructs. Here to, further improvements in synthetic construct design (Tang and Chilkoti 2016) and in the DNA assembly process will likely enable the successful synthesis of even the most difficult target sequences in the not-to-distant future. What started with a synthetic tRNA gene more than 40 years ago has led to the recent synthesis of the first minimal bacterial genome (Hutchison et al. 2016) and a deep appreciation of the complexities and challenges of engineering biology. Much of which still remains to be explored. In this, the century of DNA, the question remains. What is next?

REFERENCES

Agarwal KL, Buchi H, Caruthers MH, Gupta N, Khorana HG, Kleppe K, Kumar A, Ohtsuka E, Rajbhandary UL, Van de Sande JH, et al. 1970. Total synthesis of the gene for an alanine transfer ribonucleic acid from yeast. *Nature* **227:** 27–34.

Annaluru N, Muller H, Mitchell LA, Ramalingam S, Stracquadanio G, Richardson SM, Dymond JS, Kuang Z, Scheifele LZ, Cooper EM, et al. 2014. Total synthesis of a functional designer eukaryotic chromosome. *Science* **344:** 55–58.

Au LC, Yang FY, Yang WJ, Lo SH, Kao CF. 1998. Gene synthesis by an LCR-based approach: High-level production of leptin-l54 using synthetic gene in *Escherichia coli*. *Biochem Biophys Res Commun* **248:** 200–203.

Binkowski BF, Richmond KE, Kaysen J, Sussman MR, Belshaw PJ. 2005. Correcting errors in synthetic DNA through consensus shuffling. *Nucleic Acids Res* **33:** e55.

Brophy JA, Voigt CA. 2016. Antisense transcription as a tool to tune gene expression. *Mol Syst Biol* **12:** 854.

Callura JM, Cantor CR, Collins JJ. 2012. Genetic switchboard for synthetic biology applications. *Proc Natl Acad Sci* **109:** 5850–5855.

Carr PA, Park JS, Lee YJ, Yu T, Zhang S, Jacobson JM. 2004. Protein-mediated error correction for de novo DNA synthesis. *Nucleic Acids Res* **32:** e162.

Caruthers MH. 2011. A brief review of DNA and RNA chemical synthesis. *Biochem Soc Trans* **39:** 575–580.

Cheng JY, Chen HH, Kao YS, Kao WC, Peck K. 2002. High throughput parallel synthesis of oligonucleotides with 1536 channel synthesizer. *Nucleic Acids Res* **30:** e93.

Cleary MA, Kilian K, Wang Y, Bradshaw J, Cavet G, Ge W, Kulkarni A, Paddison PJ, Chang K, Sheth N, et al. 2004. Production of complex nucleic acid libraries using highly parallel in situ oligonucleotide synthesis. *Nat Methods* **1:** 241–248.

Czar MJ, Anderson JC, Bader JS, Peccoud J. 2009. Gene synthesis demystified. *Trends Biotechnol* **27:** 63–72.

de Kok S, Stanton LH, Slaby T, Durot M, Holmes VF, Patel KG, Platt D, Shapland EB, Serber Z, Dean J, et al. 2014. Rapid and reliable DNA assembly via ligase cycling reaction. *ACS Synth Biol* **3:** 97–106.

Dillon PJ, Rosen CA. 1990. A rapid method for the construction of synthetic genes using the polymerase chain reaction. *BioTechniques* **9:** 298–300.

Dormitzer PR, Suphaphiphat P, Gibson DG, Wentworth DE, Stockwell TB, Algire MA, Alperovich N, Barro M, Brown DM, Craig S, et al. 2013. Synthetic generation of influenza vaccine viruses for rapid response to pandemics. *Sci Transl Med* **5:** 185ra168.

Doudna JA, Charpentier E. 2014. Genome editing. The new frontier of genome engineering with CRISPR-Cas9. *Science* **346:** 1258096.

Efcavitch JW, Heiner C. 1985. Depurination as a yield decreasing mechanism in oligodeoxynucleotide synthesis. *Nucleosides Nucleotides* **4:** 267–267.

Engler C, Kandzia R, Marillonnet S. 2008. A one pot, one step, precision cloning method with high throughput capability. *PLoS ONE* **3:** e3647.

Fodor SP, Read JL, Pirrung MC, Stryer L, Lu AT, Solas D. 1991. Light-directed, spatially addressable parallel chemical synthesis. *Science* **251:** 767–773.

Fuhrmann M, Oertel W, Berthold P, Hegemann P. 2005. Removal of mismatched bases from synthetic genes by enzymatic mismatch cleavage. *Nucleic Acids Res* **33:** e58.

Gao X, LeProust E, Zhang H, Srivannavit O, Gulari E, Yu P, Nishiguchi C, Xiang Q, Zhou X. 2001. A flexible light-directed DNA chip synthesis gated by deprotection using solution photogenerated acids. *Nucleic Acids Res* **29:** 4744–4750.

Gibson DG. 2009. Synthesis of DNA fragments in yeast by one-step assembly of overlapping oligonucleotides. *Nucleic Acids Res* **37:** 6984–6990.

Gibson DG. 2011a. Enzymatic assembly of overlapping DNA fragments. *Methods Enzymol* **498:** 349–361.

Gibson DG. 2011b. Gene and genome construction in yeast. In *Current protocols in molecular biology* (ed. Ausubel FM et al.), Chapter 3, Unit 3.22. Wiley, Hoboken, NJ.

Gibson DG. 2012. Oligonucleotide assembly in yeast to produce synthetic DNA fragments. *Methods Mol Biol* **852:** 11–21.

Gibson DG, Benders GA, Andrews-Pfannkoch C, Denisova EA, Baden-Tillson H, Zaveri J, Stockwell TB, Brownley A, Thomas DW, Algire MA, et al. 2008a. Complete chemical synthesis, assembly, and cloning of a *Mycoplasma genitalium* genome. *Science* **319:** 1215–1220.

Gibson DG, Benders GA, Axelrod KC, Zaveri J, Algire MA, Moodie M, Montague MG, Venter JC, Smith HO, Hutchison CA III. 2008b. One-step assembly in yeast of 25 overlapping DNA fragments to form a complete synthetic Mycoplasma genitalium genome. *Proc Natl Acad Sci* **105:** 20404–20409.

Gibson DG, Young L, Chuang RY, Venter JC, Hutchison CA III, Smith HO. 2009. Enzymatic assembly of DNA molecules up to several hundred kilobases. *Nat Method* **6:** 343–345.

Gibson DG, Smith HO, Hutchison CA III, Venter JC, Merryman C. 2010. Chemical synthesis of the mouse mitochondrial genome. *Nat Method* **7:** 901–903.

Guye P, Li Y, Wroblewska L, Duportet X, Weiss R. 2013. Rapid, modular and reliable construction of complex mammalian gene circuits. *Nucleic Acids Res* **41:** e156.

Hughes RA, Miklos AE, Ellington AD. 2011. Gene synthesis: Methods and applications. *Methods Enzymol* **498:** 277–309.

Hutchison CA 3rd, Chuang RY, Noskov VN, Assad-Garcia N, Deerinck TJ, Ellisman MH, Gill J, Kannan K, Karas BJ, Ma L, et al. 2016. Design and synthesis of a minimal bacterial genome. *Science* **351:** aad6253.

Itakura K, Hirose T, Crea R, Riggs AD, Heyneker HL, Bolivar F, Boyer HW. 1977. Expression in *Escherichia coli* of a chemically synthesized gene for the hormone somatostatin. *Science* **198:** 1056–1063.

Jayaraman K, Fingar SA, Shah J, Fyles J. 1991. Polymerase chain reaction–mediated gene synthesis: Synthesis of a gene coding for isozyme c of horseradish peroxidase. *Proc Natl Acad Sci* **88:** 4084–4088.

Khorana HG, Buchi H, Caruthers MH, Chang SH, Gupta NK, Kumar A, Ohtsuka E, Sgaramella V, Weber H. 1968. Progress in the total synthesis of the gene for ala-tRNA. *Cold Spring Harb Symp Quant Biol* **33:** 35–44.

Khorana HG, Agarwal KL, Buchi H, Caruthers MH, Gupta NK, Kleppe K, Kumar A, Otsuka E, RajBhandary UL, Van de Sande JH, et al. 1972. Studies on polynucleotides. 103. Total synthesis of the structural gene for an alanine transfer ribonucleic acid from yeast. *J Mol Biol* **72:** 209–217.

Kim H, Han H, Ahn J, Lee J, Cho N, Jang H, Kim H, Kwon S, Bang D. 2012. "Shotgun DNA synthesis" for the high-throughput construction of large DNA molecules. *Nucleic Acids Res* **40:** e140.

Klein JC, Lajoie MJ, Schwartz JJ, Strauch EM, Nelson J, Baker D, Shendure J. 2016. Multiplex pairwise assembly of array-derived DNA oligonucleotides. *Nucleic Acids Res* **44:** e43.

Kodumal SJ, Patel KG, Reid R, Menzella HG, Welch M, Santi DV. 2004. Total synthesis of long DNA sequences: Synthesis of a contiguous 32-kb polyketide synthase gene cluster. *Proc Natl Acad Sci* **101:** 15573–15578.

Kosuri S, Church GM. 2014. Large-scale de novo DNA synthesis: Technologies and applications. *Nat Method* **11:** 499–507.

Kosuri S, Eroshenko N, Leproust EM, Super M, Way J, Li JB, Church GM. 2010. Scalable gene synthesis by selective amplification of DNA pools from high-fidelity microchips. *Nat Biotechnol* **28:** 1295–1299.

Lee TS, Krupa RA, Zhang F, Hajimorad M, Holtz WJ, Prasad N, Lee SK, Keasling JD. 2011. BglBrick vectors and datasheets: A synthetic biology platform for gene expression. *J Biol Eng* **5:** 12.

LeProust EM, Peck BJ, Spirin K, McCuen HB, Moore B, Namsaraev E, Caruthers MH. 2010. Synthesis of high-quality libraries of long (150mer) oligonucleotides by a novel depurination controlled process. *Nucleic Acids Res* **38:** 2522–2540.

Li MZ, Elledge SJ. 2007. Harnessing homologous recombination in vitro to generate recombinant DNA via SLIC. *Nat Method* **4:** 251–256.

Ma S, Tang N, Tian J. 2012. DNA synthesis, assembly and applications in synthetic biology. *Curr Opin Chem Biol* **16:** 260–267.

Matzas M, Stahler PF, Kefer N, Siebelt N, Boisguerin V, Leonard JT, Keller A, Stahler CF, Haberle P, Gharizadeh B, et al. 2010. High-fidelity gene synthesis by retrieval of sequence-verified DNA identified using high-throughput pyrosequencing. *Nat Biotechnol* **28:** 1291–1294.

Nielsen AA, Der BS, Shin J, Vaidyanathan P, Paralanov V, Strychalski EA, Ross D, Densmore D, Voigt CA. 2016. Genetic circuit design automation. *Science* **352:** aac7341.

Patwardhan RP, Hiatt JB, Witten DM, Kim MJ, Smith RP, May D, Lee C, Andrie JM, Lee SI, Cooper GM, et al. 2012. Massively parallel functional dissection of mammalian enhancers in vivo. *Nat Biotechnol* **30:** 265–270.

Pease AC, Solas D, Sullivan EJ, Cronin MT, Holmes CP, Fodor SP. 1994. Light-generated oligonucleotide arrays for rapid DNA sequence analysis. *Proc Natl Acad Sci* **91:** 5022–5026.

Quan J, Tian J. 2009. Circular polymerase extension cloning of complex gene libraries and pathways. *PLoS ONE* **4:** e6441.

Quan J, Saaem I, Tang N, Ma S, Negre N, Gong H, White KP, Tian J. 2011. Parallel on-chip gene synthesis and application to optimization of protein expression. *Nat Biotechnol* **29:** 449–452.

Rayner S, Brignac S, Bumeister R, Belosludtsev Y, Ward T, Grant O, O'Brien K, Evans GA, Garner HR. 1998. MerMade: An oligodeoxyribonucleotide synthesizer for high throughput oligonucleotide production in dual 96-well plates. *Genome Res* **8:** 741–747.

Richmond KE, Li MH, Rodesch MJ, Patel M, Lowe AM, Kim C, Chu LL, Venkataramaian N, Flickinger SF, Kaysen J, et al. 2004. Amplification and assembly of chip-eluted DNA (AACED): A method for high-throughput gene synthesis. *Nucleic Acids Res* **32:** 5011–5018.

Roy S, Caruthers M. 2013. Synthesis of DNA/RNA and their analogs via phosphoramidite and H-phosphonate chemistries. *Molecules* **18:** 14268–14284.

Rubens JR, Selvaggio G, Lu TK. 2016. Synthetic mixed-signal computation in living cells. *Nat Commun* **7:** 11658.

Saaem I, Ma KS, Marchi AN, LaBean TH, Tian J. 2010. In situ synthesis of DNA microarray on functionalized cyclic olefin copolymer substrate. *ACS Appl Mater Interfaces* **2:** 491–497.

Saaem I, Ma S, Quan J, Tian J. 2012. Error correction of microchip synthesized genes using Surveyor nuclease. *Nucleic Acids Res* **40:** e23.

Salis HM, Mirsky EA, Voigt CA. 2009. Automated design of synthetic ribosome binding sites to control protein expression. *Nat Biotechnol* **27:** 946–950.

Schatz MC, Phillippy AM. 2012. The rise of a digital immune system. *GigaScience* **1:** 4.

Scheuerbrandt G, Duffield AM, Nussbaum AL. 1963. Stepwise synthesis of deoxyribo-oligonucleotides. *Biochem Biophys Res Commun* **11:** 152–155.

Schwartz JJ, Lee C, Shendure J. 2012. Accurate gene synthesis with tag-directed retrieval of sequence-verified DNA molecules. *Nat Method* **9:** 913–915.

Sekiya T, Besmer P, Takeya T, Khorana HG. 1976. Total synthesis of the structural gene for the precursor of a tyrosine suppressor transfer RNA from *Escherichia coli*. 7. Enzymatic joining of the chemically synthesized segments to form a DNA duplex corresponding to the nucleotide sequence 1-26. *J Biol Chem* **251:** 634–641.

Sekiya T, Takeya T, Brown EL, Belagaje R, Contreras R, Fritz HJ, Gait MJ, Lees RG, Ryan MJ, Khorana HG. 1979. Total synthesis of a tyrosine suppressor transfer RNA gene. XVI: Enzymatic joinings to form the total 207-base pair-long DNA. *J Biol Chem* **254:** 5787–5801.

Septak M. 1996. Kinetic studies on depurination and detritylation of CPG-bound intermediates during oligonucleotide synthesis. *Nucleic Acids Res* **24:** 3053–3058.

Shetty RP, Endy D, Knight TF Jr. 2008. Engineering BioBrick vectors from BioBrick parts. *J Biol Eng* **2:** 5.

Singh-Gasson S, Green RD, Yue Y, Nelson C, Blattner F, Sussman MR, Cerrina F. 1999. Maskless fabrication of light-directed oligonucleotide microarrays using a digital micromirror array. *Nat Biotechnol* **17:** 974–978.

Smith HO, Hutchison CA III, Pfannkoch C, Venter JC. 2003. Generating a synthetic genome by whole genome assembly: φX174 bacteriophage from synthetic oligonucleotides. *Proc Natl Acad Sci* **100:** 15440–15445.

Sowa SW, Gelderman G, Contreras LM. 2015. Advances in synthetic dynamic circuits design: Using novel synthetic parts to engineer new generations of gene oscillations. *Curr Opin Biotechnol* **36:** 161–167.

Stemmer WP, Crameri A, Ha KD, Brennan TM, Heyneker HL. 1995. Single-step assembly of a gene and entire plasmid from large numbers of oligodeoxyribonucleotides. *Gene* **164:** 49–53.

Tang NC, Chilkoti A. 2016. Combinatorial codon scrambling enables scalable gene synthesis and amplification of repetitive proteins. *Nat Mater* **15:** 419–424.

Temme K, Zhao D, Voigt CA. 2012. Refactoring the nitrogen fixation gene cluster from *Klebsiella oxytoca*. *Proc Natl Acad Sci* **109:** 7085–7090.

Tian J, Gong H, Sheng N, Zhou X, Gulari E, Gao X, Church G. 2004. Accurate multiplex gene synthesis from programmable DNA microchips. *Nature* **432:** 1050–1054.

Torella JP, Lienert F, Boehm CR, Chen JH, Way JC, Silver PA. 2014. Unique nucleotide sequence-guided assembly of repetitive DNA parts for synthetic biology applications. *Nat Protoc* **9:** 2075–2089.

Trubitsyna M, Michlewski G, Cai Y, Elfick A, French CE. 2014. PaperClip: Rapid multi-part DNA assembly from existing libraries. *Nucleic Acids Res* **42:** e154.

Wan W, Li L, Xu Q, Wang Z, Yao Y, Wang R, Zhang J, Liu H, Gao X, Hong J. 2014. Error removal in microchip-synthesized DNA using immobilized MutS. *Nucleic Acids Res* **42:** e102.

Wang HH, Isaacs FJ, Carr PA, Sun ZZ, Xu G, Forest CR, Church GM. 2009. Programming cells by multiplex genome engineering and accelerated evolution. *Nature* **460:** 894–898.

Warner JR, Reeder PJ, Karimpour-Fard A, Woodruff LB, Gill RT. 2010. Rapid profiling of a microbial genome using mixtures of barcoded oligonucleotides. *Nat Biotechnol* **28:** 856–862.

Xiong AS, Peng RH, Zhuang J, Gao F, Li Y, Cheng ZM, Yao QH. 2008a. Chemical gene synthesis: Strategies, softwares, error corrections, and applications. *FEMS Microbiol Rev* **32:** 522–540.

Xiong AS, Peng RH, Zhuang J, Liu JG, Gao F, Chen JM, Cheng ZM, Yao QH. 2008b. Non-polymerase-cycling-assembly-based chemical gene synthesis: Strategies, methods, and progress. *Biotechnol Adv* **26:** 121–134.

Young L, Dong Q. 2004. Two-step total gene synthesis method. *Nucleic Acids Res* **32:** e59.

Zhou X, Cai S, Hong A, You Q, Yu P, Sheng N, Srivannavit O, Muranjan S, Rouillard JM, Xia Y, et al. 2004. Microfluidic PicoArray synthesis of oligodeoxynucleotides and simultaneous assembling of multiple DNA sequences. *Nucleic Acids Res* **32:** 5409–5417.

Minimal Cells—Real and Imagined

John I. Glass, Chuck Merryman, Kim S. Wise, Clyde A. Hutchison III, and Hamilton O. Smith

Synthetic Biology and Bioenergy Group, J. Craig Venter Institute, La Jolla, California 92037

Correspondence: jglass@jcvi.org

A minimal cell is one whose genome only encodes the minimal set of genes necessary for the cell to survive. Scientific reductionism postulates the best way to learn the first principles of cellular biology would be to use a minimal cell in which the functions of all genes and components are understood. The genes in a minimal cell are, by definition, essential. In 2016, synthesis of a genome comprised of only the set of essential and quasi-essential genes encoded by the bacterium *Mycoplasma mycoides* created a near-minimal bacterial cell. This organism performs the cellular functions common to all organisms. It replicates DNA, transcribes RNA, translates proteins, undergoes cell division, and little else. In this review, we examine this organism and contrast it with other bacteria that have been used as surrogates for a minimal cell.

In 2016, the construction of a bacterial cell that encoded only the genes necessary for cell life and rapid cell growth in rich laboratory media was announced. The creation of this "minimal cell" was widely heralded as a scientific milestone. The organism was called JCVI-Syn3.0. It has a smaller genome (531,490 bp) than that of any known organism that can be grown in axenic culture. There are only 438 protein-coding genes and 35 RNA-coding genes (Hutchison et al. 2016). Although there have been more than 2000 scientific papers published that mentioned minimal bacterial cells, and creating one has been an ambition of the scientific community since the 1930s, the bacterium JCVI-Syn3.0 is the first real minimal cell. This review discusses the concept of a minimal cell, why they are important, the evolution of that concept, and how the creation of a living minimal cell has changed that concept.

THE MINIMAL CELL CONCEPT

In the 19th and 20th centuries, physicists used the hydrogen atom to gain deep understanding about atomic structure and the nature of matter. Their reasoning was what is true for hydrogen, the simplest of all atoms, is also true for the other more complex elements. Even today, the hydrogen atom is still a tool for the investigation matter because some calculations about atomic structure are just too complex for the other atoms. By analogy, the minimal cell is the hydrogen atom of biology.

Even though cells are the simplest units of life, a huge number of different interacting parts comprise even the simplest cells. Motivated by the same reductionist view that led physicists to study the hydrogen atom, in the 1930s, physicist Max Delbrück founded the American Phage Group. This was a group of physicists, chemists,

Cite this article as *Cold Spring Harb Perspect Biol* doi: 10.1101/cshperspect.a023861

and biologists who reasoned that an understanding of the first principles of cellular life would come through study of the simplest biological systems (Morange 2000). They adopted a reductionist approach to understand how life works by identifying and determining how each essential element in a living cell functions (obviously, the technology of the day was not up to the task): hence, the minimal cell concept. A minimal cell would contain only the essential genes for independent growth under ideal laboratory conditions. It would live in a stress-free environment where all necessary nutrients would be provided in the growth media. No single gene can be removed without loss of viability. A minimal cell has all of the machinery for independent cellular life. There would be no unnecessary redundancy. From there, one can imagine that if the function(s) of every gene is determined, then it may be possible to achieve a complete understanding of what it takes to be alive. Additionally, it may be possible to model the minimal cell's behavior on a computer and eventually predict the effects of environmental variations or the effects of added metabolic pathways, etc. From that understanding of the first principles of cellular life, one may be able to build cells that are more complex by addition of genes and metabolic pathways to the minimal cell.

ESSENTIAL, QUASI-ESSENTIAL, AND NONESSENTIAL GENES

In discussion of minimal cells, the term "essential genes" is often used. Genes code for proteins or RNAs with precise functions that may be enzymatic, regulatory, or structural. In the case of essential genes, they code for essential functions. Part of establishing what genes to include in a minimal cell involves classifying them as to whether their functions are essential or not. In a given bacterium, a gene is essential (E) if the cell in which that gene is inactivated cannot be indefinitely propagated. The cell may continue to divide for a time, but once the E gene product is sufficiently diluted among the progeny, no more growth occurs. In this case, the gene specifies an essential function that is not supplied by any other gene in the cell. However, if this same function is supplied by two different either homologous or nonhomologous genes, then in the usual single-gene knockout studies, these two genes are both classed as nonessential, even though the genetic function they specify is an essential function. The pair of genes comprises what is called a "synthetic lethal." Essentiality is a measure of genetic redundancy. If a genome is completely redundant, there are no E genes. On the other hand, if there is no genetic functional redundancy, then E genes correspond one-to-one with essential genetic functions.

A gene is nonessential (NE) if it can be inactivated without affecting the viability or growth rate of the cell for a specific environment. It can specify either a nonessential genetic function or an essential genetic function. In the latter case, it is dispensable because another gene in the cell supplies the same essential function.

A third class of genes, quasi-essential (QE) genes, which when disrupted result in impaired growth, has been observed in several studies when pools of transposon mutagenized cells are grown competitively (Smith et al. 1996; Badarinarayana et al. 2001; Lluch-Senar et al. 2015; Hutchison et al. 2016). The degree of growth impairment can vary from modest to severe. QE genes specify genetic functions that are important for robust growth, but are not strictly essential. As an example, several QE genes could be collectively involved in supplying a critical essential function such as transport of an amino acid. Deletion of any one gene might make the transport less efficient, but two or more deletions might eliminate the transport and produce a lethal effect. In general, if several QE genes are involved in a similar function, some combinations may act as synthetic lethals when simultaneously disrupted.

MYCOPLASMAS ARE NEAR-MINIMAL CELLS

Mycoplasmas are a group of bacteria characterized their lack of a cell wall, obligate parasitic lifestyle, metabolic simplicity, and most impor-

 Cite this article as *Cold Spring Harb Perspect Biol* doi: 10.1101/cshperspect.a023861

tantly, for this review, small genomes. Long before the advent of the genomic era, it was clear to many biologists that the mycoplasmas were already near-minimal cells. In 1984, physicist Harold Morowitz, recognizing that the mycoplasmas were the simplest cells capable of autonomous growth, proposed that these bacteria be used as models for understanding the basic principles of life (Morowitz 1984). Although the mycoplasmas are often called atypical bacteria because of their salient characteristics, for minimal cell purposes, the exact opposite is true. Mycoplasmas are excellent embodiments of what is constant in all bacteria and indeed in all cellular life. Although mycoplasmas are the simplest cells capable of independent growth in laboratory media, they did not evolve as simpler forms of life. They are not ancient bacteria found at the base of the tree of life. Mycoplasmas are great models for minimal cells because of how they evolved. They did not originally evolve with small genomes and the other unusual features. Rather, they descended from conventional bacteria of the Firmicutes class (e.g., *Bacillus subtilis* or *Staphylococcus aureus*) through a process of massive gene loss, which was presumably allowed because they adopted parasitic lifestyles constraining them to live in highly nutrient-rich, stable environments (Woese et al. 1980). Indeed, the three bacteria with the smallest genomes that can be grown in axenic culture are all parasites of the human urogenital tract (this suggests that the mammalian urogenital system may be the most stress-free, hospitable-to-bacteria environment on earth). One of those three species, *Mycoplasma genitalium*, has the smallest genome at 580,076 (Fraser et al. 1995). It has long been referred to as a minimal or near-minimal cell and used as a surrogate for a minimal cell in many studies. The genes that remain in modern mycoplasmas likely consist of only the minimum number of genes needed for life in their natural habitat. This is because mycoplasmas are under some evolutionary pressure to discard NE genes. Thus, the mycoplasmas have very few redundant genes. Furthermore, their obligate parasitic lifestyle enables easy access to almost all their needed nutrients. These are imported rather than synthesized, which can be done with fewer genes. Essentially, all that mycoplasmas do is synthesize DNA, RNA, and proteins from imported precursors and replicate themselves. Those functions are done the same way in all living cells. Although the idea of using mycoplasmas as starting points for minimal cell studies predates the first genome sequences, it was the availability of whole-genome sequences for mycoplasma species (as well as other bacteria) that really jump-started minimal cell studies.

COMPUTATIONAL DETERMINATION OF MINIMAL GENE SETS

In 1995, the era of genome sequencing began with the shotgun sequencing and computational assembly of the *M. genitalium* and *Haemophilus influenza* Rd bacterial genomes. This ushered in a revolution in microbial genome sequencing. Not long after there were two genomes to compare, comparative genomics became a new field. In 1996, Mushegian and Koonin compared orthologs between the two genomes, one Gram- negative and one Gram-positive, and determined a set of 256 genes thought to specify the core functions of a minimal cell. To supply a few apparently missing essential functions not represented by orthologs, they proposed the concept of "nonorthologous gene displacements," representing recruitment of unrelated or distantly related proteins for the same function (Mushegian and Koonin 1996). Nonorthologous gene displacements may be independently evolved proteins or simply proteins diverged too far to have recognizable relatedness.

Koonin proposed one possible mechanism for establishing nonorthologous gene displacements (Koonin et al. 1996). Suppose a cell has an essential genetic function that is specified by two separate nonhomologous genes. These form an essential function redundancy. As rounds of cell division take place, an occasional daughter cell may lose one member of the pair. Eventually, two lineages will be established, each with a single gene providing the essential function. When the genomes of the two lineages are compared, one gene will be identified as supplying the essential function, but in the other lineage, no orthologous gene can be identified as supplying that essential function. It is necessary to postulate that a nonorthologous displacement has taken place. The apparently missing essential genetic function is actually present, but is unrecognized because it is provided by a gene that is not orthologous to the known gene. Over evolutionary time, this happens many times and gradually the overlap of recognizable orthologs between different bacterial species decreases. Koonin (2003) found that the intersection of orthologs common to a collection of organisms decreases to a low of 65 when comparing all sequenced organisms available in 2003. Charlebois and Doolittle (2004) compared 147 prokaryotic genomes available in 2004 and found less than 50 orthologous genes in common, most of which were involved in translation. Even in the relatively narrow mycoplasma family, comparison of 20 sequenced strains yields a core of only 196 orthologs (Liu et al. 2012). Nonorthologous gene displacements seriously impact the usefulness of comparative genomics for determining minimal gene sets.

Still, through rigorous curation, multi-bacteria comparative genomics studies can be useful. In one of the most often cited comparative genomics studies, Gil and collaborators present a set of 206 protein-coding genes that are highly conserved in both near-minimal organisms such as *M. genitalium* and endosymbiotic bacteria such as *Buchnera aphidicola,* as well as in the biotechnology workhorse species *Escherichia coli* and *B. subtilis.* The genes are described as the core elements of a possible minimum gene set necessary to sustain life (Gil et al. 2004). Numerous instances of nonorthologous gene displacement are explained showing how some E bacterial gene sets only appear to differ. These genes are almost all present in both the minimal cell JCVI-Syn3.0 and in the near-minimal species *M. genitalium.* Members of the E gene set in Gil et al. (2004) study not present in the mycoplasmas generally have the necessary cellular tasks performed by different genes.

EXPERIMENTAL DETERMINATION OF ESSENTIAL GENE SETS BY SINGLE-GENE INACTIVATION STUDIES

Hutchison and collaborators introduced global transposon mutagenesis for identifying NE genes in bacterial genomes (Hutchison et al. 1999). Their work was done in two closely related *Mycoplasma* species. The Tn4001 transposon carried on an *E. coli* plasmid, which did not replicate in mycoplasma, was introduced by electroporation into an *M. genitalium* culture. Then, cells with transposon insertions were selected on gentamycin. The principle here is that the transposon, which contains an antibiotic resistance marker, randomly inserts in the *M. genitalium* genome. Insertions in NE genes result in antibiotic resistant mutants. Insertions in E genes result in nonviable mutants. After more than 30 generations of growth, the locations of insertions in the genome were determined by inverse polymerase chain reaction (PCR) and sequencing from the transposon ends to determine junction points between Tn4001 and the mycoplasma genome. In parallel experiments, *Mycoplasma pneumoniae,* which contains orthologs of all but one of the 482 *M. genitalium* genes, plus an additional 197 genes, was mutagenized. For the combined data, 1354 unique insertions were obtained. Insertions were found in 140 *M. genitalium* genes and 179 *M. pneumoniae* genes. Combined analysis based on shared orthologs between the two species led to an estimate of between 265 and 351 E genes in *M. genitalium* (Hutchison et al. 1999). In a follow-up study, Glass and collaborators reanalyzed *M. genitalium,* this time actually isolating individual clones with insertions in different genes. The number of clones with inserts in NE genes reached a level of 101 insertions asymptotically (Table 1) (Glass et al. 2006). Likely, many of the NE genes identified only in the initial study were not found in 2006 because they were actually QE mutants, and grew too slowly to be recovered.

In 2008, a Tn4001 transposon study was done in *Mycoplasma pulmonis,* which identified insertion sites in 1700 clones (French et al. 2008). The 963-kb genome has 789 annotated

 Cite this article as *Cold Spring Harb Perspect Biol* doi: 10.1101/cshperspect.a023861

Table 1. Essential genes in *Mycoplasma* species as determined by transposon bombardment

Mycoplasma species	Genome size (kb)	Total genes	NE genes	Total–NE
M. mycoides JCVI-Syn1.0[a]	1080	901	432	469
M. pulmonis[b]	963	789	321	468
M. pneumoniae[c]	816	739	259	480
M. genitalium[d]	580	507	101	406
M. mycoides JCVI-Syn2.0[e]	576	516	90	426
M. mycoides JCVI-Syn3.0[e]	531	473	48	425

NE, Nonessential.

[a]Gibson et al. 2010.

[b]French et al. 2008.

[c]Lluch-Senar et al. 2015. Note that the small previously unannotated noncoding RNAs found by these researchers were not included in these calculations.

[d]Glass et al. 2006.

[e]Hutchison et al. 2016.

protein coding and RNA-coding genes, of which 321 were inactivated (NE) and thus 468 could be classed as E and QE. Because of the small number of insertions analyzed, 321 might be somewhat lower than the true NE gene value.

As a follow-up to the Hutchison et al. study, preliminary determination of NE genes in *M. pneumoniae* used as a control in to judge when they had achieved saturation of their *M. genitalium* transposon bombardment (Hutchison et al. 1999), Lluch-Senar and collaborators performed a much more extensive and insightful transposon mutagenesis study of *M. pneumoniae* (Lluch-Senar et al. 2015). Cells with mini-transposon inserts were grown in culture for 12 days to identify NE genes. The only cells that persisted were those with transposon insertions in NE genes with little to no growth impairment. They found 259 NE genes and 332 E genes. They also identified 93 fitness (QE) genes using a statistical criterion to identify slow growing mutants that persisted but decreased relative to the overall population with increasing passages.

In the most recent mycoplasma study by Hutchison et al. (2016), the synthetic cell *Mycoplasma mycoides* JCVI-Syn1.0 (Gibson et al. 2010), whose genome was an almost exact copy of the wild-type *M. mycoides*, subspecies *capri* genome was subjected to global Tn5 mini-transposon mutagenesis. Mutagenized cells were plated and selected for puromycin resistance conferred by Tn5 insertions. Colonies were counted and pooled. Inserts were mapped to the genome using inverse PCR (Hutchison et al. 1999, 2016), thus providing a zero passage, P_0 data set. A sample of the pool was subjected to four serial passages of approximately 40 generations and Tn5 inserts were again mapped to the genome giving a P_4 data set. Leaving out RNA-coding genes, the P_4 data set showed disrupting Tn5 insertions in 432 genes. These were classified as NE by the criteria used above. There were 205 E genes without Tn5 inserts in P_0 and 229 growth-impaired QE genes with P_0 inserts, but few if any P_4 inserts.

In addition, designing and synthesizing the genome with most of the NE genes removed produced two reduced versions of Syn1.0. The first design resulted in JCVI-Syn2.0 (576 kb) with 516 genes. Transposon mutagenesis in this new genetic background yielded 90 apparent NE genes. Some of these had been classified as QE, but in the new slightly slower growing JCVI-Syn2.0, now appeared NE. The second reduced genome, JCVI-Syn3.0 (531 kb), had 508 total genes and 48 apparent NE genes by Tn5 mutagenesis. Again, many QE genes converted to apparent NE genes in this even slower growing cell (180 min doubling rate vs. 60 min for JVCI-Syn1.0 and 90 min for JCVI-Syn2.0).

Data for the four mycoplasmas (above and in Fig. 1 and Table 1) show that the sum of E and QE genes is approximately constant at around 400 genes, whereas the number of NE genes varies directly with genome size. It is interesting

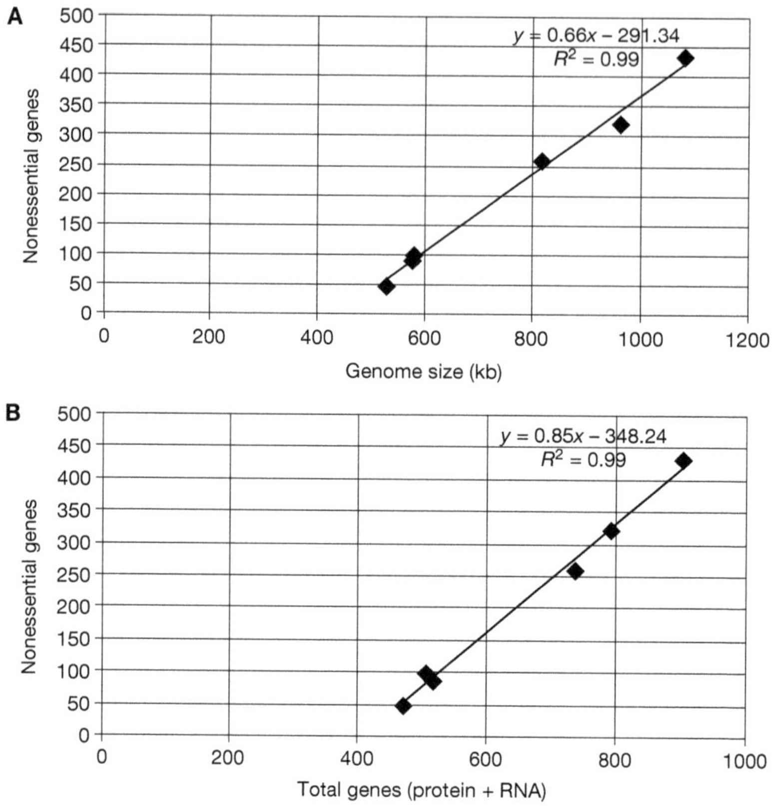

Figure 1. Plots of nonessential (NE) gene number versus genome size and total genes in *Mycoplasma* species and reduced strains. NE genes were determined by transposon mutagenesis in *Mycoplasma genitalium* (580 kb), *Mycoplasma pneumoniae* (816 kb), *Mycoplasma pulmonis* (963 kb), *Mycoplasma mycoides* JCVI-syn1.0 (1080 kb), *M. mycoides* JCVI-syn2.0 (576 kb), and *M. mycoides* JCVI-Syn3.0 (531 kb). (*A*) Plotting NE genes versus genome size yields an extrapolated value of 423 kb for the minimal genome size when NE = 0. (*B*) Plotting NE genes versus total genes yields an extrapolated value of 413 genes when NE = 0.

that when NE genes are plotted against genome size, extrapolation to NE gene to zero yields a minimal genome size of ~423 kb. Similarly, plotting NE genes versus total genes yields an extrapolated value of 413 genes when NE = 0.

GENOME MINIMIZATION BY TOP-DOWN SERIAL DELETIONS

Two different approaches to genome minimization, bottom-up and top-down, have been used. The bottom-up approach involves design, synthesis, and installation of minimal or reduced genomes based on knowledge of gene sets that supply the essential functions for life. In the top-down approach, one starts with a viable "natural" cell and sequentially removes genes, testing for viability and appropriate phenotypic properties at each step. This process can be continued until the genome is substantially reduced, or until no more genes can be removed without severe growth impairment or loss of viability.

Top-down reduction of the *M. mycoides* JCVI-Syn1.0 genome (Gibson et al. 2010) by consecutive scar-less deletion of NE gene clusters has been recently reported (Hutchison et al. 2016). The 1078-kb JCVI-Syn1.0 genome that has 901 protein and RNA-coding genes was subjected to 22 consecutive deletions of multigene clusters resulting in removal of 255 genes (28%) and 357 kb of DNA (33%) based on knowledge of individual gene essentiality without significantly affecting the growth rate. As more and more genes were removed from the genome, the rate of reduction markedly slowed as NE gene cluster sizes decreased toward single genes. Removal of some genes, because one member of a synthetic lethal pair had already been removed, proved impossible. In other instances, gene removal resulted in an unacceptable slowing of growth rate.

Although efforts to produce a minimal cell whose genome only encoded the minimal set of genes necessary for cell survival have focused on mycoplasmas, there are genome minimization projects that focus on other bacteria. The biotechnology workhorse species, *E. coli* and *B. subtilis*, are the subjects of several minimization efforts that seek to produce superior organisms for research and industry. Progress on reduction of *E. coli* and *B. subtilis* genomes was recently reviewed (Juhas et al. 2014).

In the first attempt at reduction of the *E. coli* genome in 2002 (Kolisnychenko et al. 2002), 12 strain-specific genomic islands (KI-islands) representing 8.1% of the genome were removed by scar-less deletion. Also, in 2002, Yu et al. (2002) introduced the Cre/loxP excision method for random removal of nonessential regions and reported a 6.7% reduction of *E. coli*. Hashimoto et al. (2005) constructed a series of deleted *E. coli* strains ranging up to 29% genome reduction. Significant cell morphological changes were observed in the 29% deleted strain, but this study did show that large parts of the genome could be removed without marked effects on cell growth. Posfai et al. (2006) used scar-less deletion to remove mobile elements, pathogenicity islands, and certain other nonessential regions totaling 15.3% of the E. *coli* genome. The resulting strain had good growth characteristics, improved electroporation efficiency, and decreased mutation rates. In 2008, Mizoguchi et al. (2008) and Kato and Hashimoto (2008) produced strains of *E. coli* lacking 22% and 30% of the genome, respectively. More recently, two *E. coli* strains lacking 29% and 35.2% of the genome were constructed (Hirokawa et al. 2013).

Several groups have achieved sizable reductions of the *B. subtilis* genome ranging from 20% to 35%. In 2008, Morimoto et al. (2008) constructed a reduced *B. subtilis* strain MGB874 genome lacking 874 kb (20.7%) of genomic DNA by consecutive deletion of 11 dispensable regions ranging from 11 kb to 195 kb in size with little effect on growth rate. Recently, Tanaka et al. (2013) and Reuss et al. (2017) produced *B. subtilis* strains lacking ~35% of the genome by sequential deletions.

Although these strains of *E. coli* and *B. subtilis* with reduced genomes have some interesting properties, this is probably not a practical approach to achieving a minimal genome nor was this the intention of these studies. In the next section, we present work showing that the bottom-up synthetic path can yield an approximately minimal genome when performed in the low genetic redundancy mycoplasmas.

GENOME MINIMIZATION BY BOTTOM-UP DESIGN AND SYNTHESIS

In 1999, Hutchison and coworkers proposed building a cassette-based minimal cell based on a mycoplasma chassis, but this was not feasible at the time (Cho et al. 1999; Hutchison et al. 1999). In 2006, Forster and Church proposed a minimal set of 151 genes (113 proteins and 38 RNAs) encoded by a 113-kb genome that could constitute an in vitro system capable of replication and evolution. The system would be fed by needed small molecules through a membrane (Forster and Church 2006). Although this system was not cellular and was never successfully tested, it was a concrete proposal of a bottom-up synthetic genome design with some of the attributes of a living cell. Only very recently, Hutchison et al. (2016) successfully accomplished building a minimal cell using a bottom-up approach. A number of advances were required to make the synthetic approach feasible. Methods for synthesis of large DNA molecules and for installation and booting up of a naked synthetic DNA genome in recipient cell cytoplasm (genome transplantation) were developed (Lartigue et al. 2007; Gibson et al. 2008; Benders et al. 2010). Better methods for classification of genes as E, QE, or NE were developed (Hutchison et al. 2016). These new methods allowed minimal genome designs to be made, built, and tested. A design-build-test cycle was developed so that various designs could be tested for viability, modified as necessary, and then rebuilt and retested. For troubleshooting, genomes were built in eight overlapping segments, each representing about 1/8th of the genome. Individual synthetic reduced segments could then be tested for viability in the context of the other seven wild-type segments. Viable reduced segments could be put together in various combinations with wild-type segments to obtain partially reduced genomes for testing, and then finally viable genomes made only from reduced segments of various designs could be assembled and tested. Using these methods, Hutchison and coworkers started with *M. mycoides* JCVI-Syn1.0 (1078 kb, 863 protein, and 38 RNA-coding genes), the synthetic genome published in 2010 (Gibson et al. 2010), and in a series of three design, synthesis, and test cycles, constructed an ~50% reduced version of the *M. mycoides* JCVI-Syn1.0 genome with >90% of the NE genes removed. This new "minimal" cell, JCVI-Syn3.0 (531 kb, 438 protein-coding, and 35 RNA-coding genes) grows with a doubling time of 180 min compared with 60 min for syn1.0. Syn3.0 and has a substantially smaller genome than that of any known natural cell capable of independent growth. How close to "minimal" is JCVI-Syn3.0? Hutchison and colleagues performed Tn5 mutagenesis on JCVI-Syn3.0 as reported in the supplementary materials and found 53 genes that did not affect growth rate in the slower growing JCVI-Syn3.0 cells. Most of these had been classified as QE genes in *M. mycoides* JCVI-Syn1.0 cells previously and had been retained in JCVI-Syn3.0. Adding the data for JCVI-Syn3.0 to the graph in Figure 1 suggests that additional genes could be removed to achieve a fully minimal cell. Total genes extrapolate to 413 when NE genes equals zero. Because JCVI-Syn3.0 has 473 genes (including the 35 RNA genes), as many as 60 genes might be removable while still retaining a viable cell. However, as the investigators point out, growth rate would likely be further impacted.

THE GENES IN MINIMAL CELL JCVI-Syn3.0

The precise functions of 35 RNA genes and 240 protein-coding genes are known. Not unexpectedly, the largest category of genes in the minimal cell is tasked with the expression of genome information (195, 41%). Similarly, the cytosolic metabolism (81, 17%), cell membrane (79, 17%), and preservation of genome information (34, 7%) categories comprise significant fractions of the total gene content as would be expected. A more detailed breakdown of how many JCVI-Syn3.0 genes are involved with different cellular processes is shown in Table 2. What was unexpected was that 65 genes have unknown functions and 84 have only generic

Table 2. JCVI-Syn3.0 genes listed by functional category

Functional category	Number of genes
Translation	89
Unassigned	79
RNA (rRNAs, tRNAs, small RNAs)	35
Membrane transport	31
Cofactor transport and salvage	21
Lipid salvage and biogenesis	21
Nucleotide salvage	19
tRNA modification	17
DNA replication	16
Glucose transport and glycolysis	15
Lipoprotein	15
Ribosome biogenesis	14
rRNA modification	12
Metabolic processes	10
Protein export	10
Proteolysis	10
Transcription	9
Regulation	9
Efflux	7
RNA metabolism	7
DNA repair	6
DNA topology	5
Redox homeostasis	4
Chromosome segregation	3
DNA metabolism	3
Protein folding	3
Transport and catabolism of nonglucose carbon sources	2
Cell division	1
TOTAL GENES	373

assignments such as hydrolase, esterase, ATPase, etc. That we have no clear idea of the functions performed by 149 of 473 genes in the minimal gene set makes it clear how incomplete our knowledge of cellular biology really is. To fully understand how life is specified by the genes of a minimal cell, one must know all of the functions of the cell components.

COMPARISON OF MINIMAL CELL JCVI-Syn3.0 AND *Mycoplasma genitalium* GENE SETS BASED ON COMPARATIVE GENOMICS AND TRANSPOSON BOMBARDMENT

JCVI-Syn3.0 was designed to contain only the minimal set of genes needed in an environment where all needed nutrients were provided in the growth media and there were no environmental stressors. The genome encodes 438 protein-coding genes and 35 RNA-coding genes. Given that *M. genitalium* is often used as a surrogate for a minimal cell and given that extensive transposon bombardment studies have been reported for *M. genitalium* and JCVI-Syn3.0, we compared those two E gene sets. There are 29 disrupted *M. genitalium* genes that have homologs in JCVI-Syn3.0. For five of those homologous pairs, the JCVI-Syn3.0 gene was NE but just had not been deleted from the genome. For 21, the JCVI-Syn3.0 gene was QE. Because no growth rate data for *the M. genitalium* gene disruption mutants are available, we cannot tell if those mutations cause impaired growth, but it is likely. There were 3 JCVI-Syn3.0 E genes that could be disrupted in *M. genitalium*: chromosome segregation and condensation protein B (MG_214 and JCVISYN3_0328), DNA polymerase III, delta subunit (MG_315 and JCVISYN3_0826), and an ATP-binding cassette (ABC) transporter, ATP-binding protein (MG_290 and JCVISYN3_0707). The fact that JCVI-Syn3.0 divides every 3 hours and *M. genitalium* divides every 12–15 hours under the same growth conditions could be a reason these genes are essential in one species and not in the other. Of 438 protein-coding genes in JCVI-Syn3.0, 288 have reciprocal BLASTp homologs in *M. genitalium,* leaving 150 genes unpaired. Similarly, of the 383 protein coding *M. genitalium* genes that were not disrupted in the Glass et al. (2006) transposon bombardment study, 89 had no homolog in JCVI-Syn3.0. The vast majority of the genes in each organism lacking a homolog in the other encode hypothetical proteins and membrane associated proteins. Many more homologous pairs would likely be found if less stringent criteria for homology were applied. Plus, it is worth noting that *M. mycoides* (the organism minimized to produce JCVI-Syn3.0) and *M. genitalium* come from completely different branches of mycoplasma phylogeny. By the Koonin model for nonorthologous gene displacement (Koonin et al. 1996), it is likely that the two species have many genes performing the same functions but which have evolved so that there is no longer sequence similarity. Likely, as the functions of more genes of unknown function in these species are determined, the apparent difference in their gene content will diminish.

MINIMAL CELL COMPUTATIONAL MODELS

As stated at the beginning of this review, one of the aspirations for the minimal cell is to learn the functions of all the gene products and cell components. Those data would be used to create a computational model of the cell that would mimic cell behavior and accurately predict how the cell would respond to genome changes or environmental stress. Richard Feynman's statement "What I cannot create, I do not understand" (quoted in Gleick 1992), which has become a mantra for the synthetic biology community, is applicable here. Until we understand all aspects of the minimal cell well enough to build computational models that can replicate minimal cell biology, we do not understand the cell. Thus, computational models that consider all of the minimal cell systems will become the yardstick for measuring how well we understand the cell. Already, computational models have been constructed based on the annotated genome of *M. genitalium* that showed promise in embodying several aspects of the biology of the organism (Tomita et al. 1999; Karr et al. 2012); however, they are far from perfect. Efforts are now underway by several groups to develop computational models for JCVI-Syn3.0.

EPILOGUE

Recently, Bruce Alberts, former president of the United States National Academy of Sciences, wrote about the astonishing finding that 149 genes in the JCVI-Syn3.0 minimal cell were of unknown function. "Hundreds of talented young scientists should be leaping to fill this huge gap in understanding of fundamental biological mechanisms, perhaps earning several Nobel Prizes along the way. Over the long term, such results are certain to lead to powerful new approaches for improving human health and welfare" (Alberts 2016). Having a minimal bacterial cell has been a long-standing goal of cell biologists. No longer are we limited to working with imaginary minimal cells or naturally occurring or naturally occurring organisms with small genomes as surrogates. A minimal cell has now been constructed. Clearly, there is much about its biology that biologists do not understand. First principles of cellular life are waiting to be discovered.

ACKNOWLEDGMENTS

We thank Synthetic Genomics (SGI), the Defense Advanced Research Projects Agency's Living Foundries program (contract HR0011-12-C-0063), and the J. Craig Venter Institute for funding this work. We also thank the members of the J. Craig Venter Institute and SGI who contributed to the creation of the minimal cell, JCVI-Syn3.0.

REFERENCES

Alberts B. 2016. Ensuring an innovative and productive future for the next generation of scientists: The 2016 Lasker-Koshland special achievement award in medical science. *JAMA* **316:** 1256–1257.

Badarinarayana V, Estep PWIII, Shendure J, Edwards J, Tavazoie S, Lam F, Church GM. 2001. Selection analyses of insertional mutants using subgenic-resolution arrays. *Nat Biotechnol* **19:** 1060–1065.

Benders GA, Noskov VN, Denisova EA, Lartigue C, Gibson DG, Assad-Garcia N, Chuang RY, Carrera W, Moodie M, Algire MA, et al. 2010. Cloning whole bacterial genomes in yeast. *Nucleic Acids Res* **38:** 2558–2569.

Charlebois RL, Doolittle WF. 2004. Computing prokaryotic gene ubiquity: Rescuing the core from extinction. *Genome Res* **14:** 2469–2477.

Cho MK, Magnus D, Caplan AL, McGee D. 1999. Policy forum: Genetics. Ethical considerations in synthesizing a minimal genome. *Science* **286:** 2087, 2089–2090.

Forster AC, Church GM. 2006. Towards synthesis of a minimal cell. *Mol Syst Biol* **2:** 45.

Fraser CM, Gocayne JD, White O, Adams MD, Clayton RA, Fleischmann RD, Bult CJ, Kerlavage AR, Sutton G, Kelley JM, et al. 1995. The minimal gene complement of *Mycoplasma genitalium*. *Science* **270:** 397–403.

French CT, Lao P, Loraine AE, Matthews BT, Yu H, Dybvig K. 2008. Large-scale transposon mutagenesis of *Mycoplasma pulmonis*. *Mol Microbiol* **69:** 67–76.

Gibson DG, Benders GA, Andrews-Pfannkoch C, Denisova EA, Baden-Tillson H, Zaveri J, Stockwell TB, Brownley A, Thomas DW, Algire MA, et al. 2008. Complete chemical synthesis, assembly, and cloning of a *Mycoplasma genitalium* genome. *Science* **319:** 1215–1220.

Gibson DG, Glass JI, Lartigue C, Noskov VN, Chuang RY, Algire MA, Benders GA, Montague MG, Ma L, Moodie MM, et al. 2010. Creation of a bacterial cell controlled by a chemically synthesized genome. *Science* **329:** 52–56.

Gil R, Silva FJ, Pereto J, Moya A. 2004. Determination of the core of a minimal bacterial gene set. *Microbiol Mol Biol Rev* **68:** 518–537.

Glass JI, Assad-Garcia N, Alperovich N, Yooseph S, Lewis MR, Maruf M, Hutchison CAIII, Smith HO, Venter JC. 2006. Essential genes of a minimal bacterium. *Proc Natl Acad Sci* **103:** 425–430.

Gleick J. 1992. *Genius: The life and science of Richard Feynman*. Random House, New York.

Hashimoto M, Ichimura T, Mizoguchi H, Tanaka K, Fujimitsu K, Keyamura K, Ote T, Yamakawa T, Yamazaki Y, Mori H, et al. 2005. Cell size and nucleoid organization of engineered *Escherichia coli* cells with a reduced genome. *Mol Microbiol* **55:** 137–149.

Hirokawa Y, Kawano H, Tanaka-Masuda K, Nakamura N, Nakagawa A, Ito M, Mori H, Oshima T, Ogasawara N. 2013. Genetic manipulations restored the growth fitness of reduced-genome *Escherichia coli*. *J Biosci Bioeng* **116:** 52–58.

Hutchison CA, Peterson SN, Gill SR, Cline RT, White O, Fraser CM, Smith HO, Venter JC. 1999. Global transposon mutagenesis and a minimal Mycoplasma genome. *Science* **286:** 2165–2169.

Hutchison CA3rd, Chuang RY, Noskov VN, Assad-Garcia N, Deerinck TJ, Ellisman MH, Gill J, Kannan K, Karas BJ, Ma L, et al. 2016. Design and synthesis of a minimal bacterial genome. *Science* **351:** aad6253.

Juhas M, Reuss DR, Zhu B, Commichau FM. 2014. *Bacillus subtilis* and *Escherichia coli* essential genes and minimal cell factories after one decade of genome engineering. *Microbiology* **160:** 2341–2351.

Karr JR, Sanghvi JC, Macklin DN, Gutschow MV, Jacobs JM, Bolival BJr, Assad-Garcia N, Glass JI, Covert MW. 2012. A whole-cell computational model predicts phenotype from genotype. *Cell* **150:** 389–401.

Kato J, Hashimoto M. 2008. Construction of long chromosomal deletion mutants of *Escherichia coli* and minimization of the genome. *Methods Mol Biol* **416:** 279–293.

Kolisnychenko V, Plunkett GIII, Herring CD, Feher T, Posfai J, Blattner FR, Posfai G. 2002. Engineering a reduced *Escherichia coli* genome. *Genome Res* **12:** 640–647.

Koonin EV. 2003. Comparative genomics, minimal gene-sets and the last universal common ancestor. *Nat Rev Microbiol* **1:** 127–136.

Koonin EV, Mushegian AR, Bork P. 1996. Non-orthologous gene displacement. *Trends Genet* **12:** 334–336.

Lartigue C, Glass JI, Alperovich N, Pieper R, Parmar PP, Hutchison CAIII, Smith HO, Venter JC. 2007. Genome transplantation in bacteria: Changing one species to another. *Science* **317:** 632–638.

Liu W, Fang L, Li M, Li S, Guo S, Luo R, Feng Z, Li B, Zhou Z, Shao G, et al. 2012. Comparative genomics of Mycoplasma: Analysis of conserved essential genes and diversity of the pan-genome. *PLoS ONE* **7:** e35698.

Lluch-Senar M, Delgado J, Chen WH, Llorens-Rico V, O'Reilly FJ, Wodke JA, Unal EB, Yus E, Martinez S, Nichols RJ, et al. 2015. Defining a minimal cell: Essentiality of small ORFs and ncRNAs in a genome-reduced bacterium. *Mol Syst Biol* **11:** 780.

Mizoguchi H, Sawano Y, Kato J, Mori H. 2008. Superpositioning of deletions promotes growth of *Escherichia coli* with a reduced genome. *DNA Res* **15:** 277–284.

Morange M. 2000. *A history of molecular biology.* Harvard University Press, Cambridge.

Morimoto T, Kadoya R, Endo K, Tohata M, Sawada K, Liu S, Ozawa T, Kodama T, Kakeshita H, Kageyama Y, et al. 2008. Enhanced recombinant protein productivity by genome reduction in *Bacillus subtilis. DNA Res* **15:** 73–81.

Morowitz HJ. 1984. The completeness of molecular biology. *Isr J Med Sci* **20:** 750–753.

Mushegian AR, Koonin EV. 1996. A minimal gene set for cellular life derived by comparison of complete bacterial genomes. *Proc Natl Acad Sci* **93:** 10268–10273.

Posfai G, Plunkett GIII, Feher T, Frisch D, Keil GM, Umenhoffer K, Kolisnychenko V, Stahl B, Sharma SS, de Arruda M, et al. 2006. Emergent properties of reduced-genome *Escherichia coli. Science* **312:** 1044–1046.

Reuss DR, Altenbuchner J, Mader U, Rath H, Ischebeck T, Sappa PK, Thurmer A, Guerin C, Nicolas P, Steil L, et al. 2017. Large-scale reduction of the *Bacillus subtilis* genome: Consequences for the transcriptional network, resource allocation, and metabolism. *Genome Res* **27:** 289–299.

Smith V, Chou KN, Lashkari D, Botstein D, Brown PO. 1996. Functional analysis of the genes of yeast chromosome V by genetic footprinting. *Science* **274:** 2069–2074.

Tanaka K, Henry CS, Zinner JF, Jolivet E, Cohoon MP, Xia F, Bidnenko V, Ehrlich SD, Stevens RL, Noirot P. 2013. Building the repertoire of dispensable chromosome regions in *Bacillus subtilis* entails major refinement of cognate large-scale metabolic model. *Nucleic Acids Res* **41:** 687–699.

Tomita M, Hashimoto K, Takahashi K, Shimizu TS, Matsuzaki Y, Miyoshi F, Saito K, Tanida S, Yugi K, Venter JC, et al. 1999. E-CELL: Software environment for whole-cell simulation. *Bioinformatics* **15:** 72–84.

Woese CR, Maniloff J, Zablen LB. 1980. Phylogenetic analysis of the mycoplasmas. *Proc Natl Acad Sci* **77:** 494–498.

Yu BJ, Sung BH, Koob MD, Lee CH, Lee JH, Lee WS, Kim MS, Kim SC. 2002. Minimization of the *Escherichia coli* genome using a Tn5-targeted Cre/loxP excision system. *Nat Biotechnol* **20:** 1018–1023.

Genome-Editing Technologies: Principles and Applications

Thomas Gaj,[1,4] Shannon J. Sirk,[2] Sai-lan Shui,[3] and Jia Liu[3,4]

[1]Department of Bioengineering, University of California, Berkeley, California 94720

[2]Department of Chemical Engineering, Stanford University, Stanford, California 94305

[3]Shanghai Institute for Advanced Immunochemical Studies, ShanghaiTech University, Shanghai, China

Correspondence: gaj@berkeley.edu; liujia@shanghaitech.edu.cn

Targeted nucleases have provided researchers with the ability to manipulate virtually any genomic sequence, enabling the facile creation of isogenic cell lines and animal models for the study of human disease, and promoting exciting new possibilities for human gene therapy. Here we review three foundational technologies—clustered regularly interspaced short palindromic repeats (CRISPR)-CRISPR-associated protein 9 (Cas9), transcription activator-like effector nucleases (TALENs), and zinc-finger nucleases (ZFNs). We discuss the engineering advances that facilitated their development and highlight several achievements in genome engineering that were made possible by these tools. We also consider artificial transcription factors, illustrating how this technology can complement targeted nucleases for synthetic biology and gene therapy.

In recent years, the emergence of highly versatile genome-editing technologies has provided investigators with the ability to rapidly and economically introduce sequence-specific modifications into the genomes of a broad spectrum of cell types and organisms. The core technologies now most commonly used to facilitate genome editing, shown in Figure 1, are (1) clustered regularly interspaced short palindromic repeats (CRISPR)-CRISPR-associated protein 9 (Cas9), (2) transcription activator-like effector nucleases (TALENs), (3) zinc-finger nucleases (ZFNs), and (4) homing endonucleases or meganucleases.

In particular, the ease with which CRISPR-Cas9 and TALENs can be configured to recognize new genomic sequences has driven a revolution in genome editing that has accelerated scientific breakthroughs and discoveries in disciplines as diverse as synthetic biology, human gene therapy, disease modeling, drug discovery, neuroscience, and the agricultural sciences.

The diverse array of genetic outcomes made possible by these technologies is the result, in large part, of their ability to efficiently induce targeted DNA double-strand breaks (DSBs). These DNA breaks then drive activation of cellular DNA repair pathways and facilitate the introduction of site-specific genomic modifications (Rouet et al. 1994; Choulika et al. 1995). This process is most often used to achieve gene knockout via random base insertions and/or

[4]These authors contributed equally to this work.

Cite this article as *Cold Spring Harb Perspect Biol* doi: 10.1101/cshperspect.a023754

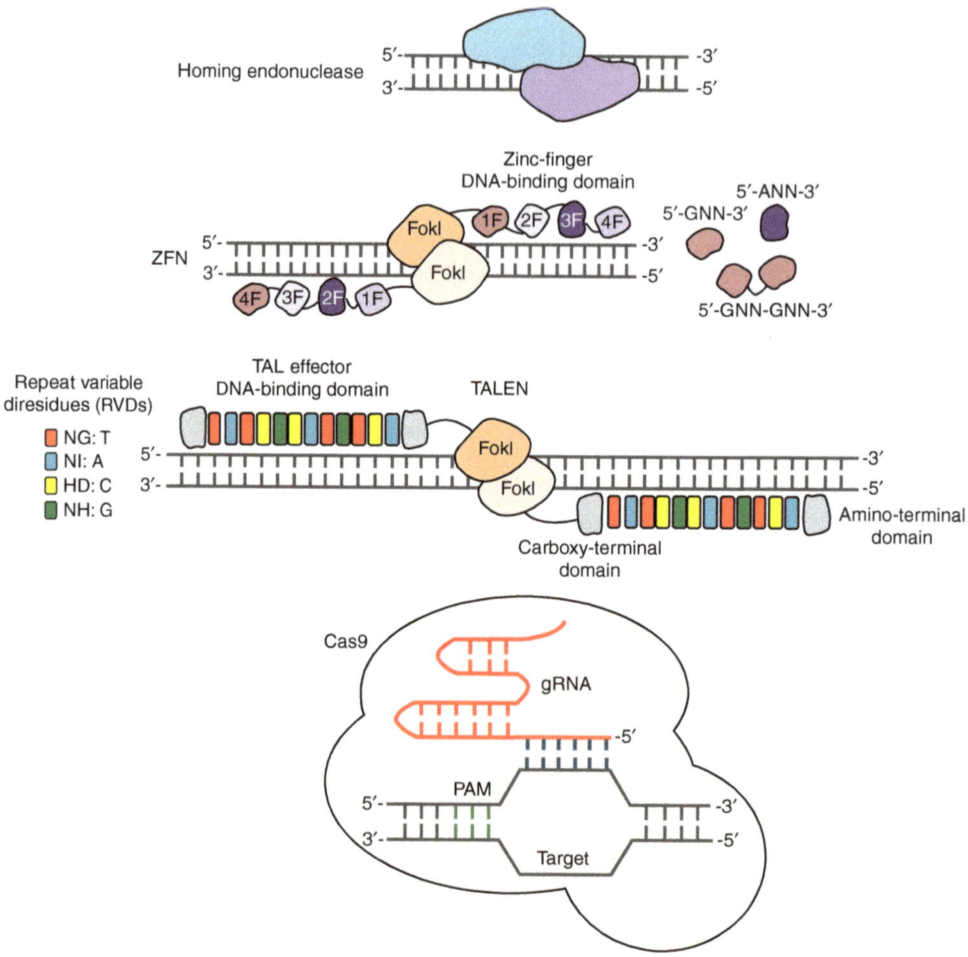

Figure 1. Genome-editing technologies. Cartoons illustrating the mechanisms of targeted nucleases. From *top* to *bottom*: homing endonucleases, zinc-finger nucleases (ZFNs), transcription activator-like effector (TALE) nucleases (TALENs), and clustered regularly interspaced short palindromic repeats (CRISPR)-CRISPR-associated protein 9 (Cas9). Homing endonucleases generally cleave their DNA substrates as dimers, and do not have distinct binding and cleavage domains. ZFNs recognize target sites that consist of two zinc-finger binding sites that flank a 5- to 7-base pair (bp) spacer sequence recognized by the FokI cleavage domain. TALENs recognize target sites that consist of two TALE DNA-binding sites that flank a 12- to 20-bp spacer sequence recognized by the FokI cleavage domain. The Cas9 nuclease is targeted to DNA sequences complementary to the targeting sequence within the single guide RNA (gRNA) located immediately upstream of a compatible protospacer adjacent motif (PAM). DNA and protein are not drawn to scale.

deletions that can be introduced by nonhomologous end joining (NHEJ) (Fig. 2A) (Bibikova et al. 2002). Alternatively, in the presence of a donor template with homology to the targeted chromosomal site, gene integration, or base correction via homology-directed repair (HDR) can occur (HDR) (Fig. 2B) (see Fig. 2 for an overview of other possible genome-editing outcomes) (Bibikova et al. 2001, 2003; Porteus and Baltimore 2003; Urnov et al. 2005). Indeed, the broad versatility of these genome-modifying enzymes is evidenced by the fact that they also serve as the foundation for artificial transcription factors, a class of tools capable of modulating the expression of nearly any gene within a genome.

Cite this article as *Cold Spring Harb Perspect Biol* doi: 10.1101/cshperspect.a023754

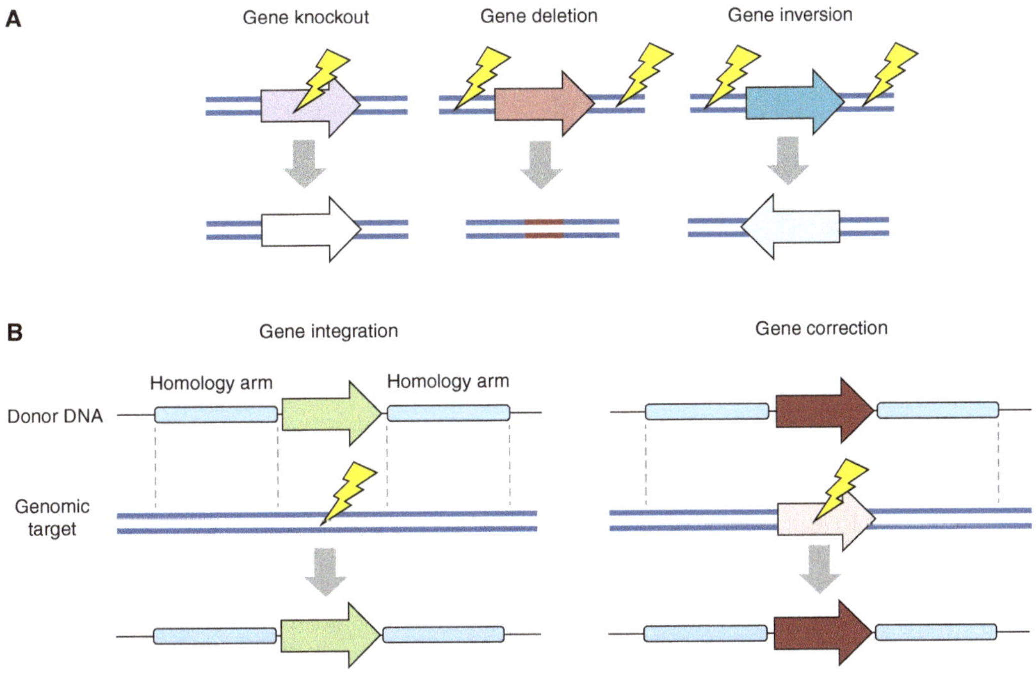

Figure 2. Genome-editing outcomes. Targeted nucleases induce DNA double-strand breaks (DSBs) that are repaired by nonhomologous end joining (NHEJ) or, in the presence of donor template, homology-directed repair (HDR). (*A*) In the absence of a donor template, NHEJ introduces small base insertions or deletions that can result in gene disruption. When two DSBs are induced simultaneously, the intervening genomic sequence can be deleted or inverted. (*B*) In the presence of donor DNA (plasmid or single-stranded oligonucleotide), recombination between homologous DNA sequences present on the donor template and a specific chromosomal site can facilitate targeted integration. Lightning bolts indicate DSBs.

Here we review key principles of genome editing, emphasizing many of the engineering advances that have laid the groundwork for the creation, refinement, and implementation of the current suite of genome-modifying tools. We also provide an overview of the achievements made possible by genome editing, illustrating how this technology can enable advances throughout the life sciences.

TARGETED NUCLEASES

Zinc-Finger Nucleases

ZFNs, which are fusions between a custom-designed Cys_2-His_2 zinc-finger protein and the cleavage domain of the FokI restriction endonuclease (Kim et al. 1996), were the first targeted nuclease to achieve widespread use (Porteus and Carroll 2005; Urnov et al. 2010). ZFNs function as dimers, with each monomer recognizing a specific "half site" sequence—typically nine to 18 base pairs (bps) of DNA—via the zinc-finger DNA-binding domain (Fig. 1). Dimerization of the ZFN proteins is mediated by the FokI cleavage domain, which cuts DNA within a five- to seven-bp spacer sequence that separates two flanking zinc-finger binding sites (Smith et al. 2000). Each ZFN is typically composed of three or four zinc-finger domains, with each individual domain composed of ~30 amino acid residues that are organized in a ββα motif (Pavletich and Pabo 1991). The residues that facilitate DNA recognition are located within the α-helical domain and typically interact with three bps of DNA, with occasional overlap from an adjacent domain (Wolfe et al. 2000). Using methods such as phage display (Choo and Klug 1994; Jamieson et al. 1994; Wu et al. 1995), a large number of zinc-finger domains recognizing distinct DNA triplets have been identified (Segal et al. 1999; Dreier et al. 2001, 2005; Bae et al. 2003). These domains can be fused together in tandem using a canonical linker peptide (Liu et al. 1997) to generate polydactyl zinc-finger proteins that can target a wide range of possible DNA sequences (Beerli et al. 1998, 2000a; Kim et al. 2009). In addition to this "modular assembly" approach to zinc-finger construction, selection-based methods for constructing zinc-finger proteins have also been reported (Greisman and Pabo 1997; Isalan et al. 2001; Hurt et al. 2003; Magnenat et al. 2004), including those that consider context-dependent interactions between adjacent zinc-finger domains, such as oligomerized pool engineering (OPEN) (Maeder et al. 2008). In addition, specialized sets of validated two-finger, zinc-finger modules have been used to assemble zinc-finger arrays (Kim et al. 2009; Bhakta et al. 2013), including those that take context-dependent effects into account (Sander et al. 2011b; Gupta et al. 2012).

One major concern associated with the use of ZFNs for genome editing (in addition to all targeted nucleases) is off-target mutations (Gabriel et al. 2011; Pattanayak et al. 2011). As a result, several approaches have been undertaken to enhance their specificity. Among the most successful of these has been the creation of obligate heterodimeric ZFN architectures that rely on charge–charge repulsion to prevent unwanted homodimerization of the FokI cleavage domain (Miller et al. 2007; Doyon et al. 2011), thereby minimizing the potential for ZFNs to dimerize at off-target sites. Additionally, protein-engineering methods have been used to enhance the cleavage efficiency of the FokI cleavage domain (Guo et al. 2010). One particularly promising approach for improving ZFN specificity is to deliver them into cells as protein. Because of the intrinsic cell-penetrating activity of zinc-finger domains (Gaj et al. 2014a), ZFN proteins themselves are inherently cell-permeable and can facilitate gene editing with fewer off-target effects when applied directly onto cells as purified protein compared to when expressed within cells from nucleic acids (Gaj et al. 2012). Modified ZFN proteins endowed with improved cell-penetrating activity have since been described (Liu et al. 2015a). ZFNickases can also facilitate gene correction in the absence of a DSB (Kim et al. 2012; Ramirez et al. 2012; Wang et al. 2012). These enzymes, which consist of one catalytically inactivated ZFN monomer in combination with a second native ZFN monomer, can stimulate HDR by nicking or cleaving one strand of DNA and are derived from a concept first illustrated by Stoddard and colleagues using homing endonucleases (McConnell Smith et al. 2009).

Unlike TALENs and CRISPR-Cas9, the difficulty associated with constructing zinc-finger arrays has hindered their widespread adoption in unspecialized laboratories. In particular, it remains challenging to create zinc-finger domains that can effectively recognize all DNA triplets, especially those of the 5′-CNN-3′ and 5′-TNN-3′ variety. As a result, ZFNs lack the target flexibility inherent to more recent genome-editing platforms. Nevertheless, the potential for ZFNs to mediate specific and efficient genome editing is evidenced by ongoing clinical trials based on ZFN-mediated knockout of the human immunodeficiency virus (HIV)-1 coreceptor CCR5 for treatment of HIV/acquired immune deficiency syndrome (AIDS) (Tebas et al. 2014) and a planned clinical trial based on site-specific integration of the factor IX gene into the albumin locus to treat hemophilia B (Clinical Trial ID: NCT02695160) (Sharma et al. 2015).

TALE Nucleases

TALE proteins are bacterial effectors. In 2009, the code used by TALE proteins to recognize DNA was uncovered (Boch et al. 2009; Moscou and Bogdanove 2009). In a matter of months, this discovery enabled the creation of custom TALENs capable of modifying nearly any gene. Like ZFNs, TALENs are modular in form and function, comprised of an amino-terminal TALE DNA-binding domain fused to a carboxy-terminal FokI cleavage domain (Christian et al. 2010; Miller et al. 2011). Also like ZFNs, dimerization of TALEN proteins is mediated by the FokI cleavage domain, which cuts within a 12- to 19-bp spacer sequence that separates each TALE binding site (Fig. 1) (Miller et al. 2011). TALEs are typically assembled to recognize between 12- to 20-bps of DNA, with more bases typically leading to higher genome-editing specificity (Guilinger et al. 2014a). The TALE-binding domain consists of a series of repeat domains, each ~34 residues in length. Each repeat contacts DNA via the amino acid residues at positions 12 and 13, known as the repeat variable diresidues (RVDs) (Boch et al. 2009; Moscou and Bogdanove 2009). Unlike zinc fingers, which recognize DNA triplets, each TALE repeat recognizes only a single bp, with little to no target site overlap from adjacent domains (Deng et al. 2012; Mak et al. 2012). The most commonly used RVDs for assembling synthetic TALE arrays are: NI for adenine, HD for cytosine, NG for thymine, and NN or HN for guanine or adenine (Boch et al. 2009; Moscou and Bogdanove 2009; Cong et al. 2012; Streubel et al. 2012). TALE DNA-binding domains can be constructed using a variety of methods, with the most straightforward approach being Golden Gate assembly (Cermak et al. 2011). However, high-throughput TALE assembly methods have also been developed, including FLASH assembly (Reyon et al. 2012), iterative capped assembly (Briggs et al. 2012), and ligation independent cloning (Schmid-Burgk et al. 2013), among others. More recent advances in TALEN assembly, though, have focused on the development of methods that can enhance their performance, including specificity profiling to uncover nonconventional RVDs that improve TALEN activity (Guilinger et al. 2014a; Yang et al. 2014; Juillerat et al. 2015; Miller et al. 2015), directed evolution as means to refine TALE specificity (Hubbard et al. 2015), and even fusing TALE domains to homing endonuclease variants to generate chimeric nucleases with extended targeting specificity (discussed in more detail below) (Boissel et al. 2014).

Compared to ZFNs, TALENs offer two distinct advantages for genome editing. First, no selection or directed evolution is necessary to engineer TALE arrays, dramatically reducing the amount of time and experience needed to assemble a functional nuclease. Second, TALENs have been reported to show improved specificity and reduced toxicity compared to some ZFNs (Mussolino et al. 2014), potentially because of their increased affinity for target DNA (Meckler et al. 2013) or perhaps a greater energetic penalty for associating with base mismatches. However, TALENs are substantially larger than ZFNs, and have a highly repetitive structure, making their efficient delivery into cells through the use of lentivirus (Holkers et al. 2013) or a single adeno-associated virus (AAV) particle challenging. Methods for overcoming these limitations have emerged as TALENs can be readily delivered into cells as mRNA (Mahiny et al. 2015; Mock et al. 2015) and even protein (Cai et al. 2014; Liu et al. 2014a), although alternative codon usage and amino acid degeneracy can also be leveraged to express RVD arrays that might be less susceptible to recombination (Kim et al. 2013a). In addition, adenoviral vectors have also proven particularly useful for mediating TALEN delivery to hard-to-transfect cell types (Holkers et al. 2014; Maggio et al. 2016).

CRISPR-Cas9

The CRISPR-Cas9 system, which has a role in adaptive immunity in bacteria (Horvath and Barrangou 2010; Marraffini and Sontheimer 2010), is the most recent addition to the genome-editing toolbox. In bacteria, the type-II CRISPR system provides protection against DNA from invading viruses and plasmids via RNA-guided DNA cleavage by Cas proteins (Wiedenheft et al. 2012; Sorek et al. 2013). Short segments of foreign DNA are integrated within the CRISPR locus and transcribed into CRISPR RNA (crRNA), which then anneal to *trans*-activating crRNA (tracrRNA) to direct sequence-specific degradation of pathogenic DNA by the Cas9 protein (Jinek et al. 2012). In 2012, Charpentier, Doudna, and co-workers reported that target recognition by the Cas9 protein only requires a seed sequence within the crRNA and a conserved protospacer-adjacent motif (PAM) upstream of the crRNA binding site (Jinek et al. 2012). This system has since been simplified for genome engineering (Cho et al. 2013; Cong et al. 2013; Jinek et al. 2013; Mali et al. 2013b) and now consists of only the Cas9 nuclease and a single guide RNA (gRNA) containing the essential crRNA and tracrRNA elements (Fig. 1). Because target site recognition is mediated entirely by the gRNA, CRISPR-Cas9 has emerged as the most flexible and user-friendly platform for genome editing, eliminating the need for engineering new proteins to recognize each new target site. The only major restriction for Cas9 target site recognition is that the PAM motif—which is recognized by the Cas9 nuclease and is essential for DNA cleavage—be located immediately downstream of the gRNA target site. The PAM sequence for the *Streptococcus pyogenes* Cas9, for example, is 5′-NGG-3′ (although in some cases 5′-NAG-3′ can be tolerated) (Hsu et al. 2013; Jiang et al. 2013; Mali et al. 2013a). Several studies have now shed light on the structural basis of DNA recognition by Cas9, revealing that the heteroduplex formed by the gRNA and its complementary strand of DNA is housed in a positively charged groove between the two nuclease domains (RuvC and HNH) within the Cas9 protein (Nishimasu et al. 2014), and that PAM recognition is mediated by an arginine-rich motif present in Cas9 (Anders et al. 2014). Doudna and colleagues have since proposed that DNA strand displacement induces a structural rearrangement within the Cas9 protein that directs the nontarget DNA strand into the RuvC active site, which then positions the HNH domain near target DNA (Jiang et al. 2016), enabling Cas9-mediated cleavage of both DNA strands.

The Cas9 nuclease and its gRNA can be delivered into cells for genome editing on the same or separate plasmids, and numerous resources have been developed to facilitate target site selection and gRNA construction, including E-CRISP (Heigwer et al. 2014), among others. Although Cas9 boasts the highest ease of use among the targeted nuclease platforms, several reports have indicated that it could be prone to inducing off-target mutations (Cradick et al. 2013; Fu et al. 2013). To this end, considerable effort has been devoted to improving the specificity of this system, including using paired Cas9 nickases (Mali et al. 2013a; Ran et al. 2013), which increase gene-editing specificity by requiring the induction of two sequential and adjacent nicking events for DSB formation, or truncated gRNA that are more sensitive to mismatches at the genomic target site than a full-length gRNA (Fu et al. 2014). Off-target cleavage has also been reduced by controlling the dosage of either the Cas9 protein or gRNA within the cell (Hsu et al. 2013), or even by using Cas9 variants configured to enable conditional genome editing, such as a rapamycin-inducible split-Cas9 architecture (Zetsche et al. 2015b) or a Cas9 variant that contains a strategically placed small-molecule-responsive intein domain (Davis et al. 2015). Nucleofection (Kim et al. 2014) or transient transfection (Zuris et al. 2015) of a preformed Cas9 ribonucleoprotein complex has also been shown to reduce off-target effects, enabling DNA-free gene editing in primary human T cells (Schumann et al. 2015), embryonic stem cells (Liu et al. 2015b), *Caenorhabditis elegans* gonads (Paix et al. 2015), mouse (Menoret et al. 2015; Wang et al. 2015a) and zebrafish embryos (Sung et al. 2014), and even plant protoplasts (Woo et al. 2015). The incorporation of specific chemical modifications known to protect RNA from nuclease degradation and stabilize secondary structure can further enhance Cas9 ribonucleoprotein activity (Hendel et al. 2015; Rahdar et al. 2015). In a clever marriage of genome-editing platforms, the FokI cleavage domain has even been fused to an inactivated Cas9 variant to generate hybrid nucleases that require protein dimerization for DNA cleavage (Guilinger et al. 2014b; Tsai et al. 2014), theoretically increasing CRISPR-Cas9 specificity. Similarly, fusing Cas9 to DNA-binding domains has also proven effective at improving its specificity (Bolukbasi et al. 2015). Finally, several studies have recently showed that protein engineering can broadly enhance Cas9 specificity (Kleinstiver et al. 2016; Slaymaker et al. 2016) and even alter its PAM requirements (Kleinstiver et al. 2015), the latter having the potential to enable creation of customized variants of Cas9 for allele-specific gene editing, although Cas9 orthologs (Cong

 Cite this article as *Cold Spring Harb Perspect Biol* doi: 10.1101/cshperspect.a023754

et al. 2013; Esvelt et al. 2013; Hou et al. 2013; Ran et al. 2015) or alternative CRISPR systems (Zetsche et al. 2015a) with unique PAM specificities have been uncovered in nature.

Homing Endonucleases

Homing endonucleases, also known as meganucleases, represent the final member of the targeted nuclease family. These enzymes have been reviewed at length elsewhere (Silva et al. 2011; Stoddard 2014) but, briefly, members of the LAGLIDADG family of endonucleases—so named for the conserved amino acid motif present within these enzymes that interacts with DNA—are a collection of naturally occurring enzymes that recognize and cleave long DNA sequences (14–40 bps) (Fig. 1). These enzymes make extensive sequence-specific contacts with their DNA substrate (Stoddard 2011), and thus typically show exquisite specificity. However, unlike ZFNs and TALENs, the binding and cleavage domains in homing endonucleases are not modular. This overlap in form and function make their repurposing challenging, and limits their utility for more routine applications of genome editing. More recently megaTALs—fusions of a rare-cleaving homing endonuclease to a TALE-binding domain—have been reported to induce highly specific gene modifications (Boissel et al. 2014; Lin et al. 2015a). These enzymes have enabled integration of antitumor and anti-HIV factors into the human *CCR5* gene in both primary T cells and hematopoietic stem/progenitor cells (Sather et al. 2015), as well as disruption of endogenous T-cell receptor elements in T cells (Osborn et al. 2016), indicating their potential for enabling and enhancing immunotherapies.

GENOME-EDITING APPLICATIONS

Engineering Cell Lines and Organisms

Before the emergence of engineered nucleases, genetically modifying mammalian cell lines was labor intensive, costly, and often times limited to laboratories with specialized expertise. However, with the advent of cost-effective and user-friendly gene-editing technologies, custom cell lines carrying nearly any genomic modification can now be generated in simply a matter of weeks. Examples of some of the outcomes that have become routine because of the emergence of targeted nucleases include gene knockout (Santiago et al. 2008; Mali et al. 2013b), gene deletion (Lee et al. 2010), gene inversion (Xiao et al. 2013), gene correction (Urnov et al. 2005; Ran et al. 2013), gene addition (Moehle et al. 2007; Hockemeyer et al. 2011; Hou et al. 2013), and even chromosomal translocation (Fig. 2) (Torres et al. 2014). In addition to cell line engineering, targeted nucleases have also expedited the generation of genetically modified organisms, facilitating the rapid creation of transgenic zebrafish (Doyon et al. 2008; Sander et al. 2011a; Hwang et al. 2013), mice (Cui et al. 2011; Wang et al. 2013; Wu et al. 2013), rats (Geurts et al. 2009; Tesson et al. 2011; Li et al. 2013), monkeys (Liu et al. 2014c), and livestock (Hauschild et al. 2011; Carlson et al. 2012), which together have the capacity to accelerate human disease modeling and the discovery of new therapeutics.

Targeted nucleases have also emerged as powerful tools for plant engineering (Baltes and Voytas 2015). Both TALENs and CRISPR-Cas9 have been used to modify multiple alleles within hexaploid bread wheat to confer heritable resistance to powdery mildew (Wang et al. 2014b). In another study, TALENs were used to knock out nonessential genes in the fatty acid metabolic pathway in soybean to generate a simplified plant cell with reduced metabolic components (Haun et al. 2014). Of special note, two recent reports showed that purified nuclease proteins can be introduced directly into plant protoplasts, enabling the introduction of germline-transmissible modifications that are virtually indistinguishable from naturally occurring (Luo et al. 2015; Woo et al. 2015). This technical advance could help to overcome certain regulatory hurdles associated with the use of transgenic crops. Finally, targeted nucleases have also been used to inactivate pathogenic genes to prevent viral (Lin et al. 2014) or parasitic (Ghorbal et al. 2014) infection, as well as to introduce knockin-specific factors capable of imparting pathogen resistance (Wu et al. 2015).

Intriguingly, targeted nucleases could also serve as conduits for curbing mosquito- or insect-borne diseases through a technique known as gene drive (Burt 2003; Sinkins and Gould 2006), which harnesses genome editing to facilitate the introduction of a specific gene or mutation that can then confer a particular phenotype into a host and also be transmitted to its progeny (Windbichler et al. 2011). Gene drives have now been tested in the malaria vector mosquitos *Anopheles stephensi* (Gantz et al. 2015) and *Anopheles gambiae* (Hammond et al. 2016) as a means for achieving population control and to prevent disease transmission, respectively. However, owing to the ease with which CRISPR-Cas9 can be programmed (Gantz and Bier 2015), debate has ignited on the potential societal and environmental impact of this technology (Esvelt et al. 2014; Akbari et al. 2015), spurring the development of safeguard elements that could help to minimize the risk of gene-edited organisms escaping from the laboratory (DiCarlo et al. 2015).

Synthetic Biology and Genome-Scale Engineering

Targeted nucleases also offer a facile means for generating modified bacterial and yeast strains for synthetic biology, including metabolic pathway engineering. Bacterial species of the order *Actinomycetales*, for instance, are one of the most important sources of industrially relevant secondary metabolites. However, many *Actinomycetales* species are recalcitrant to genetic manipulation, a fact that has severely hampered their use for metabolic engineering. CRISPR-Cas9 has been used to inactivate multiple genes in actinomycetes (Tong et al. 2015), indicating its ability to enable the creation of designer bacterial strains with enhanced metabolite production capabilities. CRISPR has also facilitated multiplexed metabolic pathway engineering in yeast at high efficiencies (Jakociunas et al. 2015a,b), as well as random mutagenesis of yeast chromosomal DNA for phenotypic screening of desired mutants (Ryan et al. 2014). Indeed, genome-wide CRISPR-based knockout screens hold tremendous potential for functional genomics (Hilton and Gersbach 2015), having facilitated the discovery of genomic loci that confer drug resistance to cells (Koike-Yusa et al. 2014; Shalem et al. 2014; Wang et al. 2014a; Zhou et al. 2014), uncovered how cells can control induction of the host immune response (Parnas et al. 2015), provided new insights into the genetic basis of cellular fitness (Hart et al. 2015; Wang et al. 2015b), and even shed light on how certain viruses induce cell death (Ma et al. 2015). Genome-wide CRISPR screens can also facilitate the discovery of functional noncoding elements (Kim et al. 2013b; Korkmaz et al. 2016), and provide a means for studying the structure and evolution of the human genome. In a remarkable example of the latter, Shendure and colleagues used Cas9 to mediate integration of short randomized DNA sequences into the *BRCA1* and *DBR1* genes (Findlay et al. 2014). They then measured the functional consequences of these mutations on fitness, achieving an unprecedented look at some of the factors driving genome and disease evolution. Finally, CRISPR screens have even proven effective in vivo, enabling the identification of factors involved in zebrafish development (Shah et al. 2015) and disease progression in mice (Chen et al. 2015).

Therapeutic Genome Editing

Genome editing itself also holds tremendous potential for treating the underlying genetic causes of certain diseases (Cox et al. 2015; Porteus 2015; Maeder and Gersbach 2016). In one of the most successful examples of this to date, ZFN-mediated disruption of the HIV coreceptor CCR5 was used to engineer HIV resistance into both $CD4^+$ T cells (Perez et al. 2008) and $CD34^+$ hematopoietic stem/progenitor cells (HSPCs) (Holt et al. 2010), proving safe and well-tolerated in a phase I clinical trial that infused these gene-modified T cells into individuals with HIV/AIDS (Tebas et al. 2014). In addition to enabling the introduction of gene modification that can enhance autologous cell therapies, targeted nucleases can also be combined with viral vectors—including AAV—to mediate genome editing in situ (Gaj et al.

2016). For instance, delivery of an AAV vector encoding a ZFN pair designed to target a defective copy of the factor IX gene, along with its repair template, led to efficient gene correction in mouse liver, increasing factor IX protein production in both neonatal (Li et al. 2011) and adult (Anguela et al. 2013) models of the disease. In vivo genome editing also recently enabled the restoration of dystrophin gene expression and the rescue of muscle function in mouse models of Duchenne muscular dystrophy (Long et al. 2015; Nelson et al. 2015; Tabebordbar et al. 2015). Therapeutic gene editing in a mouse model of human hereditary tyrosinemia has also been reported using both hydrodynamic injection of plasmid DNA encoding CRISPR-Cas9 (Yin et al. 2014) and by combining nanoparticle-mediated delivery of Cas9-encoding mRNA with AAV-mediated delivery of the DNA template for gene correction (Yin et al. 2016). More recently, a dual particle AAV system, wherein one AAV vector carried the Cas9 nuclease and a second harbored the gRNA and donor repair template, was able to mediate correction of a disease-causing mutation in the ornithine transcarbamylase gene in the liver of a neonatal model of the disease (Yang et al. 2016). This work, in particular, showed that therapeutic levels of gene correction could be achieved in a regenerating tissue even when using multiple AAV particles. Although highly promising, numerous hurdles still need to be overcome for in vivo applications of genome editing to reach its full potential. Chief among these are methods that can facilitate nuclease delivery or expression to only diseased cells or tissues, and the development of new strategies that can enhance HDR in disease-associated postmitotic cells in vivo.

TARGETED TRANSCRIPTION FACTORS

Tools for Modulating Gene Expression

The modular qualities of zinc-finger and TALE proteins, in addition to the highly flexible DNA recognition ability of CRISPR-Cas9, also provide investigators with the ability to modulate the expression of nearly any gene from its promoter or enhancer sequences via their fusion to transcriptional activator and repressor protein domains. Among the first fully synthetic transcriptional effector proteins to be generated (Beerli et al. 1998) were those based on the fusion of engineered zinc-finger proteins with either the Herpes simplex–derived transactivation domain (Sadowski et al. 1988) or the Krüppel-associated box (KRAB) repression protein (Margolin et al. 1994). Over the course of the next 15 years, zinc-finger-based transcriptional modulators were expanded and featured several other types of effector domains (Beerli and Barbas 2002), including, for example, the Dnmt3a methyltransferase domain (Rivenbark et al. 2012; Siddique et al. 2013) and the ten-eleven translocation methylcytosine dioxygenase 1 (TET1) (Chen et al. 2014), which can modulate transcription via targeted methylation or demethylation, respectively. As a natural extension of zinc-finger transcription factors, and further drawing on the parallels with zinc-finger proteins, TALE transcription factors have also emerged as an especially effective platform for achieving targeted transcriptional modulation (Miller et al. 2011; Zhang et al. 2011). Effector domains are generally fused to the carboxyl terminus of the synthetic TALE array and, contrary to the longer sequence typically required for efficient modulation by zinc-finger transcription factors, TALEs have been reported to regulate gene expression with as few as 10.5 repeats (Boch et al. 2009). Like zinc fingers, TALEs are also compatible with numerous epigenetic modifiers, including the TET1 hydroxylase catalytic domain (Maeder et al. 2013b) and the lysine-specific histone demethylase 1 (LSD1) (Mendenhall et al. 2013) domains, which have been used for targeted CpG demethylation and histone demethylation, respectively. In particular, the ease with which a large number of TALEs can be constructed has enabled the discovery that tiling a promoter sequence with combinations of synthetic transcription factors can lead to a synergistic increase in gene expression (Maeder et al. 2013b; Perez-Pinera et al. 2013). And, like zinc fingers (Beerli et al. 2000b; Pollock et al. 2002; Magnenat et al. 2008; Polstein and Gersbach 2012), TALE activators have also been successfully engineered to regulate gene expression in response to external (Mercer et al. 2014) or endogenous (Li et al. 2012) chemical stimuli, optical signals (Konermann et al. 2013), and even proteolytic cues (Copeland et al. 2016; Lonzaric et al. 2016).

Because of the exquisite ease with which it can be programmed, the CRISPR-Cas9 system has also been adapted for transcriptional modulation through fusion of specific effector domains to a catalytically inactivated variant of the Cas9 protein (Qi et al. 2013). Deactivation is achieved by introducing two amino acid substitutions, D10A and H840A, into the RuvC and NHN endonuclease domains of Cas9, respectively. Although unable to cleave DNA, this mutant, referred to as dCas9, retains its ability to bind DNA in an RNA-directed manner. Effector domains are fused to the carboxyl terminus of the dCas9 protein and can modulate gene expression from either strand of the targeted DNA sequence (Farzadfard et al. 2013; Maeder et al. 2013a; Perez-Pinera et al. 2013). Genome-scale activation studies have indicated that the most robust levels of activation are generally observed when dCas9 activators are targeted to -400 to -50 bp upstream from the transcriptional start site (Gilbert et al. 2014; Hu et al. 2014). Additionally, dCas9 can inhibit gene expression by simply blocking transcriptional initiation or elongation through a process known as CRISPR interference (Qi et al. 2013), although fusing dCas9 to transcriptional repressor domains can also lead to efficient silencing from the promoter (Gilbert et al. 2013; Zalatan et al. 2015). Much like zinc fingers and TALEs, methods for achieving conditional gene modulation using dCas9 have also been reported, including the fusion of a dihydrofolate reductase destabilization domain to dCas9, which can provide chemical control over activation, enabling cellular reprogramming or differentiation (Balboa et al. 2015). Light-inducible dCas9-based systems capable of providing optical control of gene expression provide another means for achieving conditional control of gene expression (Nihongaki et al. 2015; Polstein and Gersbach 2015).

Although flexible, first-generation dCas9 activators were routinely found to display suboptimal levels of activation. As a result, the development of second-generation CRISPR activators quickly emerged as a highly active area of research. One particularly elegant approach for overcoming the low activation thresholds inherent within first-generation systems was by strategically inserting an RNA aptamer within a functionally inert region of the gRNA. This aptamer recruits specific activation helper proteins that work in concert with a dCas9 activator to enhance transcription (Konermann et al. 2015; Zalatan et al. 2015). Other strategies based on directly fusing additional helper activation domains to dCas9 have also been shown to enhance transcription (Chavez et al. 2015). Targeted acetylation of histone proteins within a promoter or enhancer sequence via epigenome editing using the catalytic core of the human acetyltransferase p300 fused to dCas9 can also lead to robust levels of gene activation (Hilton et al. 2015). Similarly, dCas9 repressor proteins targeted to distal regulatory elements have been found to facilitate chromatin remodeling and gene repression via epigenomic modification (Thakore et al. 2015). Finally, by simply reducing the length of the gRNA, catalytically active variants of Cas9 can stimulate transcription without inducing DNA breaks (Dahlman et al. 2015; Kiani et al. 2015), enabling orthogonal gene knockout and activation with the same Cas9 variant in a single cell.

Applications of Targeted Transcriptional Regulation

Early work on the use of engineered zinc-finger transcription factors revealed that synthetic transcriptional modulators are effective tools for a broad range of applications, enabling such tasks as inhibiting viral replication (Papworth et al. 2003; Reynolds et al. 2003; Segal et al. 2004; Eberhardy et al. 2006), modulating the expression of disease-associated loci (Graslund et al. 2005; Wilber et al. 2010), inducing angiogenesis for accelerated wound healing (Rebar et al. 2002), and genomic screening of cellular targets for cancer progression and drug resistance (Park et al. 2003; Blancafort et al. 2005, 2008). Facilitated by many of the insights gained from zinc-finger transcription factor

 Cite this article as *Cold Spring Harb Perspect Biol* doi: 10.1101/cshperspect.a023754

technology, both TALEs and CRISPR-Cas9 have now further expanded the possibilities of engineered transcriptional activators and repressors. For example, TALEs and CRISPR-Cas9 have enabled rapid construction of custom genetic circuits and logic gates (Gaber et al. 2014; Lebar et al. 2014; Liu et al. 2014b), complex gene regulation networks (Nielsen and Voigt 2014; Nissim et al. 2014), and even facilitated cellular reprogramming (Gao et al. 2013) and the differentiation of mouse embryonic fibroblasts to skeletal myocytes (Chakraborty et al. 2014). dCas9 transcriptional effectors have even been used to efficiently mediate repression and activation of endogenous genes in *Drosophila* (Lin et al. 2015b) and in plant cells (Piatek et al. 2015). Both TALE and Cas9 activators have also been configured to stimulate transcription of latent HIV (Zhang et al. 2015; Ji et al. 2016; Limsirichai et al. 2016; Perdigao et al. 2016; Saayman et al. 2016), indicating their potential to work in concert with antiretroviral therapy for eradicating HIV infection. Importantly, because of the ease with which the CRISPR-Cas9 system can be used, genome-wide screens using Cas9 transcriptional activators (Gilbert et al. 2014; Konermann et al. 2015) and repressors (Gilbert et al. 2014) can be easily implemented to discover genes involved in a number of diverse processes, including drug resistance and cancer metastasis. In particular, CRISPR-based genome-scale screening methods have the potential to overcome many of the technical hurdles associated with other contemporary screening technologies, such as cDNA libraries and RNAi, indicating its potential for facilitating drug discovery and basic biological research.

CONCLUSIONS

Despite the successes already achieved, many challenges remain before the full potential of genome editing can be realized. First and foremost are the development of new tools capable of introducing genomic modifications in the absence of DNA breaks. Targeted recombinases (Akopian et al. 2003; Mercer et al. 2012), which can be programmed to recognize specific DNA sequences (Gaj et al. 2013; Sirk et al. 2014; Wallen et al. 2015) and even integrate therapeutic factors into the human genome (Gaj et al. 2014b), are one such option. More recent work has indicated that single-base editing without DNA breaks can be achieved using an engineered Cas9 nickase complex (Komor et al. 2016), although it remains unknown how effective this technology is in therapeutically relevant settings. By linking genomic modifications induced by targeted nucleases to their own self-degradation, self-inactivating vectors are also poised to improve the specificity of genome editing, especially because the frequency of off-target modifications can be directly proportional to the duration of cellular exposure to a nuclease (Pruett-Miller et al. 2009). In addition, much of the knowledge behind genome engineering has been obtained in immortalized cell lines. However, in the case of regenerative medicine, it is highly desirable to genetically manipulate progenitor or stem-cell populations, both of which can differ markedly from transformed cell lines with respect to their epigenome or three-dimensional organization of their genomic DNA. These differences can have profound effects on the ability of genome-editing tools to either modify a specific sequence or regulate gene expression. Although the union between genome engineering and regenerative medicine is still in its infancy, realizing the full potential of these technologies in stem/progenitor cells requires that their functional landscape be fully explored in these genetic backgrounds. Only then will genome editing technologies truly be able to reprogram cell fate and behavior for the next generation of advances in synthetic biology and gene therapy.

ACKNOWLEDGMENTS

We gratefully acknowledge the support and mentorship of the late Carlos F. Barbas, III (1964–2014). This work is supported by the National Institutes of Health (F32GM113446 to T.G.) and ShanghaiTech University (to J.L.). We apologize to those whose important contributions were not cited because of space constraints.

REFERENCES

Akbari OS, Bellen HJ, Bier E, Bullock SL, Burt A, Church GM, Cook KR, Duchek P, Edwards OR, Esvelt KM, et al. 2015. Safeguarding gene drive experiments in the laboratory. *Science* **349:** 927–929.

Akopian A, He J, Boocock MR, Stark WM. 2003. Chimeric recombinases with designed DNA sequence recognition. *Proc Natl Acad Sci* **100:** 8688–8691.

Anders C, Niewoehner O, Duerst A, Jinek M. 2014. Structural basis of PAM-dependent target DNA recognition by the Cas9 endonuclease. *Nature* **513:** 569–573.

Anguela XM, Sharma R, Doyon Y, Miller JC, Li H, Haurigot V, Rohde ME, Wong SY, Davidson RJ, Zhou S, et al. 2013. Robust ZFN-mediated genome editing in adult hemophilic mice. *Blood* **122:** 3283–3287.

Bae KH, Kwon YD, Shin HC, Hwang MS, Ryu EH, Park KS, Yang HY, Lee DK, Lee Y, Park J, et al. 2003. Human zinc fingers as building blocks in the construction of artificial transcription factors. *Nat Biotechnol* **21:** 275–280.

Balboa D, Weltner J, Eurola S, Trokovic R, Wartiovaara K, Otonkoski T. 2015. Conditionally stabilized dCas9 activator for controlling gene expression in human cell reprogramming and differentiation. *Stem Cell Rep* **5:** 448–459.

Baltes NJ, Voytas DF. 2015. Enabling plant synthetic biology through genome engineering. *Trends Biotechnol* **33:** 120–131.

Beerli RR, Barbas CF III. 2002. Engineering polydactyl zinc-finger transcription factors. *Nat Biotechnol* **20:** 135–141.

Beerli RR, Segal DJ, Dreier B, Barbas CF III. 1998. Toward controlling gene expression at will: Specific regulation of the erbB-2/HER-2 promoter by using polydactyl zinc finger proteins constructed from modular building blocks. *Proc Natl Acad Sci* **95:** 14628–14633.

Beerli RR, Dreier B, Barbas CF III. 2000a. Positive and negative regulation of endogenous genes by designed transcription factors. *Proc Natl Acad Sci* **97:** 1495–1500.

Beerli RR, Schopfer U, Dreier B, Barbas CF III. 2000b. Chemically regulated zinc finger transcription factors. *J Biol Chem* **275:** 32617–32627.

Bhakta MS, Henry IM, Ousterout DG, Das KT, Lockwood SH, Meckler JF, Wallen MC, Zykovich A, Yu Y, Leo H, et al. 2013. Highly active zinc-finger nucleases by extended modular assembly. *Genome Res* **23:** 530–538.

Bibikova M, Carroll D, Segal DJ, Trautman JK, Smith J, Kim YG, Chandrasegaran S. 2001. Stimulation of homologous recombination through targeted cleavage by chimeric nucleases. *Mol Cell Biol* **21:** 289–297.

Bibikova M, Golic M, Golic KG, Carroll D. 2002. Targeted chromosomal cleavage and mutagenesis in *Drosophila* using zinc-finger nucleases. *Genetics* **161:** 1169–1175.

Bibikova M, Beumer K, Trautman JK, Carroll D. 2003. Enhancing gene targeting with designed zinc finger nucleases. *Science* **300:** 764.

Blancafort P, Chen EI, Gonzalez B, Bergquist S, Zijlstra A, Guthy D, Brachat A, Brakenhoff RH, Quigley JP, Erdmann D, et al. 2005. Genetic reprogramming of tumor cells by zinc finger transcription factors. *Proc Natl Acad Sci* **102:** 11716–11721.

Blancafort P, Tschan MP, Bergquist S, Guthy D, Brachat A, Sheeter DA, Torbett BE, Erdmann D, Barbas CF III. 2008. Modulation of drug resistance by artificial transcription factors. *Mol Cancer Ther* **7:** 688–697.

Boch J, Scholze H, Schornack S, Landgraf A, Hahn S, Kay S, Lahaye T, Nickstadt A, Bonas U. 2009. Breaking the code of DNA binding specificity of TAL-type III effectors. *Science* **326:** 1509–1512.

Boissel S, Jarjour J, Astrakhan A, Adey A, Gouble A, Duchateau P, Shendure J, Stoddard BL, Certo MT, Baker D, et al. 2014. megaTALs: A rare-cleaving nuclease architecture for therapeutic genome engineering. *Nucleic Acids Res* **42:** 2591–2601.

Bolukbasi MF, Gupta A, Oikemus S, Derr AG, Garber M, Brodsky MH, Zhu LJ, Wolfe SA. 2015. DNA-binding-domain fusions enhance the targeting range and precision of Cas9. *Nat Methods* **12:** 1150–1156.

Briggs AW, Rios X, Chari R, Yang L, Zhang F, Mali P, Church GM. 2012. Iterative capped assembly: rapid and scalable synthesis of repeat-module DNA such as TAL effectors from individual monomers. *Nucleic Acids Res* **40:** e117.

Burt A. 2003. Site-specific selfish genes as tools for the control and genetic engineering of natural populations. *Proc Biol Sci* **270:** 921–928.

Cai Y, Bak RO, Mikkelsen JG. 2014. Targeted genome editing by lentiviral protein transduction of zinc-finger and TAL-effector nucleases. *eLife* **3:** e01911.

Carlson DF, Tan W, Lillico SG, Stverakova D, Proudfoot C, Christian M, Voytas DF, Long CR, Whitelaw CB, Fahrenkrug SC. 2012. Efficient TALEN-mediated gene knockout in livestock. *Proc Natl Acad Sci* **109:** 17382–17387.

Cermak T, Doyle EL, Christian M, Wang L, Zhang Y, Schmidt C, Baller JA, Somia NV, Bogdanove AJ, Voytas DF. 2011. Efficient design and assembly of custom TALEN and other TAL effector-based constructs for DNA targeting. *Nucleic Acids Res* **39:** e82.

Chakraborty S, Ji H, Kabadi AM, Gersbach CA, Christoforou N, Leong KW. 2014. A CRISPR/Cas9-based system for reprogramming cell lineage specification. *Stem Cell Rep* **3:** 940–947.

Chavez A, Scheiman J, Vora S, Pruitt BW, Tuttle M, Iyer EPR, Lin S, Kiani S, Guzman CD, Wiegand DJ, et al. 2015. Highly efficient Cas9-mediated transcriptional programming. *Nat Methods* **12:** 326–328.

Chen H, Kazemier HG, de Groote ML, Ruiters MH, Xu GL, Rots MG. 2014. Induced DNA demethylation by targeting Ten-Eleven Translocation 2 to the human ICAM-1 promoter. *Nucleic Acids Res* **42:** 1563–1574.

Chen S, Sanjana NE, Zheng K, Shalem O, Lee K, Shi X, Scott DA, Song J, Pan JQ, Weissleder R, et al. 2015. Genome-wide CRISPR screen in a mouse model of tumor growth and metastasis. *Cell* **160:** 1246–1260.

Cho SW, Kim S, Kim JM, Kim JS. 2013. Targeted genome engineering in human cells with the Cas9 RNA-guided endonuclease. *Nat Biotechnol* **31:** 230–232.

Choo Y, Klug A. 1994. Toward a code for the interactions of zinc fingers with DNA: Selection of randomized fingers displayed on phage. *Proc Natl Acad Sci* **91:** 11163–11167.

Choulika A, Perrin A, Dujon B, Nicolas JF. 1995. Induction of homologous recombination in mammalian chromosomes by using the I-SceI system of *Saccharomyces cerevisiae*. *Mol Cell Biol* **15:** 1968–1973.

Christian M, Cermak T, Doyle EL, Schmidt C, Zhang F, Hummel A, Bogdanove AJ, Voytas DF. 2010. Targeting DNA double-strand breaks with TAL effector nucleases. *Genetics* **186:** 757–761.

Cong L, Zhou R, Kuo YC, Cunniff M, Zhang F. 2012. Comprehensive interrogation of natural TALE DNA-binding modules and transcriptional repressor domains. *Nat Commun* **3:** 968.

Cong L, Ran FA, Cox D, Lin S, Barretto R, Habib N, Hsu PD, Wu X, Jiang W, Marraffini LA, et al. 2013. Multiplex genome engineering using CRISPR/Cas systems. *Science* **339:** 819–823.

Copeland MF, Politz MC, Johnson CB, Markley AL, Pfleger BF. 2016. A transcription activator-like effector (TALE) induction system mediated by proteolysis. *Nat Chem Biol* **12:** 254–260.

Cox DB, Platt RJ, Zhang F. 2015. Therapeutic genome editing: Prospects and challenges. *Nat Med* **21:** 121–131.

Cradick TJ, Fine EJ, Antico CJ, Bao G. 2013. CRISPR/Cas9 systems targeting β-globin and CCR5 genes have substantial off-target activity. *Nucleic Acids Res* **41:** 9584–9592.

Cui X, Ji D, Fisher DA, Wu Y, Briner DM, Weinstein EJ. 2011. Targeted integration in rat and mouse embryos with zinc-finger nucleases. *Nat Biotechnol* **29:** 64–67.

Dahlman JE, Abudayyeh OO, Joung J, Gootenberg JS, Zhang F, Konermann S. 2015. Orthogonal gene knockout and activation with a catalytically active Cas9 nuclease. *Nat Biotechnol* **33:** 1159–1161.

Davis KM, Pattanayak V, Thompson DB, Zuris JA, Liu DR. 2015. Small molecule-triggered Cas9 protein with improved genome-editing specificity. *Nat Chem Biol* **11:** 316–318.

Deng D, Yan C, Pan X, Mahfouz M, Wang J, Zhu JK, Shi Y, Yan N. 2012. Structural basis for sequence-specific recognition of DNA by TAL effectors. *Science* **335:** 720–723.

DiCarlo JE, Chavez A, Dietz SL, Esvelt KM, Church GM. 2015. Safeguarding CRISPR-Cas9 gene drives in yeast. *Nat Biotechnol* **33:** 1250–1255.

Doyon Y, McCammon JM, Miller JC, Faraji F, Ngo C, Katibah GE, Amora R, Hocking TD, Zhang L, Rebar EJ, et al. 2008. Heritable targeted gene disruption in zebrafish using designed zinc-finger nucleases. *Nat Biotechnol* **26:** 702–708.

Doyon Y, Vo TD, Mendel MC, Greenberg SG, Wang J, Xia DF, Miller JC, Urnov FD, Gregory PD, Holmes MC. 2011. Enhancing zinc-finger-nuclease activity with improved obligate heterodimeric architectures. *Nat Methods* **8:** 74–79.

Dreier B, Beerli RR, Segal DJ, Flippin JD, Barbas CF III. 2001. Development of zinc finger domains for recognition of the 5′-ANN-3′ family of DNA sequences and their use in the construction of artificial transcription factors. *J Biol Chem* **276:** 29466–29478.

Dreier B, Fuller RP, Segal DJ, Lund CV, Blancafort P, Huber A, Koksch B, Barbas CF III. 2005. Development of zinc finger domains for recognition of the 5′-CNN-3′ family DNA sequences and their use in the construction of artificial transcription factors. *J Biol Chem* **280:** 35588–35597.

Eberhardy SR, Goncalves J, Coelho S, Segal DJ, Berkhout B, Barbas CF III. 2006. Inhibition of human immunodeficiency virus type 1 replication with artificial transcription factors targeting the highly conserved primer-binding site. *J Virol* **80:** 2873–2883.

Esvelt KM, Mali P, Braff JL, Moosburner M, Yaung SJ, Church GM. 2013. Orthogonal Cas9 proteins for RNA-guided gene regulation and editing. *Nat Methods* **10:** 1116–1121.

Esvelt KM, Smidler AL, Catteruccia F, Church GM. 2014. Concerning RNA-guided gene drives for the alteration of wild populations. *eLife* doi: 10.7554/eLife.03401.

Farzadfard F, Perli SD, Lu TK. 2013. Tunable and multifunctional eukaryotic transcription factors based on CRISPR/Cas. *ACS Synth Biol* **2:** 604–613.

Findlay GM, Boyle EA, Hause RJ, Klein JC, Shendure J. 2014. Saturation editing of genomic regions by multiplex homology-directed repair. *Nature* **513:** 120–123.

Fu Y, Foden JA, Khayter C, Maeder ML, Reyon D, Joung JK, Sander JD. 2013. High-frequency off-target mutagenesis induced by CRISPR-Cas nucleases in human cells. *Nat Biotechnol* **31:** 822–826.

Fu Y, Sander JD, Reyon D, Cascio VM, Joung JK. 2014. Improving CRISPR-Cas nuclease specificity using truncated guide RNAs. *Nat Biotechnol* **32:** 279–284.

Gaber R, Lebar T, Majerle A, Ster B, Dobnikar A, Bencina M, Jerala R. 2014. Designable DNA-binding domains enable construction of logic circuits in mammalian cells. *Nat Chem Biol* **10:** 203–208.

Gabriel R, Lombardo A, Arens A, Miller JC, Genovese P, Kaeppel C, Nowrouzi A, Bartholomae CC, Wang J, Friedman G, et al. 2011. An unbiased genome-wide analysis of zinc-finger nuclease specificity. *Nat Biotechnol* **29:** 816–823.

Gaj T, Guo J, Kato Y, Sirk SJ, Barbas CF III. 2012. Targeted gene knockout by direct delivery of zinc-finger nuclease proteins. *Nat Methods* **9:** 805–807.

Gaj T, Mercer AC, Sirk SJ, Smith HL, Barbas CF III. 2013. A comprehensive approach to zinc-finger recombinase customization enables genomic targeting in human cells. *Nucleic Acids Res* **41:** 3937–3946.

Gaj T, Liu J, Anderson KE, Sirk SJ, Barbas CF III. 2014a. Protein delivery using Cys2-His2 zinc-finger domains. *ACS Chem Biol* **9:** 1662–1667.

Gaj T, Sirk SJ, Tingle RD, Mercer AC, Wallen MC, Barbas CF III. 2014b. Enhancing the specificity of recombinase-mediated genome engineering through dimer interface redesign. *J Am Chem Soc* **136:** 5047–5056.

Gaj T, Epstein BE, Schaffer DV. 2016. Genome engineering using adeno-associated virus: Basic and clinical research applications. *Mol Ther* **24:** 458–464.

Gantz VM, Bier E. 2015. The mutagenic chain reaction: A method for converting heterozygous to homozygous mutations. *Science* **348:** 442–444.

Gantz VM, Jasinskiene N, Tatarenkova O, Fazekas A, Macias VM, Bier E, James AA. 2015. Highly efficient Cas9-mediated gene drive for population modification of the malaria vector mosquito *Anopheles stephensi*. *Proc Natl Acad Sci* **112:** E6736–E6743.

Gao X, Yang J, Tsang JC, Ooi J, Wu D, Liu P. 2013. Reprogramming to pluripotency using designer TALE transcription factors targeting enhancers. *Stem Cell Rep* **1:** 183–197.

Geurts AM, Cost GJ, Freyvert Y, Zeitler B, Miller JC, Choi VM, Jenkins SS, Wood A, Cui X, Meng X, et al. 2009. Knockout rats via embryo microinjection of zinc-finger nucleases. *Science* **325:** 433.

Ghorbal M, Gorman M, Macpherson CR, Martins RM, Scherf A, Lopez-Rubio JJ. 2014. Genome editing in the human malaria parasite *Plasmodium falciparum* using the CRISPR-Cas9 system. *Nat Biotechnol* **32:** 819–821.

Gilbert LA, Larson MH, Morsut L, Liu Z, Brar GA, Torres SE, Stern-Ginossar N, Brandman O, Whitehead EH, Doudna JA, et al. 2013. CRISPR-mediated modular RNA-guided regulation of transcription in eukaryotes. *Cell* **154:** 442–451.

Gilbert LA, Horlbeck MA, Adamson B, Villalta JE, Chen Y, Whitehead EH, Guimaraes C, Panning B, Ploegh HL, Bassik MC, et al. 2014. Genome-scale CRISPR-mediated control of gene repression and activation. *Cell* **159:** 647–661.

Graslund T, Li X, Magnenat L, Popkov M, Barbas CF III. 2005. Exploring strategies for the design of artificial transcription factors: Targeting sites proximal to known regulatory regions for the induction of γ-globin expression and the treatment of sickle cell disease. *J Biol Chem* **280:** 3707–3714.

Greisman HA, Pabo CO. 1997. A general strategy for selecting high-affinity zinc finger proteins for diverse DNA target sites. *Science* **275:** 657–661.

Guilinger JP, Pattanayak V, Reyon D, Tsai SQ, Sander JD, Joung JK, Liu DR. 2014a. Broad specificity profiling of TALENs results in engineered nucleases with improved DNA-cleavage specificity. *Nat Methods* **11:** 429–435.

Guilinger JP, Thompson DB, Liu DR. 2014b. Fusion of catalytically inactive Cas9 to FokI nuclease improves the specificity of genome modification. *Nat Biotechnol* **32:** 577–582.

Guo J, Gaj T, Barbas CF III. 2010. Directed evolution of an enhanced and highly efficient FokI cleavage domain for zinc finger nucleases. *J Mol Biol* **400:** 96–107.

Gupta A, Christensen RG, Rayla AL, Lakshmanan A, Stormo GD, Wolfe SA. 2012. An optimized two-finger archive for ZFN-mediated gene targeting. *Nat Methods* **9:** 588–590.

Hammond A, Galizi R, Kyrou K, Simoni A, Siniscalchi C, Katsanos D, Gribble M, Baker D, Marois E, Russell S, et al. 2016. A CRISPR-Cas9 gene drive system targeting female reproduction in the malaria mosquito vector Anopheles gambiae. *Nat Biotechnol* **34:** 78–83.

Hart T, Chandrashekhar M, Aregger M, Steinhart Z, Brown KR, MacLeod G, Mis M, Zimmermann M, Fradet-Turcotte A, Sun S, et al. 2015. High-resolution CRISPR screens reveal fitness genes and genotype-specific cancer liabilities. *Cell* **163:** 1515–1526.

Haun W, Coffman A, Clasen BM, Demorest ZL, Lowy A, Ray E, Retterath A, Stoddard T, Juillerat A, Cedrone F, et al. 2014. Improved soybean oil quality by targeted mutagenesis of the fatty acid desaturase 2 gene family. *Plant Biotechnol J* **12:** 934–940.

Hauschild J, Petersen B, Santiago Y, Queisser AL, Carnwath JW, Lucas-Hahn A, Zhang L, Meng X, Gregory PD, Schwinzer R, et al. 2011. Efficient generation of a biallelic knockout in pigs using zinc-finger nucleases. *Proc Natl Acad Sci* **108:** 12013–12017.

Heigwer F, Kerr G, Boutros M. 2014. E-CRISP: Fast CRISPR target site identification. *Nat Methods* **11:** 122–123.

Hendel A, Bak RO, Clark JT, Kennedy AB, Ryan DE, Roy S, Steinfeld I, Lunstad BD, Kaiser RJ, Wilkens AB, et al. 2015. Chemically modified guide RNAs enhance CRISPR-Cas genome editing in human primary cells. *Nat Biotechnol* **33:** 985–989.

Hilton IB, Gersbach CA. 2015. Enabling functional genomics with genome engineering. *Genome Res* **25:** 1442–1455.

Hilton IB, D'Ippolito AM, Vockley CM, Thakore PI, Crawford GE, Reddy TE. 2015. Epigenome editing by a CRISPR-Cas9-based acetyltransferase activates genes from promoters and enhancers. *Nat Biotechnol* **33:** 510–517.

Hockemeyer D, Wang H, Kiani S, Lai CS, Gao Q, Cassady JP, Cost GJ, Zhang L, Santiago Y, Miller JC, et al. 2011. Genetic engineering of human pluripotent cells using TALE nucleases. *Nat Biotechnol* **29:** 731–734.

Holkers M, Maggio I, Liu J, Janssen JM, Miselli F, Mussolino C, Recchia A, Cathomen T, Goncalves MA. 2013. Differential integrity of TALE nuclease genes following adenoviral and lentiviral vector gene transfer into human cells. *Nucleic Acids Res* **41:** e63.

Holkers M, Maggio I, Henriques SF, Janssen JM, Cathomen T, Goncalves MA. 2014. Adenoviral vector DNA for accurate genome editing with engineered nucleases. *Nat Methods* **11:** 1051–1057.

Holt N, Wang J, Kim K, Friedman G, Wang X, Taupin V, Crooks GM, Kohn DB, Gregory PD, Holmes MC, et al. 2010. Human hematopoietic stem/progenitor cells modified by zinc-finger nucleases targeted to CCR5 control HIV-1 in vivo. *Nat Biotechnol* **28:** 839–847.

Horvath P, Barrangou R. 2010. CRISPR/Cas, the immune system of bacteria and archaea. *Science* **327:** 167–170.

Hou Z, Zhang Y, Propson NE, Howden SE, Chu LF, Sontheimer EJ, Thomson JA. 2013. Efficient genome engineering in human pluripotent stem cells using Cas9 from *Neisseria meningitidis*. *Proc Natl Acad Sci* **110:** 15644–15649.

Hsu PD, Scott DA, Weinstein JA, Ran FA, Konermann S, Agarwala V, Li Y, Fine EJ, Wu X, Shalem O, et al. 2013. DNA targeting specificity of RNA-guided Cas9 nucleases. *Nat Biotechnol* **31:** 827–832.

Hu J, Lei Y, Wong WK, Liu S, Lee KC, He X, You W, Zhou R, Guo JT, Chen X, et al. 2014. Direct activation of human and mouse Oct4 genes using engineered TALE and Cas9 transcription factors. *Nucleic Acids Res* **42:** 4375–4390.

Hubbard BP, Badran AH, Zuris JA, Guilinger JP, Davis KM, Chen L, Tsai SQ, Sander JD, Joung JK, Liu DR. 2015. Continuous directed evolution of DNA-binding proteins to improve TALEN specificity. *Nat Methods* **12:** 939–942.

Hurt JA, Thibodeau SA, Hirsh AS, Pabo CO, Joung JK. 2003. Highly specific zinc finger proteins obtained by directed domain shuffling and cell-based selection. *Proc Natl Acad Sci* **100:** 12271–12276.

Hwang WY, Fu Y, Reyon D, Maeder ML, Tsai SQ, Sander JD, Peterson RT, Yeh JR, Joung JK. 2013. Efficient genome editing in zebrafish using a CRISPR-Cas system. *Nat Biotechnol* **31:** 227–229.

Isalan M, Klug A, Choo Y. 2001. A rapid, generally applicable method to engineer zinc fingers illustrated by targeting the HIV-1 promoter. *Nat Biotechnol* **19:** 656–660.

Jakociunas T, Bonde I, Herrgard M, Harrison SJ, Kristensen M, Pedersen LE, Jensen MK, Keasling JD. 2015a. Multiplex metabolic pathway engineering using CRISPR/Cas9 in *Saccharomyces cerevisiae. Metab Eng* **28:** 213–222.

Jakociunas T, Rajkumar AS, Zhang J, Arsovska D, Rodriguez A, Jendresen CB, Skjodt ML, Nielsen AT, Borodina I, Jensen MK, et al. 2015b. CasEMBLR: Cas9-facilitated multiloci genomic integration of in vivo assembled DNA parts in *Saccharomyces cerevisiae. ACS Synth Biol* **4:** 1226–1234.

Jamieson AC, Kim SH, Wells JA. 1994. In vitro selection of zinc fingers with altered DNA-binding specificity. *Biochemistry* **33:** 5689–5695.

Ji H, Jiang Z, Lu P, Ma L, Li C, Pan H, Fu Z, Qu X, Wang P, Deng J, et al. 2016. Specific reactivation of latent HIV-1 by dCas9-SunTag-VP64-mediated guide RNA targeting the HIV-1 promoter. *Mol Ther* **24:** 508–521.

Jiang W, Bikard D, Cox D, Zhang F, Marraffini LA. 2013. RNA-guided editing of bacterial genomes using CRISPR-Cas systems. *Nat Biotechnol* **31:** 233–239.

Jiang F, Taylor DW, Chen JS, Kornfeld JE, Zhou K, Thompson AJ, Nogales E, Doudna JA. 2016. Structures of a CRISPR-Cas9 R-loop complex primed for DNA cleavage. *Science* **351:** 867–871.

Jinek M, Chylinski K, Fonfara I, Hauer M, Doudna JA, Charpentier E. 2012. A programmable dual-RNA-guided DNA endonuclease in adaptive bacterial immunity. *Science* **337:** 816–821.

Jinek M, East A, Cheng A, Lin S, Ma E, Doudna J. 2013. RNA-programmed genome editing in human cells. *eLife* **2:** e00471.

Juillerat A, Pessereau C, Dubois G, Guyot V, Marechal A, Valton J, Daboussi F, Poirot L, Duclert A, Duchateau P. 2015. Optimized tuning of TALEN specificity using nonconventional RVDs. *Sci Rep* **5:** 8150.

Kiani S, Chavez A, Tuttle M, Hall RN, Chari R, Ter-Ovanesyan D, Qian J, Pruitt BW, Beal J, Vora S, et al. 2015. Cas9 gRNA engineering for genome editing, activation and repression. *Nat Methods* **12:** 1051–1054.

Kim YG, Cha J, Chandrasegaran S. 1996. Hybrid restriction enzymes: Zinc finger fusions to Fok I cleavage domain. *Proc Natl Acad Sci* **93:** 1156–1160.

Kim HJ, Lee HJ, Kim H, Cho SW, Kim JS. 2009. Targeted genome editing in human cells with zinc finger nucleases constructed via modular assembly. *Genome Res* **19:** 1279–1288.

Kim E, Kim S, Kim DH, Choi BS, Choi IY, Kim JS. 2012. Precision genome engineering with programmable DNA-nicking enzymes. *Genome Res* **22:** 1327–1333.

Kim Y, Kweon J, Kim A, Chon JK, Yoo JY, Kim HJ, Kim S, Lee C, Jeong E, Chung E, et al. 2013a. A library of TAL effector nucleases spanning the human genome. *Nat Biotechnol* **31:** 251–258.

Kim YK, Wee G, Park J, Kim J, Baek D, Kim JS, Kim VN. 2013b. TALEN-based knockout library for human microRNAs. *Nat Struct Mol Biol* **20:** 1458–1464.

Kim S, Kim D, Cho SW, Kim J, Kim JS. 2014. Highly efficient RNA-guided genome editing in human cells via delivery of purified Cas9 ribonucleoproteins. *Genome Res* **24:** 1012–1019.

Kleinstiver BP, Prew MS, Tsai SQ, Topkar VV, Nguyen NT, Zheng Z, Gonzales AP, Li Z, Peterson RT, Yeh JR, et al. 2015. Engineered CRISPR-Cas9 nucleases with altered PAM specificities. *Nature* **523:** 481–485.

Kleinstiver BP, Pattanayak V, Prew MS, Tsai SQ, Nguyen NT, Zheng Z, Joung JK. 2016. High-fidelity CRISPR-Cas9 nucleases with no detectable genome-wide off-target effects. *Nature* **529:** 490–495.

Koike-Yusa H, Li Y, Tan EP, Velasco-Herrera Mdel C, Yusa K. 2014. Genome-wide recessive genetic screening in mammalian cells with a lentiviral CRISPR-guide RNA library. *Nat Biotechnol* **32:** 267–273.

Komor AC, Kim YB, Packer MS, Zuris JA, Liu DR. 2016. Programmable editing of a target base in genomic DNA without double-stranded DNA cleavage. *Nature* doi: 10.1038/nature17946.

Konermann S, Brigham MD, Trevino AE, Hsu PD, Heidenreich M, Cong L, Platt RJ, Scott DA, Church GM, Zhang F. 2013. Optical control of mammalian endogenous transcription and epigenetic states. *Nature* **500:** 472–476.

Konermann S, Brigham MD, Trevino AE, Joung J, Abudayyeh OO, Barcena C, Hsu PD, Habib N, Gootenberg JS, Nishimasu H, et al. 2015. Genome-scale transcriptional activation by an engineered CRISPR-Cas9 complex. *Nature* **517:** 583–588.

Korkmaz G, Lopes R, Ugalde AP, Nevedomskaya E, Han R, Myacheva K, Zwart W, Elkon R, Agami R. 2016. Functional genetic screens for enhancer elements in the human genome using CRISPR-Cas9. *Nat Biotechnol* **34:** 192–198.

Lebar T, Bezeljak U, Golob A, Jerala M, Kadunc L, Pirs B, Strazar M, Vucko D, Zupancic U, Bencina M, et al. 2014. A bistable genetic switch based on designable DNA-binding domains. *Nat Commun* **5:** 5007.

Lee HJ, Kim E, Kim JS. 2010. Targeted chromosomal deletions in human cells using zinc finger nucleases. *Genome Res* **20:** 81–89.

Li H, Haurigot V, Doyon Y, Li T, Wong SY, Bhagwat AS, Malani N, Anguela XM, Sharma R, Ivanciu L, et al. 2011. In vivo genome editing restores haemostasis in a mouse model of haemophilia. *Nature* **475:** 217–221.

Li Y, Moore R, Guinn M, Bleris L. 2012. Transcription activator-like effector hybrids for conditional control and rewiring of chromosomal transgene expression. *Sci Rep* **2:** 897.

Li W, Teng F, Li T, Zhou Q. 2013. Simultaneous generation and germline transmission of multiple gene mutations in rat using CRISPR-Cas systems. *Nat Biotechnol* **31:** 684–686.

Limsirichai P, Gaj T, Schaffer DV. 2016. CRISPR-mediated activation of latent HIV-1 expression. *Mol Ther* **24:** 499–507.

Lin SR, Yang HC, Kuo YT, Liu CJ, Yang TY, Sung KC, Lin YY, Wang HY, Wang CC, Shen YC, et al. 2014. The CRISPR/

Cas9 system facilitates clearance of the intrahepatic HBV templates in vivo. *Mol Ther Nucleic Acids* **3:** e186.

Lin J, Chen H, Luo L, Lai Y, Xie W, Kee K. 2015a. Creating a monomeric endonuclease TALE-I-SceI with high specificity and low genotoxicity in human cells. *Nucleic Acids Res* **43:** 1112–1122.

Lin S, Ewen-Campen B, Ni X, Housden BE, Perrimon N. 2015b. In vivo transcriptional activation using CRISPR/ Cas9 in *Drosophila. Genetics* **201:** 433–442.

Liu Q, Segal DJ, Ghiara JB, Barbas CF III. 1997. Design of polydactyl zinc-finger proteins for unique addressing within complex genomes. *Proc Natl Acad Sci* **94:** 5525–5530.

Liu J, Gaj T, Patterson JT, Sirk SJ, Barbas CF III. 2014a. Cell-penetrating peptide-mediated delivery of TALEN proteins via bioconjugation for genome engineering. *PLoS ONE* **9:** e85755.

Liu Y, Zeng Y, Liu L, Zhuang C, Fu X, Huang W, Cai Z. 2014b. Synthesizing AND gate genetic circuits based on CRISPR-Cas9 for identification of bladder cancer cells. *Nat Commun* **5:** 5393.

Liu Z, Zhou X, Zhu Y, Chen ZF, Yu B, Wang Y, Zhang CC, Nie YH, Sang X, Cai YJ, et al. 2014c. Generation of a monkey with MECP2 mutations by TALEN-based gene targeting. *Neurosci Bull* **30:** 381–386.

Liu J, Gaj T, Wallen MC, Barbas CF III. 2015a. Improved cell-penetrating zinc-finger nuclease proteins for precision genome engineering. *Mol Ther Nucleic Acids* **4:** e232.

Liu J, Gaj T, Yang Y, Wang N, Shui S, Kim S, Kanchiswamy CN, Kim JS, Barbas CF III. 2015b. Efficient delivery of nuclease proteins for genome editing in human stem cells and primary cells. *Nat Protoc* **10:** 1842–1859.

Long C, Amoasii L, Mireault AA, McAnally JR, Li H, Sanchez-Ortiz E, Bhattacharyya S, Shelton JM, Bassel-Duby R, Olson EN. 2015. Postnatal genome editing partially restores dystrophin expression in a mouse model of muscular dystrophy. *Science* **351:** 400–403.

Lonzaric J, Lebar T, Majerle A, Mancek-Keber M, Jerala R. 2016. Locked and proteolysis-based transcription activator-like effector (TALE) regulation. *Nucleic Acids Res* **44:** 1471–1481.

Luo S, Li J, Stoddard TJ, Baltes NJ, Demorest ZL, Clasen BM, Coffman A, Retterath A, Mathis L, Voytas DF, et al. 2015. Non-transgenic plant genome editing using purified sequence-specific nucleases. *Mol Plant* **8:** 1425–1427.

Ma H, Dang Y, Wu Y, Jia G, Anaya E, Zhang J, Abraham S, Choi JG, Shi G, Qi L, et al. 2015. A CRISPR-based screen identifies genes essential for West-Nile-virus-induced cell death. *Cell Rep* **12:** 673–683.

Maeder ML, Gersbach CA. 2016. Genome-editing technologies for gene and cell therapy. *Mol Ther* **24:** 430–446.

Maeder ML, Thibodeau-Beganny S, Osiak A, Wright DA, Anthony RM, Eichtinger M, Jiang T, Foley JE, Winfrey RJ, Townsend JA, et al. 2008. Rapid "open-source" engineering of customized zinc-finger nucleases for highly efficient gene modification. *Mol Cell* **31:** 294–301.

Maeder ML, Linder SJ, Cascio VM, Fu Y, Ho QH, Joung JK. 2013a. CRISPR RNA-guided activation of endogenous human genes. *Nat Methods* **10:** 977–979.

Maeder ML, Linder SJ, Reyon D, Angstman JF, Fu Y, Sander JD, Joung JK. 2013b. Robust, synergistic regulation of human gene expression using TALE activators. *Nat Methods* **10:** 243–245.

Maggio I, Stefanucci L, Janssen JM, Liu J, Chen X, Mouly V, Goncalves MA. 2016. Selection-free gene repair after adenoviral vector transduction of designer nucleases: rescue of dystrophin synthesis in DMD muscle cell populations. *Nucleic Acids Res* **44:** 1449–1470.

Magnenat L, Blancafort P, Barbas CF III. 2004. In vivo selection of combinatorial libraries and designed affinity maturation of polydactyl zinc finger transcription factors for ICAM-1 provides new insights into gene regulation. *J Mol Biol* **341:** 635–649.

Magnenat L, Schwimmer LJ, Barbas CF III. 2008. Drug-inducible and simultaneous regulation of endogenous genes by single-chain nuclear receptor-based zinc-finger transcription factor gene switches. *Gene Ther* **15:** 1223–1232.

Mahiny AJ, Dewerth A, Mays LE, Alkhaled M, Mothes B, Malaeksefat E, Loretz B, Rottenberger J, Brosch DM, Reautschnig P, et al. 2015. In vivo genome editing using nuclease-encoding mRNA corrects SP-B deficiency. *Nat Biotechnol* **33:** 584–586.

Mak AN, Bradley P, Cernadas RA, Bogdanove AJ, Stoddard BL. 2012. The crystal structure of TAL effector PthXo1 bound to its DNA target. *Science* **335:** 716–719.

Mali P, Aach J, Stranges PB, Esvelt KM, Moosburner M, Kosuri S, Yang L, Church GM. 2013a. CAS9 transcriptional activators for target specificity screening and paired nickases for cooperative genome engineering. *Nat Biotechnol* **31:** 833–838.

Mali P, Yang L, Esvelt KM, Aach J, Guell M, DiCarlo JE, Norville JE, Church GM. 2013b. RNA-guided human genome engineering via Cas9. *Science* **339:** 823–826.

Margolin JF, Friedman JR, Meyer WK, Vissing H, Thiesen HJ, Rauscher FJ III. 1994. Kruppel-associated boxes are potent transcriptional repression domains. *Proc Natl Acad Sci* **91:** 4509–4513.

Marraffini LA, Sontheimer EJ. 2010. CRISPR interference: RNA-directed adaptive immunity in bacteria and archaea. *Nat Rev Genet* **11:** 181–190.

McConnell Smith A, Takeuchi R, Pellenz S, Davis L, Maizels N, Monnat RJ Jr, Stoddard BL. 2009. Generation of a nicking enzyme that stimulates site-specific gene conversion from the I-AniI LAGLIDADG homing endonuclease. *Proc Natl Acad Sci* **106:** 5099–5104.

Meckler JF, Bhakta MS, Kim MS, Ovadia R, Habrian CH, Zykovich A, Yu A, Lockwood SH, Morbitzer R, Elsaesser J, et al. 2013. Quantitative analysis of TALE-DNA interactions suggests polarity effects. *Nucleic Acids Res* **41:** 4118–4128.

Mendenhall EM, Williamson KE, Reyon D, Zou JY, Ram O, Joung JK, Bernstein BE. 2013. Locus-specific editing of histone modifications at endogenous enhancers. *Nat Biotechnol* **31:** 1133–1136.

Menoret S, De Cian A, Tesson L, Remy S, Usal C, Boule JB, Boix C, Fontaniere S, Creneguy A, Nguyen TH, et al. 2015. Homology-directed repair in rodent zygotes using Cas9 and TALEN engineered proteins. *Sci Rep* **5:** 14410.

Mercer AC, Gaj T, Fuller RP, Barbas CF 3rd. 2012. Chimeric TALE recombinases with programmable DNA sequence specificity. *Nucleic Acids Res* **40:** 11163–11172.

Mercer AC, Gaj T, Sirk SJ, Lamb BM, Barbas CF III. 2014. Regulation of endogenous human gene expression by ligand-inducible TALE transcription factors. *ACS Synth Biol* **3:** 723–730.

Miller JC, Holmes MC, Wang J, Guschin DY, Lee YL, Rupniewski I, Beausejour CM, Waite AJ, Wang NS, Kim KA, et al. 2007. An improved zinc-finger nuclease architecture for highly specific genome editing. *Nat Biotechnol* **25:** 778–785.

Miller JC, Tan S, Qiao G, Barlow KA, Wang J, Xia DF, Meng X, Paschon DE, Leung E, Hinkley SJ, et al. 2011. A TALE nuclease architecture for efficient genome editing. *Nat Biotechnol* **29:** 143–148.

Miller JC, Zhang L, Xia DF, Campo JJ, Ankoudinova IV, Guschin DY, Babiarz JE, Meng X, Hinkley SJ, Lam SC, et al. 2015. Improved specificity of TALE-based genome editing using an expanded RVD repertoire. *Nat Methods* **12:** 465–471.

Mock U, Machowicz R, Hauber I, Horn S, Abramowski P, Berdien B, Hauber J, Fehse B. 2015. mRNA transfection of a novel TAL effector nuclease (TALEN) facilitates efficient knockout of HIV co-receptor CCR5. *Nucleic Acids Res* **43:** 5560–5571.

Moehle EA, Rock JM, Lee YL, Jouvenot Y, DeKelver RC, Gregory PD, Urnov FD, Holmes MC. 2007. Targeted gene addition into a specified location in the human genome using designed zinc finger nucleases. *Proc Natl Acad Sci* **104:** 3055–3060.

Moscou MJ, Bogdanove AJ. 2009. A simple cipher governs DNA recognition by TAL effectors. *Science* **326:** 1501.

Mussolino C, Alzubi J, Fine EJ, Morbitzer R, Cradick TJ, Lahaye T, Bao G, Cathomen T. 2014. TALENs facilitate targeted genome editing in human cells with high specificity and low cytotoxicity. *Nucleic Acids Res* **42:** 6762–6773.

Nelson CE, Hakim CH, Ousterout DG, Thakore PI, Moreb EA, Rivera RMC, Madhavan S, Pan X, Ran FA, Yan WX, et al. 2015. In vivo genome editing improves muscle function in a mouse model of Duchenne muscular dystrophy. *Science* **351:** 403–407.

Nielsen AA, Voigt CA. 2014. Multi-input CRISPR/Cas genetic circuits that interface host regulatory networks. *Mol Syst Biol* **10:** 763.

Nihongaki Y, Yamamoto S, Kawano F, Suzuki H, Sato M. 2015. CRISPR-Cas9-based photoactivatable transcription system. *Chem Biol* **22:** 169–174.

Nishimasu H, Ran FA, Hsu PD, Konermann S, Shehata SI, Dohmae N, Ishitani R, Zhang F, Nureki O. 2014. Crystal structure of Cas9 in complex with guide RNA and target DNA. *Cell* **156:** 935–949.

Nissim L, Perli SD, Fridkin A, Perez-Pinera P, Lu TK. 2014. Multiplexed and programmable regulation of gene networks with an integrated RNA and CRISPR/Cas toolkit in human cells. *Mol Cell* **54:** 698–710.

Osborn MJ, Webber BR, Knipping F, Lonetree CL, Tennis N, DeFeo AP, McElroy AN, Starker CG, Lee C, Merkel S, et al. 2016. Evaluation of TCR gene editing achieved by TALENs, CRISPR/Cas9, and megaTAL nucleases. *Mol Ther* **24:** 570–581.

Paix A, Folkmann A, Rasoloson D, Seydoux G. 2015. High efficiency, homology-directed genome editing in *Caenorhabditis elegans* using CRISPR-Cas9 ribonucleoprotein complexes. *Genetics* **201:** 47–54.

Papworth M, Moore M, Isalan M, Minczuk M, Choo Y, Klug A. 2003. Inhibition of herpes simplex virus 1 gene expression by designer zinc-finger transcription factors. *Proc Natl Acad Sci* **100:** 1621–1626.

Park KS, Lee DK, Lee H, Lee Y, Jang YS, Kim YH, Yang HY, Lee SI, Seol W, Kim JS. 2003. Phenotypic alteration of eukaryotic cells using randomized libraries of artificial transcription factors. *Nat Biotechnol* **21:** 1208–1214.

Parnas O, Jovanovic M, Eisenhaure TM, Herbst RH, Dixit A, Ye CJ, Przybylski D, Platt RJ, Tirosh I, Sanjana NE, et al. 2015. A genome-wide CRISPR screen in primary immune cells to dissect regulatory networks. *Cell* **162:** 675–686.

Pattanayak V, Ramirez CL, Joung JK, Liu DR. 2011. Revealing off-target cleavage specificities of zinc-finger nucleases by in vitro selection. *Nat Methods* **8:** 765–770.

Pavletich NP, Pabo CO. 1991. Zinc finger-DNA recognition: Crystal structure of a Zif268-DNA complex at 2.1 A. *Science* **252:** 809–817.

Perdigao P, Gaj T, Santa-Marta M, Barbas CF III, Goncalves J. 2016. Reactivation of latent HIV-1 expression by engineered TALE transcription factors. *PLoS ONE* **11:** e0150037.

Perez EE, Wang J, Miller JC, Jouvenot Y, Kim KA, Liu O, Wang N, Lee G, Bartsevich VV, Lee YL, et al. 2008. Establishment of HIV-1 resistance in $CD4^+$ T cells by genome editing using zinc-finger nucleases. *Nat Biotechnol* **26:** 808–816.

Perez-Pinera P, Ousterout DG, Brunger JM, Farin AM, Glass KA, Guilak F, Crawford GE, Hartemink AJ, Gersbach CA. 2013. Synergistic and tunable human gene activation by combinations of synthetic transcription factors. *Nat Methods* **10:** 239–242.

Piatek A, Ali Z, Baazim H, Li L, Abulfaraj A, Al-Shareef S, Aouida M, Mahfouz MM. 2015. RNA-guided transcriptional regulation in planta via synthetic dCas9-based transcription factors. *Plant Biotechnol J* **13:** 578–589.

Pollock R, Giel M, Linher K, Clackson T. 2002. Regulation of endogenous gene expression with a small-molecule dimerizer. *Nat Biotechnol* **20:** 729–733.

Polstein LR, Gersbach CA. 2012. Light-inducible spatiotemporal control of gene activation by customizable zinc finger transcription factors. *J Am Chem Soc* **134:** 16480–16483.

Polstein LR, Gersbach CA. 2015. A light-inducible CRISPR-Cas9 system for control of endogenous gene activation. *Nat Chem Biol* **11:** 198–200.

Porteus MH. 2015. Towards a new era in medicine: Therapeutic genome editing. *Genome Biol* **16:** 286.

Porteus MH, Baltimore D. 2003. Chimeric nucleases stimulate gene targeting in human cells. *Science* **300:** 763.

Porteus MH, Carroll D. 2005. Gene targeting using zinc finger nucleases. *Nat Biotechnol* **23:** 967–973.

Pruett-Miller SM, Reading DW, Porter SN, Porteus MH. 2009. Attenuation of zinc finger nuclease toxicity by small-molecule regulation of protein levels. *PLoS Genet* **5:** e1000376.

Qi LS, Larson MH, Gilbert LA, Doudna JA, Weissman JS, Arkin AP, Lim WA. 2013. Repurposing CRISPR as an

RNA-guided platform for sequence-specific control of gene expression. *Cell* **152:** 1173–1183.

Rahdar M, McMahon MA, Prakash TP, Swayze EE, Bennett CF, Cleveland DW. 2015. Synthetic CRISPR RNA-Cas9-guided genome editing in human cells. *Proc Natl Acad Sci* **112:** E7110–E7117.

Ramirez CL, Certo MT, Mussolino C, Goodwin MJ, Cradick TJ, McCaffrey AP, Cathomen T, Scharenberg AM, Joung JK. 2012. Engineered zinc finger nickases induce homology-directed repair with reduced mutagenic effects. *Nucleic Acids Res* **40:** 5560–5568.

Ran FA, Hsu PD, Lin CY, Gootenberg JS, Konermann S, Trevino AE, Scott DA, Inoue A, Matoba S, Zhang Y, et al. 2013. Double nicking by RNA-guided CRISPR Cas9 for enhanced genome editing specificity. *Cell* **154:** 1380–1389.

Ran FA, Cong L, Yan WX, Scott DA, Gootenberg JS, Kriz AJ, Zetsche B, Shalem O, Wu X, Makarova KS, et al. 2015. In vivo genome editing using *Staphylococcus aureus* Cas9. *Nature* **520:** 186–191.

Rebar EJ, Huang Y, Hickey R, Nath AK, Meoli D, Nath S, Chen B, Xu L, Liang Y, Jamieson AC, et al. 2002. Induction of angiogenesis in a mouse model using engineered transcription factors. *Nat Med* **8:** 1427–1432.

Reynolds L, Ullman C, Moore M, Isalan M, West MJ, Clapham P, Klug A, Choo Y. 2003. Repression of the HIV-1 5′ LTR promoter and inhibition of HIV-1 replication by using engineered zinc-finger transcription factors. *Proc Natl Acad Sci* **100:** 1615–1620.

Reyon D, Tsai SQ, Khayter C, Foden JA, Sander JD, Joung JK. 2012. FLASH assembly of TALENs for high-throughput genome editing. *Nat Biotechnol* **30:** 460–465.

Rivenbark AG, Stolzenburg S, Beltran AS, Yuan X, Rots MG, Strahl BD, Blancafort P. 2012. Epigenetic reprogramming of cancer cells via targeted DNA methylation. *Epigenetics* **7:** 350–360.

Rouet P, Smih F, Jasin M. 1994. Introduction of double-strand breaks into the genome of mouse cells by expression of a rare-cutting endonuclease. *Mol Cell Biol* **14:** 8096–8106.

Ryan OW, Skerker JM, Maurer MJ, Li X, Tsai JC, Poddar S, Lee ME, DeLoache W, Dueber JE, Arkin AP, et al. 2014. Selection of chromosomal DNA libraries using a multiplex CRISPR system. *eLife* doi: 10.7554/eLife.03703.

Saayman SM, Lazar DC, Scott TA, Hart JR, Takahashi M, Burnett JC, Planelles V, Morris KV, Weinberg MS. 2016. Potent and targeted activation of latent HIV-1 using the CRISPR/dCas9 activator complex. *Mol Ther* **24:** 488–498.

Sadowski I, Ma J, Triezenberg S, Ptashne M. 1988. GAL4-VP16 is an unusually potent transcriptional activator. *Nature* **335:** 563–564.

Sander JD, Cade L, Khayter C, Reyon D, Peterson RT, Joung JK, Yeh JR. 2011a. Targeted gene disruption in somatic zebrafish cells using engineered TALENs. *Nat Biotechnol* **29:** 697–698.

Sander JD, Dahlborg EJ, Goodwin MJ, Cade L, Zhang F, Cifuentes D, Curtin SJ, Blackburn JS, Thibodeau-Beganny S, Qi Y, et al. 2011b. Selection-free zinc-finger-nuclease engineering by context-dependent assembly (CoDA). *Nat Methods* **8:** 67–69.

Santiago Y, Chan E, Liu PQ, Orlando S, Zhang L, Urnov FD, Holmes MC, Guschin D, Waite A, Miller JC, et al. 2008. Targeted gene knockout in mammalian cells by using engineered zinc-finger nucleases. *Proc Natl Acad Sci* **105:** 5809–5814.

Sather BD, Romano Ibarra GS, Sommer K, Curinga G, Hale M, Khan IF, Singh S, Song Y, Gwiazda K, Sahni J, et al. 2015. Efficient modification of CCR5 in primary human hematopoietic cells using a megaTAL nuclease and AAV donor template. *Sci Transl Med* **7:** 307ra156.

Schmid-Burgk JL, Schmidt T, Kaiser V, Honing K, Hornung V. 2013. A ligation-independent cloning technique for high-throughput assembly of transcription activator-like effector genes. *Nat Biotechnol* **31:** 76–81.

Schumann K, Lin S, Boyer E, Simeonov DR, Subramaniam M, Gate RE, Haliburton GE, Ye CJ, Bluestone JA, Doudna JA, et al. 2015. Generation of knock-in primary human T cells using Cas9 ribonucleoproteins. *Proc Natl Acad Sci* **112:** 10437–10442.

Segal DJ, Dreier B, Beerli RR, Barbas CF III. 1999. Toward controlling gene expression at will: Selection and design of zinc finger domains recognizing each of the 5′-GNN-3′ DNA target sequences. *Proc Natl Acad Sci* **96:** 2758–2763.

Segal DJ, Goncalves J, Eberhardy S, Swan CH, Torbett BE, Li X, Barbas CF III. 2004. Attenuation of HIV-1 replication in primary human cells with a designed zinc finger transcription factor. *J Biol Chem* **279:** 14509–14519.

Shah AN, Davey CF, Whitebirch AC, Miller AC, Moens CB. 2015. Rapid reverse genetic screening using CRISPR in zebrafish. *Nat Methods* **12:** 535–540.

Shalem O, Sanjana NE, Hartenian E, Shi X, Scott DA, Mikkelsen TS, Heckl D, Ebert BL, Root DE, Doench JG, et al. 2014. Genome-scale CRISPR-Cas9 knockout screening in human cells. *Science* **343:** 84–87.

Sharma R, Anguela XM, Doyon Y, Wechsler T, DeKelver RC, Sproul S, Paschon DE, Miller JC, Davidson RJ, Shivak D, et al. 2015. In vivo genome editing of the albumin locus as a platform for protein replacement therapy. *Blood* **126:** 1777–1784.

Siddique AN, Nunna S, Rajavelu A, Zhang Y, Jurkowska RZ, Reinhardt R, Rots MG, Ragozin S, Jurkowski TP, Jeltsch A. 2013. Targeted methylation and gene silencing of VEGF-A in human cells by using a designed Dnmt3a-Dnmt3L single-chain fusion protein with increased DNA methylation activity. *J Mol Biol* **425:** 479–491.

Silva G, Poirot L, Galetto R, Smith J, Montoya G, Duchateau P, Paques F. 2011. Meganucleases and other tools for targeted genome engineering: perspectives and challenges for gene therapy. *Curr Gene Ther* **11:** 11–27.

Sinkins SP, Gould F. 2006. Gene drive systems for insect disease vectors. *Nat Rev Genet* **7:** 427–435.

Sirk SJ, Gaj T, Jonsson A, Mercer AC, Barbas CF III. 2014. Expanding the zinc-finger recombinase repertoire: Directed evolution and mutational analysis of serine recombinase specificity determinants. *Nucleic Acids Res* **42:** 4755–4766.

Slaymaker IM, Gao L, Zetsche B, Scott DA, Yan WX, Zhang F. 2016. Rationally engineered Cas9 nucleases with improved specificity. *Science* **351:** 84–88.

Smith J, Bibikova M, Whitby FG, Reddy AR, Chandrasegaran S, Carroll D. 2000. Requirements for double-strand cleavage by chimeric restriction enzymes with zinc finger

DNA-recognition domains. *Nucleic Acids Res* **28:** 3361–3369.

Sorek R, Lawrence CM, Wiedenheft B. 2013. CRISPR-mediated adaptive immune systems in bacteria and archaea. *Annu Rev Biochem* **82:** 237–266.

Stoddard BL. 2011. Homing endonucleases: From microbial genetic invaders to reagents for targeted DNA modification. *Structure* **19:** 7–15.

Stoddard BL. 2014. Homing endonucleases from mobile group I introns: Discovery to genome engineering. *Mob DNA* **5:** 7.

Streubel J, Blucher C, Landgraf A, Boch J. 2012. TAL effector RVD specificities and efficiencies. *Nat Biotechnol* **30:** 593–595.

Sung YH, Kim JM, Kim HT, Lee J, Jeon J, Jin Y, Choi JH, Ban YH, Ha SJ, Kim CH, et al. 2014. Highly efficient gene knockout in mice and zebrafish with RNA-guided endonucleases. *Genome Res* **24:** 125–131.

Tabebordbar M, Zhu K, Cheng JKW, Chew WL, Widrick JJ, Yan WX, Maesner C, Wu EY, Xiao R, Ran FA, et al. 2015. In vivo gene editing in dystrophic mouse muscle and muscle stem cells. *Science* **351:** 407–411.

Tebas P, Stein D, Tang WW, Frank I, Wang SQ, Lee G, Spratt SK, Surosky RT, Giedlin MA, Nichol G, et al. 2014. Gene editing of CCR5 in autologous CD4 T cells of persons infected with HIV. *N Engl J Med* **370:** 901–910.

Tesson L, Usal C, Menoret S, Leung E, Niles BJ, Remy S, Santiago Y, Vincent AI, Meng X, Zhang L, et al. 2011. Knockout rats generated by embryo microinjection of TALENs. *Nat Biotechnol* **29:** 695–696.

Thakore PI, D'Ippolito AM, Song L, Safi A, Shivakumar NK, Kabadi AM, Reddy TE, Crawford GE, Gersbach CA. 2015. Highly specific epigenome editing by CRISPR-Cas9 repressors for silencing of distal regulatory elements. *Nat Methods* **12:** 1143–1149.

Tong Y, Charusanti P, Zhang L, Weber T, Lee SY. 2015. CRISPR-Cas9 based engineering of actinomycetal genomes. *ACS Synth Biol* **4:** 1020–1029.

Torres R, Martin MC, Garcia A, Cigudosa JC, Ramirez JC, Rodriguez-Perales S. 2014. Engineering human tumour-associated chromosomal translocations with the RNA-guided CRISPR-Cas9 system. *Nat Commun* **5:** 3964.

Tsai SQ, Wyvekens N, Khayter C, Foden JA, Thapar V, Reyon D, Goodwin MJ, Aryee MJ, Joung JK. 2014. Dimeric CRISPR RNA-guided FokI nucleases for highly specific genome editing. *Nat Biotechnol* **32:** 569–576.

Urnov FD, Miller JC, Lee YL, Beausejour CM, Rock JM, Augustus S, Jamieson AC, Porteus MH, Gregory PD, Holmes MC. 2005. Highly efficient endogenous human gene correction using designed zinc-finger nucleases. *Nature* **435:** 646–651.

Urnov FD, Rebar EJ, Holmes MC, Zhang HS, Gregory PD. 2010. Genome editing with engineered zinc finger nucleases. *Nat Rev Genet* **11:** 636–646.

Wallen MC, Gaj T, Barbas CF III. 2015. Redesigning recombinase specificity for safe harbor sites in the human genome. *PLoS ONE* **10:** e0139123.

Wang J, Friedman G, Doyon Y, Wang NS, Li CJ, Miller JC, Hua KL, Yan JJ, Babiarz JE, Gregory PD, et al. 2012. Targeted gene addition to a predetermined site in the human genome using a ZFN-based nicking enzyme. *Genome Res* **22:** 1316–1326.

Wang H, Hu YC, Markoulaki S, Welstead GG, Cheng AW, Shivalila CS, Pyntikova T, Dadon DB, Voytas DF, Bogdanove AJ, et al. 2013. TALEN-mediated editing of the mouse Y chromosome. *Nat Biotechnol* **31:** 530–532.

Wang T, Wei JJ, Sabatini DM, Lander ES. 2014a. Genetic screens in human cells using the CRISPR-Cas9 system. *Science* **343:** 80–84.

Wang Y, Cheng X, Shan Q, Zhang Y, Liu J, Gao C, Qiu JL. 2014b. Simultaneous editing of three homoeoalleles in hexaploid bread wheat confers heritable resistance to powdery mildew. *Nat Biotechnol* **32:** 947–951.

Wang L, Shao Y, Guan Y, Li L, Wu L, Chen F, Liu M, Chen H, Ma Y, Ma X, et al. 2015a. Large genomic fragment deletion and functional gene cassette knock-in via Cas9 protein mediated genome editing in one-cell rodent embryos. *Sci Rep* **5:** 17517.

Wang T, Birsoy K, Hughes NW, Krupczak KM, Post Y, Wei JJ, Lander ES, Sabatini DM. 2015b. Identification and characterization of essential genes in the human genome. *Science* **350:** 1096–1101.

Wiedenheft B, Sternberg SH, Doudna JA. 2012. RNA-guided genetic silencing systems in bacteria and archaea. *Nature* **482:** 331–338.

Wilber A, Tschulena U, Hargrove PW, Kim YS, Persons DA, Barbas CF III, Nienhuis AW. 2010. A zinc-finger transcriptional activator designed to interact with the γ-globin gene promoters enhances fetal hemoglobin production in primary human adult erythroblasts. *Blood* **115:** 3033–3041.

Windbichler N, Menichelli M, Papathanos PA, Thyme SB, Li H, Ulge UY, Hovde BT, Baker D, Monnat RJ Jr, Burt A, et al. 2011. A synthetic homing endonuclease-based gene drive system in the human malaria mosquito. *Nature* **473:** 212–215.

Wolfe SA, Nekludova L, Pabo CO. 2000. DNA recognition by Cys2His2 zinc finger proteins. *Annu Rev Biophys Biomol Struct* **29:** 183–212.

Woo JW, Kim J, Kwon SI, Corvalan C, Cho SW, Kim H, Kim SG, Kim ST, Choe S, Kim JS. 2015. DNA-free genome editing in plants with preassembled CRISPR-Cas9 ribonucleoproteins. *Nat Biotechnol* **33:** 1162–1164.

Wu H, Yang WP, Barbas CF III. 1995. Building zinc fingers by selection: Toward a therapeutic application. *Proc Natl Acad Sci* **92:** 344–348.

Wu Y, Liang D, Wang Y, Bai M, Tang W, Bao S, Yan Z, Li D, Li J. 2013. Correction of a genetic disease in mouse via use of CRISPR-Cas9. *Cell Stem Cell* **13:** 659–662.

Wu H, Wang Y, Zhang Y, Yang M, Lv J, Liu J, Zhang Y. 2015. TALE nickase-mediated SP110 knockin endows cattle with increased resistance to tuberculosis. *Proc Natl Acad Sci* **112:** E1530–E1539.

Xiao A, Wang Z, Hu Y, Wu Y, Luo Z, Yang Z, Zu Y, Li W, Huang P, Tong X, et al. 2013. Chromosomal deletions and inversions mediated by TALENs and CRISPR/Cas in zebrafish. *Nucleic Acids Res* **41:** e141.

Yang J, Zhang Y, Yuan P, Zhou Y, Cai C, Ren Q, Wen D, Chu C, Qi H, Wei W. 2014. Complete decoding of TAL effectors for DNA recognition. *Cell Res* **24:** 628–631.

Yang Y, Wang L, Bell P, McMenamin D, He Z, White J, Yu H, Xu C, Morizono H, Musunuru K, et al. 2016. A dual AAV system enables the Cas9-mediated correction of a metabolic liver disease in newborn mice. *Nat Biotechnol* **34:** 334–338.

Yin H, Xue W, Chen S, Bogorad RL, Benedetti E, Grompe M, Koteliansky V, Sharp PA, Jacks T, Anderson DG. 2014. Genome editing with Cas9 in adult mice corrects a disease mutation and phenotype. *Nat Biotechnol* **32:** 551–553.

Yin H, Song CQ, Dorkin JR, Zhu LJ, Li Y, Wu Q, Park A, Yang J, Suresh S, Bizhanova A, et al. 2016. Therapeutic genome editing by combined viral and non-viral delivery of CRISPR system components in vivo. *Nat Biotechnol* **34:** 328–333.

Zalatan JG, Lee ME, Almeida R, Gilbert LA, Whitehead EH, La Russa M, Tsai JC, Weissman JS, Dueber JE, Qi LS, et al. 2015. Engineering complex synthetic transcriptional programs with CRISPR RNA scaffolds. *Cell* **160:** 339–350.

Zetsche B, Gootenberg JS, Abudayyeh OO, Slaymaker IM, Makarova KS, Essletzbichler P, Volz SE, Joung J, van der Oost J, Regev A, et al. 2015a. Cpf1 is a single RNA-guided endonuclease of a class 2 CRISPR-Cas system. *Cell* **163:** 759–771.

Zetsche B, Volz SE, Zhang F. 2015b. A split-Cas9 architecture for inducible genome editing and transcription modulation. *Nat Biotechnol* **33:** 139–142.

Zhang F, Cong L, Lodato S, Kosuri S, Church GM, Arlotta P. 2011. Efficient construction of sequence-specific TAL effectors for modulating mammalian transcription. *Nat Biotechnol* **29:** 149–153.

Zhang Y, Yin C, Zhang T, Li F, Yang W, Kaminski R, Fagan PR, Putatunda R, Young WB, Khalili K, et al. 2015. CRISPR/gRNA-directed synergistic activation mediator (SAM) induces specific, persistent and robust reactivation of the HIV-1 latent reservoirs. *Sci Rep* **5:** 16277.

Zhou Y, Zhu S, Cai C, Yuan P, Li C, Huang Y, Wei W. 2014. High-throughput screening of a CRISPR/Cas9 library for functional genomics in human cells. *Nature* **509:** 487–491.

Zuris JA, Thompson DB, Shu Y, Guilinger JP, Bessen JL, Hu JH, Maeder ML, Joung JK, Chen ZY, Liu DR. 2015. Cationic lipid-mediated delivery of proteins enables efficient protein-based genome editing in vitro and in vivo. *Nat Biotechnol* **33:** 73–80.

Alternative Watson–Crick Synthetic Genetic Systems

Steven A. Benner, Nilesh B. Karalkar, Shuichi Hoshika, Roberto Laos, Ryan W. Shaw, Mariko Matsuura, Diego Fajardo, and Patricia Moussatche

The Westheimer Institute for Science and Technology, The Foundation for Applied Molecular Evolution, Alachua, Florida 32615

Correspondence: sbenner@ffame.org

In its "grand challenge" format in chemistry, "synthesis" as an activity sets out a goal that is substantially beyond current theoretical and technological capabilities. In pursuit of this goal, scientists are forced across uncharted territory, where they must answer unscripted questions and solve unscripted problems, creating new theories and new technologies in ways that would not be created by hypothesis-directed research. Thus, synthesis drives discovery and paradigm changes in ways that analysis cannot. Described here are the products that have arisen so far through the pursuit of one grand challenge in synthetic biology: Recreate the genetics, catalysis, evolution, and adaptation that we value in life, but using genetic and catalytic biopolymers different from those that have been delivered to us by natural history on Earth. The outcomes in technology include new diagnostic tools that have helped personalize the care of hundreds of thousands of patients worldwide. In science, the effort has generated a fundamentally different view of DNA, RNA, and how they work.

On the occasion of the 90th birthday of Albert Eschenmoser, a master of synthesis.

Many have noted that the phrase "synthetic biology" has had no consistent meaning among the communities that have used it over the past 40 years (Brent 2004). This inconsistency is reflected in the literature. To some, synthetic biology means simply "synthesizing a lot of DNA," perhaps even entire genomes (Glass et al. 2017; Hughes and Ellington 2017). To others, synthetic biology is a new name for the much older field of metabolic engineering, but on a grander scale than modestly constructing a microbe that manufactures a single natural product using a single heterologously expressed gene (Lechner et al. 2016). To others, "synthetic biology" is the redirecting of information in living systems, perhaps to create a microbial platform for further engineering (Gaj et al. 2016). To these can be added concepts not represented in this series, such as the construction of devices that use natural nucleic acids and proteins as biobricks (Smolke 2009; Win et al. 2009), perhaps to test a theory about how those parts work together naturally (Prehoda et al. 2000; Dueber et al. 2004). Others use "synthetic biology" as suggested by Eric Kool, to mean the use of unnatural molecules in the context of natural biological systems (Rawls 2000). In its original definition, synthetic biology meant the creation of artificial life (Leduc 1912).

Cite this article as *Cold Spring Harb Perspect Biol* doi: 10.1101/cshperspect.a023770

We have ourselves long held the view that synthesis is not a field, but rather an activity, an activity that derives value not only from products that it creates, but also from what is learned as the syntheses are attempted (Sismour and Benner 2005a). Indeed, the heroes of synthesis in classical organic chemistry often chose targets that had no product value at all (Woodward 1968; Kishi 1989). Instead they chose targets that would present a "grand challenge," a molecule whose synthesis was beyond current capabilities, whose pursuit would therefore challenge molecular theory.

Here, synthesis does something that "hypothesis-directed research" cannot. When scientists control the hypotheses that they test, they often strategically limit their activities to safe hypotheses that are likely to be true. If they do not, then their funding agencies will. Thus, "hypothesis-based research" tends to not challenge core convictions.

In contrast, synthesis in pursuit of a "grand challenge" forces scientists across uncharted grounds, where they must ask and answer unscripted questions. Thus, a well-selected synthetic grand challenge tests all of the theories and assumptions that go into any strategic synthetic plan. Many of these are unstated; the scientists involved might not even realize that they are making them. As a result, synthesis can drive discovery and paradigm change in ways that hypothesis cannot.

For example, in his classic defense of grand challenge synthesis, Woodward (1968) discussed his choice, with Albert Eschenmoser, of vitamin B_{12} as a synthetic target. In 1965 (and in some sense still), B_{12} was the most complicated nonpolymeric natural product known. The product of the synthesis itself had no commercial value; fermentation was already generating B_{12} for pennies per unit. However, the effort led to the discovery of the intimate relation between molecular reactivity and molecular orbital structure. That discovery, captured as the Woodward–Hoffman rules (1970), was later recognized by a Nobel Prize (Fukui 1982; Hoffmann 1982).

The celebrated total synthesis of the genome of a bacterium by Venter and his coworkers (Gibson et al. 2010) was, of course, nothing more (and nothing less) than the total synthesis of a natural product, one that happens to be involved in microbial inheritance. Venter reportedly said that he undertook this challenge because someone told him it "could not be done." This "someone" was certainly not a synthetic organic chemist. The credo of chemistry, for at least 50 years, has held that if a molecular structure can be drawn, and if the arrangement of atoms that it represents is an energy minimum, then the molecule can be synthesized given enough effort and money. That credo rests on many successful syntheses; some examples include tetrodotoxin (which is barely an energy minimum) (Kishi et al. 1972) and palytoxin (whose complexity is almost boring) (Fig. 1) (Kishi 1989).

Despite that credo, chemists in the early 1980s fully understood that "structure theory" in chemistry had broad deficiencies. In particular, that theory could not tell us "what" molecules to synthesize to create a desired molecular behavior. This was especially true if the grand challenge goal related to artificial life (Leduc 1912). Chemical theory could not then, and still cannot, meet the following easy-to-express challenge: "Draw me structures of some molecules that, if synthesized, will together have the properties that we value in living systems."

Indeed, even simpler tasks were (in 1980) and remain (today) beyond the power of chemical theory. Despite advanced computers, advanced theory (Pople 1999), and decades of effort, we still cannot predict the solubility of salts in water, the packing of organic crystals (Dunitz and Bernstein 1995), or the freezing point of water. Yet these "simple" processes are the elements of the molecular interactions that generate the molecular behaviors that we value in biology. We cannot create medicines by direct design. We cannot create molecular electronic devices by direct design. We cannot engineer self-assembling materials by direct design. We get these today only by trial, error, intuition, and experiment.

However, DNA has long appeared to be special. At first glance, the Watson–Crick model for the double helix did appear to allow design of molecules that bound to other molecules in wa-

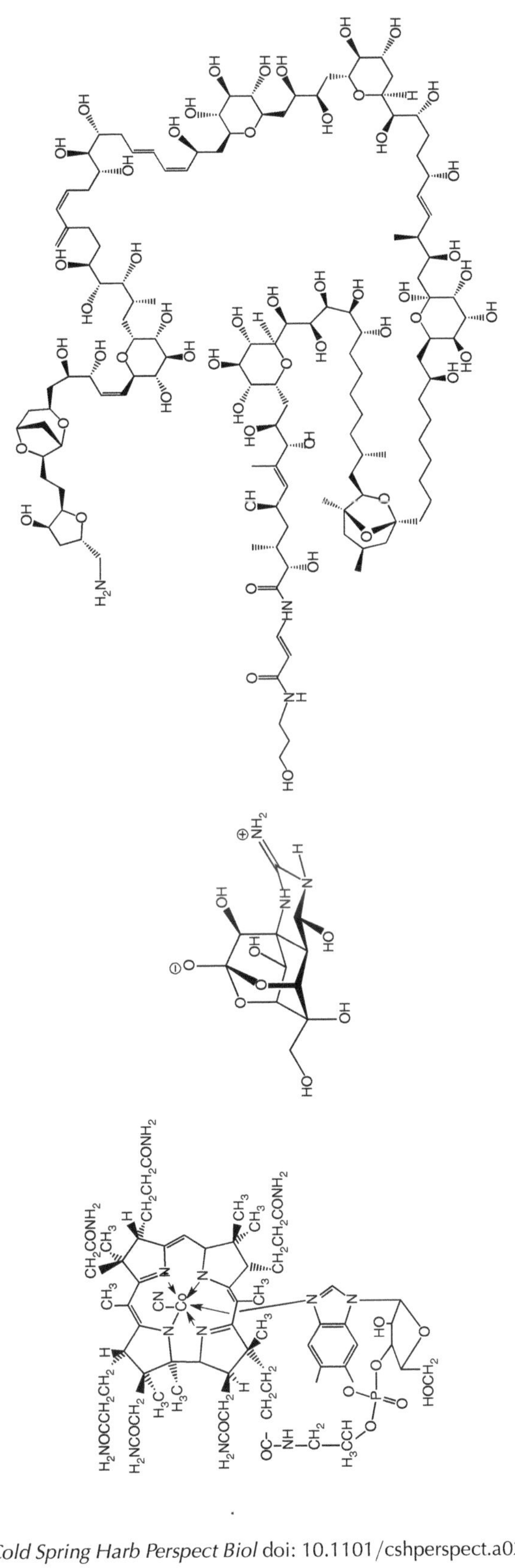

Figure 1. The "grand challenge" target. Vitamin B_{12} (*left*) was synthesized to drive chemical theory, not because it was a multistep tour de force or because the product itself was valuable. Tetrodotoxin (*middle*) is barely stable, but was nevertheless synthesized (Kishi et al. 1972). Palytoxin (*right*) is not a biopolymer, although its structure reveals a biosynthetic pathway that involves repetitive assembly of building blocks; chemists could synthesize this also (Kishi 1989).

ter (a problematic solvent). Indeed, such design was being performed almost routinely by molecular biologists who had no training in the fraternity of organic chemistry. An entire industry ("antisense therapeutics") (Miller and T'so 1988) was based on the notion that the Watson–Crick model of DNA–DNA binding could be easily transferred to other backbones, in particular, to nonionic backbones. Millions of dollars were spent to show that this transfer was not easy.

In contrast with models for other molecular systems in chemistry, the Watson–Crick model is almost trivially simple. First, the model holds that molecular recognition depends entirely on two very simple rules of complementarity: size complementarity (big purines pair with small pyrimidines) and hydrogen-bonding complementarity (hydrogen bond donors pair with hydrogen bond acceptors) (Fig. 2). High school students (ourselves included) were taught "how genetics works" using paper cutouts to illustrate these rules. Generations of future molecular biologists came to view the structure of DNA as "obviously correct."

Yet, even a newly minted Ph.D. in 1980 could see the multiple absurdities in this model. One hardly expects to get good molecular recognition from complementary hydrogen bonding in water; water as a solvent is "nothing but" competing hydrogen bonds. Further, the size complementarity required for Watson–Crick pairing is, at first glance, not likely to be enforced by the flexible backbone. Indeed, as Kool et al. (2000), Romesberg and collaborators (Malyshev et al.

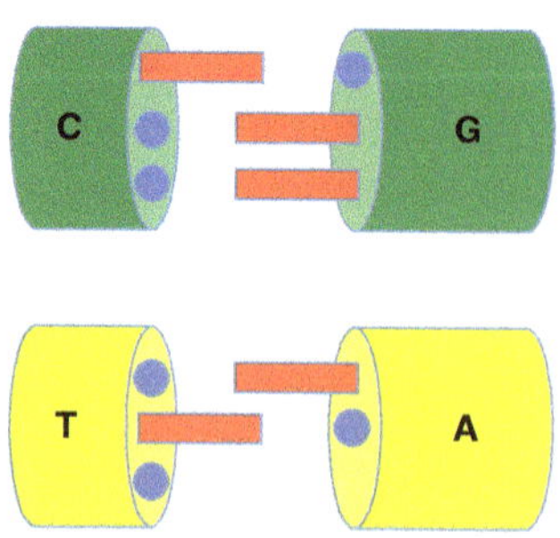

Figure 2. How genetics works. The cartoon that "explains everything"—paper cutouts that taught generations of schoolchildren that molecular genetics were simple chemistry.

2009, 2014), Hirao and collaborators. (Kimoto et al. 2011), Heuberger and Switzer (2008), and others (Doi et al. 2008) later showed, the backbone can easily adjust itself to accommodate geometries other than edge-on contacts (Fig. 3). Further, as "obvious" as nucleobase stacking seems (pennies stacked in a roll is a common analogy) (Bowman and Williams 2011), simple aromatic solids (benzene is an archetype) do not stack like pennies in a roll (Fig. 3).

Once one begins down this path of reasoning, horrors pile on top of horrors. Adenine is "missing" a hydrogen-bonding group, perhaps because of how it emerged in prebiotic processes (Fig. 3) (Orgel 2004); the resulting instability of the A:T pair creates unending problems in biotechnology (Wei et al. 2012). Cytosine suffers spontaneous deamination in water, requiring constant repair in our genomes; so do adenine and guanine, at slower rates (Shapiro 1987). DNA sequences with consecutive dGs are so pathological that, at some point, commercial supply houses hesitated to make them (Mizusawa et al. 1986). And things get worse with RNA sequences that have consecutive Gs (Davis 2004).

These realizations helped make the 1980s a good time to expand synthesis past natural bioproducts to include unnatural bioproducts. Here, the "grand challenge" centered on questions "why?" and "why not?" Why did DNA have these molecular perplexities? What other molecular systems can "do" genetics? Could we synthesize alternative genetic molecules that perform better? The grand challenge question could even be put fancifully: If we encountered an alien species capable of Darwinian evolution, would their "DNA" be DNA or some other molecular system?

This article discusses the synthesis of alternative Watson–Crick systems (Benner et al. 1998; Benner 2004). As it turns out, the alternative molecular recognition systems do themselves have value as products. However, the major value of the synthetic effort to create unnatural genetic systems came from what was learned as the challenge was undertaken, under terms that did not allow failure to be an option (Bostick 2010).

 Cite this article as *Cold Spring Harb Perspect Biol* doi: 10.1101/cshperspect.a023770

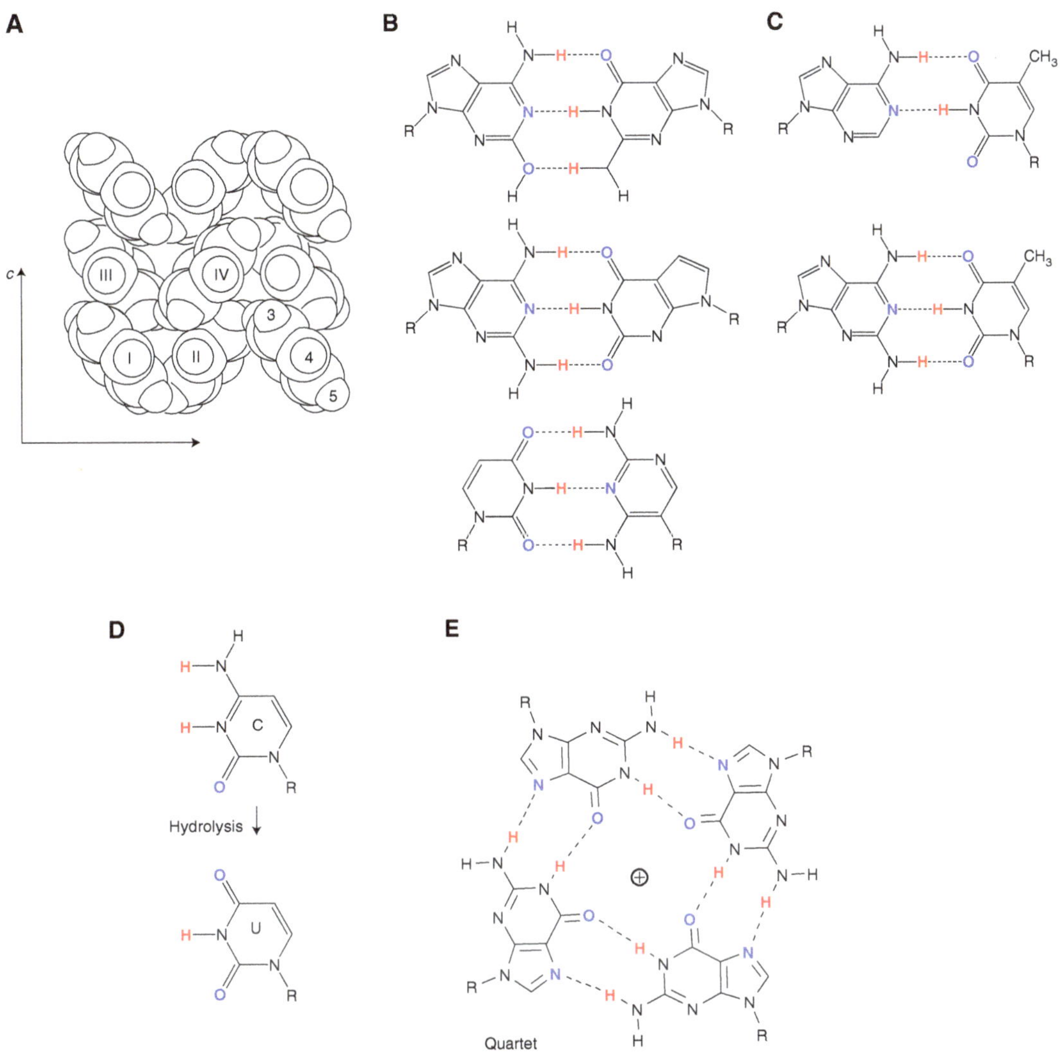

Figure 3. Some of the features of "noncartoon" natural DNA that make the Watson–Crick model less self-evident. (*A*) Unlike the "pennies in a roll" model for nucleobase stacking, the aromatic rings of benzene do not stack in a crystal (Cox et al. 1958). (*B*) Some synthetic nucleobase pairs that contribute to duplexes similar to the T:A pair, showing the flexibility of the backbone (Geyer et al. 2003; Gao et al. 2004; Heuberger and Switzer 2008; Winnacker and Kool 2013). (*C*) Adenine is missing a hydrogen-bonding group, creating many problems when using DNA to assemble nanostructures or to construct large DNA assemblies; diaminopurine forms a stronger pair. (*D*) Deamination of cytosine (shown), adenine, and guanine requires continuous repair. (*E*) G quartets are an example of a noncanonical structure that can dominate Watson–Crickery.

REARRANGING THE SYNTHESIS OF HYDROGEN-BONDING UNITS—GENERATIONS OF ARTIFICIAL WATSON–CRICKERY

Even 30 years ago, one could easily draw nucleobase pairs that retained the Watson–Crick pairing "concept," but had hydrogen-bonding units rearranged to give, at first glance, new Watson–Crick pairs. As shown in Figure 4, this rearrangement could readily generate 12 hypothetical nucleobases that might form a total of six orthogonal Watson–Crick pairs.

The structures in Figure 4 came to be called "artificially expanded genetic information systems" (AEGIS). This term covers pairs that fully retain the Watson–Crick pairing concept; it distinguishes this concept from strategies from the Kool laboratory (in which interstrand hydrogen bonding is dispensed with entirely), the Hirao laboratory (in which size complementarity is key) (Hirao et al. 2002; Hirao 2006; Kimoto et al. 2009, 2013), and the Romesberg and Schultz laboratories (in which combinatorics defined the scope of polymerase–DNA interactions in a particularly interesting way) (McMinn et al. 1999; Malyshev et al. 2009).

We were not the first to hypothesize that an expanded genetic alphabet might be obtained by shuffling hydrogen-bonding units. Alex Rich, some 20 years earlier, had recognized that isoguanine (a natural product) and isocytosine might possibly form a third pair (Rich 1962) (the first-generation S:B pair in Fig. 1). Independently, Geoffrey Zubay (1988) proposed another alternative pair (Fig. 5), not recognizing that his hypothetical structure for the small component of the new pair lacked the aromatic planar geometry that the Watson–Crick model suggested was necessary for nucleobase stacking. Interestingly, even the writers of the movie *E.T. the Extra-Terrestrial* understood the possibility of an expanded genetic alphabet; E.T. has DNA built from six nucleotides, "inosine and a pyrimidine we cannot identify" (Mathison 1982).

However, synthesis as an activity was necessary to determine whether nucleobase pairing was as simple as the Watson–Crick model implied. In this undertaking, it soon became clear that more than one heterocyclic system would support, or "implement," any particular hydrogen-bonding pattern. For example, among natural nucleobases, uridine and pseudouridine both present a hydrogen bond acceptor–donor–acceptor hydrogen-bonding pattern (Fig. 5). Those seeking to meet the grand challenge of creating an artificial genetic system needed to decide which heterocyclic system to synthesize to implement each of the orthogonal hydrogen-bonding patterns.

In many cases, the first heterocyclic system prepared to implement each of the four additional hydrogen-bonding patterns turned out not to be the best heterocycle to support genetics. Several first-generation AEGIS components suffered from chemical defects, indicated in magenta in Figure 4. For example, the pyrazine heterocycles used first to implement the pyADD and pyDDA hydrogen-bonding patterns epimerized rapidly (Fig. 5) (Voegel et al. 1993a,b; Voegel and Benner 1994, 1996a,b; von Krosigk and Benner 2004). The purine ring system used to implement the puDDA hydrogen-bonding patterns had a substantial amount of a minor tautomer that created nucleobase-pairing ambiguity (Sepiol et al. 1976; Sismour et al. 2004). In a long process documented in the literature, second-generation implementations of various bonding patterns were then synthesized to fix problems in the new DNA (Benner 2009).

These second-generation improvements are summarized in Figure 4. The effort produced much new knowledge in heterocyclic chemistry. Within the framework of the Watson–Crick pair, a rather comprehensive view of what heterocycles can support genetics has now emerged. We can now even make a good guess about the "pyrimidine that we cannot identify" in E.T.'s DNA. We also know that if E.T. indeed had inosine in his/her genome, (s)he would have had difficulty surviving to the point where his/her species could attempt interplanetary travel.

NUCLEOBASE PAIRING IS CONSTRUCTIVELY AS SIMPLE AS THE WATSON–CRICK MODEL SUGGESTS

Once synthesis delivered heterocycles that adequately implemented the four extra hydrogen-bonding pairing patterns, it was relatively easy to adapt the then-emerging phosphoramidite-based solid-phase-synthesis chemistry (the key enabling technology for synthetic biology) to create DNA containing AEGIS components. These supported experiments to determine how AEGIS pairing contributes to overall duplex stability.

Remarkably, these studies found the Watson–Crick concept to be quite robust with re-

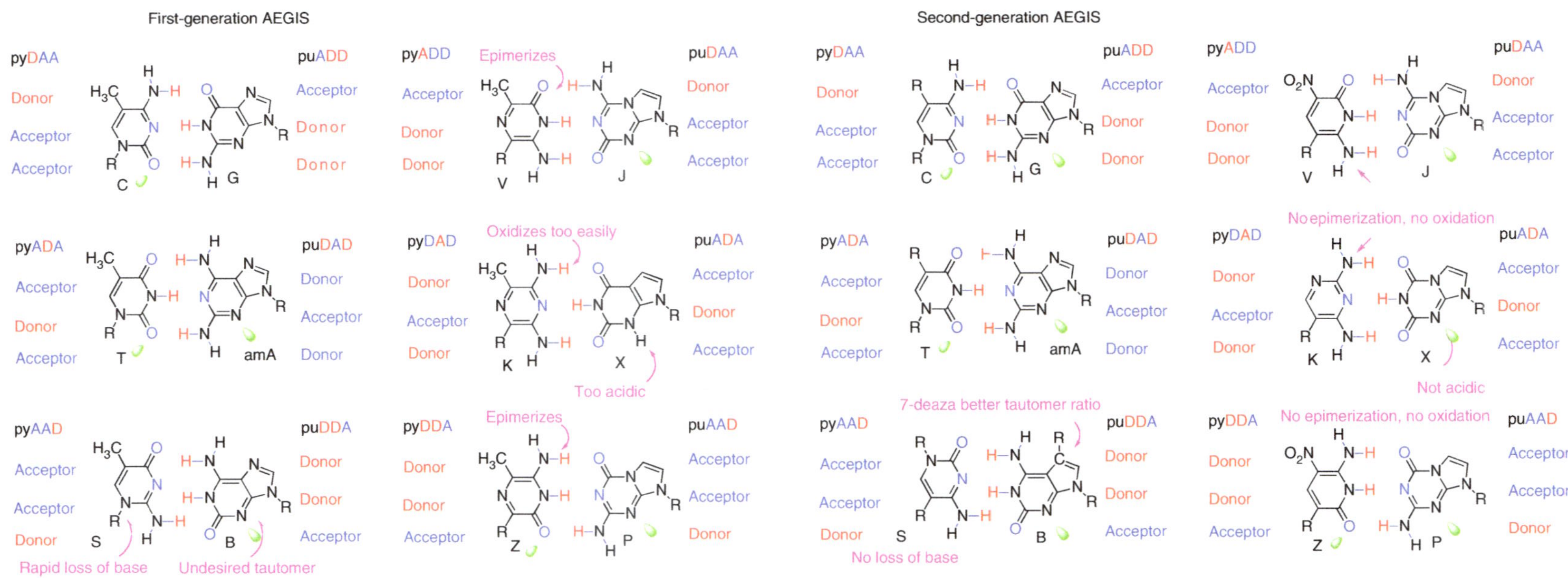

Figure 4. Two rules of complementarity guide base pairing in DNA and RNA (collectively xeno nucleic acid [XNA]). The rules are (1) size complementarity (large purines pair with small pyrimidines), and (2) hydrogen-bonding complementarity (hydrogen bond donors, D, pair with hydrogen bond acceptors, A). Rearranging the nucleobase D and A groups gives artificially expanded genetic information systems (AEGIS). Chemical issues in the "first-generation" AEGIS (*left* pairs) are indicated in magenta. These motivated the synthesis of a second-generation AEGIS (*right* pairs). Electron density presented to the minor groove (green lobes) is recognized by polymerases (Benner et al. 1998).

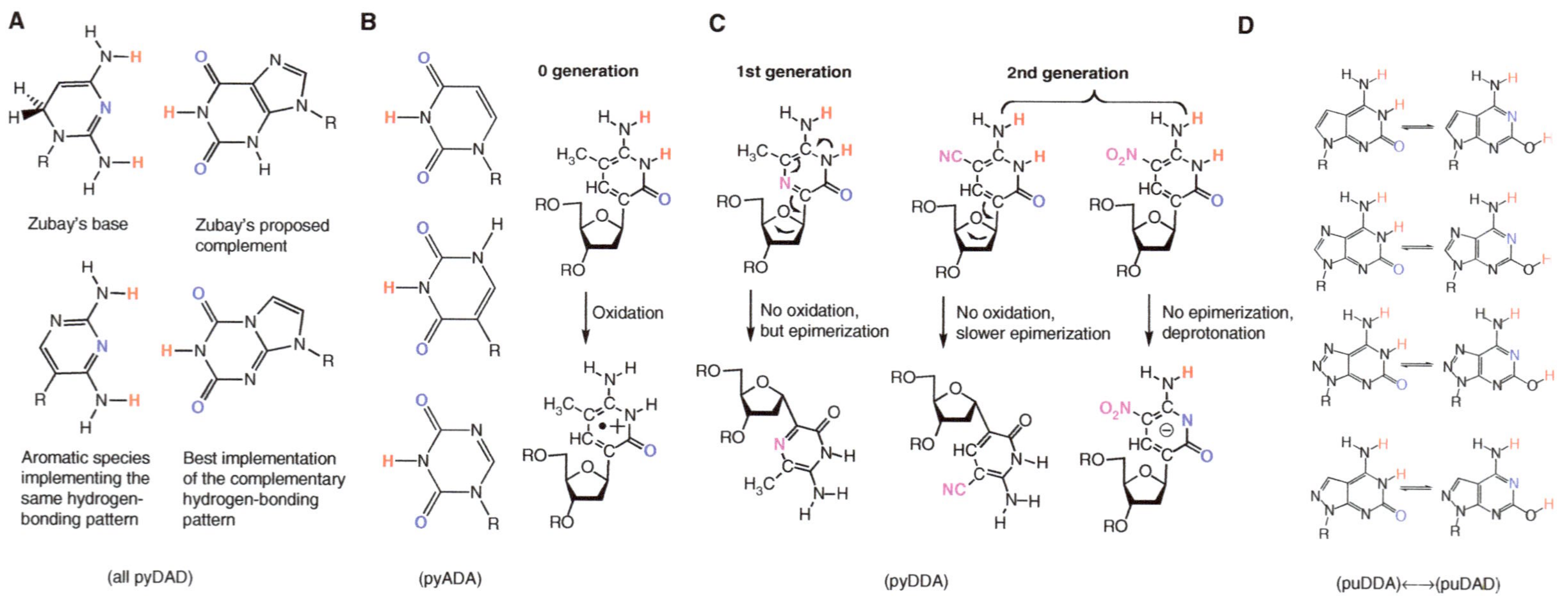

Figure 5. Different ring systems having different properties can implement the same Watson–Crick hydrogen-bonding pattern. (*A*) Zubay's ring system formally implements a hydrogen bond donor–acceptor–donor pattern, but without the planar aromatic system that the Watson–Crick model implies (Zubay 1988); to get both, one must make a C-glycoside (Benner et al. 1987). (*B*) In nature, uridine and pseudouridine implement the same hydrogen-bonding pattern, the first as an N-glycoside, the second as a C-glycoside. (*C*) Obtaining a heterocycle to implement the puDDA hydrogen-bonding pattern was especially challenging, as various C-glycosides are easy to oxidize or easily epimerized. (*D*) Various different 5,6-ring systems implement the pu(DDA) hydrogen-bonding pattern, with different amounts of a minor tautomeric form, which implements a different pu(DAD) hydrogen-bonding pattern complementary to thymidine.

Cite this article as *Cold Spring Harb Perspect Biol* doi: 10.1101/cshperspect.a023770

spect to the shuffling of hydrogen-bonding units. In a study that looked at some 300 duplexes, pairs joined by three hydrogen bonds generally contributed more to duplex stability than pairs joined by just two; pairs joined by just one hydrogen bond contributed no stability to a duplex in competition with alternative interactions with bulk solvent (Geyer et al. 2003).

A set of second-order rules could also be discerned. For example, an uncompensated C=O group was not particularly disfavored in a pair (Fig. 6). However, an uncompensated $-NH_2$ group was. Further, adding a proton and a positive charge to a nucleobase to create formal Watson–Crick complementarity was generally accepted. In contrast, losing a proton to put a negative charge on the nucleobase destabilized a formally Watson–Crick complement. These observations were used to develop a "self-avoiding molecular recognition system" (Hoshika et al. 2010), another unnatural DNA that is gaining in multiplex diagnostic systems and isothermal DNA amplification architectures (Sharma et al. 2014; Glushakova et al. 2015; Yang et al. 2015).

The flexibility of the DNA backbone, and by implication its inability to enforce size complementarity, was also evident in these studies. For example, pairing of two small nucleobase analogs could stabilize the duplex if they were joined by three hydrogen bonds (Fig. 3B). Indeed, a small:small pair joined by three hydrogen bonds stabilized the duplex as much as a small:large pair joined by just two hydrogen bonds (Geyer et al. 2003). The observation that these pairs might compete with standard small:large pairs remains an important constraint on the design of artificial genetic systems.

These biophysical studies were followed by crystallographic studies that showed that AEGIS components did in fact pair with Watson–Crick geometry. For example, introduction of a single Z:P pair into the stem of an RNA riboswitch marginally increased the stability of that stem; a crystal structure showed essentially no geometric perturbation (Fig. 7, right) (Hernandez et al. 2015). In DNA, the crystal structure of a single Z:P pair likewise shows no substantial deviation from Watson–Crick geometry (Zhang et al. 2015) (Fig. 7, left). Indeed, duplexes with four or six Z:P pairs retain their overall Watson–Crick geometry (Fig. 7, center) (Georgiadis et al. 2015).

THE USE OF ORTHOGONAL AEGIS BINDING IN DIAGNOSTICS

Even as "why?" and "why not?" questions were being pursued in a grand challenge effort for discovery, not for technology, the intrinsic ability of AEGIS to fit within the canonical Watson–Crick structure was proving to be important for many applications. For example, Mickey Urdea and Thomas Horn at Chiron (Emeryville, CA) sought to create a branched DNA (b-DNA) assay kit to measure viral loads in patients infected with HIV, hepatitis B, and hepatitis C viruses (Bushnell et al. 1999). Viral load measurements are critical to determining when the management of a viral infection starts to fail because of mutation of the virus to create resistance to a drug being

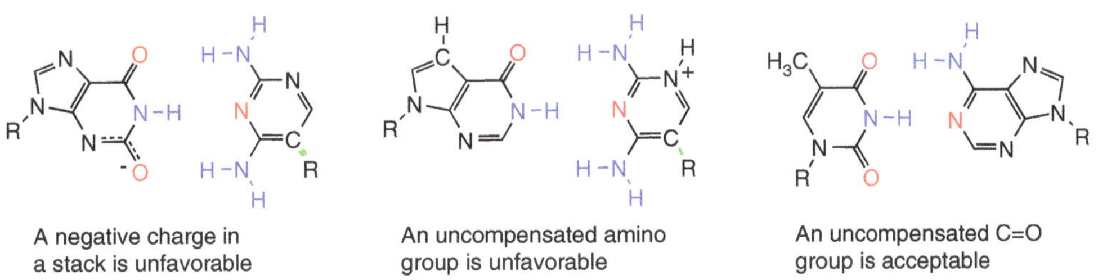

Figure 6. Some second-order Watson–Crick pairing rules obtained via artificially expanded genetic information systems (AEGIS) development (Geyer et al. 2003). A negative charge in the nucleobase stack destabilizing (*left*). An uncompensated $C-NH_2$ unit is destabilizing (*center*). An uncompensated C=O unit is acceptable (*right*).

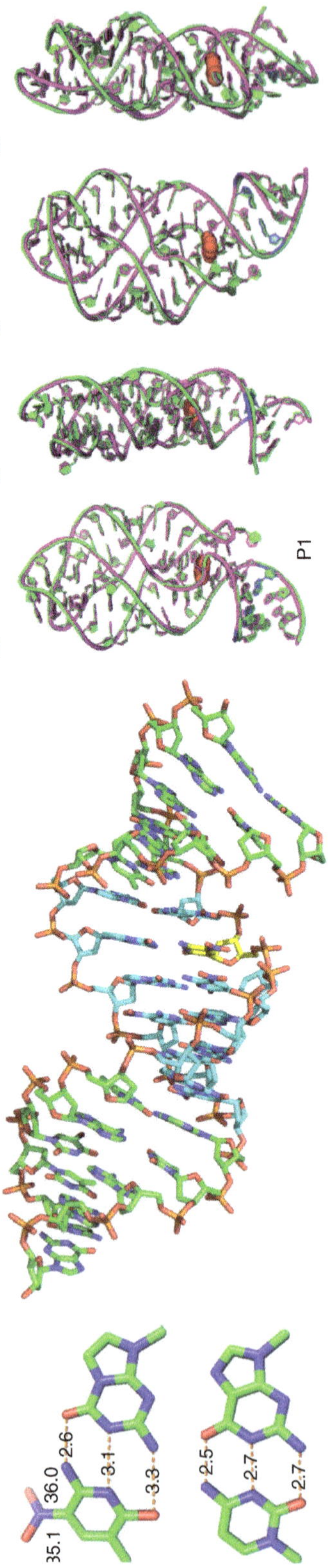

Figure 7. Three crystal structures with Z:P artificially expanded genetic information systems (AEGIS) pairs. An isolated pair in a short, A-form DNA duplex crystallized with the aid of a selenium substitution (*left*) (Zhang et al. 2015). A 16-mer duplex with six consecutive Z:P pairs (*center*) (Georgiadis et al. 2015). A single Z:P pair in an RNA riboswitch in four views *A*, *B*, *C*, and *D* each rotated 90° (*right*) (Hernandez et al. 2015).

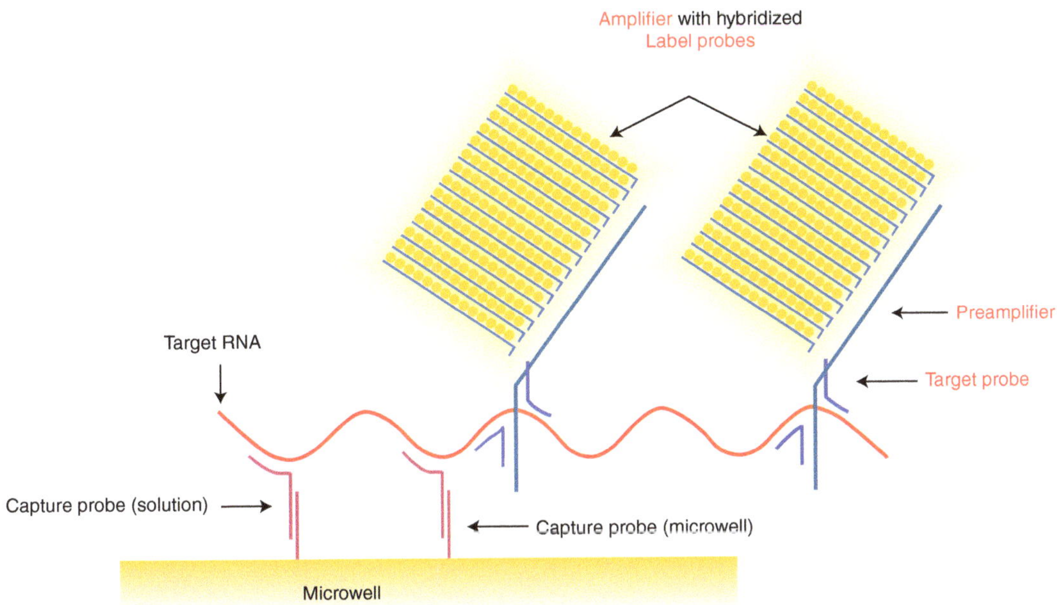

Figure 8. The branched DNA assay (Bushnell et al. 1999). When artificially expanded genetic information systems (AEGIS) nucleotides (in this case, first-generation S and B) are placed in the amplifier nanostructure, the noise is dramatically decreased (Elbeik et al. 2004a,b).

administered. Here, instead of amplifying the nucleic acid target, the b-DNA assay uses AEGIS components to assemble a signaling nanostructure (Fig. 8).

Standard Watson–Crick pairing was used to capture the viral sequence on the solid support. Then, in a "sandwich" format, standard Watson–Crick pairing allowed the immobilized viral sequence to capture a second DNA molecule. The second DNA molecule then captured a b-DNA molecule, which, in turn, captured multiple signaling molecules. The b-DNA assay did not amplify the target viral nucleic acid sequence, creating a downstream contamination problem. Thus, the b-DNA assay became an alternative to a polymerase chain reaction (PCR) that required less skill to perform and less expertise to interpret.

Unfortunately, the b-DNA architecture with only natural nucleic acids failed to detect target nucleic acid sequence with low noise. Natural biological samples (e.g., blood) contain many nucleic acids, some of which partially complement "any" DNA sequences that might be used to assemble the signaling nanostructure. These can interact with the signaling molecules, perhaps mismatched, perhaps via concatenation, to immobilize signaling molecules on the surface of the support even in the absence of the target virus molecule.

This background noise was mitigated by putting AEGIS components (first-generation S and B) into the sequences that self-assembled to give the signaling nanostructure. AEGIS oligonucleotides cannot complement any natural oligonucleotides. Therefore, they cannot form structures that generate background noise. This allowed the b-DNA assay to become FDA approved with a level of detection at 30 molecules and serve millions of patients (Elbeik et al. 2004a,b). This represents the first examples of using DNA to construct commercially useful nanostructures.

The nonenzymatic hybridization of oligonucleotides containing AEGIS components showed that the simple Watson–Crick model could be generalized to increase the number of orthogonal pairs. In parallel, synthesis going on at the same time was showing the importance of the repeating charge in the backbone to molecular recognition (Steinbeck and Richert 1998; Benner and Hutter 2002). Here, various neutral backbone analogs of DNA, including polyamide nucleic acid (PNA) (Nielsen et al. 1991), were made and studied. The outcome of these studies was to show that to support Darwinian evolution, a linear genetic biopolymer was likely to require a repeating backbone charge. This "polyelectrolyte theory of the gene" (Benner and Hutter 2002) allowed the physical properties of the polymer to be largely independent of sequence, a property that is unusual in all molecular systems. Indeed, the repeating backbone charge is the reason why DNA molecules with expected properties are so easy to design.

At the same time, synthesis was showing the delicate and unpredictable impact of changing the structure of the sugar ring (Schneider and Benner 1990; Freier and Altmann 1997; Wilds et al. 2002). Here, a wide diversity of DNA analogs with different backbone carbohydrates (Eschenmoser 1999; Declercq et al. 2002; Wilds et al. 2002; Horhota et al. 2005) were synthesized and studied. Some had prebiotic significance (Krishnamurthy 2015).

Discussion of alternative backbone units for DNA and RNA is regrettably beyond the scope of this article. Nevertheless, the point was analogous: By the activity of synthesis, nucleobase pairing, at the core of genetics, was found to be more malleable than the phosphates and carbohydrates backbone units, the "uprights" in the "ladder" that the standard model had relegated as largely incidental to the performance of DNA in genetics.

CREATING A MOLECULAR BIOLOGY TO SUPPORT AEGIS

But could AEGIS components do more than just bind to other AEGIS components? Could an AEGIS DNA strand also dynamically participate in the enzymatic synthesis of its complement, a key property that we value in natural biomolecules?

Here, the grandness of the challenge arose from natural history. To serve in a genetic system, a polymerase must take instruction only from the template with extraordinary fidelity. Achieving this was no small trick. Polymerases have four natural substrates: A(template):T-triphosphate, T(template):A-triphosphate, G(template):C-triphosphate, and C(template):G-triphosphate. Thus, any direct contact between a polymerase and the nucleobase is "dangerous." For example, contact in the major groove might cause a polymerase to prefer some nucleotides over others, disregarding the instructions from the template, as natural nucleobases differ greatly in the moieties that they present to the major groove.

The minor groove is different in this respect. As noted by Joyce and Steitz (1994), all of the four standard nucleobases present electron density to the minor groove. This density is delivered by the exocyclic C=O moieties of C and T and by the N3 nitrogens of A and G (Fig. 4). Crystallographic analysis of many polymerases identifies side chains that contact this electron density in both the template and primer in primer–template–enzyme complexes.

For just one AEGIS pair, both components present analogous electron density to the minor groove. Z has an exocyclic C=O moiety; its partner P carries electron density on its N3. In the three other second-generation pairs, the small component presents a hydrogen bond donating moiety (-NH_2) to the minor groove. Thus, the Joyce–Steitz "minor groove scanning hypothesis" suggested that polymerases would readily synthesize duplex DNA containing Z:P pairs from templates and triphosphates, whereas relatively few would synthesize the S:B, K:X, and V:J pairs; those pairs would be synthesized best by polymerases that had been mutated.

This proved to be the case empirically. Nearly all polymerases examined over the past decade do a reasonable, sometimes acceptable, and occasionally an excellent job replicating Z:P pairs. In contrast, the S:B (Sismour and Benner 2005b) and K:X (Sismour et al. 2004) pairs, which lack this electron density, are often best replicated by polymerases that are first mutated.

To create polymerase variants that accept AEGIS components, a directed evolution tool developed by Tawfik and Griffiths (1998) and

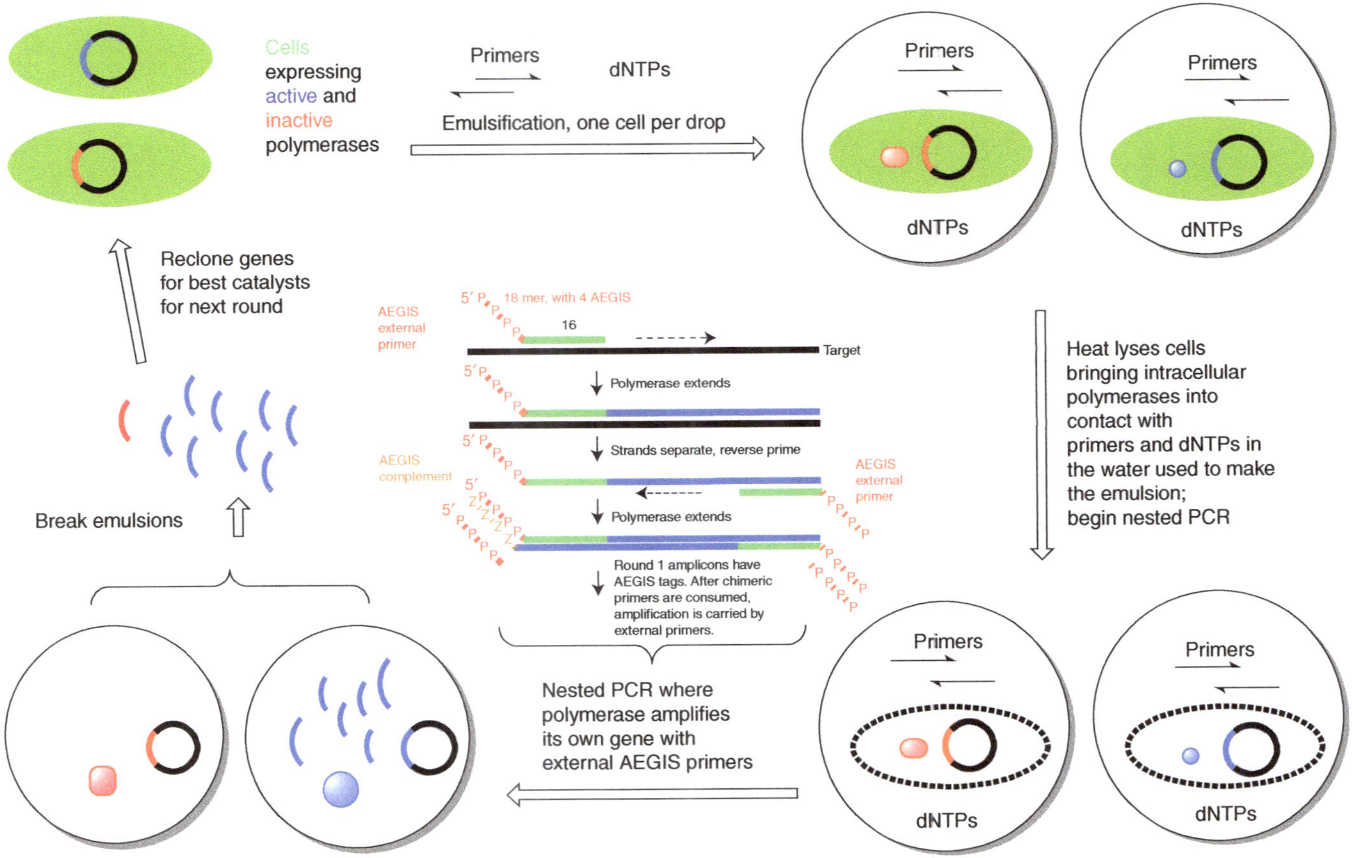

Figure 9. Schematic of colony self-replication (CSR) (Ghadessy et al. 2001) for the selection of polymerases that amplify AEGIS pairs, here in a nested polymerase chain reaction (PCR) architecture. A library of genes encoding active (blue) and inactive (red) polymerases is cloned into bacterial cells, which are dispersed (one cell per droplet) into an emulsion where the extracellular buffer contains PCR primers and triphosphates. The initial heat cycle places the polymerases in contact with the PCR primers and triphosphates. The droplets keep the polymerase variant associated with its own encoding gene; if the gene is to be amplified, it must be amplified by its encoded polymerase. After the target-specific primers are consumed, the nested PCR is "carried" by external primers containing AEGIS components. The genes encoding polymerases that are able to copy AEGIS nucleotides (the blue arcs) are enriched in the product pool. The products are then recloned, and the selection is repeated.

adapted to polymerases by Holliger (Ghadessy et al. 2001) proved to be useful. Called compartmentalized self-replication (CSR) (Fig. 9), bacteria containing plasmids encoding polymerase variants are placed in water droplets emulsified in oil. The water droplets carry buffer, primers, and triphosphates necessary for PCR. The bacteria biosynthesize the encoded variant polymerase, which is released with its encoding gene to the extracellular PCR mixture in the first heat step. If the variant can replicate its own gene, then that gene is amplified and the amplified gene presented to subsequent cycles of laboratory evolution. If, however, the variant cannot replicate its own gene, it does not survive. By placing AEGIS components strategically into a CSR experiment, polymerases that copy them with increased efficiency are evolved in the laboratory.

Polymerases contain too many important amino acid residues to expect good results to emerge via random variation. Evidence of this comes, for example, from the fact that a library of 10^8 polymerase variants having 4–6 amino-acid replacements does not generate any that can be recovered in a laboratory selection experiment that has useful activity (Laos et al. 2014).

However, the natural history of polymerase evolution can rationally improve polymerase libraries. Sites that display unusual evolutionary behavior (such as heterotachy, homoplasy, and parallelism) (Fig. 10) are more productively altered in a nonrandom library (Chen et al. 2010). As a consequence, many polymerase systems are now available that replicate many different six-letter AEGIS alphabets (Sismour et al. 2004, 2005; Yang et al. 2010; Laos et al. 2014; Sefah et al. 2014).

USING AEGIS MOLECULAR BIOLOGY

With the availability of polymerases that replicate AEGIS pairs, the orthogonality of AEGIS pairing seen in the b-DNA assay can be exploited in useful processes and products that include AEGIS PCR. Consider, for example, the multiplexed PCR problem. In most cases, PCR amplicons can be extracted from a complex biological mixture by adding a single pair of primers. However, as the number of primer pairs is increased to target more and more amplicons, the primers interact with each other, find off-target sites in a complex genomic environment to bind, and create spurious amplicons.

AEGIS proved able to manage this problem in a nested PCR format (Fig. 11). Here, the PCR is initiated with low concentrations of chimeric primers with 3′-ends complementary to the target (natural) sequence, and a 5′-AEGIS tag. The initial rounds of PCR create amplicons carrying complementary AEGIS sequences at their ends. Therefore, after the small amounts of chimeric primers are consumed, the PCR is carried by large amounts of "external" AEGIS primers complementary to these tags. Even though they are present at high concentration, the AEGIS external primers cannot find any natural DNA to bind off-target, even in very complex biological mixtures. Therefore, AEGIS nested PCR is very clean, even in a multiplexed form (Fig. 11) (Yang et al. 2010).

AEGIS orthogonality gained further use with the invention of "conversion" (Yang et al. 2013). Conversion copies a standard DNA molecule in a solution that lacks, for example, dCTP. Instead, the solution contains dZTP. With the correct polymerase, the correct buffer components, and the correct pH, the polymerase is forced to put in Z opposite G in the template (Yang et al. 2013). This creates a GAZT product, which cannot complement any natural xeno nucleic acid (XNA) sequence in any biological sample, no matter how complex. This allows clean and uniform capture of AEGIS DNA.

The use of conversion is shown in Figure 12, in a single assay that targets 22 different arboviral RNA sequences that might be present in a single mosquito carcass (Glushakova et al. 2015). Self-avoiding primers (Hoshika et al. 2010) are used with AEGIS external primers to cleanly create amplicons. Then, primers are extended with conversion to create an AEGIS tag that is the only oligonucleotide in the mixture complementary to a CTPA (computed tomography pulmonary angiogram) probe carried by a Luminex bead. This gives high and uniform detection of the amplicons arising from what-

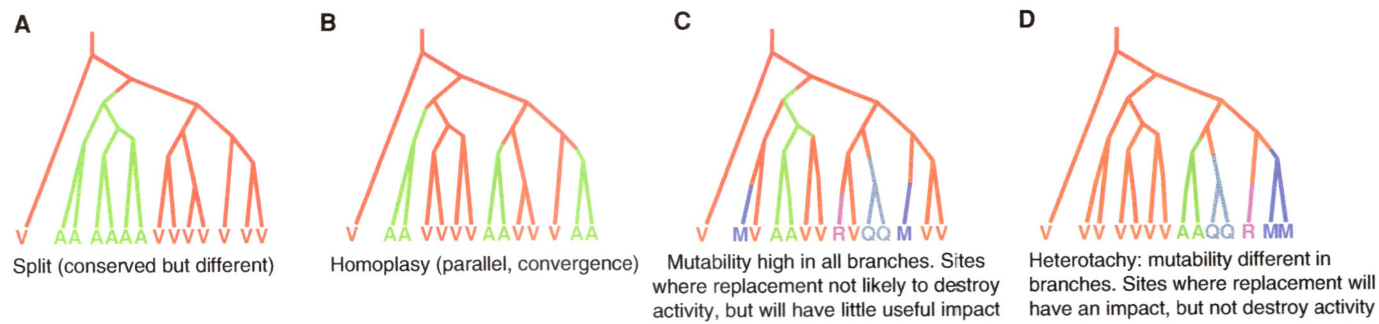

Figure 10. The pattern of amino acid replacement in natural history. Represented here by a phylogenetic tree for an individual site, the pattern can be used to guide protein engineering and directed evolution experiments. (*A*) Amino acid replacement at sites with low-level of replacements typically leads to inactivation of the polymerase. (*B*) Homoplasy in the form of parallel evolution indicating purifying selection removes variation that moves beyond a small set of amino acids; sites displaying this variation may be replaced in a controlled way. (*C*) Substantial variation indicates a site experiencing little purifying selective pressure; replacement at this site is unlikely to have little interesting impact on polymerase behavior. (*D*) Heterotachy, in which different branches have different rates of amino acid replacement, indicates changing functional constraints at a site; these sites are the foci of the most successfully designed protein engineering libraries (Chen et al. 2010).

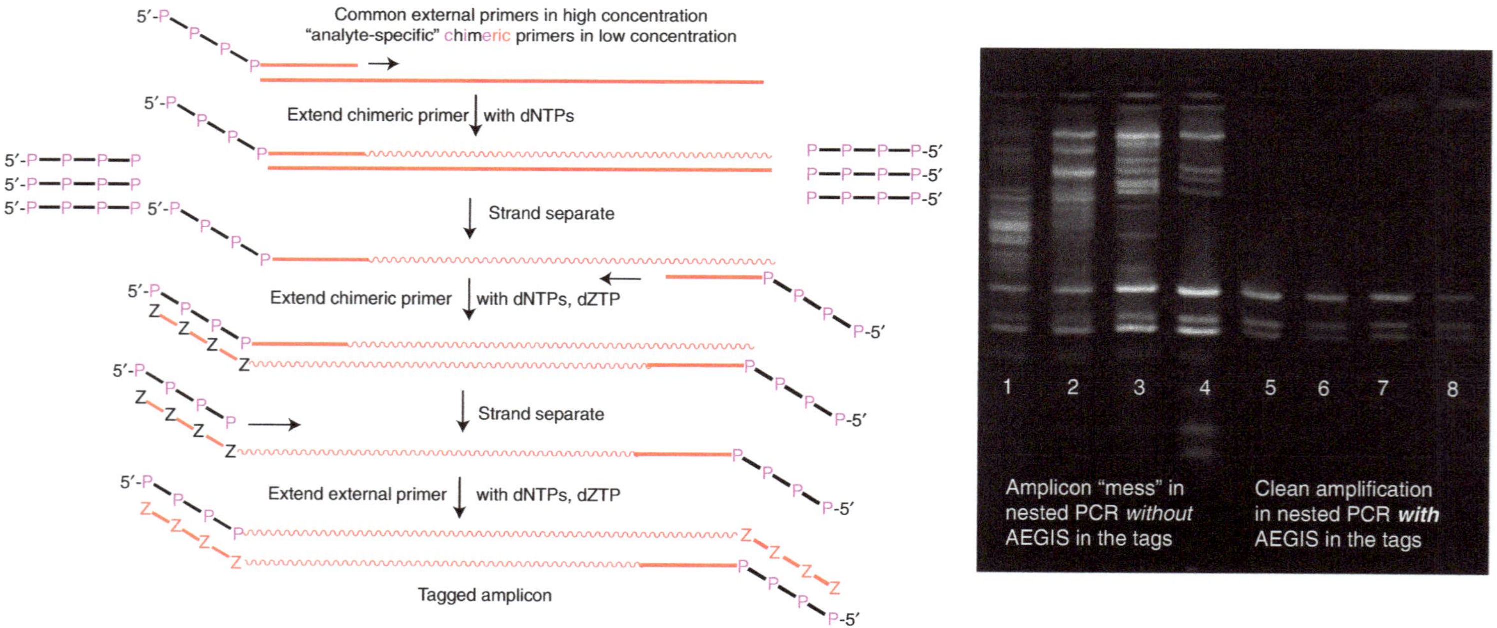

Figure 11. Although synthesized in a "grand challenge" to ask "what if?" and "why not?" questions, artificially expanded genetic information systems (AEGIS) components turned out to have practical value. Here, AEGIS orthogonality creates very clean multiplexed polymerase chain reaction (PCR) (*right*) in a nested PCR architecture (*left*). With AEGIS nucleotides in the external primers, PCR is clean, even in complex biological mixtures compared with the same PCR but with external primers having only standard nucleotides (Yang et al. 2010).

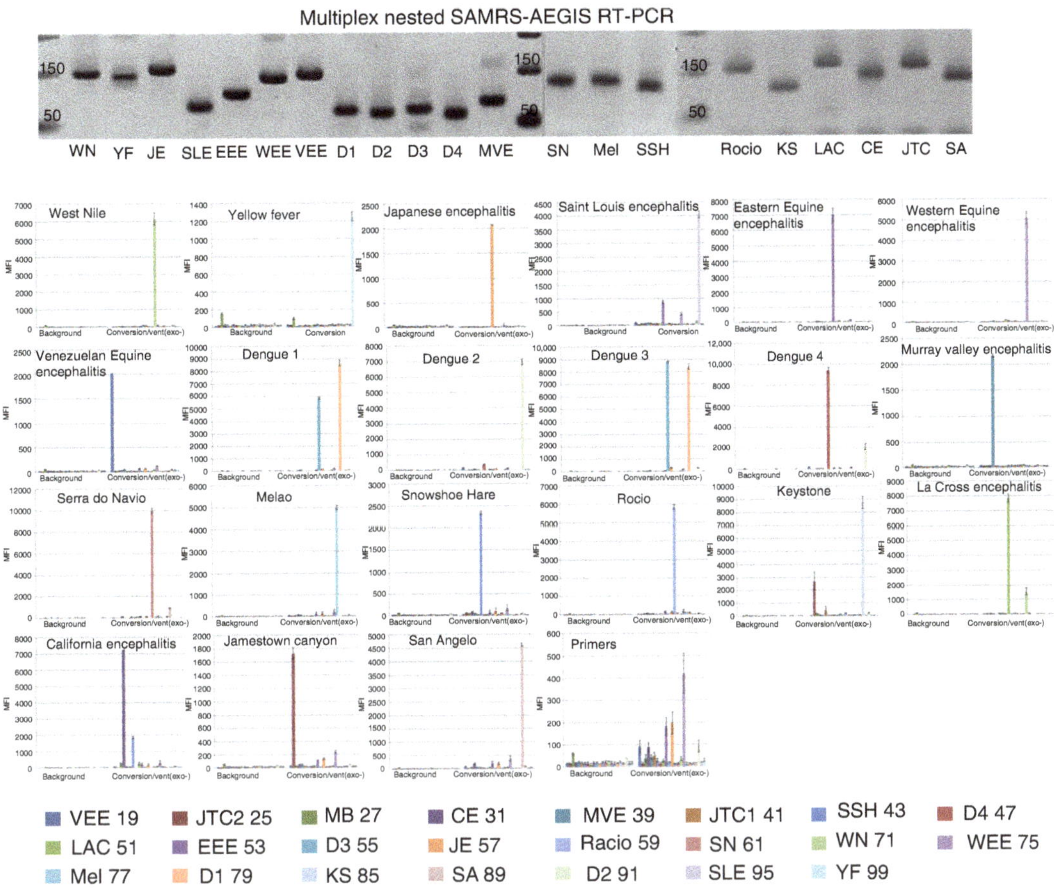

Figure 12. Artificially expanded genetic information systems (AEGIS) conversion supports an assay that allows 22 mosquito-borne viruses to be sought in a single mosquito carcass (for details, see Glushakova et al. 2015).

ever RNA virus might be present. These combinations are now being used in assays to detect many infectious disease agents, from NiV (Nipah virus) and SARS (severe acute respiratory syndrome) virus to MERS (Middle East respiratory syndrome) virus and Zika virus (Benner et al. 2015; Yang et al. 2015).

USING AEGIS INSTEAD OF NATURAL NUCLEOTIDES

This kind of synthetic biology shows that technology need not be constrained by the structure of the biopolymers that prebiotic chemistry (and four billion years of subsequent historical accident examined by natural selection) on Earth has delivered to us. Peter Schultz and others have made the parallel point with respect to the protein lexicon (Chatterjee et al. 2013).

The two can be joined. More nucleotide "letters" in a genetic alphabet should allow the writing of more amino acid "words" in the protein "lexicon." Indeed, using the first-generation AEGIS S and B "letters" to support the codon:anticodon interaction in mRNA and tRNA molecules allowed an AEGIS mRNA molecule to encode a 21st amino acid in in vitro translation (Bain et al. 1992). One outcome of this synthesis was a deeper understanding of the role played by release factors in preventing frame shifting during translation termination. Parallel experiments using the second-generation Z:P pair, in which the AEGIS tRNA was charged by flexizymes (Morimoto et al. 2011) did the same. Flexizymes are RNA catalysts that charge tRNA molecules with amino acids; the inspiration for their design was the RNA enzymes that presumably charged tRNA in the RNA world that invented ribosomal translation.

AEGIS IN LARGE-SCALE DNA SYNTHESIS

AEGIS can also be used for another goal discussed in this volume: the synthesis of large DNA constructs from short fragments. Here, AEGIS components are placed in the overhangs of synthetic fragments that are mixed in a multicomponent ligation to create a large DNA molecule (Fig. 13). Because AEGIS nucleotides do not pair with standard nucleotides, these overhangs cannot form hairpins or other undesired secondary structures, either within their fragment or between fragments; they therefore are free to hybridize as designed. After ligation, conversion in the reverse direction replaces the AEGIS nucleotides by standard nucleotides, completing the synthesis of a large DNA construct. This was shown in a "one-pot" synthesis of a gene encoding kanamycin resistance (Merritt et al. 2014).

AEGIS AS A PLATFORM FOR EVOLUTION

Natural DNA and RNA (collectively XNA) can perform functions beyond genetics (Ellington and Szostak 1992; Bartel and Szostak 1993; Breaker and Joyce 1994; Schneider et al. 1995; Kraemer et al. 2011). RNA catalysis may have supported the first forms of life on Earth; a current model for natural history holds that an earlier episode of life on Earth (the "RNA world") used RNA as its only genetically encoded catalytic component (Benner et al. 1989). Indeed, the design of flexizymes to charge AEGIS tRNA with unnatural amino acids was based on the RNA-world model (Morimoto et al. 2011).

Adding replicable nucleobases should increase the binding and catalytic potential of the XNA libraries. However, adding nucleobases also expands the "sequence space" accessible to a biopolymer, from 4^n species in a library n nucleotides in length to 6^n in an XNA species with six building blocks, and 12^n if the AEGIS is completed. It remains open whether AEGIS helps or hurts the search for functional nucleic acids, as the success of that search depends on the "density" of functional behavior in that space. Here, we might be advised to follow the Perrin or Silverman strategy of simply functionalizing the four standard nucleotides (Hollenstein et al. 2009a,b; Zhou et al. 2016).

The availability of polymerases that replicate AEGIS nucleotides makes it possible to apply laboratory in vitro evolution (LIVE) experiments to address that question (Fig. 14). The approach follows in vitro selection experiments applied to four-letter nucleic acids by Szostak, Ellington, Gold, Joyce, and others (Ellington and Szostak 1992; Bartel and Szostak 1993; Breaker and Joyce 1994; Schneider et al. 1995; Kraemer et al. 2011). These experiments have shown that natural nucleic acids are relatively poor reservoirs of receptors, ligands, and catalysts, perhaps because of their limited number and functionality of their monomers.

AEGIS-LIVE is in its infancy, now with just four examples. Nevertheless, it appears as if adding functionalized AEGIS nucleotides to a DNA library delivers better and more specific receptors and ligands. In one example (Zhang et al. 2015), a GACTZP DNA library was used in an AEGIS-LIVE experiment to find AEGIS molecules that bind to HepG2 liver cancer cells. A counterselection against untransformed liver cells (Hu1545V) was used to remove nonspecific binders (Fig. 14) (Zhang et al. 2015). Binding was seen in the bulk pools after 12 rounds of affirmative selection. Four rounds of negative selection were embedded in the process, which also included 200 cycles of PCR. Sequencing recovered 17 motifs that contributed from 0.14% to 26% of the total surviving population. These were resynthesized and their affinities for HepG2 cells were measured (specificity data are shown in Fig. 15).

Several features of the data are striking. First, AEGIS-LIVE delivered cell-binding molecules that had more than one Z and/or P. These include several that had Z and P nearby (ZnP and PnnZ, where "n" is any nucleotide), multiple Z

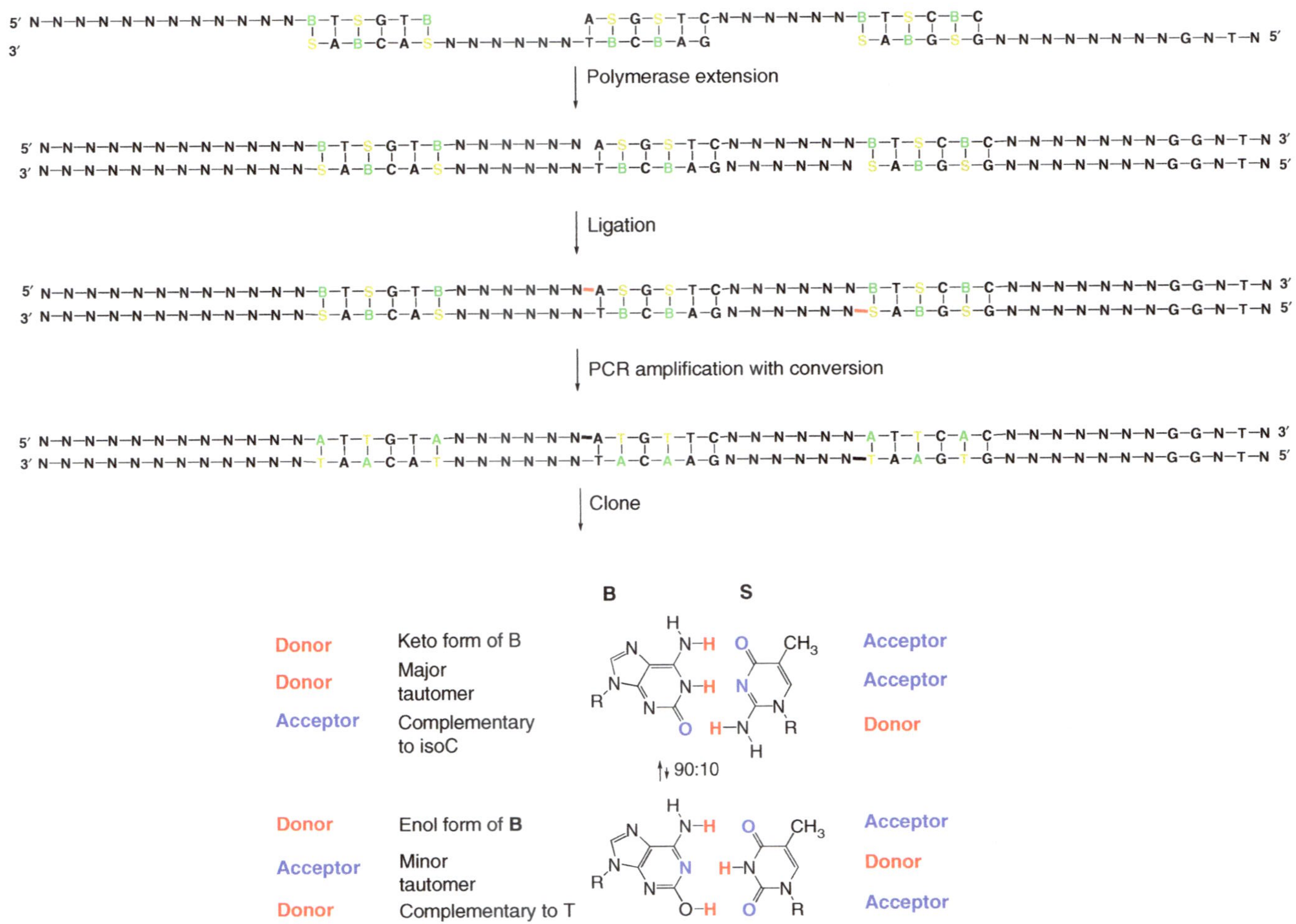

Figure 13. Assembly of large DNA constructs using artificially expanded genetic information systems (AEGIS) pairs to assemble the strands without their forming hairpins or other undesired structures. Here, the added information density of a six-letter polymer diminishes assembly ambiguity. After the assembly is complete, AEGIS nucleotides are converted to standard nucleotides, giving an entirely natural construct (Merritt et al. 2014). This allows shorter fragments to be used, which, in turn, recognizes that the longer the synthetic fragment, the greater the chance of error.

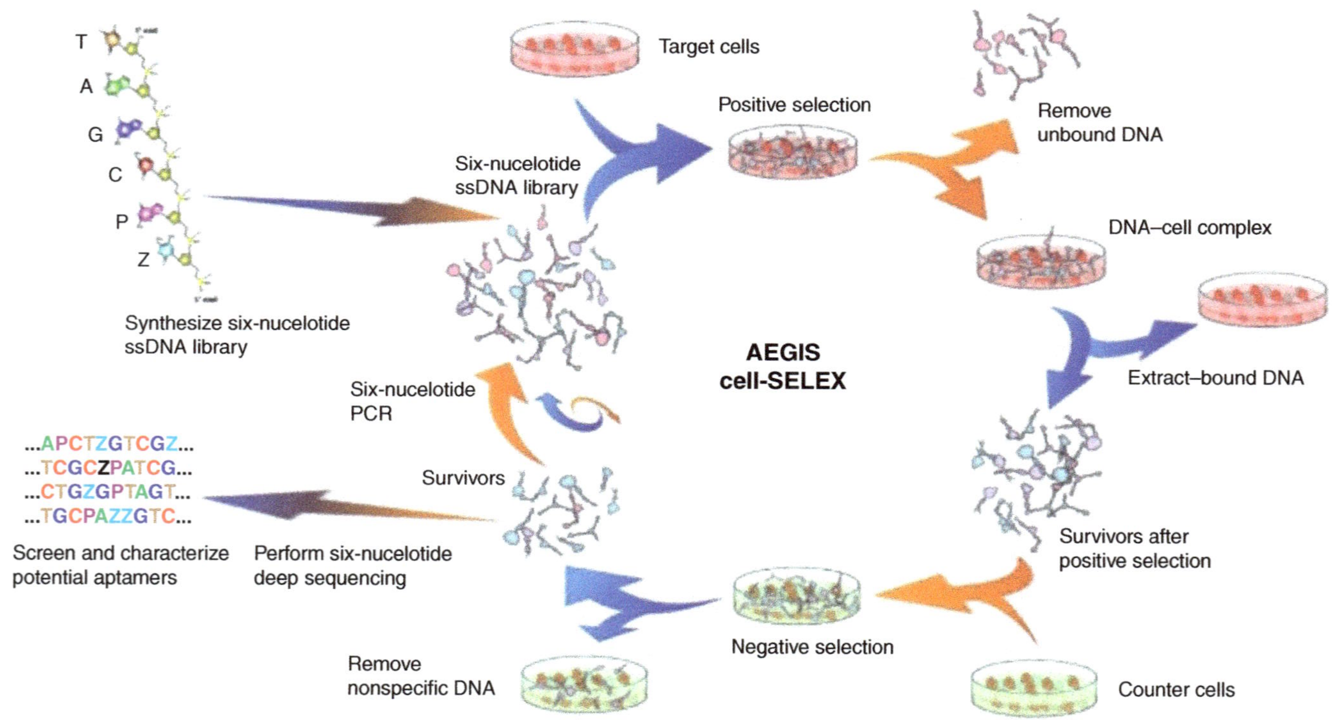

Figure 14. Schematic of the cell-targeted artificially expanded genetic information systems–laboratory in vitro evolution (AEGIS-LIVE) reported in Zhang et al. (2015). A GACTZP DNA library with a randomized region flanked by primer binding sites was incubated with the target cells. Unbound sequences were then washed away. Bound sequences were eluted from the cells, and the supernatant enriched in AEGIS DNA molecules ("survivors") having affinity for the cells was collected. Counterselections were performed against untransformed liver cells. The survivors were polymerase chain reaction (PCR)-amplified using a fluorescein isothiocyanate (FITC)-labeled primer and a biotinylated primer. Single-stranded DNA was made from the PCR products using the biotin handle and entered into the next round of selection. In each round, survivor pools were monitored for the appearance of "bulk" binding. After this was seen, survivors were sequenced and resynthesized for study.

Cite this article as *Cold Spring Harb Perspect Biol* doi: 10.1101/cshperspect.a023770

A

Name	Sequences	K_d (nM)	Percentage in pool
LZH1	~CCAATAAATCTPGCZGPGGTATCGG~	14±3	0.40%
LZH2	~GGAAGTGACGGTAGCPTTTTGGAGG~	24±3	0.28%
LZH3	~CGACCZGACTTTTAGCPTCGAATAG~	24±5	7.84%
LZH4	~GGATAAGTCTPACZGPGGTATCATG~	36±5	0.50%
LZH5	~GCTACPTGGGCCCTGGTPTCTGTGC~	41±7	0.66%
LZH6	~TATCAGCCCGATTTAACTCPZATGG~	47±5	2.15%
LZH7	~CAATAATTCTPGCZGCGGTATTGGG~	55±8	23.04%
LZH8	~TATTAGTACGGCTTAACCCPCATGG~	96±10	0.65%
LZH9	~TATCPGTTGCCCTTAAAGGCTATGG~	214±26	25.67%
LZH10	~CGCCCACGGAAGAGTCTCTGCGGCC~	234±53	1.35%
LZH11	~CGGCCGCZGAGCAGGPCCCCCCCCG~	247±25	0.27%
LZH12	~CGGCTTGACAGACPGCATZGATCAG~	298±37	0.14%
LZH13	~GTGCGGCCACCATACCCTCCTGGGC~	326±53	0.91%
LZH14	~CCAACCTGCGACCCACAACCCTATG~	346±50	1.27%
LZH15	~TTGCGCATGCCACCACCTACCAGGC~	727±84	1.25%
LZH16	~TCCCTACATGCGAGTACCACCCCTG~	>1000	0.74%
LZH17	~CCACCTAAGCTCTGGTTTCCCGTGG~	>1000	0.54%

B

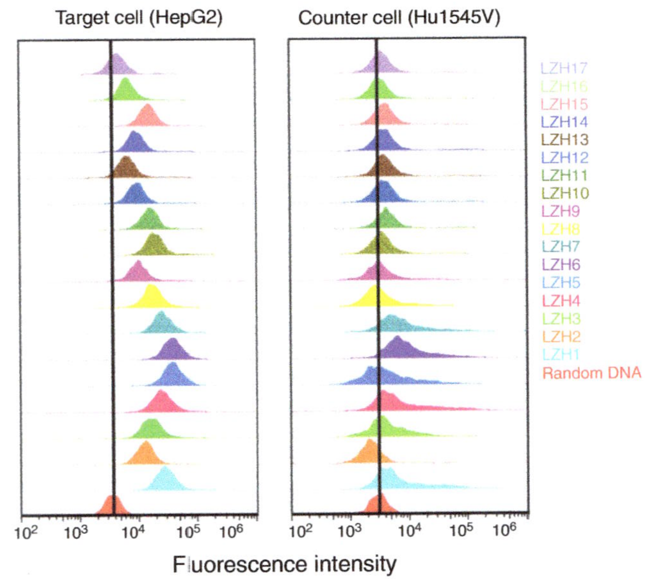

Figure 15. DNA aptamers from artificially expanded genetic information systems–laboratory in vitro evolution (AEGIS-LIVE) (Zhang et al. 2015). (*A*) Sequences (only randomized region), dissociation constants (K_d), and their percentage in the pool of binders after 12 rounds of positive selection and four rounds of negative selection (Z and P in red). Sequences are arranged in order of increasing K_d. (*B*) Binding and specificity of AEGIS aptamers, arranged from the weakest binding (*top*) to the tightest (*bottom*). Binding is measured by incubating cells with fluorescently tagged binding species in cell sorter to give fluorescence intensity per cell (*x*-axis, log scale; the *y*-axis indicates the number of cells having the indicated intensity). (*Left*) Binding to transformed cancer cells, the cells that served as the target in the aptamer selection process. (*Right*) Binding to untransformed liver cells, which were the counterselection cells. Red distribution at bottom of each panel is the binding displayed by the DNA library before selection.

and P units nearby (e.g., PnnZnP), and even one with an adjacent Z and P (PZ). This suggests that the AEGIS-LIVE experiment searched much of the substantially larger "sequence space" of the GACTZP system. The binders were also more specific (Fig. 15).

Binders also emerged from the same experiment that had neither Z nor P. However, these had systematically weaker affinity for the cells, all with K_{diss} values that were >200 nM. Those with Z and P typically had K_{diss} values that were < 50 nM. These indicate that the additional functionality presented on Z, or possibly the additional information density of a six-letter nucleic acid, helped improve the quality of a DNA library as a reservoir of functional molecules. The nitro group is a "universal binding moiety" (remembering the ability of nitrocellulose to bind many proteins). Increasing the information density in an oligonucleotide manages folding ambiguity, a major problem in DNA catalysts that have been studied in detail (Carrigan et al. 2004).

MOVING AEGIS INTO LIVING CELLS

These results showed that the rather simple Watson–Crick model not only supported an enlarged molecular recognition system and an enlarged molecular biology, but also a genetic system that has enlarged functional potential. Further, if combined with natural enzymes, AEGIS provides many of the properties that we value in a genetic system. AEGIS forms duplexes with sequence specificity, directs its own replication, can adapt under selective pressure, and can evolve. Further, AEGIS has value, not only for what it has taught us about the intimate connection between molecular structure and genetics, but also in human diagnostics, pathogen surveillance, and in a platform to create receptors, ligands, and catalysts "on demand."

Thus, AEGIS fits closely the Kool definition for synthetic biology (unnatural parts working in the context of natural systems) (Rawls 2000). However, AEGIS has not been placed into living cells. Here, the much more "unnatural" pair described by Romesberg has been replicated for ~15 hr in *Escherichia coli* (Malyshev et al. 2014), as a single exemplar. Unfortunately, this tour de force required that an engineered *E. coli* cell be fed presynthesized triphosphates presented in the growth medium.

Fortunately, efforts to meet the grander challenge, a robust bacterial system that replicates AEGIS-containing plasmids, are well underway. Mutants of kinases that add a single phosphate to a nucleoside to make the nucleoside monophosphate, and then further transform the monophosphate to the di- and triphosphates inside of living cells, have been prepared (Matsuura et al. 2016). Cells have been engineered to manage the unnatural functional groups that AEGIS carries. The construction of a cell that makes a fifth and sixth triphosphate is teaching us much about how cells regulate this key machinery required for life. Perhaps someday E.T. will be present on Earth, not by interstellar travel, but rather by the hands of the synthetic biologist.

CONCLUSIONS

"Synthetic biology" means different things in different communities. Most communities, however, remain focused on activities that involve only rearranging natural bioparts. This is certainly pragmatic. Natural DNA can be ordered inexpensively from Integrated DNA Technologies (IDT). Natural polymerases that manage natural DNA can be likewise obtained inexpensively. Moving beyond "the natural" requires that one reinvent much of molecular biology, including polymerases, restriction enzymes (Chen et al. 2011), and other tools that biotechnologists take for granted. Moving beyond the natural also requires the development of new synthetic pipelines to create triphosphates and phosphoramidites and new analytical chemistry tools, including AEGIS sequencing technology (Yang et al. 2013). Thus, there is little wonder that most synthetic biology uses natural biological parts.

However, moving synthetic biology beyond "the natural" is not without benefits. As with classical synthesis, the effort to recreate the properties that we value in life, but with unnatural products, has provided (and is providing)

 Cite this article as *Cold Spring Harb Perspect Biol* doi: 10.1101/cshperspect.a023770

unexpected insights into how biomolecules work in terran systems, and how biomolecular structure is intimately connected to biomolecular behavior.

But moving synthetic biology beyond the natural also benefits technology. Natural DNA is the product of prebiotically constrained starting molecules whose molecular structures have been conserved despite their multiple defects (Fig. 3). Those who do not move beyond the natural are condemned to suffer (in perpetuity) from these defects, which lead to failed microarrays (Wei et al. 2012), PCR multiplexed messes (Yang et al. 2010), expensive diagnostics (Glushakova et al. 2015), laborious DNA construct synthesis (Merritt et al. 2014), poor quality aptamers (Sefah et al. 2014), slow DNAzymes (Zhang et al. 2015), ambiguously folded DNA nanostructures, and proteins with only 20 (or 21) different kinds of amino acids (Chatterjee et al. 2013).

In return for its added effort, the unnatural kind of synthetic biology solves these problems. Indeed, given the advantages of redesigned DNA that "fixes God's mistakes," it might soon be surprising to find anyone who uses the natural stuff anymore.

ACKNOWLEDGMENTS

We are indebted to the National Aeronautics and Space Administration (NASA) (Experimental Approaches to Potential Alien Molecular Biologies: A Two-Biopolymer Darwinian System, Grant No. NNX14AK37G and Expanded Alphabets for Constructing Evolutionary Machines, Grant No. NNX15AF46G), The Defense Threat Reduction Agency (DTRA) (Aptamers from Artificial Genetic Systems, Grant No. HDTRA1-13-1-0004), and the Templeton World Charity Foundation (TWCF) (Doing Bioinformation Differently: A Two-Biopolymer Synthetic Life Form without Encoding or Instruction, with Bidirectional Information Flow, Grant No. TWCF0092/AB57) for their support of this work. This notwithstanding, the opinions expressed herein are those of the authors, and not of the Federal Government, NASA, the Department of Defense, or the TWCF.

REFERENCES

*Reference is also in this collection.

Bain JD, Chamberlin AR, Switzer CY, Benner SA. 1992. Ribosome-mediated incorporation of non-standard amino acids into a peptide through expansion of the genetic code. *Nature* **356:** 537–539.

Bartel DP, Szostak JW. 1993. Isolation of new ribozymes from a large pool of random sequences. *Science* **261:** 1411–1418.

Benner SA. 2004. Understanding nucleic acids using synthetic chemistry. *Accounts Chem Res* **37:** 784–797.

Benner SA. 2009. *The life, the universe and the scientific method.* FfAME, Gainesville, FL.

Benner SA, Hutter D. 2002. Phosphates, DNA, and the search for nonterrean life. A second generation model for genetic molecules. *Bioorg Chem* **30:** 62–80.

Benner SA, Allemann RK, Ellington AD, Ge L, Glasfeld A, Leanz GF, Krauch T, Macpherson LJ, Moroney SE, Piccirilli JA. 1987. Natural selection, protein engineering and the last riboorganism. Rational model building in biochemistry. *Cold Spring Harbor Symp Quant Biol* **52:** 53–63.

Benner SA, Ellington AD, Tauer A. 1989. Modern metabolism as a palimpsest of the RNA world. *Proc Natl Acad Sci* **86:** 7054–7058.

Benner SA, Battersby TR, Kodra J, Switzer CY, Moroney SE, Voegel J, MacPherson L, von Krosigk U, Hammer C, Richert C, et al. 1998. Redesigning nucleic acids. *Pure Appl Chem* **70:** 263–266.

Benner SA, Kim HJ, Merritt KB, Yang Z, McLendon C, Hoshika S, Hutter D. 2015. Next-generation DNA in pathogen detection, surveillance, and CLIA-waivable diagnostics. *Advances in Global Health through Sensing Technologies 2015*, doi: 10.1117/12.2183481.

Bostick J. 2010. Origin of Apollo 13 quote: "Failure is not an option," www.thespacereview.com/article/357/1.

Bowman JC, Williams LD. 2011. Nucleic acids. In *Encyclopedia of astrobiology*, pp. 1141–1147. Springer, Berlin.

Breaker RR, Joyce GF. 1994. A DNA enzyme that cleaves RNA. *Chem Biol* **1:** 223–229.

Brent RL. 2004. A partnership between biology and engineering. *Nat Biotechnol* **22:** 1211–1214.

Bushnell S, Budde J, Catino T, Cole J, Derti A, Kelso R, Collins ML, Molino G, Sheridan P, Monahan J, et al. 1999. ProbeDesigner for the design of probesets for branched DNA (bDNA) signal amplification assays. *Bioinformatics* **15:** 348–355.

Carrigan MA, Ricardo A, Ang DN, Benner SA. 2004. Quantitative analysis of a deoxyribonucleotide catalyst obtained via in vitro selection. A DNA ribonuclease. *Biochemistry* **43:** 11446–11459.

Chatterjee A, Sun SB, Furman JL, Xiao H, Schultz PG. 2013. A versatile platform for single- and multiple-unnatural amino acid mutagenesis in *Escherichia coli*. *Biochemistry* **52:** 1828–1837.

Chen F, Gaucher EA, Leal NA, Hutter D, Havemann SA, Govindarajan S, Ortlund EA, Benner SA. 2010. Reconstructed evolutionary adaptive paths give polymerases accepting reversible terminators for sequencing and SNP detection. *Proc Natl Acad Sci* **107:** 1948–1953.

Chen F, Yang ZY, Yang MC, Alvarado JB, Wang GG, Benner SA. 2011. Recognition of an expanded genetic alphabet by type-II restriction endonucleases and their application to analyze polymerase fidelity. *Nucleic Acids Res* **39:** 3949–3961.

Cox EG, Cruickshank DWJ, Smith JAS. 1958. The crystal structure of benzene at −3°C. *Proc R Soc A* **247:** 1–21.

Davis JT. 2004. G quartets 40 years later: From 5′-GMP to molecular biology and supramolecular chemistry. *Angew Chem Int Ed Engl* **43:** 668–898.

Declercq R, Van Aerschot A, Read RJ, Herdewijn P, Van Meervelt L. 2002. Crystal structure of double helical hexitol nucleic acids. *J Am Chem Soc* **124:** 928–933.

Doi Y, Chiba J, Morikawa T, Inouye M. 2008. Artificial DNA made exclusively of nonnatural C-nucleosides with four types of nonnatural bases. *J Am Chem Soc* **130:** 8762–8768.

Dueber JE, Yeh BJ, Bhattacharyya RP, Lim WA. 2004. Rewiring cell signaling: The logic and plasticity of eukaryotic protein circuitry. *Curr Opin Struct Biol* **14:** 690–699.

Dunitz JD, Bernstein J. 1995. Disappearing polymorphs. *Acc Chem Res* **28:** 193–200.

Elbeik T, Markowitz N, Nassos P, Kumar U, Beringer S, Haller B, Ng V. 2004a. Simultaneous runs of the Bayer VERSANT HIV-1 version 3.0 and HCV bDNAversion 3.0 quantitative assays on the system 340 platform provide reliable quantitation and improved work flow. *J Clin Microbiol* **42:** 3120–3127.

Elbeik T, Surtihadi J, Destree M, Gorlin J, Holodniy M, Jortani SA, Kuramoto K, Ng V, Valdes R, Valsamakis A, et al. 2004b. Multicenter evaluation of the performance characteristics of the Bayer VERSANT HCV RNA 3.0 assay (bDNA). *J Clin Microbiol* **42:** 563–569.

Ellington AD, Szostak JW. 1992. Selection in vitro of single-stranded molecules that fold into specific ligand-binding structures. *Nature* **355:** 850–852.

Eschenmoser A. 1999. Chemical etiology of nucleic acid structure. *Science* **284:** 2118–2124.

Freier SM, Altmann KH. 1997. The ups and downs of nucleic acid duplex stability. Structure-stability studies on chemically-modified DNA:RNA duplexes. *Nucleic Acids Res* **25:** 4429–4443.

Fukui K. 1982. Role of frontier orbitals in chemical reactions. *Science* **218:** 747–754.

* Gaj T, Sirk SJ, Shui SI, Liu J. 2016. Genome-editing technologies: Principles and applications. *Cold Spring Harb Perspect Biol* **8:** a023754.

Gao JM, Liu HB, Kool ET. 2004. Expanded-size bases in naturally sized DNA: Evaluation of steric effects in Watson–Crick pairing. *J Am Chem Soc* **126:** 11826–11831.

Georgiadis MM, Singh I, Kellett WF, Hoshika S, Benner SA, Richards NGJ. 2015. Crystal structures of non-natural nucleobase pairs in A- and B-DNA. *J Am Chem Soc* **137:** 6947–6955.

Geyer CR, Battersby TR, Benner SA. 2003. Nucleobase pairing in expanded Watson–Crick like genetic information systems. The nucleobases. *Structure* **11:** 1485–1498.

Ghadessy FJ, Ong JL, Holliger P. 2001. Directed evolution of polymerase function by compartmentalized self-replication. *Proc Natl Acad Sci* **98:** 4552–4557.

Gibson DG, Glass JI, Lartigue C, Noskov VN, Chuang RY, Algire MA, Benders GA, Montague MG, Ma L, Moodie MM, et al. 2010. Creation of a bacterial cell controlled by a chemically synthesized genome. *Science* **329:** 52–56.

* Glass JI, Merryman C, Wise KS, Hutchison III CA, Smith HO. 2017. Minimal cells—Real and imagined. *Cold Spring Harb Perspect Biol* doi: 10.1101/cshperspect.a023861.

Glushakova LG, Bradley A, Bradley KM, Alto BW, Hoshika S, Hutter D, Sharma N, Yang Benner SA. 2015. High-throughput multiplexed xMAP Luminex array panel for detection of twenty two medically important mosquito-borne arboviruses based on innovations in synthetic biology. *J Virol Meth* **214:** 60–74.

Hernandez AR, Shao Y, Hoshika S, Yang Z, Shelke SA, Herrou J, Kim HJ, Kim MJ, Piccirilli JA, Benner SA. 2015. A crystal structure of a functional RNA molecule containing an artificial nucleobase pair. *Angew Chem Int Ed Engl* **54:** 9853–9856.

Heuberger BD, Switzer C. 2008. An alternative nucleobase code: Characterization of purine–purine DNA double helices bearing guanine–isoguanine and diaminopurine 7-deazaxanthine base pairs. *ChemBioChem* **9:** 2779–2783.

Hirao I. 2006. Unnatural base pair systems for DNA/RNA-based biotechnology. *Curr Opin Chem Biol* **10:** 622–627.

Hirao I, Ohtsuki T, Fujiwara T, Mitsui T, Yokogawa T, Okuni T, Nakayama H, Takie K, Yabuki T, Kigawa T, et al. 2002. An unnatural base pair for incorporating amino acid analogs into proteins. *Nat Biotechnol* **20:** 177–182.

Hoffmann R. 1982. Building bridges between inorganic and organic chemistry (Nobel Lecture). *Angew Chem Int Ed Engl* **21:** 711–724.

Hollenstein M, Hipolito CJ, Lam CH, Perrin DM. 2009a. A DNAzyme with three protein-like functional groups. Enhancing catalytic efficiency of M^{2+}-independent RNA cleavage. *ChemBioChem* **10:** 1988–1992.

Hollenstein M, Hipolito CJ, Lam CH, Perrin DM. 2009b. A self-cleaving DNA enzyme modified with amines, guanidines and imidazoles operates independently of divalent metal cations (M^{2+}). *Nucleic Acids Res* **37:** 1638–1649.

Horhota A, Zou KY, Ichida JK, Yu B, McLaughlin LW, Szostak JW, Chaput JC. 2005. Kinetic analysis of an efficient DNA-dependent TNA polymerase. *J Amer Chem Soc* **127:** 7427–7434.

Hoshika S, Leal N, Chen F, Benner SA. 2010. Artificial genetic systems. Self-avoiding DNA in PCR and multiplexed PCR. *Angew Chem Int Ed Engl* **49:** 5554–5557.

* Hughes RA, Ellington AD. 2017. Synthetic DNA synthesis and assembly: Putting the synthetic in synthetic biology. *Cold Spring Harb Perspect Biol* **9:** a023812.

Joyce CM, Steitz TA. 1994. Function and structure relationships in DNA polymerases. *Ann Rev Biochem* **63:** 777–822.

Kimoto M, Kawai R, Mitsui T, Yokoyama S, Hirao I. 2009. An unnatural base pair system for efficient PCR amplification and functionalization of DNA molecules. *Nucleic Acids Res* **37:** e14.

Kimoto M, Cox RS, Hirao I. 2011. Unnatural base pair systems for sensing and diagnostic applications. *Expert Rev Mol Diagn* **11:** 321–331.

Kimoto M, Yamashige R, Matsunaga K, Yokoyama S, Hirao I. 2013. Generation of high-affinity DNA aptamers using an expanded genetic alphabet. *Nat Biotechnol* **31:** 453–457.

Kishi Y. 1989. Natural products synthesis. Palytoxin. *Pure Appl Chem* **61:** 313–324.

Kishi Y, Fujuyama T, Aratani M, Nakatsubo F, Goto T, Inoue S, Tanini H, Sugiura S, Kakoi H. 1972. Synthetic studies on tetrodotoxin and related compounds. IV: Stereospecific total synthesis of DL-tetrodotoxin. *J Am Chem Soc* **94:** 9219–9221.

Kool ET, Morales JC, Guckian KM. 2000. Mimicking the structure and function of DNA. Insights into DNA stability and replication. *Angew Chem Int Ed Engl* **39:** 990–1009.

Kraemer S, Vaught JD, Bock C, Gold L, Katilius E, Keeney TR, Kim N, Saccomano NA, Wilcox SK, Zichi D, et al. 2011. From SOMAmer-based biomarker discovery to diagnostic and clinical applications: A SOMAmer-based, streamlined multiplex proteomic assay. *PLoS ONE* **6:** e26332.

Krishnamurthy R. 2015. On the emergence of RNA. *Israel J Chem* **55:** 837–850.

Laos R, Thomson JM, Benner SA. 2014. DNA polymerases engineered by directed evolution to incorporate non-standard nucleotides. *Front Microbiol* **5:** 565.

* Lechner A, Brunk E, Keasling JD. 2016. The need for integrated approaches in metabolic engineering. *Cold Spring Harb Perspect Biol* **8:** a023903.

Leduc S. 1912. *La biologie synthetique*. Poinat, Paris.

Malyshev DA, Seo YJ, Ordoukhanian P, Romesberg FE. 2009. PCR with an expanded genetic alphabet. *J Am Chem Soc* **131:** 14620–14621.

Malyshev DA, Dhami K, Lavergne T, Chen T, Dai N, Foster JM, Correa IR Jr, Romesberg FE. 2014. A semi-synthetic organism with an expanded genetic alphabet. *Nature* **509:** 385–388.

Mathison M. 1982. *E.T. the Extra-Terrestrial.* Directed by Steven Spielberg, Universal Pictures.

Matsuura MF, Shaw RW, Moses JD, Kim HJ, Kim MJ, Kim MS, Benner SA. 2016. Assays to detect the formation of triphosphates of unnatural nucleotides: Application to *Escherichia coli* nucleoside diphosphate kinase. *ACS Synth Biol* **5:** 234–240.

McMinn DL, Ogawa AK, Wu Y, Liu J, Schultz PG, Romesberg FE. 1999. Efforts toward expansion of the genetic alphabet. DNA polymerase recognition of a highly stable, self-pairing hydrophobic base. *J Am Chem Soc* **121:** 11585–11586.

Merritt KB, Bradley KM, Hutter D, Benner SA. 2014. Whole genes from autonomous self-assembly of synthetic oligonucleotides incorporating artificial nucleotides. *Beilstein J Org Chem* **10:** 2348–2360.

Miller PS, Ts'o POP. 1988. Oligonucleotide inhibitors of gene-expression in living cells. New opportunities in drug design. *Ann Rep Med Chem* **23:** 295–304.

Mizusawa S, Nishimur S, Seela F. 1986. Improvement of the dideoxy chain termination method of DNA sequencing by use of deoxy-7-deazaguanosine triphosphate in place of dGTP. *Nucleic Acids Res* **14:** 1319–1324.

Morimoto J, Hayashi Y, Iwasaki K, Suga H. 2011. Flexizymes: Their evolutionary history and the origin of catalytic function. *Acc Chem Res* **44:** 1359–1368.

Nielsen PE, Egholm M, Berg RH, Buchardt O. 1991. Sequence-selective recognition of DNA by strand displacement with a thymine-substituted polyamide. *Science* **254:** 1497–1500.

Orgel LE. 2004. Prebiotic adenine revisited: Eutectics and photochemistry. *Orig Life Evol Biosph* **34:**: 361–369.

Pople JA. 1999. Nobel lecture. Quantum mechanical models. *Rev Mod Phys* **71:**: 1267–1274.

Prehoda KE, Scott JA, Mullins RD, Lim WA. 2000. Integration of multiple signals through cooperative regulation of the N-WASP-Arp2/3 complex. *Science* **290:** 801–806.

Rawls RL. 2000. "Synthetic biology" makes its debut. *Chem Eng News* **78:** 49–53.

Rich A. 1962. On the problems of evolution and biochemical information transfer. In *Horizons in biochemistry* (ed. Kasha M, Pullmann B), pp. 103–126. Academic, New York.

Schneider KC, Benner SA. 1990. Oligonucleotides containing flexible nucleoside analogs. *J Am Chem Soc* **112:** 453–455.

Schneider DJ, Feigon J, Hostomsky Z, Gold L. 1995. High-affinity ssDNA inhibitors of the reverse transcriptase of type 1 human immunodeficiency virus. *Biochemistry* **34:** 9599–9610.

Sefah K, Yang Z, Bradley KM, Hoshika S, Jimenez E, Zhu G, Shanker S, Yu F, Tan WH, Benner SA. 2014. In vitro selection with artificial expanded genetic information systems (AEGIS). *Proc Natl Acad Sci* **111:** 1449–1456.

Sepiol J, Kazimierczuk Z, Shugar DZ. 1976. Tautomerism of iso-guanosine and solvent-induced keto-enol equilibrium. *Z Naturforsch* **31:** 361–370.

Shapiro R. 1987. *Origins: A skeptic's guide to the creation of life on Earth.* Bantam Dell, New York.

Sharma N, Hoshika S, Hutter D, Bradley KM, Benner SA. 2014. Recombinase-based isothermal amplification of nucleic acids with self-avoiding molecular recognition systems (SAMRS). *ChemBioChem* **15:** 2268–2274.

Sismour AM, Benner SA. 2005a. Synthetic biology. *Expert Opin Biol Ther* **5:** 1409–1414.

Sismour AM, Benner SA. 2005b. The use of thymidine analogs to improve the replication of an extra DNA base pair. A synthetic biological system. *Nucleic Acids Res* **33:** 5640–5646.

Sismour AM, Lutz S, Park J-H, Lutz MJ, Boyer PL, Hughes SH, Benner SA. 2004. PCR amplification of DNA containing non-standard base pairs by variants of reverse transcriptase from human immunodeficiency virus-1. *Nucleic Acids Res* **32:** 728–735.

Smolke CD. 2009. Building outside of the box: iGEM and the BioBricks Foundation. *Nat Biotechnol* **27:** 1099–1102.

Steinbeck C, Richert C. 1998. The role of ionic backbones in RNA structure. An unusually stable non-Watson–Crick duplex of a nonionic analog in an apolar medium. *J Am Chem Soc* **120:** 11576–11580.

Stoker HS. 1988. *General, organic and biological chemistry*, p. 745. McGraw-Hill, New York.

Tawfik DS, Griffiths AD. 1998. Man-made cell-like compartments for molecular evolution. *Nat Biotechnol* **16:** 652–656.

Voegel JJ, Benner SA. 1994. Non-standard hydrogen bonding in duplex oligonucleotides. The base pair between an acceptor–donor–donor pyrimidine analog and a donor–acceptor–acceptor purine analog. *J Am Chem Soc* **116:** 6929–6930.

Voegel JJ, Benner SA. 1996a. Synthesis and characterization of non-standard nucleosides and nucleotides bearing the acceptor–donor–donor pyrimidine analog 6-amino-3-methylpyrazin-2(1*H*)-one. *Helv Chim Acta* **79:** 1863–1880.

Voegel JJ, Benner SA. 1996b. Synthesis, molecular recognition and enzymology of oligonucleotides containing the non-standard base pair between 5-aza-7-deaza-isoguanine and 6-amino-3-methylpyrazin-2(1*H*)-one, a donor–acceptor–acceptor purine analog and an acceptor–donor–donor pyrimidine analog. *Helv Chim Acta* **79:** 1881–1898.

Voegel JJ, Altorfer MM, Benner SA. 1993a. The donor–acceptor–acceptor purine analog: Transformation of 5-aza-7-deaza-1*H*-isoguanine (=4 aminoimidazo-[1,2-*a*]-1,3,5-triazin-2(1*H*)-one) to 2′-deoxy-5-aza-7-deaza-isoguanosine using purine nucleoside phosphorylase. *Helv Chim Acta* **76:** 2061–2069.

Voegel JJ, von Krosigk U, Benner SA. 1993b. Synthesis and tautomeric equilibrium of 6-amino-5-benzyl-3-methylpyrazin-2-one. An acceptor–donor–donor nucleoside base analog. *J Org Chem* **58:** 7542–7547.

von Krosigk U, Benner SA. 2004. Expanding the genetic alphabet. Pyrazine nucleosides that support a donor–donor–acceptor hydrogen bonding pattern. *Helv Chim Acta* **87:** 1299–1324.

Wei T, Pearson MN, Armstrong K, Blohm D, Liu J. 2012. Analysis of crucial factors resulting in microarray hybridization failure. *Mol Biosyst* **8:** 1325–1339.

Wilds CJ, Wawrzak Z, Krishnamurthy R, Eschenmoser A, Egli M. 2002. Crystal structure of a B-form DNA duplex containing (L)-α-threofuranosyl (3′→2′) nucleosides: A four-carbon sugar is easily accommodated into the backbone of DNA. *J Am Chem Soc* **124:** 13716–13721.

Win MN, Liang JC, Smolke CD. 2009. Frameworks for programming biological function through RNA parts and devices. *Chem Biol* **16:** 298–310.

Winnacker M, Kool ET. 2013. Artificial genetic sets composed of size-expanded base pairs. *Angew Chem Int Ed Engl* **52:** 12498–12508.

Woodward RB. 1968. Recent advances in the chemistry of natural products. *Pure Appl Chem* **17:** 519–547.

Woodward RB, Hoffmann R. 1970. *The conservation of orbital symmetry*. Springer, Heidelberg.

Yang Z, Chen F, Chamberlin SG, Benner SA. 2010. Expanded genetic alphabets in the polymerase chain reaction. *Angew Chem Int Ed Engl* **49:** 177–180.

Yang Z, Durante M, Glushakova LG, Sharma N, Leal NA, Bradley KM, Chen F, Benner SA. 2013. Conversion strategy using an expanded genetic alphabet to assay nucleic acids. *Anal Chem* **85:** 4705–4712.

Yang Z, McLendon C, Hutter D, Bradley KM, Frye C, Benner SA. 2015. Helicase dependent isothermal amplification of DNA and RNA using self-avoiding molecular recognition systems. *ChemBioChem* **16:** 1365–1370.

Zhang L, Yang Z, Sefah K, Bradley KM, Hoshika S, Zhu G, Cansiz S, Teng IT, Champanhac C, McLendon C, et al. 2015. Evolution of functional six-nucleotide DNA. *J Am Chem Soc* **137:** 6734–6737.

Zhou C, Avins JL, Klauser PC, Brandsen BM, Lee Y, Silverman SK. 2016. DNA-catalyzed amide hydrolysis. *J Am Chem Soc* **138:** 2106–2109.

Zubay G. 1988. *The roots of modern biochemistry* (ed. Kleinkauf H, von Döehren H, Jaenicke L), pp. 911–916. Walter de Gruyter, Berlin.

At the Interface of Chemical and Biological Synthesis: An Expanded Genetic Code

Han Xiao[1] and Peter G. Schultz[1,2]

[1]Department of Chemistry and the Skaggs Institute for Chemical Biology, The Scripps Research Institute, La Jolla, California 92037

[2]California Institute for Biomedical Research, La Jolla, California 92037

Correspondence: schultz@scripps.edu

The ability to site specifically incorporate noncanonical amino acids (ncAAs) with novel structures into proteins in living cells affords a powerful tool to investigate and manipulate protein structure and function. More than 200 ncAAs with diverse biological, chemical, and physical properties have been genetically encoded in response to nonsense or frameshift codons in both prokaryotic and eukaryotic organisms with high fidelity and efficiency. In this review, recent advances in the technology and its application to problems in protein biochemistry, cellular biology, and medicine are highlighted.

With the rare exceptions of pyrrolysine and selenocysteine, the genetic codes of all known organisms consist of the same 20 canonical amino acids. However, additional functional groups, including organic and inorganic cofactors and posttranslational modifications (PTMs), are required for many of the functions performed by proteins. This observation suggests that an expanded genetic code might allow one to either rationally design or evolve proteins with new or enhanced physical, chemical, and biological properties. To this end, methodology has been developed to exploit the endogenous protein biosynthetic machinery to site-specifically incorporate noncanonical amino acids (ncAAs) into proteins in living organisms with high translational fidelity and efficiency (up to 5 g/L on a commercial scale). The desired ncAA is encoded by a nonsense or frameshift codon using an orthogonal aminoacyl-tRNA synthetase (aaRS)/transfer RNA (tRNA) pair that is specific for the ncAA but does not cross-react with endogenous host aaRSs, tRNAs, or amino acids (Fig. 1A) (Wang et al. 2006b; Liu and Schultz 2010). Using this system, over 200 structurally distinct ncAAs have been genetically encoded in both prokaryotic and eukaryotic organisms. These ncAAs include spectroscopic probes, metal ion chelators, photo-affinity probes and photocaged amino acids, posttranslational modifications, and amino acids with orthogonal chemical activity for the site-specific modification of proteins (Wang et al. 2006b; Liu and Schultz 2010; Wan et al. 2014). They have been used as probes of protein structure and function, both in vitro and in vivo, and in the rational design of proteins with new or enhanced biological or pharmacological activities. Most recently, in vitro evolution experiments have begun to show that additional amino acid building

Cite this article as *Cold Spring Harb Perspect Biol* doi: 10.1101/cshperspect.a023945

blocks can lead to novel protein structures and activities. Herein, we review a number of recent advances in the field, both in the development and application of this technology.

GENETICALLY ENCODING ncAAs

Evolved *Methanococcus jannaschii* tyrosyl aaRS/tRNA pairs and *Methanosarcina barkeri* and *mazei* pyrrolysyl aaRS/tRNA pairs have enabled the incorporation of a large number of ncAAs into proteins in *Escherichia coli* (Fig. 1B) (Wang et al. 2006b; Liu and Schultz 2010; Wan et al. 2014). To further expand the number and nature of genetically encoded ncAAs, aaRS/tRNA pairs from other archeal and eukaryotic organisms have been recently developed, including *Pyrococcus horikoshii* lysyl aaRS/tRNA, *P. horikoshii* glutamyl aaRS/tRNA, *Saccharomyces cerevisiae* tryptophanyl aaRS/tRNA, heterogeneous leucyl *Mt*-tRNA/*Hs*-aaRS, and proly *Af*-tRNA/*Ph*-aaRS pairs (Anderson and Schultz 2003; Santoro et al. 2003; Anderson et al. 2004; Chatterjee et al. 2012, 2013d; Xiao et al. 2014). The structurally distinct active sites of these aaRSs allow one to encode chemically diverse amino acid side chains; however, the efficiencies with which the ncAAs can be incorporated at a given site in the proteome vary—ranging from milligrams to 5+ g/L of mutant protein, most likely because of the differing degrees to which the aaRS/tRNA pair is optimized (of course, for any given protein, efficiency is also affected by the mutation site). Platform vectors, for example, pEvol and pUltra, have been established that allow high-level protein expression of a desired mutant protein in bacteria, and large-scale fermentation (>10,000 L) has yielded mutant proteins on a 5 g/L scale (Young et al. 2010; Chatterjee et al. 2013b). A current focus is the efficient incorporation of multiple identical or distinct ncAAs into the same protein, or the synthesis of an entirely “unnatural” biopolymer. One hypothesis is that the presence of endogenous release factors leads to low overall incorporation efficiency for multiple ncAAs encoded by nonsense codons. Huang et al. (2010) first showed that up to three ncAAs could be genetically incorporated into the same protein by overexpression of the carboxy-terminal domain of ribosomal protein L11 to reduce release factor 1 (RF1)-mediated termination of protein trans-

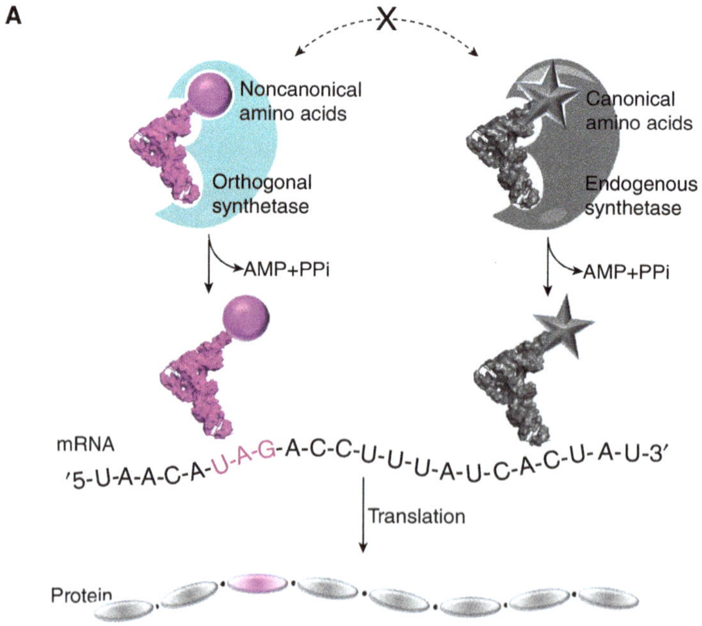

Figure 1. Expanding the genetic code. (*A*) The site-specific incorporation of noncanonical amino acids (ncAAs) using an orthogonal aminoacyl-tRNA synthetase (aaRS)/transfer RNA (tRNA) pair. (*Figure continues on following page.*)

Figure 1. (*Continued*) (*B*) ncAAs described in this review. mRNA, Messenger RNA.

lation. Using an RF1 knockout strain, Johnson et al. (2011) showed that up to 10 *p*-acetylphenylalanines (pAcF; Fig. 1B1) could be incorporated into the same protein, albeit in relatively low yield (Fig. 2A). Knockout of RF1 improves ncAA incorporation in response to the amber codon; however, impaired growth rates resulting from RF1 deletion may limit the utility of this strain.

Recently, multiplex automated genome engineering (MAGE) and conjugative assembly genome engineering (CAGE) were used to replace all of the TAG stop codons in the *E. coli* MG1655 genome with the TAA ochre nonsense codon (Fig. 2A) (Lajoie et al. 2013); subsequently, RF1 was knocked out to generate *E. coli* strain C321.ΔA. In the resulting genomically recoded organism (C321.ΔA), UAG was reassigned from a nonsense codon to a sense codon specific for an ncAA in the presence of the cognate orthogonal tRNA/aaRS pair. As a result of genome recoding, RF1 knockout is no longer deleterious and robust ncAA incorporation was achieved in response to the amber codon (Lajoie et al. 2013). In addition, the suppression efficiency of the four-base codon UAGA, which starts with UAG, was significantly increased in the C321.ΔA strain (Fig. 2B) (Chatterjee et al. 2014). This genomically recoded strain was also used to generate a library of chromosomally integrated aaRSs for the selection of more efficient aaRS variants. The evolved aaRSs showed increased ncAA-containing protein production, enabling genetic incorporation of 30 ncAAs into a single protein (Amiram et al. 2015). An alternative approach to minimize the effects of RF1-mediated translational termination is the generation of ribosome-mRNA pairs that function in the presence of, but are orthogonal to, the endogenous translational machinery. The use of orthogonal ribosome–mRNA pairs in *E. coli* significantly increases the suppression efficiency of ncAA incorporation, and facilitates the use of frameshift codons and incorporation of multiple ncAAs into proteins (Wang et al. 2007; Neumann et al. 2010a,b).

In mammalian cells, the pyrrolysyl pair from archaea, and the tyrosyl and leucyl pairs from *E. coli* have been widely used to genetically encode novel amino acids, including amino acids with bio-orthogonal chemical reactivities (e.g., azido, alkynyl, and tetrazinyl), fluorescent probes, PTMs, and photocaged amino acids (Fig. 1B) (Wang et al. 2006b; Liu and Schultz 2010; Wan et al. 2014). These orthogonal tRNA/aaRS pairs are either evolved in yeast (for *E. coli*–derived pairs) or bacteria (for archaea-derived pairs) and then transferred to vectors with appropriate promoters for transient expression in the host cell, typically, HEK293 or CHO cells. Yields of transiently expressed proteins on the order of 5–20 mg/L are common, and yields up to 1 g/L have been reported for stable cell lines expressing mutant full-length antibodies (Tian et al. 2014). However, the application of this technology in mammalian cells is limited by the inherent drawbacks of transient transfection, including the relatively low transfection efficiency of certain cell types (e.g., neurons and embryonic stem cells) and the toxicity of transfection reagents. To overcome these challenges, an enhanced suppression system delivered by a hybrid baculovirus has been reported (Chatterjee et al. 2013c). This baculovirus-based system significantly increases the incorporation efficiency of ncAA into proteins in mammalian cells and allows ncAAs to be encoded in embryonic stem cells, rat cardiac fibroblasts, and neurons. This enhanced suppression system has been used to incorporate up to three identical ncAAs or two distinct ncAAs into a single protein in good yield (Xiao et al. 2013). For example, in a full-length antibody, pAcF (Fig. 1B1) was incorporated in response to a TAG codon in the heavy chain and an azidolysine analog (AzK; Fig. 1B34) was incorporated in response to a TAA codon in the light chain, allowing the mutant antibody to be selectively conjugated to a drug and a fluorophore at these two distinct sites (Xiao et al. 2013). Recently, a *Methanosarcina mazei* pyrrolysyl aaRS/tRNA pair was stably integrated into the mammalian genome using the PiggyBac transposon system, which enables efficient incorporation of ncAAs into target proteins in diverse mammalian cells. The utility of this system was shown by genetically incorporating six ε-*N*-acetyllysines (AcK; Fig. 1B59) into histone H3 (Elsasser et al. 2016). Methodological advances have also allowed the genetic

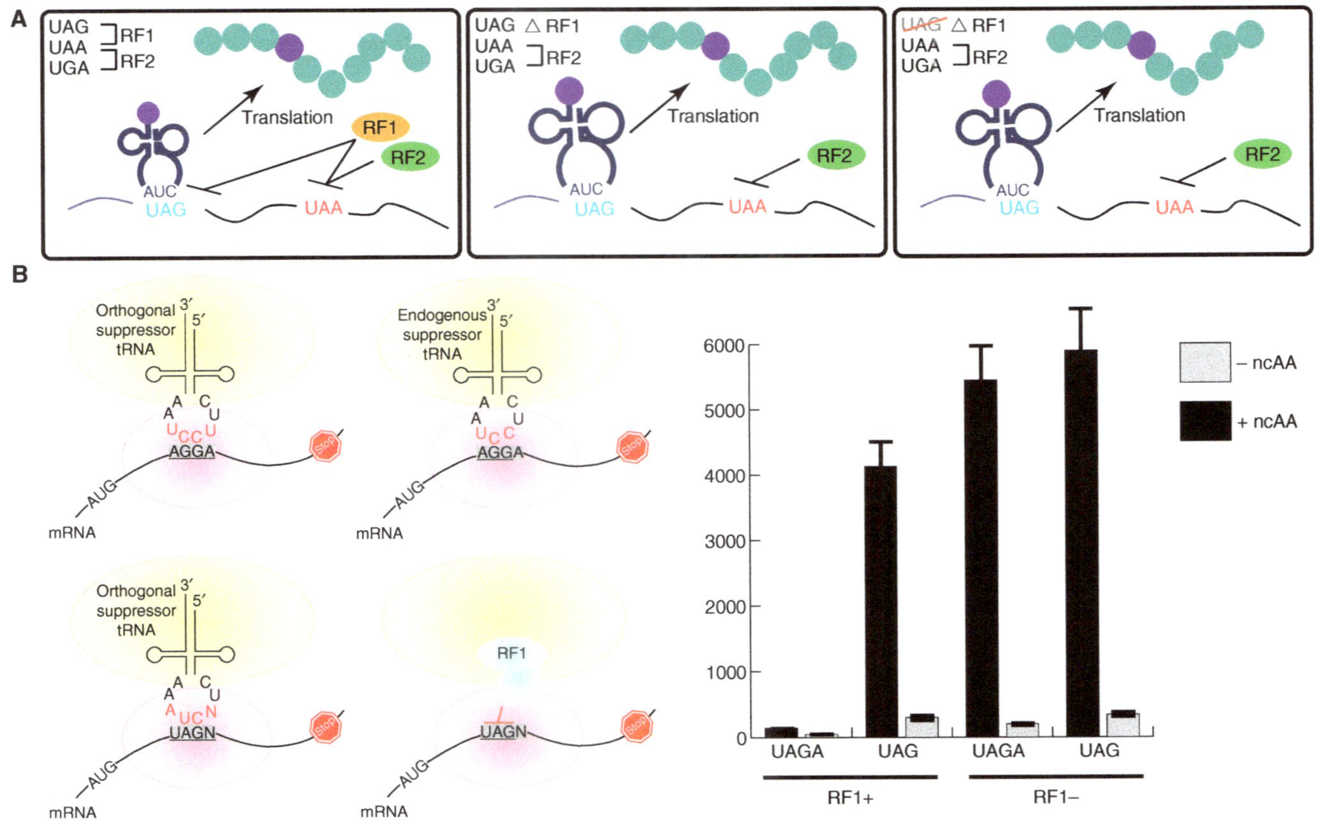

Figure 2. Engineering *Escherichia coli* strains with a reassigned codon. (*A*) *Escherichia coli* genome engineering for increasing noncanonical amino acids (ncAAs) incorporation efficiency. (*B*) Competing triplet recognition during the translation of quadruplet codons. Deletion of RF1 significantly improves UAGN-suppression levels. mRNA, Messenger RNA; tRNA, transfer RNA.

incorporation of ncAAs into proteins in various multicellular organisms, including *Caenorhabditis elegans*, *Drosophila melanogaster*, *Arabidopsis thaliana*, and *Streptomyces venezuelae* (Greiss and Chin 2011; Bianco et al. 2012; Li et al. 2013; Chin 2014; He et al. 2015). Most recently, we have been able to genetically encode ncAAs in *Bacillus cereus*, allowing the synthesis of ribosomally derived natural products from unnatural building blocks (Luo et al. 2016).

PROBES OF PROTEIN STRUCTURE

The site-specific incorporation of ncAAs with unique spectroscopic properties provides a powerful tool to explore protein structure and dynamics. $^{13}C/^{15}N$-labeled *p*-methoxyphenylalanine (Fig. 1B2) and *p*-tifluoromethoxyphenylalanine (Fig. 1B13) have been genetically incorporated into human fatty acid synthetase to probe conformational changes that occur on ligand binding by nuclear magnetic resonance (NMR) (Cellitti et al. 2008). Similarly, ^{19}F-labeled *p*-trifluoroethyl-phenylalanine (Fig. 1B14) was incorporated into nitroreductase and histidinol dehydrogenase to probe structural changes that occur on substrate binding (Jackson et al. 2007; Li et al. 2010). Furthermore, the site-specific incorporation of the metal-binding ncAA, 2-amino-3-(8-hydroxyquinolin-3-yl)propanoic acid (Fig. 1B21), into membrane proteins (CXCR1 and viroporin p7) allowed long-range paramagnetic relaxation measurements of intra- and intermolecular distance (Park et al. 2014). In related work, to facilitate electron paramagnetic resonance (EPR) studies of proteins, a ketoxime-linked spin label side chain has been site-specifically conjugated to proteins through a reactive chemical handle (Fleissner et al. 2009). However, direct incorporation of these probes is advantageous in many cases and, thus, a spin-labeled ncAA (Fig. 1B51) was recently incorporated into green fluorescent protein (GFP) and thioredoxin in *E. coli* (Schmidt et al. 2014).

Infrared (IR) spectroscopy is a valuable technique that can be used to characterize protein structure, dynamics, and microenvironment on a range of timescales. To this end, ncAAs with unique IR signatures, including *p*-cyano-phenylalanine (Fig. 1B15), *p*-azidophenylalanine (pAzF, Fig. 1B4), *p*-nitrophenylalanine (pNO_2F; Fig. 1B16), and 3-nitrotyrosine ($3\text{-}NO_2Y$, Fig. 1B19) have been efficiently incorporated into proteins (Schultz et al. 2006; Taskent-Sezgin et al. 2009; Ye et al. 2009; Smith et al. 2011). To study protein dynamics and microenvironments, photocaged ^2D-labeled *o*-nitrobenzyl-tyrosine (oNBTyr, Fig. 1B46) enabled the site-specific deuteration of Tyr100 within dihydrofolate reductase and revealed that Tyr100 stabilizes the positive charge of the nicotinamide moiety in the transition state (TS) via electrostatic interactions (Groff et al. 2009). Photocaged lysine, cysteine, and serine (Fig. 1B46–50) have all been genetically incorporated into proteins, and can be used to site-specifically deuterium-label these amino acids as well (Liu and Schultz 2010; Davis and Chin 2012).

Fluorescent proteins (FPs) have been widely used to study protein expression, localization, and conformation in living cells. By substituting core residues in the chromophores of various FPs with ncAAs, FP variants with distinct spectral properties have been generated, including FP sensors for monitoring pH, reducing and oxidizing environments, and heavy metals (Wang et al. 2003; Ai 2012; Chatterjee et al. 2013d; Niu and Guo 2013; Xiao et al. 2014). For example, a fluorogenic biosensor sensitive to oxidation was generated by substitution of Tyr66 in GFP with *p*-borono-phenylalanine (Fig. 1B11) (Wang et al. 2012; Chen et al. 2013). In the presence of oxidants, the mutant chromophore is converted to the wild-type chromophore with a change in fluorescence quantum yield. However, fusion of an FP can significantly perturb the structure and function of a target protein, and is largely limited to the amino and carboxyl termini (Zheng et al. 2014). The ability to site-specifically incorporate relatively small fluorescent amino acids at defined sites within proteins would further increase the utility of fluorescence spectroscopy to study protein structure in vitro and in living cells. Several fluorescent ncAAs have been genetically incorporated into proteins in *E. coli*,

 Cite this article as *Cold Spring Harb Perspect Biol* doi: 10.1101/cshperspect.a023945

including a hydroxycoumarin-derived amino acid (7HC, Fig. 1B42) and acridon-2-ylalanine (Fig. 1B45) (Wang et al. 2006a; Speight et al. 2013). 7HC was inserted into bacterial tubulin FtsZ and used to track FtsZ localization during cell division (Charbon et al. 2011). This fluorophore was also substituted for Trp 564 in activator of transcription 3 to monitor the phosphorylation state of Tyr705 (Lacey et al. 2011). To determine protein localization in yeast and mammalian cells, 2-amino-3-(5-(dimethylamino)naphthalene-1-sulfonamido) propanoic acid (Fig. 1B43) and 3-(6-acetylnaphthalen-2-ylamino)-2-aminopropanoic acid (Anap; Fig. 1B44) have been used by a number of groups (Summerer et al. 2006; Lee et al. 2009; Chatterjee et al. 2013a). Anap was successfully used to detect conformational changes on ligand binding to glutamine-binding protein (QBP) and dynamic internal pore opening in voltage-gated potassium channels (K_V) in oocytes (Fig. 3B,C) (Lee et al. 2009; Kalstrup and Blunck 2013). These latter experiments underscore the ability to probe local changes in structure or environment by the site-specific insertion of smaller fluorophores. An important advance in this regard will be the genetic encoding of probes with improved photostability and longer wavelengths.

PROBING PROTEIN FUNCTION WITH ncAAs

Genetically encoded ncAAs can also be used as mechanistic probes. For example, ncAAs have been used in mechanistic studies of class I ribonucleotide reductases (Cotruvo and Stubbe 2011). The site-specific incorporation of a redox active probe, 3-aminotyrosine (Fig. 1B18), in place of transiently oxidized Tyr356, Tyr730, and Tyr731 residues was used to probe the role of these amino acids in long-range proton coupled electron transfer (PCET) (Seyedsayamdost et al. 2007; Minnihan et al. 2011a). To measure perturbations to the pKas of redox-active tyrosines within the PCET pathway, 3-NO_2Y (Fig. 1B19), was substituted for Tyr122, Tyr730, and Tyr731 (Yokoyama et al. 2010). This radical transfer mechanism was also explored using a number of fluorotyrosines (Fig. 1B20) with distinct EPR properties (Minnihan et al. 2011b). Recently, ncAAs have even been used as direct structural probes of the transition states (TSs) of chemical reactions. Specifically, the X-ray crystal structure of the planar TS configuration

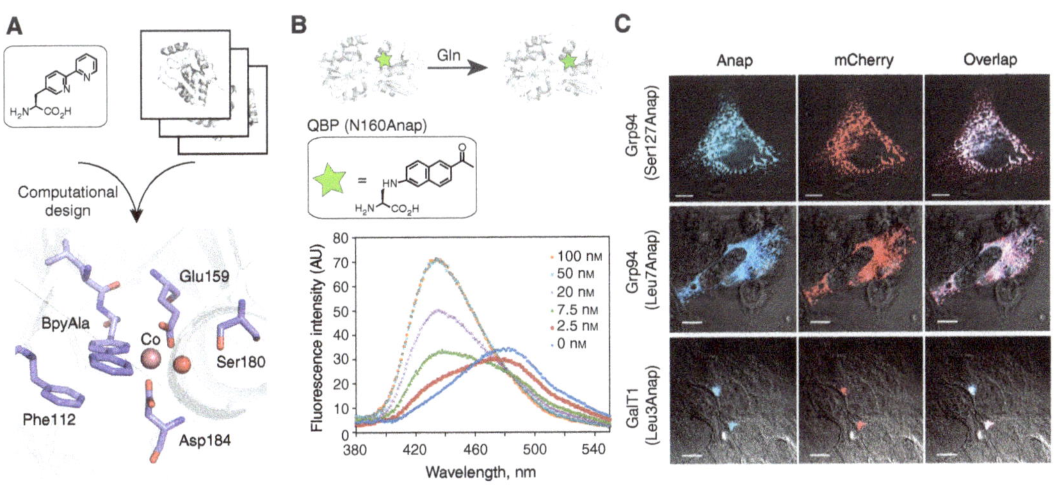

Figure 3. Some applications of noncanonical amino acids (ncAAs). (*A*) X-ray crystal structure of a BipyAla-containing metalloprotein generated by computational design (protein data bank [PDB] code 4IWW). (*B*) Fluorescence intensities of glutamine-binding protein (QBP) (Asn160Anap) with different concentrations of Gln. (*C*) Subcellular localization of endoplasmic and Golgi-residue proteins in HEK293 cells visualized using Anap. Scale bars, 10 μm.

for rotation about the central C-C bond of biphenyl was directly observed by genetically introducing *p*-biphenylalanine (BiPhe; Fig. 1B8) into a site in *Pyrococcus abyssi* threonyl-tRNA synthetase that was designed to bind BiPhe in a planar TS configuration (Pearson et al. 2015). After several rounds of optimization, the crystal structure of the designed protein at 2.05 Å resolution revealed a planar biphenyl TS configuration stabilized by van der Waals interactions with side chains in the pocket.

Photocaged ncAAs can be used to control the temporal and spatial activity of proteins in living cells. To this end, photocaged tyrosine, lysine, cysteine, and serine have been genetically encoded in *E. coli*, yeast, and mammalian cells (Fig. 1B46–50) (Liu and Schultz 2010; Davis and Chin 2012). In an early example, 4,5-dimethoxy-2-nitrobenzyl-serine (Fig. 1B50) was substituted for Ser114 and Ser128 in the yeast transcription factor Pho4 (Lemke et al. 2007). Uncaging and subsequent phosphorylation of these two sites, which are involved in nuclear export of Pho4, was triggered by a laser pulse and monitored in real time. Similarly, a photocaged lysine, ε-*N*-[(1-(6-nitrobenzo[*d*][1,3]-dioxol-5-yl)ethoxy)-carbonyl]lysine (Fig. 1B47), was used to control the nuclear import of p53 (Gautier et al. 2010). This same lysine analog was also used to control the activation of mitogen-associated protein kinase (MAPK) signaling in mammalian cells by photocaging the MAPK kinase MEK1 (Gautier et al. 2011); and a photocaged tyrosine, oNBTyr (Fig. 1B46), has been used to control phosphorylation of Tyr 701 in human signal transducer and activator of transcription 1 (STAT1) and thereby activate the Janus kinase/STAT signal transduction pathway (Arbely et al. 2012). Recently, a photocaged cysteine (Fig. 1B48) was genetically incorporated into the pore of the inwardly rectifying potassium channel Kir2.1 to generate a light-activated ion channel in the embryonic mouse neocortex (Kang et al. 2013). Photocaged ncAAs have been also used in other studies, including development of light-responsive FPs, light-induced transcription using a light-activated RNA polymerase, photocleavage of the polypeptide backbone, and photoactivatable intein-mediated protein splicing (Peters et al. 2009; Groff et al. 2010; Hemphill et al. 2013; Ren et al. 2015).

To map protein interactions, several ncAAs containing UV-induced cross-linking moieties, including benzophenone, azides, and diazirines (Fig. 1B52, 4, and 53), have been site-specifically incorporated into proteins in both prokaryotic and eukaryotic cells (Liu and Schultz 2010; Davis and Chin 2012). The formation of covalent cross-links by UV-irradiation is especially important for studying relatively weak protein interactions that are not stable to immunoprecipitation or cell lysis, and also allows one to probe intra- or interprotein contacts between specific residues. Numerous studies have been performed using these photo-cross-linking ncAAs, including probing interactions of chaperones with substrates, conformational changes in RNA polymerases, interactions among components within signal transduction pathways, and membrane protein interactions (Schlieker et al. 2004; Mori and Ito 2006; Chen et al. 2007; Mohibullah and Hahn 2008; Ieva and Bernstein 2009; Okuda and Tokuda 2009; Tagami et al. 2010). The use of ncAAs to selectively form covalent bonds not only provides valuable information regarding molecular interactions, but also allows the design of covalent protein-based agonists, antagonists, or inhibitors. For example, *p*-vinylsulfonamido-phenylalanine (Fig. 1B54), which contains an aza-Michael acceptor that can alkylate nearby cysteine or lysine residues, was site-specifically introduced into Herceptin; the mutant antibody selectively formed a covalent complex with a specific surface lysine of the oncogenic receptor tyrosine kinase ErbB2 on live cells (Furman et al. 2014). In addition, a number of ncAAs containing electrophilic reactive handles (Fig. 1B55 and 56) have been reported by Lei and his coworkers and used to form specific inter- or intramolecular cross-links in proteins (Chen et al. 2014; Xiang et al. 2014).

PTMs can alter protein function, folding, and stability, and thereby regulate numerous biological processes in the cell. The ability to site-specifically incorporate these PTMs into target proteins provides a valuable tool to study their function. Several tyrosine analogs, including *p*-sulfo-tyrosine (pSY; Fig. 1B58) and *p*-carboxymethylphenylalanine (pCMF; Fig. 1B57), have been genetically encoded to mimic tyrosine PTMs (Liu and Schultz 2006; Xie et al. 2007). Site-specific incorporation of pSY allowed for the preparation of high-affinity sulfo-hirudin and various sulfated antibodies to the human immunodeficiency virus (HIV) protein GP120 (Liu et al. 2007). Recently, pCMF was used to mimic phosphoTyr291 in protein arginine methyltransferase 1, which plays a role in substrate recognition (Rust et al. 2014). A similar strategy has been used to study the function of lysine PTMs within histones. For example, ε-*N*-acetyllysine (AcK; Fig. 1B59), ε-*N*-methyllysine (MeK; Fig. 1B60), ε-*N*,*N*-dimethyllysine (Fig. 1B61), ε-*N*-crotonyllysine (Fig. 1B62), and ε-*N*-2-hydroxyisobutyryllysine (Fig. 1B63) have all been site-specifically incorporated into histone proteins (Neumann et al. 2008; Nguyen et al. 2009, 2010; Kim et al. 2012b; Xiao et al. 2015b). Site-specific substitution of AcK for H3K56 revealed that H3K56 has no direct effect on the compaction of chromatin, but leads to an increase in DNA breathing (Neumann et al. 2009); and the substitution of MeK for H3K9 allowed determination of its effects on the binding specificity of heterochromatin protein 1 (Nguyen et al. 2009).

THE DESIGN OF PROTEINS WITH NOVEL PROPERTIES USING ncAAs

There is considerable interest in the site-specific modification of proteins with biological probes, polyethylene glycols (PEGs), drugs, and other polypeptides/proteins for a range of applications, including the generation of antibody–drug conjugates (ADCs) and bispecific antibodies. Unfortunately, it is difficult to do chemistry on proteins with the same degree of control over structure that chemists have with small molecules. Typically, the modification of proteins involves the electrophilic modification of cysteine or lysine residues, which in the case of large, disulfide cross-linked proteins like antibodies, results in chemically heterogeneous products with varying stoichiometries and sites of modification. This heterogeneity makes it difficult to optimize the properties of the resulting conjugates. The site-specific incorporation of ncAAs with bioorthogonal chemical reactivity enables the preparation of homogenous, chemically defined macromolecules. For example, homogenous therapeutic proteins with defined PEG modifications have been prepared by the site-specific incorporation of pAcF (Fig. 1B1), followed by efficient conjugation to alkoxyamine-derivatized PEG by oxime ligation (Cho et al. 2011). Using this strategy, PEG-conjugated human growth hormone, human fibroblast growth factor 21, and bovine granulocyte colony–stimulating factor (GCSF) were prepared and shown in clinical studies to have significantly enhanced serum half-lives while preserving a high degree of biological activity, with little or no observed immunogenicity (Cho et al. 2011; Mu et al. 2012).

ADCs have emerged as a promising new class of drugs, as evidenced by the clinical efficacy of trastuzumab emtansine for the treatment of Her2-positive breast cancer (Sun et al. 2014; Wals and Ovaa 2014). ADCs allow highly cytotoxic drugs to be selectively delivered to the target tumor, thus, increasing therapeutic efficacy and reducing adverse systemic side effects (Fig. 4). Most ADCs are based on the reactivity of cysteine and lysine residues, which often result in heterogeneous products. Recently, we prepared homogenous ADCs by the site-specific incorporation of pAcF (Fig. 1B1) into Herceptin and conjugation of the mutant antibody to an alkoxy-amine auristatin derivative in good yield (Axup et al. 2012) (a similar ADC developed by Ambrx (San Diego, CA) has entered clinical trials; Tian et al. 2014). Complete regression of Her2-positive tumors in mouse xenograft models was observed without overt toxicity. Importantly, the site and stoichiometry of conjugation were found to affect the pharmacokinetics and efficacy of the ADC. Similarly, site-specific ADCs have been generated that target CXCR4-positive metastatic cancer cells, and this technology is now being applied outside of oncology in the preparation of ADCs for the treatment of inflammatory, autoimmune, and metabolic disease (Kularatne et al. 2014). Other constructs, including antibodies conjugated to

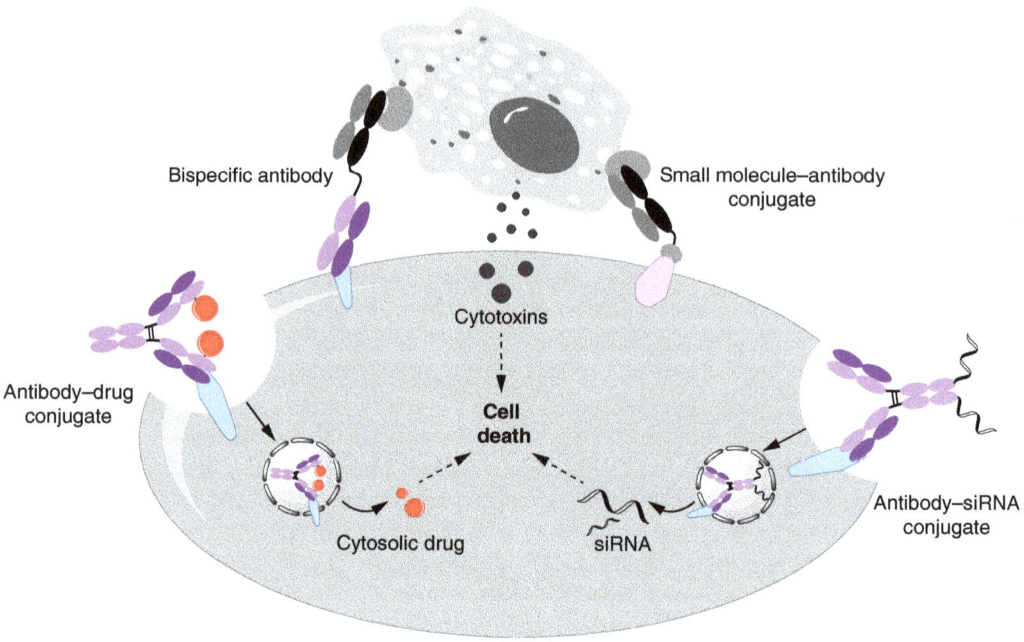

Figure 4. Overview of noncanonical amino acid (ncAA)-containing antibody conjugates. siRNA, Small interfering RNA.

imaging probes or two different drugs with synergistic effects, are being explored. Furthermore, the utility of site-specific antibody conjugation was also shown in antibody small-interfering RNA (siRNA) conjugation, which significantly increases the selectivity and efficiency of siRNA delivery (Fig. 4) (Lu et al. 2013).

Another example of use of ncAAs in the preparation of therapeutic proteins is the generation of bispecific antibodies that selectively deliver cytotoxic T cells to tumor cells (Fig. 4) (Byrne et al. 2013). Bispecific antibodies are typically prepared as genetic fusions or heterodimers, which limits the possible relative spatial orientations of the two antibody variable domains and may affect stability, efficacy, and selectivity (Bluemel et al. 2010; Moore et al. 2011). Again, the use of genetically encoded ncAAs with bio-orthogonal reactivity allows far more control over the relative orientation and distance of the two antibody fragments (Kim et al. 2012a). To illustrate this notion, pAcF-containing anti-Her2 and anti-CD3 Fabs were expressed in *E. coli* and each was conjugated to short polyethylene linkers terminated with either an azide or cyclooctyne group (Cao et al. 2015). These two Fabs were then conjugated via a copper-free click reaction to afford a bispecific antibody that binds the CD3 protein on T cells and Her2 on tumor cells to form an immunological synapse. The resulting anti-HER2/anti-CD3 bispecific antibody showed good selectivity and high potency with Her2-positive tumor cells in vitro and showed excellent efficacy against both high and low Her2-positive tumors in rodent xenografts in the presence of purified peripheral blood mononuclear cells. Similarly, bispecifics have been made that target CD33 and CLL1 for the treatment of acute myelogenous leukemia (Lu et al. 2014). Tumor-specific small molecules can also be used to recruit cytotoxic T cells to tumor cells (Murelli et al. 2009; Kularatne et al. 2010). For example, 2-[3-(1,3-dicarboxy propyl)-ureido] pentanedioic acid, a small molecule that selectively binds PSMA, which is overexpressed in prostate cancer, was site-specifically conjugated to anti-CD3 to generate a PSMA-targeting small molecule–antibody conjugate (SMAC) (Fig. 4) (Kim et al. 2013). The optimized SMAC showed excellent selectivity and cytotoxicity ($EC_{50} \sim$ 100 pM) against patient-derived PSMA-positive prostate cancer cells both in vitro and in vivo and is being advanced to clinical development. A similar strategy was used to recruit T cells to ovarian cancer cells using a folate-anti-CD3 bispecific construct (Kularatne et al. 2013).

Despite the impressive clinical efficacy of chimeric antigen receptor (CAR)-T-cell therapy, the lack of control in the activation and expansion and in the termination of conventional CAR-T cells has greatly limited its application. To overcome this challenge, a CAR was engineered that binds the bio-orthogonal ligand fluorescein isothiocyanate (FITC). The resulting transduced CAR-T cells are inactive until a switch molecule, consisting of an FITC-conjugated site specifically to a tumor-targeting antibody, are dosed. It was shown that the ability to control the site and stoichiometry of FITC conjugation using ncAAs dramatically impacts effective immunological synapse formation and, as a consequence, in vitro and in vivo activity. These switch molecules allowed dose titration of the CAR-T-cell response and resulted in complete clearance of tumors in rodent xenografts with significantly reduced cytokine levels; the CAR-T-cell response could also be terminated by simply removing the switch. Finally, this single universal CAR can be used for heterogeneous tumors with distinct antigens (or reoccurring tumors with altered antigen expression) by simply using multiple switch molecules. Thus, this platform (which also includes a recombinant version using peptide neoepitopes) should decrease the acute and chronic toxicity associated with this promising new cell therapy, simplify its application, and allow it to be used with both liquid and solid tumors (Ma et al. 2016).

ncAA mutagenesis has also been used for the generation of therapeutic vaccines to break immunological tolerance against a native protein (Sun et al. 2014; Wals and Ovaa 2014). Replacement of a single tyrosine residue (Tyr86) of murine tumor necrosis factor α (mTNF-α) with pNO_2F (Fig. 1B16) induced a high-titer, long-lived antibody response to both the mutant and native mTNF-α in vaccinated mice (Grunewald et al. 2008). Mechanistic studies revealed that the ncAA mutation generated a T-cell neoepitope, which led to a polyclonal antibody response to the autologous antigen (Gauba et al. 2011). Similar results were obtained with other mutant antigens containing ncAAs, including complement component 5α and proprotein convertase subtilisin/kexin type 9. It was also shown that the naturally posttranslationally modified amino acids nitrotyrosine and sulfotyrosine also break immunological tolerance, supporting the hypothesis that such PTMs, induced by viruses or inflammation, could be an underlying cause of autoimmune disease. Recently, an HIV-1 virus that requires ncAA translational machinery for replication has been reported by Wang et al. (2014). Multiple amber codons were introduced at permissive sites of the Gag gene; in the presence of an amber suppressor tRNA/aaRS pair and specific ncAA, full-length Gag protein is generated leading to the assembly of infective virus. However, none of these ncAA translational components are naturally present in humans, thus making the virus unable to reproduce in human cells. This strategy is also being used to create other vaccines—several pyrrolysine analogs have been incorporated into the surface protein of human hepatitis D virus, and *Mj*TyrRS/tRNA has been used to encode a number of tyrosine analogs in *Mycobacterium smegmatis* and *tuberculosis* (Wang et al. 2010; Lin et al. 2013). Using a similar strategy, a live-attenuated tuberculosis vaccine is being explored in our laboratory that is dependent on the presence of an ncAA for growth ex vivo, but dies in the absence of the ncAA. Surprisingly, it has not yet been possible to isolate a revertant for some sites of suppression. Alternatively, one could exploit a strategy developed by Church and coworkers, who used an ncAA-based synthetic system to control unintended proliferation of genetically modified organisms (Mandell et al. 2015). A synthetic essential gene that requires the ncAA BiPhe (Fig. 1B8) for translation, folding, and function was generated by computational design. Substitution of the wild-type gene with this synthetic gene in the *E. coli* genome provides a novel biocontainment mechanism against mutational escape, environmental supplementation, and horizontal gene transfer.

ncAAs have also been used in the preparation of biomaterials with improved or novel properties. For example, a reversible biomechano-responsive surface was generated via covalently coating pAzF (Fig. 1B4)-containing GFPs onto polydimethylsiloxane with PEG brushes (Longo et al. 2015). A reversible decrease in fluorescence intensity was observed on applying mechanical stress to the surface. To prepare homogenous protein–polymer conjugates, an ncAA with a polymer initiator side chain was genetically incorporated into proteins in *E. coli*. The designed amino acid, 4-(2′-bromoisobutyramido)-phenylalanine (Fig. 1B24) can serve as an initiator in atom-transfer radical polymerization and provides a stable link between target proteins and polymers (Peeler et al. 2010). Metal ion-binding amino acids may allow the generation of enzymes with novel functions, including oxygen transport, electron transfer, and catalytic activity. One such example is the computational design of a metal ion-binding site in the protein using (2,2′-bipyridin-5yl)-alanine (BpyAla; Fig. 1B22) (Fig. 3A) (Mills et al. 2013). Using the software Rosetta, Mills et al. (2013) generated a BpyAla-containing metalloprotein with high affinity (K_d for Zn^{2+} is ~ 37 pM) against a number of divalent cations. The three-dimensional structure determined by X-ray crystallography was very close to that of the design model (Fig. 3A).

PROTEIN EVOLUTION WITH ncAAs

Advances in methods for incorporation ncAAs now make it possible to ask whether additional amino acids can confer an evolutionary advantage to a host organism in response to selective pressure. For example, an in vitro antibody phage display system was developed in which ncAAs were available for directed evolution. A germline antibody library containing six randomized residues in CDR3H was subjected to affinity-based phage panning for binding to gp120, an HIV protein that naturally binds the sulfated cytokine receptor CCR5 (Liu et al. 2008). When sulfotyrosine (pSY; Fig. 1B58) was genetically encoded in the host strain for phage production, the highest affinity antibodies identified in the panning experiment all contained sulfotyrosine in CDRH3. A similar phage display system was used to select a Zn(II)-finger-like domain in which two histidine residues involved in Zn binding were substituted with the ncAA, bipyridylalanine (BpyAla; Fig. 1B22). To alter the metal ion-binding site, five residues in the amino terminus of Zif268 zinc-finger protein, including the two histidines, were randomized. Panning this library against the DNA recognition site revealed a mutant in which the wild-type Zn(II) site was replaced with a unique 4-bipyridylalanine·Fe(II) site with no impact on DNA affinity and selectivity (Kang et al. 2014). More recently, a bacterial selection scheme was used to identify cyclic peptides that inhibit HIV protease. Using a split intein system, a library of ncAA-containing ribosomal cyclic peptides was generated in *E. coli*. This library was further subjected to a selection based on survival of *E. coli* in response to proteolytic stress. A *p*-benzoylphenylalanine)-containing (Fig. 1B52) peptide that forms a covalent bond with the ε-amino group of Lys14 on the HIV protease and destabilizes the soluble cytoplasmic protein was isolated as the most potent hit (Young et al. 2011). And recently, Hammerling et al. (2014) generated bacteriophage T7 with an expanded genetic code by passaging phage on an *E. coli* host in which the amber codon was reassigned to 3-iodotyrosine (Fig. 1B17). After serial transfers of lytic phage, it was found that substitution of 3-iodotyrosine in the type II holing protein of bacteriophage T7 leads to increased phage fitness (Hammerling et al. 2014).

However, a more general rapid in vitro directed evolution for novel functional proteins with ncAAs has remained elusive. To this end, we recently developed a novel in vitro system with ncAAs in which a number of different ncAAs were randomly substituted at each of 144 of the 286 residues in TEM-1 β-lactamase based on the X-ray crystal structure (Fig. 5A) (Xiao et al. 2015a). To simplify construction of a β-lactamase library containing a variety of structurally distinct ncAAs, a polyspecific aminoacyl-tRNA synthetase was used that is orthog-

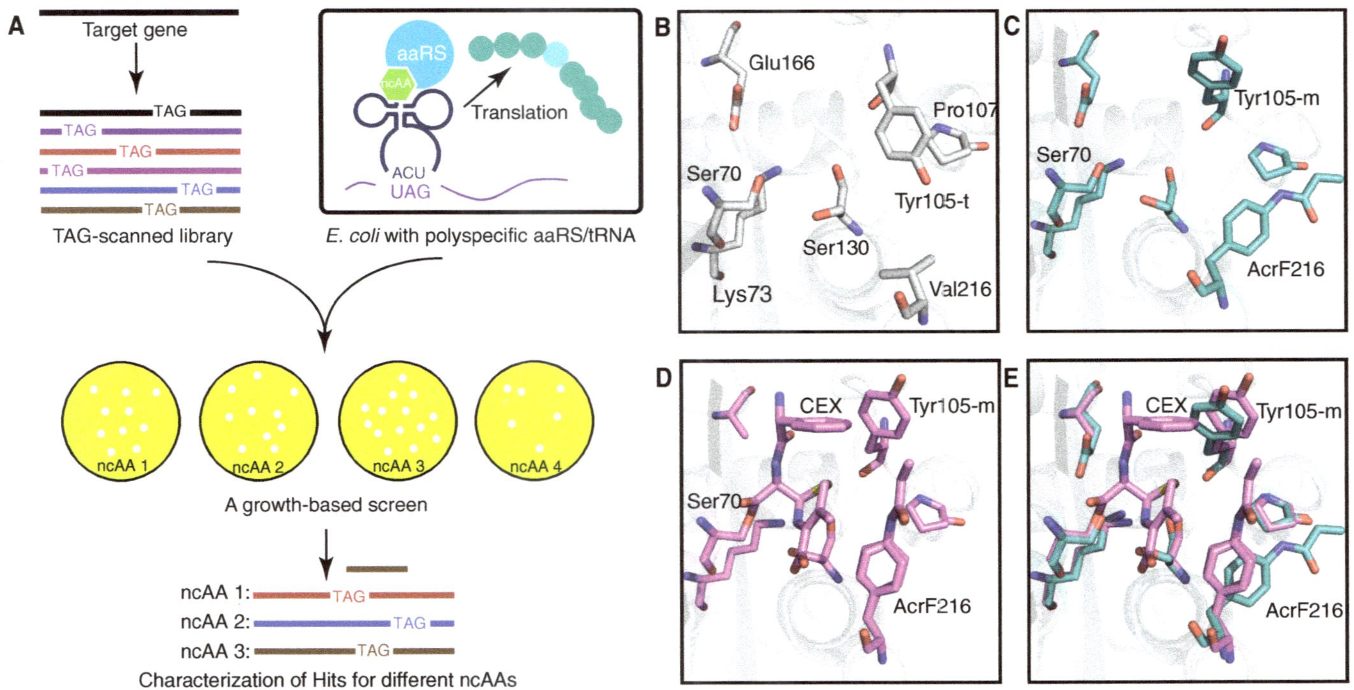

Figure 5. Directed evolution by noncanonical amino acid (ncAA) mutagenesis. (*A*) A TAG-scanned library was transformed into cells expressing a polyspecific amber suppressor aminoacyl-tRNA synthetase (aaRS)/transfer RNA (tRNA) pair and a growth selection was performed with various ncAAs. (*B*) X-ray crystal structure of ligand-free wild-type β-lactamase (protein data bank [PDB] code 1BTL). (*C*) X-ray crystal structure of ligand-free Val216AcrF mutant enzyme (PDB code 4ZJ1). (*D*) X-ray crystal structure of cephalexin (CEX) acyl-enzyme intermediate for the Val216AcrF mutant enzyme (PDB code 4ZJ3). (*E*) Overlay of active site residues of cephalexin-bound and ligand-free Val216AcrF mutant enzyme.

onal to the endogenous translational machinery of *E. coli*, and selectively incorporates 10 distinct ncAAs in response to the amber nonsense codon (Fig. 5A). A screen for growth on the β-lactam antibiotic cephalexin afforded a unique *p*-acrylamidophenylalanine mutation at Val216 that leads to an increase in catalytic efficiency by increasing k_{cat} without significantly affecting K_M. This increase in activity is not recapitulated by substitution with any of the canonical amino acids at residue 216. A comparison of the X-ray crystal structures of the cephalexin complexes of deacylation-defective wild-type and Val216*p*-AcrPhe mutant β-lactamases suggests that this mutation leads to changes in active site conformation with increased complementarity to the putative reaction TS (Fig. 5B–E). More recently, the impact of ncAA substitution on Phe66, Leu162, and Thr189 of TEM-1 β-lactamase have been systematically studied by Tack et al. (2016), and it has been shown that enzyme activity can be made strictly dependent on the presence of a ncAA. Using a temperature-dependent growth selection system, ncAAs containing long-side-chain thiols that can pair with cysteines to afford extended disulfide bonds have been used to identify enzymes showing enhanced thermostability. A mutant enzyme was identified that is cross-linked by one such extended disulfide bond and is stabilized by 9°C. In addition, a protein has been isolated that appears to form a covalent adduct between a Cys- and keto-containing ncAA and is stabilized by >25°C. These experiments provide more evidence that additional amino acids can confer an evolutionary advantage to a host organism in response to selective pressure.

CONCLUSIONS AND PERSPECTIVES

The canonical 20 amino acid genetic code has significantly limited the chemical diversity of proteins throughout evolution. Nature has added additional chemistries to proteins by the use of cofactors and PTMs and, in a small number of cases, pyrrolysine and selenocysteine. Using a relative simple strategy, we and others have been able to augment the natural protein biosynthetic machinery with a large number of ncAAs with novel chemical, physical, and biological properties. These new amino acids have been genetically encoded in both prokaryotic and eukaryotic organisms with excellent fidelity and, in some cases, efficiencies that allow mutant proteins to be produced on a commercial scale. This expanded genetic code has provided powerful new tools to probe protein structure and function both in vitro and in living cells, and is allowing the rational design and evolution of proteins with useful new properties for a variety of applications, including the creation of new medicines. Many opportunities exist to further expand and exploit this technology, including methods for the generation of proteins consisting of a large number (or entirely) of distinct ncAAs, methods to evolve whole proteomes with novel properties rather than individual proteins, the use of computational methods together with ncAAs to generate new function, and the use of this technology to create a new generation of therapeutic proteins whose pharmacology can be optimized with medicinal chemistry like control over chemical structure and homogeneity. Clearly, the synergistic use of tools and principles from both chemistry and biology has made it possible to remove the constraints of nature's 20 amino acid genetic code.

ACKNOWLEDGMENTS

We acknowledge Kristen Williams for her assistance in manuscript preparation. This work is supported by DOE Grant DE-SC0011787 (P.G.S.). This is manuscript number 29056 of The Scripps Research Institute.

REFERENCES

Ai HW. 2012. Biochemical analysis with the expanded genetic lexicon. *Anal Bioanal Chem* **403:** 2089–2102.

Amiram M, Haimovich AD, Fan C, Wang YS, Aerni HR, Ntai I, Moonan DW, Ma NJ, Rovner AJ, Hong SH, et al. 2015. Evolution of translation machinery in recoded bacteria enables multi-site incorporation of nonstandard amino acids. *Nat Biotechnol* **33:** 1272–1279.

Anderson JC, Schultz PG. 2003. Adaptation of an orthogonal archaeal leucyl-tRNA and synthetase pair for four-base, amber, and opal suppression. *Biochemistry* **42:** 9598–9608.

Anderson JC, Wu N, Santoro SW, Lakshman V, King DS, Schultz PG. 2004. An expanded genetic code with a functional quadruplet codon. *Proc Natl Acad Sci* **101:** 7566–7571.

Arbely E, Torres-Kolbus J, Deiters A, Chin JW. 2012. Photocontrol of tyrosine phosphorylation in mammalian cells via genetic encoding of photocaged tyrosine. *J Am Chem Soc* **134:** 11912–11915.

Axup JY, Bajjuri KM, Ritland M, Hutchins BM, Kim CH, Kazane SA, Halder R, Forsyth JS, Santidrian AF, Stafin K, et al. 2012. Synthesis of site-specific antibody-drug conjugates using unnatural amino acids. *Proc Natl Acad Sci* **109:** 16101–16106.

Bianco A, Townsley FM, Greiss S, Lang K, Chin JW. 2012. Expanding the genetic code of *Drosophila melanogaster. Nat Chem Biol* **8:** 748–750.

Bluemel C, Hausmann S, Fluhr P, Sriskandarajah M, Stallcup WB, Baeuerle PA, Kufer P. 2010. Epitope distance to the target cell membrane and antigen size determine the potency of T cell-mediated lysis by BiTE antibodies specific for a large melanoma surface antigen. *Cancer Immunol Immunother* **59:** 1197–1209.

Byrne H, Conroy PJ, Whisstock JC, O'Kennedy RJ. 2013. A tale of two specificities: Bispecific antibodies for therapeutic and diagnostic applications. *Trends Biotechnol* **31:** 621–632.

Cao Y, Axup JY, Ma JS, Wang RE, Choi S, Tardif V, Lim RK, Pugh HM, Lawson BR, Welzel G, et al. 2015. Multiformat T-cell-engaging bispecific antibodies targeting human breast cancers. *Angew Chem Int Ed Engl* **54:** 7022–7027.

Cellitti SE, Jones DH, Lagpacan L, Hao X, Zhang Q, Hu H, Brittain SM, Brinker A, Caldwell J, Bursulaya B, et al. 2008. In vivo incorporation of unnatural amino acids to probe structure, dynamics, and ligand binding in a large protein by nuclear magnetic resonance spectroscopy. *J Am Chem Soc* **130:** 9268–9281.

Charbon G, Brustad E, Scott KA, Wang JY, Lobner-Olesen A, Schultz PG, Jacobs-Wagner C, Chapman E. 2011. Subcellular protein localization by using a genetically encoded fluorescent amino acid. *Chembiochem* **12:** 1818–1821.

Chatterjee A, Xiao H, Schultz PG. 2012. Evolution of multiple, mutually orthogonal prolyl-tRNA synthetase/tRNA pairs for unnatural amino acid mutagenesis in *Escherichia coli. Proc Natl Acad Sci* **109:** 14841–14846.

Chatterjee A, Guo J, Lee HS, Schultz PG. 2013a. A genetically encoded fluorescent probe in mammalian cells. *J Am Chem Soc* **135:** 12540–12543.

Chatterjee A, Sun SB, Furman JL, Xiao H, Schultz PG. 2013b. A versatile platform for single- and multiple-unnatural amino acid mutagenesis in *Escherichia coli. Biochemistry* **52:** 1828–1837.

Chatterjee A, Xiao H, Bollong M, Ai HW, Schultz PG. 2013c. Efficient viral delivery system for unnatural amino acid mutagenesis in mammalian cells. *Proc Natl Acad Sci* **110:** 11803–11808.

Chatterjee A, Xiao H, Yang PY, Soundararajan G, Schultz PG. 2013d. A tryptophanyl–tRNA synthetase/tRNA pair for unnatural amino acid mutagenesis in *E. coli. Angew Chem Int Ed Engl* **52:** 5106–5109.

Chatterjee A, Lajoie MJ, Xiao H, Church GM, Schultz PG. 2014. A bacterial strain with a unique quadruplet codon specifying non-native amino acids. *Chembiochem* **15:** 1782–1786.

Chen HT, Warfield L, Hahn S. 2007. The positions of TFIIF and TFIIE in the RNA polymerase II transcription preinitiation complex. *Nat Struct Mol Biol* **14:** 696–703.

Chen ZJ, Ren W, Wright QE, Ai HW. 2013. Genetically encoded fluorescent probe for the selective detection of peroxynitrite. *J Am Chem Soc* **135:** 14940–14943.

Chen XH, Xiang Z, Hu YS, Lacey VK, Cang H, Wang L. 2014. Genetically encoding an electrophilic amino acid for protein stapling and covalent binding to native receptors. *ACS Chem Biol* **9:** 1956–1961.

Chin JW. 2014. Expanding and reprogramming the genetic code of cells and animals. *Annu Rev Biochem* **83:** 379–408.

Cho H, Daniel T, Buechler YJ, Litzinger DC, Maio Z, Putnam AM, Kraynov VS, Sim BC, Bussell S, Javahishvili T, et al. 2011. Optimized clinical performance of growth hormone with an expanded genetic code. *Proc Natl Acad Sci* **108:** 9060–9065.

Cotruvo JA, Stubbe J. 2011. Class I ribonucleotide reductases: Metallocofactor assembly and repair in vitro and in vivo. *Annu Rev Biochem* **80:** 733–767.

Davis L, Chin JW. 2012. Designer proteins: Applications of genetic code expansion in cell biology. *Nat Rev Mol Cell Biol* **13:** 168–182.

Elsasser SJ, Ernst RJ, Walker OS, Chin JW. 2016. Genetic code expansion in stable cell lines enables encoded chromatin modification. *Nat Methods* **13:** 158–164.

Fleissner MR, Brustad EM, Kalai T, Altenbach C, Cascio D, Peters FB, Hideg K, Peuker S, Schultz PG, Hubbell WL. 2009. Site-directed spin labeling of a genetically encoded unnatural amino acid. *Proc Natl Acad Sci* **106:** 21637–21642.

Furman JL, Kang M, Choi S, Cao Y, Wold ED, Sun SB, Smider VV, Schultz PG, Kim CH. 2014. A genetically encoded aza-Michael acceptor for covalent cross-linking of protein–receptor complexes. *J Am Chem Soc* **136:** 8411–8417.

Gauba V, Grunewald J, Gorney V, Deaton LM, Kang MC, Bursulaya B, Ou WJ, Lerner RA, Schmedt C, Geierstanger BH, et al. 2011. Loss of CD4 T-cell-dependent tolerance to proteins with modified amino acids. *Proc Natl Acad Sci* **108:** 12821–12826.

Gautier A, Nguyen DP, Lusic H, An WA, Deiters A, Chin JW. 2010. Genetically encoded photocontrol of protein localization in mammalian cells. *J Am Chem Soc* **132:** 4086–4088.

Gautier A, Deiters A, Chin JW. 2011. Light-activated kinases enable temporal dissection of signaling networks in living cells. *J Am Chem Soc* **133:** 2124–2127.

Greiss S, Chin JW. 2011. Expanding the genetic code of an animal. *J Am Chem Soc* **133:** 14196–14199.

Groff D, Thielges MC, Cellitti S, Schultz PG, Romesberg FE. 2009. Efforts toward the direct experimental characterization of enzyme microenvironments: Tyrosine100 in dihydrofolate reductase. *Angew Chem Int Ed Engl* **48:** 3478–3481.

Groff D, Wang F, Jockusch S, Turro NJ, Schultz PG. 2010. A new strategy to photoactivate green fluorescent protein. *Angew Chem Int Ed Engl* **49:** 7677–7679.

Gruenewald J, Tsao ML, Perera R, Dong LQ, Niessen F, Wen BG, Kubitz DM, Smider VV, Ruf W, Nasoff M, et al. 2008. Immunochemical termination of self-tolerance. *Proc Natl Acad Sci* **105:** 11276–11280.

Hammerling MJ, Ellefson JW, Boutz DR, Marcotte EM, Ellington AD, Barrick JE. 2014. Bacteriophages use an expanded genetic code on evolutionary paths to higher fitness. *Nat Chem Biol* **10:** 178–180.

He J, Van Treeck B, Nguyen HB, Melancon CE III. 2015. Development of an unnatural amino acid incorporation system in the actinobacterial natural product producer *Streptomyces venezuelae* ATCC 15439. *ACS Synth Biol* **5:** 125–132.

Hemphill J, Chou CJ, Chin JW, Deiters A. 2013. Genetically encoded light-activated transcription for spatiotemporal control of gene expression and gene silencing in mammalian cells. *J Am Chem Soc* **135:** 13433–13439.

Huang Y, Russell WK, Wan W, Pai PJ, Russell DH, Liu W. 2010. A convenient method for genetic incorporation of multiple noncanonical amino acids into one protein in *Escherichia coli. Mol Biosyst* **6:** 683–686.

Ieva R, Bernstein HD. 2009. Interaction of an autotransporter passenger domain with BamA during its translocation across the bacterial outer membrane. *Proc Natl Acad Sci* **106:** 19120–19125.

Jackson JC, Hammill JT, Mehl RA. 2007. Site-specific incorporation of a ^{19}F-amino acid into proteins as an NMR probe for characterizing protein structure and reactivity. *J Am Chem Soc* **129:** 1160–1166.

Johnson DB, Xu J, Shen Z, Takimoto JK, Schultz MD, Schmitz RJ, Xiang Z, Ecker JR, Briggs SP, Wang L. 2011. RF1 knockout allows ribosomal incorporation of unnatural amino acids at multiple sites. *Nat Chem Biol* **7:** 779–786.

Kalstrup T, Blunck R. 2013. Dynamics of internal pore opening in K_V channels probed by a fluorescent unnatural amino acid. *Proc Natl Acad Sci* **110:** 8272–8277.

Kang JY, Kawaguchi D, Coin I, Xiang Z, O'Leary DDM, Slesinger PA, Wang L. 2013. In vivo expression of a light-activatable potassium channel using unnatural amino acids. *Neuron* **80:** 358–370.

Kang M, Light K, Ai HW, Shen W, Kim CH, Chen PR, Lee HS, Solomon EI, Schultz PG. 2014. Evolution of iron(II)-finger peptides by using a bipyridyl amino acid. *Chembiochem* **15:** 822–825.

Kim CH, Axup JY, Dubrovska A, Kazane SA, Hutchins BA, Wold ED, Smider VV, Schultz PG. 2012a. Synthesis of bispecific antibodies using genetically encoded unnatural amino acids. *J Am Chem Soc* **134:** 9918–9921.

Kim CH, Kang M, Kim HJ, Chatterjee A, Schultz PG. 2012b. Site-specific incorporation of ε-*N*-crotonyllysine into histones. *Angew Chem Int Ed Engl* **51:** 7246–7249.

Kim CH, Axup JY, Lawson BR, Yun H, Tardif V, Choi SH, Zhou Q, Dubrovska A, Biroc SL, Marsden R, et al. 2013. Bispecific small molecule-antibody conjugate targeting prostate cancer. *Proc Natl Acad Sci* **110:** 17796–17801.

Kularatne SA, Venkatesh C, Santhapuram HK, Wang K, Vaitilingam B, Henne WA, Low PS. 2010. Synthesis and biological analysis of prostate-specific membrane antigen-targeted anticancer prodrugs. *J Med Chem* **53:** 7767–7777.

Kularatne SA, Deshmukh V, Gymnopoulos M, Biroc SL, Xia J, Srinagesh S, Sun Y, Zou N, Shimazu M, Pinkstaff J, et al. 2013. Recruiting cytotoxic T cells to folate-receptor-positive cancer cells. *Angew Chem Int Ed Engl* **52:** 12101–12104.

Kularatne SA, Deshmukh V, Ma J, Tardif V, Lim RK, Pugh HM, Sun Y, Manibusan A, Sellers AJ, Barnett RS, et al. 2014. A CXCR4-targeted site-specific antibody-drug conjugate. *Angew Chem Int Ed Engl* **53:** 11863–11867.

Lacey VK, Parrish AR, Han SL, Shen ZX, Briggs SP, Ma YG, Wang L. 2011. A fluorescent reporter of the phosphorylation status of the substrate protein STAT3. *Angew Chem Int Ed Engl* **50:** 8692–8696.

Lajoie MJ, Rovner AJ, Goodman DB, Aerni HR, Haimovich AD, Kuznetsov G, Mercer JA, Wang HH, Carr PA, Mosberg JA, et al. 2013. Genomically recoded organisms expand biological functions. *Science* **342:** 357–360.

Lee HS, Guo J, Lemke EA, Dimla RD, Schultz PG. 2009. Genetic incorporation of a small, environmentally sensitive, fluorescent probe into proteins in *Saccharomyces cerevisiae. J Am Chem Soc* **131:** 12921–12923.

Lemke EA, Summerer D, Geierstanger BH, Brittain SM, Schultz PG. 2007. Control of protein phosphorylation with a genetically encoded photocaged amino acid. *Nat Chem Biol* **3:** 769–772.

Li C, Wang GF, Wang Y, Creager-Allen R, Lutz EA, Scronce H, Slade KM, Ruf RA, Mehl RA, Pielak GJ. 2010. Protein ^{19}F NMR in *Escherichia coli. J Am Chem Soc* **132:** 321–327.

Li F, Zhang H, Sun Y, Pan Y, Zhou J, Wang J. 2013. Expanding the genetic code for photoclick chemistry in *E. coli*, mammalian cells, and *A. thaliana. Angew Chem Int Ed Engl* **52:** 9700–9704.

Lin SX, Yan H, Li L, Yang MY, Peng B, Chen S, Li WH, Chen PR. 2013. Site-specific engineering of chemical functionalities on the surface of live hepatitis D virus. *Angew Chem Int Ed Engl* **52:** 13970–13974.

Liu CC, Schultz PG. 2006. Recombinant expression of selectively sulfated proteins in *Escherichia coli. Nat Biotechnol* **24:** 1436–1440.

Liu CC, Schultz PG. 2010. Adding new chemistries to the genetic code. *Annu Rev Biochem* **79:** 413–444.

Liu CC, Brustad E, Liu W, Schultz PG. 2007. Crystal structure of a biosynthetic sulfo-hirudin complexed to thrombin. *J Am Chem Soc* **129:** 10648–10649.

Liu CC, Mack AV, Tsao ML, Mills JH, Lee HS, Choe H, Farzan M, Schultz PG, Smider VV. 2008. Protein evolution with an expanded genetic code. *Proc Natl Acad Sci* **105:** 17688–17693.

Longo J, Yao C, Rios C, Chau NT, Boulmedais F, Hemmerle J, Lavalle P, Schiller SM, Schaaf P, Jierry L. 2015. Reversible biomechano-responsive surface based on green fluorescent protein genetically modified with unnatural amino acids. *Chem Commun (Camb)* **51:** 232–235.

Lu H, Wang D, Kazane S, Javahishvili T, Tian F, Song F, Sellers A, Barnett B, Schultz PG. 2013. Site-specific antibody-polymer conjugates for siRNA delivery. *J Am Chem Soc* **135:** 13885–13891.

Lu H, Zhou Q, Deshmukh V, Phull H, Ma J, Tardif V, Naik RR, Bouvard C, Zhang Y, Choi S, et al. 2014. Targeting human C-type lectin-like molecule-1 (CLL1) with a bispecific antibody for immunotherapy of acute myeloid leukemia. *Angew Chem Int Ed Engl* **53:** 9841–9845.

Luo XZ, Zambaldo C, Liu T, Zhang YH, Xuan WM, Wang C, Reed SA, Yang PY, Wang RSE, Javahishvili T, et al. 2016. Recombinant thiopeptides containing noncanonical amino acids. *Proc Natl Acad Sc* **113:** 3615–3620.

Ma JS, Kim JY, Kazane SA, Choi SH, Yun HY, Kim MS, Rodgers DT, Pugh HM, Singer O, Sun SB, et al. 2016. Versatile strategy for controlling the specificity and activity of engineered T cells. *Proc Natl Acad Sci* **113:** E450–E458.

Mandell DJ, Lajoie MJ, Mee MT, Takeuchi R, Kuznetsov G, Norville JE, Gregg CJ, Stoddard BL, Church GM. 2015. Biocontainment of genetically modified organisms by synthetic protein design. *Nature* **518:** 55–60.

Mills JH, Khare SD, Bolduc JM, Forouhar F, Mulligan VK, Lew S, Seetharaman J, Tong L, Stoddard BL, Baker D. 2013. Computational design of an unnatural amino acid dependent metalloprotein with atomic level accuracy. *J Am Chem Soc* **135:** 13393–13399.

Minnihan EC, Seyedsayamdost MR, Uhlin U, Stubbe J. 2011a. Kinetics of radical intermediate formation and deoxynucleotide production in 3-aminotyrosine-substituted *Escherichia coli* ribonucleotide reductases. *J Am Chem Soc* **133:** 9430–9440.

Minnihan EC, Young DD, Schultz PG, Stubbe J. 2011b. Incorporation of fluorotyrosines into ribonucleotide reductase using an evolved, polyspecific aminoacyl-tRNA synthetase. *J Am Chem Soc* **133:** 15942–15945.

Mohibullah N, Hahn S. 2008. Site-specific cross-linking of TBP in vivo and in vitro reveals a direct functional interaction with the SAGA subunit Spt3. *Genes Dev* **22:** 2994–3006.

Moore PA, Zhang WJ, Rainey GJ, Burke S, Li H, Huang L, Gorlatov S, Veri MC, Aggarwal S, Yang YH, et al. 2011. Application of dual affinity retargeting molecules to achieve optimal redirected T-cell killing of B-cell lymphoma. *Blood* **117:** 4542–4551.

Mori H, Ito K. 2006. Different modes of SecY-SecA interactions revealed by site-directed in vivo photo-cross-linking. *Proc Natl Acad Sci* **103:** 16159–16164.

Mu J, Pinkstaff J, Li Z, Skidmore L, Li N, Myler H, Dallas-Yang Q, Putnam AM, Yao J, Bussell S, et al. 2012. FGF21 analogs of sustained action enabled by orthogonal biosynthesis demonstrate enhanced antidiabetic pharmacology in rodents. *Diabetes* **61:** 505–512.

Murelli RP, Zhang AX, Michel J, Jorgensen WL, Spiegel DA. 2009. Chemical control over immune recognition: A class of antibody-recruiting small molecules that target prostate cancer. *J Am Chem Soc* **131:** 17090–17092.

Neumann H, Peak-Chew SY, Chin JW. 2008. Genetically encoding N^{ε}-acetyllysine in recombinant proteins. *Nat Chem Biol* **4:** 232–234.

Neumann H, Hancock SM, Buning R, Routh A, Chapman L, Somers J, Owen-Hughes T, van Noort J, Rhodes D, Chin JW. 2009. A method for genetically installing site-specific acetylation in recombinant histones defines the effects of H3 K56 acetylation. *Mol Cell* **36:** 153–163.

Neumann H, Slusarczyk AL, Chin JW. 2010a. De novo generation of mutually orthogonal aminoacyl-tRNA synthetase/tRNA pairs. *J Am Chem Soc* **132:** 2142–2144.

Neumann H, Wang K, Davis L, Garcia-Alai M, Chin JW. 2010b. Encoding multiple unnatural amino acids via evolution of a quadruplet-decoding ribosome. *Nature* **464:** 441–444.

Nguyen DP, Garcia Alai MM, Kapadnis PB, Neumann H, Chin JW. 2009. Genetically encoding N^{ε}-methyl-L-lysine in recombinant histones. *J Am Chem Soc* **131:** 14194–14195.

Nguyen DP, Garcia Alai MM, Virdee S, Chin JW. 2010. Genetically directing ε-*N*,*N*-dimethyl-L-lysine in recombinant histones. *Chem Biol* **17:** 1072–1076.

Niu W, Guo J. 2013. Expanding the chemistry of fluorescent protein biosensors through genetic incorporation of unnatural amino acids. *Mol Biosyst* **9:** 2961–2970.

Okuda S, Tokuda H. 2009. Model of mouth-to-mouth transfer of bacterial lipoproteins through inner membrane LolC, periplasmic LolA, and outer membrane LolB. *Proc Natl Acad Sci* **106:** 5877–5882.

Park SH, Wang VS, Radoicic J, De Angelis AA, Berkamp S, Opella SJ. 2014. Paramagnetic relaxation enhancement of membrane proteins by incorporation of the metal chelating unnatural amino acid 2-amino-3-(8-hydroxyquinolin-3-yl)propanoic acid (HQA). *J Biomol NMR* **61:** 185–196.

Pearson AD, Mills JH, Song Y, Nasertorabi F, Han GW, Baker D, Stevens RC, Schultz PG. 2015. Transition states. Trapping a transition state in a computationally designed protein bottle. *Science* **347:** 863–867.

Peeler JC, Woodman BF, Averick S, Miyake-Stoner SJ, Stokes AL, Hess KR, Matyjaszewski K, Mehl RA. 2010. Genetically encoded initiator for polymer growth from proteins. *J Am Chem Soc* **132:** 13575–13577.

Peters FB, Brock A, Wang JY, Schultz PG. 2009. Photocleavage of the polypeptide backbone by 2-nitrophenylalanine. *Chem Biol* **16:** 148–152.

Ren W, Ji A, Ai HW. 2015. Light activation of protein splicing with a photocaged fast intein. *J Am Chem Soc* **137:** 2155–2158.

Rust HL, Subramanian V, West GM, Young DD, Schultz PG, Thompson PR. 2014. Using unnatural amino acid mutagenesis to probe the regulation of PRMT1. *ACS Chem Biol* **9:** 649–655.

Santoro SW, Anderson JC, Lakshman V, Schultz PG. 2003. An archaebacteria-derived glutamyl-tRNA synthetase and tRNA pair for unnatural amino acid mutagenesis of proteins in *Escherichia coli*. *Nucleic Acids Res* **31:** 6700–6709.

Schlieker C, Weibezahn J, Patzelt H, Tessarz P, Strub C, Zeth K, Erbse A, Schneider-Mergener J, Chin JW, Schultz PG, et al. 2004. Substrate recognition by the AAA^{+} chaperone ClpB. *Nat Struct Mol Biol* **11:** 607–615.

Schmidt MJ, Borbas J, Drescher M, Summerer D. 2014. A genetically encoded spin label for electron paramagnetic resonance distance measurements. *J Am Chem Soc* **136:** 1238–1241.

Schultz KC, Supekova L, Ryu Y, Xie J, Perera R, Schultz PG. 2006. A genetically encoded infrared probe. *J Am Chem Soc* **128:** 13984–13985.

Seyedsayamdost MR, Xie J, Chan CTY, Schultz PG, Stubbe J. 2007. Site-specific insertion of 3-aminotyrosine into subunit α2 of *E. coli* ribonucleotide reductase: Direct evidence for involvement of Y_{730} and Y_{731} in radical propagation. *J Am Chem Soc* **129:** 15060–15071.

Smith EE, Linderman BY, Luskin AC, Brewer SH. 2011. Probing local environments with the infrared probe: L-4-nitrophenylalanine. *J Phys Chem B* **115:** 2380–2385.

Speight LC, Muthusamy AK, Goldberg JM, Warner JB, Wissner RF, Willi TS, Woodman BF, Mehl RA, Petersson EJ. 2013. Efficient synthesis and in vivo incorporation of acridon-2-ylalanine, a fluorescent amino acid for lifetime and Förster resonance energy transfer/luminescence resonance energy transfer studies. *J Am Chem Soc* **135:** 18806–18814.

Summerer D, Chen S, Wu N, Deiters A, Chin JW, Schultz PG. 2006. A genetically encoded fluorescent amino acid. *Proc Natl Acad Sci* **103:** 9785–9789.

Sun SB, Schultz PG, Kim CH. 2014. Therapeutic applications of an expanded genetic code. *Chembiochem* **15:** 1721–1729.

Tack DS, Ellefson JW, Thyer R, Wang B, Gollihar J, Forster MT, Ellington AD. 2016. Addicting diverse bacteria to a noncanonical amino acid. *Nat Chem Biol* **12:** 138–140.

Tagami S, Sekine S, Kumarevel T, Hino N, Murayama Y, Kamegamori S, Yamamoto M, Sakamoto K, Yokoyama S. 2010. Crystal structure of bacterial RNA polymerase bound with a transcription inhibitor protein. *Nature* **468:** 978–982.

Taskent-Sezgin H, Chung J, Patsalo V, Miyake-Stoner SJ, Miller AM, Brewer SH, Mehl RA, Green DF, Raleigh DP, Carrico I. 2009. Interpretation of *p*-cyanophenylalanine fluorescence in proteins in terms of solvent exposure and contribution of side-chain quenchers: A combined fluorescence, IR and molecular dynamics study. *Biochemistry* **48:** 9040–9046.

Tian F, Lu Y, Manibusan A, Sellers A, Tran H, Sun Y, Phuong T, Barnett R, Hehli B, Song F, et al. 2014. A general approach to site-specific antibody drug conjugates. *Proc Natl Acad Sci* **111:** 1766–1771.

Wals K, Ovaa H. 2014. Unnatural amino acid incorporation in *E. coli*: Current and future applications in the design of therapeutic proteins. *Front Chem* **2:** 15.

Wan W, Tharp JM, Liu WR. 2014. Pyrrolysyl-tRNA synthetase: An ordinary enzyme but an outstanding genetic code expansion tool. *Biochim Biophys Acta* **1844:** 1059–1070.

Wang L, Xie J, Deniz AA, Schultz PG. 2003. Unnatural amino acid mutagenesis of green fluorescent protein. *J Org Chem* **68:** 174–176.

Wang JY, Xie JM, Schultz PG. 2006a. A genetically encoded fluorescent amino acid. *J Am Chem Soc* **128:** 8738–8739.

Wang L, Xie J, Schultz PG. 2006b. Expanding the genetic code. *Annu Rev Biophys Biomol Struct* **35:** 225–249.

Wang K, Neumann H, Peak-Chew SY, Chin JW. 2007. Evolved orthogonal ribosomes enhance the efficiency of synthetic genetic code expansion. *Nat Biotechnol* **25:** 770–777.

Wang F, Robbins S, Guo JT, Shen WJ, Schultz PG. 2010. Genetic incorporation of unnatural amino acids into proteins in Mycobacterium tuberculosis. *PloS ONE* **5:** e9354.

Wang F, Niu W, Guo J, Schultz PG. 2012. Unnatural amino acid mutagenesis of fluorescent proteins. *Angew Chem Int Ed Engl* **51:** 10132–10135.

Wang NX, Li Y, Niu W, Sun M, Cerny R, Li QS, Guo JT. 2014. Construction of a live-attenuated HIV-1 vaccine through genetic code expansion. *Angew Chem Int Ed Engl* **53:** 4867–4871.

Xiang Z, Ren HY, Hu YS, Coin I, Wei J, Cang H, Wang L. 2014. Adding an unnatural covalent bond to proteins through proximity-enhanced bioreactivity. *Nat Methods* **10:** 885–888.

Xiao H, Chatterjee A, Choi SH, Bajjuri KM, Sinha SC, Schultz PG. 2013. Genetic incorporation of multiple unnatural amino acids into proteins in mammalian cells. *Angew Chem Int Ed Eng* **52:** 14080–14083.

Xiao H, Peters FB, Yang PY, Reed S, Chittuluru JR, Schultz PG. 2014. Genetic incorporation of histidine derivatives using an engineered pyrrolysyl-tRNA synthetase. *ACS Chem Biol* **9:** 1092–1096.

Xiao H, Nasertorabi F, Choi SH, Han GW, Reed SA, Stevens RC, Schultz PG. 2015a. Exploring the potential impact of an expanded genetic code on protein function. *Proc Natl Acad Sci* **112:** 6961–6966.

Xiao H, Xuan W, Shao S, Liu T, Schultz PG. 2015b. Genetic incorporation of ε-*N*-2-hydroxyisobutyryl-lysine into recombinant histones. *ACS Chem Biol* **10:** 1599–1603.

Xie J, Supekova L, Schultz PG. 2007. A genetically encoded metabolically stable analogue of phosphotyrosine in *Escherichia coli*. *ACS Chem Biol* **2:** 474–478.

Ye S, Huber T, Vogel R, Sakmar TP. 2009. FTIR analysis of GPCR activation using azido probes. *Nat Chem Biol* **5:** 397–399.

Yokoyama K, Uhlin U, Stubbe J. 2010. Site-specific incorporation of 3-nitrotyrosine as a probe of pK_a perturbation of redox-active tyrosines in ribonucleotide reductase. *J Am Chem Soc* **132:** 8385–8397.

Young TS, Ahmad I, Yin JA, Schultz PG. 2010. An enhanced system for unnatural amino acid mutagenesis in *E. coli*. *J Mol Biol* **395:** 361–374.

Young TS, Young DD, Ahmad I, Louis JM, Benkovic SJ, Schultz PG. 2011. Evolution of cyclic peptide protease inhibitors. *Proc Natl Acad Sci* **108:** 11052–11056.

Zheng SQ, Zhang G, Li J, Chen PR. 2014. Monitoring endocytic trafficking of anthrax lethal factor by precise and quantitative protein labeling. *Angew Chem Int Ed Engl* **53:** 6449–6453.

Cell-Free Synthetic Biology: Engineering Beyond the Cell

Jessica G. Perez,[1,2,5] Jessica C. Stark,[1,2,5] and Michael C. Jewett[1,2,3,4]

[1]Department of Chemical and Biological Engineering, Northwestern University, Evanston, Illinois 60208-3120
[2]Chemistry of Life Processes Institute, Northwestern University, Evanston, Illinois 60208-3120
[3]Robert H. Lurie Comprehensive Cancer Center, Northwestern University, Chicago, Illinois 60611-3068
[4]Simpson Querrey Institute for BioNanotechnology, Northwestern University, Chicago, Illinois 60611-2875
Correspondence: m-jewett@northwestern.edu

Cell-free protein synthesis (CFPS) technologies have enabled inexpensive and rapid recombinant protein expression. Numerous highly active CFPS platforms are now available and have recently been used for synthetic biology applications. In this review, we focus on the ability of CFPS to expand our understanding of biological systems and its applications in the synthetic biology field. First, we outline a variety of CFPS platforms that provide alternative and complementary methods for expressing proteins from different organisms, compared with in vivo approaches. Next, we review the types of proteins, protein complexes, and protein modifications that have been achieved using CFPS systems. Finally, we introduce recent work on genetic networks in cell-free systems and the use of cell-free systems for rapid prototyping of in vivo networks. Given the flexibility of cell-free systems, CFPS holds promise to be a powerful tool for synthetic biology as well as a protein production technology in years to come.

Cell-free protein synthesis (CFPS) technology was first used more than 50 years ago by Nirenberg and Matthaei to decipher the genetic code (Nirenberg and Matthaei 1961). In the late 1960s and early 1970s, CFPS was used to help elucidate the regulatory mechanisms of the *Escherichia coli* lactose (Chambers and Zubay 1969) and tryptophan (Zalkin et al. 1974) operons. Now, in the last two decades, cell-free protein expression platforms have experienced a surge in development to meet the increasing demand for inexpensive and rapid recombinant protein expression technologies, which has resulted in the development of numerous highly active CFPS platforms (Carlson et al. 2012).

This renewed interest in CFPS technology was motivated by the advantages offered by this methodology for the production of recombinant proteins. In particular, the open reaction environment allows for the addition or removal of substrates for protein synthesis, as well as precise, online reaction monitoring. Furthermore, the CFPS reaction environment can be wholly directed toward and optimized for the production of the protein product of interest. In this way, CFPS platforms separate catalyst

[5]These authors contributed equally to this work.

Cite this article as *Cold Spring Harb Perspect Biol* doi: 10.1101/cshperspect.a023853

synthesis (cell growth) from catalyst usage (protein synthesis), representing a significant departure from cell-based processes that rely on microscopic cellular "reactors." CFPS effectively decouples the cell's objectives (growth and reproduction) from the engineer's objectives (protein overexpression and simple product purification). Overall, the nature of CFPS technology allows for shortened protein synthesis timelines and increased flexibility for the addition or removal of natural or synthetic components compared with in vivo approaches. The versatility of CFPS makes it especially attractive for fundamental discovery and high-throughput screening applications.

The ability to prioritize the engineer's objectives in CFPS has further motivated recent applications of CFPS technology to the exciting and ever-growing field of synthetic biology. For instance, cell-free synthetic biology approaches have enabled development of an in vitro prototyping environment for characterization of synthetic parts or genetic networks (Siegal-Gaskins et al. 2014; Takahashi et al. 2014; Chappell et al. 2015). The open environment and reduced complexity of cell-free systems has also made it possible to develop quantitative models describing cell-free genetic network performance and perform machine learning optimization of CFPS (Caschera et al. 2011; Siegal-Gaskins et al. 2014). Additionally, the absence of cell viability constraints has made CFPS an attractive technology for expanding the possible applications of synthetic biology. Recent advances in cell-free synthetic biology include the incorporation of nonnatural chemistries into biological polymers (Goerke and Swartz 2009; Bundy and Swartz 2010; Albayrak and Swartz 2013a; Hong et al. 2014a, 2015), in vitro assembly of complex biological machines and devices (Matthies et al. 2011), and the development of minimal cells (Shin and Noireaux 2012; Stano and Luisi 2013; Caschera and Noireaux 2014a). Excitingly, cell-free technology has also transitioned beyond the laboratory bench, both to the industrial scale for therapeutic production (Zawada et al. 2011; Yin et al. 2012) and to a low-cost, user-friendly format for diagnostic applications (Pardee et al. 2014).

In this review, we focus on the application of CFPS technology to synthetic biology. More detailed reviews on the development of CFPS technology and the types of proteins produced in cell-free systems have been published recently (Katzen et al. 2005; Carlson et al. 2012; Chong 2014; Harbers 2014; Hong et al. 2014a; Lian et al. 2014; Zemella et al. 2015). Here, we begin by introducing the various CFPS platforms and discuss their technological capabilities. We then outline the types of proteins, protein complexes, and protein modifications that have been achieved using CFPS technologies. Finally, we discuss cutting-edge cell-free synthetic biology applications.

MULTIPLE CELL-FREE PROTEIN SYNTHESIS TECHNOLOGIES ENABLE PRODUCTION OF DIVERSE PROTEINS

The recent technological renaissance has resulted in a variety of highly active CFPS platforms for expression of proteins from diverse organisms. Although *E. coli* and wheat germ extracts have been predominantly used in a high-throughput format, all CFPS platforms have the potential to be used for high-throughput screening of DNA libraries and gene products from diverse organisms for biological discovery and synthetic biology applications.

CFPS systems carry out protein synthesis by harnessing the biological catalysts for translation, protein folding, and energy generation from prokaryotic or eukaryotic cells. When combined with a DNA template, amino acids, an RNA polymerase, an adenosine triphosphate (ATP)-regeneration system, salts, and other buffers or environmental stabilizers (e.g., HEPES), these complex biological catalytic ensembles carry out sustained protein synthesis in vitro (Fig. 1) (Jewett et al. 2008).

There are two categories of CFPS platforms, which represent bottom-up and top-down approaches to in vitro protein synthesis. Protein synthesis using purified recombinant elements (PURE) systems build the protein synthesis ensemble from the bottom up. The PURE approach involves purifying the molecular components required for protein synthesis and sub-

 Cite this article as *Cold Spring Harb Perspect Biol* doi: 10.1101/cshperspect.a023853

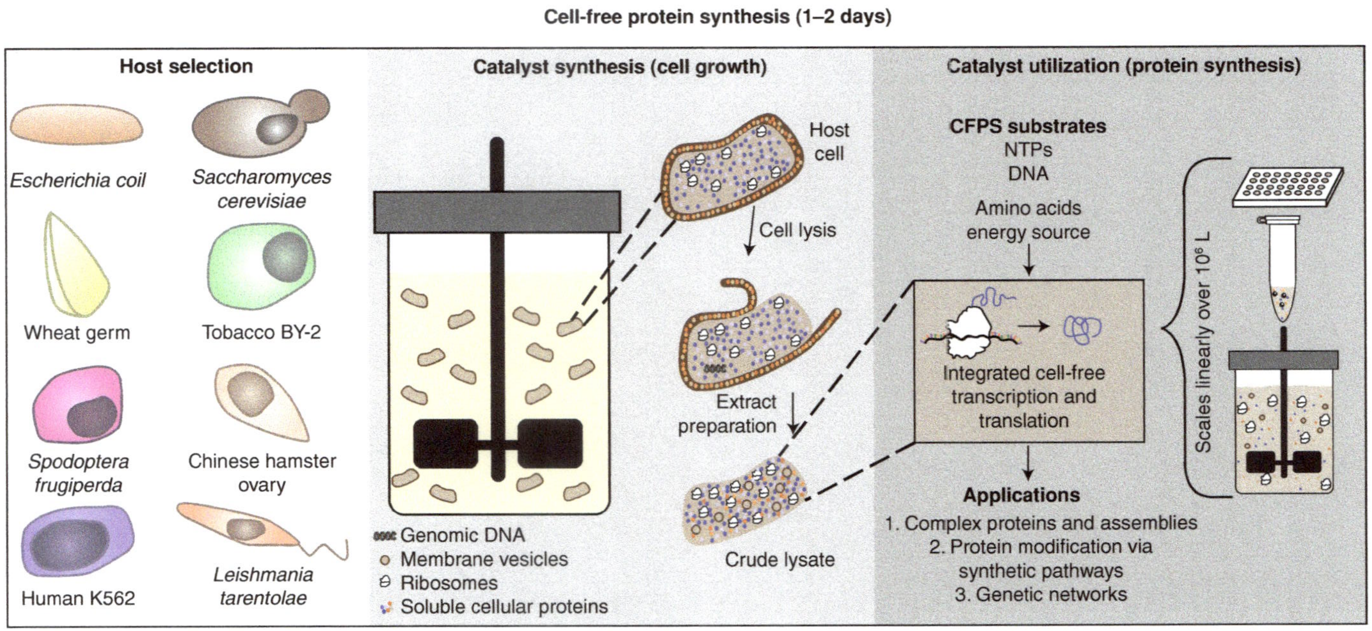

Figure 1. Cell-free protein synthesis (CFPS) platforms allow for increased flexibility and shortened process timelines to create a variety of high-value recombinant proteins. This technology provides platforms from numerous organisms with varied complexity to meet the needs of the specific target proteins. CFPS also decouples catalyst synthesis and catalyst usage, traditionally interconnected during in vivo recombinant protein production schemes. For CFPS, catalyst synthesis involves cell growth, cell lysis, and extract processing that removes genomic DNA to create crude extract. The crude extract may be frozen for future use or used directly for catalyst usage. This process requires NTPs, DNA, amino acids, and an energy source to produce various proteins with applications in the synthetic biology field.

sequently adding them to CFPS reactions (Shimizu et al. 2014). Crude extract systems represent the alternative, top-down approach. Crude cell extract is generated by clarifying whole-cell lysate via centrifugation to remove genomic DNA, insoluble biological elements, and unlysed cells. In some platforms, additional measures are taken to degrade endogenous messenger RNAs (mRNAs) (Pratt 1984). Importantly, the crude extract contains all of the biological components required for translation, protein folding, and energy regeneration (e.g., ribosomes, tRNAs, chaperones, metabolic enzymes, elongation factor-Tu [EF-Tu], translation initiation factors, etc.). Crude lysate CFPS platforms have been developed using cells from a number of organisms. Each crude extract CFPS platform has advantages and disadvantages depending on the desired application (Table 1) (Zemella et al. 2015).

PURE Systems

PURE technology was pioneered by Ueda and colleagues (Shimizu et al. 2001). The PURE approach reconstitutes the transcriptional, translational, and energy-generation machinery from *E. coli* by purifying recombinantly expressed histidine (His)-tagged components and adding them to the CFPS reaction mixture. *E. coli* ribosomes are purified using sucrose-density gradient centrifugation (Shimizu et al. 2014).

PURE CFPS systems are commercially available and widely used for both fundamental discovery and synthetic biology applications. Protein synthesis yields from the PURE system have been optimized recently (Jackson et al. 2014a) and specialized PURE systems have been developed for continuous substrate replenishment (Jackson et al. 2014b, 2015), and to produce membrane proteins (Niwa et al. 2015), as well as disulfide bond- and nonstandard amino acid (nsAA)-containing proteins (Shimizu et al. 2005; Murakami et al. 2006). PURE technology has proven particularly useful for the isolated study of recombinant proteins and protein complexes (Kuruma et al. 2010, 2012; Matsubayashi et al. 2014) because of the simplicity of the system. Furthermore, PURE technology has played a pivotal role in the development of minimal cells (Nishimura et al. 2012, 2014; Stano and Luisi 2013; Caschera and Noireaux 2014a; Matsubayashi and Ueda 2014). However, the expense and cross-reactivity of the His-tag-based component purification process limits its utility in some cases.

Microbial Crude Extract Systems

E. coli and *Saccharomyces cerevisiae* are attractive hosts for CFPS and cell-free synthetic biology because (1) they are easily fermentable, and (2) they are model organisms. As model organisms, there exists a wealth of experimental tools and genetic information for both *E. coli* and *S. cerevisiae* to aid synthetic biology efforts.

The prokaryotic *E. coli* crude extract (ECE) system is one of the most widely adopted platforms for CFPS. This technology is commercially available and is used at the industrial scale (Zawada et al. 2011; Yin et al. 2012; Groff et al. 2014). ECE has been widely adopted for two main reasons: its high batch yields, with up to 2.3 g/L of green fluorescent protein (GFP) reported (Caschera and Noireaux 2014b), and the fast, scalable, and cost-effective extract preparation process (Swartz 2006). In addition, simple and rapid methods for extract preparation have been developed (Shrestha et al. 2012; Kwon and Jewett 2015). Furthermore, ECE reactions can use nucleoside monophosphates and inexpensive energy substrates, such as glucose or starch to regenerate ATP (Calhoun and Swartz 2005a,b; Lian et al. 2014). ECE has been used to synthesize recombinant bacterial and eukaryotic proteins containing posttranslational modifications (PTMs) (Guarino and DeLisa 2012) and nsAAs (Goerke and Swartz 2009; Bundy and Swartz 2010; Ozawa et al. 2012; Albayrak and Swartz 2013a,b; Hong et al. 2014b, 2015). Furthermore, ECE has been adapted to a high-throughput format for production of antibody variants for drug development and screening studies (Stafford et al. 2014), demonstrating its use for screening or functional genomic studies. Finally, ECE has been shown to scale linearly over a 10^6 L range in reaction volumes

 Cite this article as *Cold Spring Harb Perspect Biol* doi: 10.1101/cshperspect.a023853

Table 1. Advantages and disadvantages of existing CFPS technologies

Platform	Advantages	Disadvantages	Representative yields (μg/mL)	Applications
PURE	1. No nucleases or proteases remaining after component purification 2. Flexible and modular 3. Commercially available	1. Expensive 2. Cannot activate endogenous metabolism 3. His purification	GFP: 380[a] β-galactosidase: 4400[b] (Kazuta et al. 2014)	Minimal cells (Matsubayashi and Ueda 2014) Complex proteins (Kuruma et al. 2012) nsAAs (Murakami et al. 2006)
Escherichia coli extract (ECE)	1. High-batch yields 2. Low-cost extract preparation and reagents 3. Commercially available 4. Genetic tools available for strain modifications or synthetic biology 5. CFPS scales linearly $>10^6$ L (Zawada et al. 2011)	1. Limited PTMs	GFP: 2300[a] (Caschera and Noireaux 2014b) GM-CSF: 700[a] (Zawada et al. 2011) VLP: 356[a] (Bundy et al. 2008)	High-throughput format (Stafford et al. 2014) Antibodies (Groff et al. 2014) Vaccines (Ng et al. 2012) Diagnostics (Pardee et al. 2014) nsAAs (Goerke and Swartz 2009; Ozawa et al. 2012) Genetic circuits (Noireaux et al. 2003; Chappell et al. 2013)
Saccharomyces cerevisiae extract (SCE)	1. Simple, low-cost extract preparation 2. Cotranslational folding mechanisms present 3. Genetic tools available for strain modifications or synthetic biology	1. No PTMs shown 2. Low batch yields	Luc: 8.9[a] (Choudhury et al. 2014) GFP: 17[b] (Schoborg et al. 2014) HPV L1 protein: 60[a] (Wang et al. 2008)	Complex proteins (unpubl.)
Wheat germ extract (WGE)	1. High yields 2. Proven for synthesis of many types of eukaryotic proteins, including membrane proteins 3. Continuous exchange reactions last ≤60 hours	1. Extract preparation is long and labor-intensive 2. Difficult technology transfer	GFP: 1600[a] (Harbers 2014), 9700[b] (Endo and Sawasaki 2006)	High-throughput format (Goshima et al. 2008) Vaccines (Tsuboi et al. 2008) Expressed proteomes (Nozawa and Tozawa 2014) Structural characterization (Bernhard and Tozawa 2013)
Tobacco BY-2 cell extract (BYE)	1. Greatly simplified extract preparation compared with WGE 2. Continuous exchange reactions last ≤18 hours	1. Lower yields than WGE 2. No PTMs shown	Luc: 270[a] (Buntru et al. 2014b) YFP: 400[b] Human mAb: 150[a] Hb-EGF: 25[a] (Buntru et al. 2014a)	Complex proteins (Buntru et al. 2014a)

Continued

Table 1. *Continued*

Platform	Advantages	Disadvantages	Representative yields (μg/mL)	Applications
Insect cell extract (ICE)	1. PTMs possible, including glycosylation 2. Proven for synthesis of membrane proteins 3. Commercially available	1. Low batch yields 2. Requires more extract (50% CFPS reaction volume vs. 25% for ECE)	YFP: 10[a] (Katzen and Kudlicki 2006), 50[b] (Stech et al. 2014)	Complex proteins (Katzen and Kudlicki 2006; Suzuki et al. 2007; Orth et al. 2011; Stech et al. 2012; Ezure et al. 2014; Fenz et al. 2014)
CHO, K562, and HeLa extracts	1. PTMs possible, including glycosylation 2. Proven for synthesis of membrane proteins 3. Mammalian cells are preferred expression platform for human proteins	1. Low batch yields 2. Requires more extract (50% CFPS reaction volume vs. 25% for ECE)	Luc: 50[a] (CHO), 20[a] (K562) (Brodel et al. 2015) Glutathione-*S*-transferase: 50[b] (HeLa) (Mikami et al. 2006)	Complex proteins (Mikami et al. 2006; Stech et al. 2013; Brodel et al. 2015)
Leishmania tarentolae extract (LTE)	1. Showed advantages for expressing parasitic proteins 2. Background translation mostly repressed	1. No PTMs shown 2. Limited number and types of protein synthesized	GFP: 300[a] (Mureev et al. 2009)	Complex proteins (Mureev et al. 2009) Expressed proteomes (Kovtun et al. 2010)

Hb-EGF, Heparin-binding EGF-like growth factor; EPO, erythropoietin; GFP, green fluorescent protein; Luc, luciferase; mAb, monoclonal antibody; YFP, yellow fluorescent protein; PTM, posttranslational modification; nsAAs, nonstandard amino acids; CFPS, cell-free protein synthesis.

[a]Batch reaction.

[b]Continuous exchange reaction.

(Zawada et al. 2011), which helped motivate the adoption of this technology for large-scale protein synthesis (Yin et al. 2012; Groff et al. 2014; Zimmerman et al. 2014). Although bacterial CFPS systems are able to produce proteins that would be difficult or impossible in cells, further reductions in cost would enable wider adoption for industrial biomanufacturing.

A second microbial CFPS system uses extract from *S. cerevisiae* (baker's yeast) to catalyze protein synthesis. *S. cerevisiae* cells contain eukaryotic folding machinery that can be beneficial for recombinant production of complex eukaryotic proteins. The first *S. cerevisiae* extract (SCE) system was pioneered by Iizuka et al. (1994) more than 20 years ago. Recently, the Jewett laboratory developed and optimized a scalable, low-cost SCE preparation method (Hodgman and Jewett 2013; Choudhury et al. 2014). Many efforts have been made to improve the efficiency and reduce the cost of SCE CFPS (Gan and Jewett 2014; Hodgman and Jewett 2014; Schoborg et al. 2014; Anderson et al. 2015). Together, these advances have resulted in a relatively low-cost eukaryotic CFPS technology with protein yields comparable to other eukaryotic platforms (Table 1). However, further improvements in protein synthesis yields will be necessary to encourage adoption of yeast CFPS for synthetic biology applications.

Plant, Animal, and Insect Crude Extract Systems

Crude extract CFPS technologies derived from higher eukaryotes offer advantages for complex

 Cite this article as *Cold Spring Harb Perspect Biol* doi: 10.1101/cshperspect.a023853

protein production. Importantly, these extracts enable complex cotranslational folding mechanisms and the addition of PTMs not currently possible using microbial platforms. However, batch yields from these systems are typically at least an order of magnitude lower, whereas the system cost is much higher, compared with ECE (Table 1). Additionally, applications of these technologies to synthetic biology are currently limited. One challenge preventing high-yielding eukaryotic CFPS is the difficulty of enabling endogenous translation initiation mechanisms (e.g., 5′ capping) in vitro.

Two main plant-based crude extract CFPS systems have been developed to date are wheat germ extract (WGE) and tobacco BY-2 cell extract (BYE). The WGE platform, which is derived from wheat-seed embryos (Madin et al. 2000), has achieved both high yields (Endo and Sawasaki 2006) and high-throughput expression (Sawasaki et al. 2002) of recombinant proteins. Typically, this system is used with continuous substrate replenishment (continuous exchange) (Jackson et al. 2014b, 2015). For example, Goshima et al. (2008) used WGE to synthesize ~13,000 human proteins in a single study. WGE was also used to synthesize bioactive proteins of diverse species origin, including *Arabidopsis* (Nozawa et al. 2009) and the malaria parasite genus *Plasmodium* (Tsuboi et al. 2008), without codon optimization in most cases. WGE is the highest-yielding eukaryotic CFPS platform, with yields of up to 1.6 mg/mL GFP in batch (Harbers 2014) and 9.7 mg/mL GFP with continuous exchange reported (Endo and Sawasaki 2006). Notably, the active lifetime of WGE is remarkably long—up to 60 h in continuous exchange reactions that replenish substrates (Endo and Sawasaki 2006)—because most nucleases and proteases are removed during extract preparation. However, the primary drawback of WGE is the extensive extract preparation process, which takes 4–5 days (Madin et al. 2000).

The BYE CFPS system was recently developed as an alternative plant-based CFPS platform. Preparation of BYE is greatly simplified compared with WGE and can be completed in just 4–5 h. Like WGE, BYE has also been used to synthesize proteins from diverse organisms, including a transmembrane growth factor, a glycosylated *Apergillus niger* enzyme, and a bioactive human antibody (Buntru et al. 2014b). However, unlike WGE, mg/mL yields from BYE have not yet been achieved (Table 1).

The insect cell extract (ICE) CFPS platform was developed using extract from *Spodoptera frugiperda* (fall army worm) cells and is commercially available. The first ICE system was developed by Ezure et al. (2006). A second ICE system, developed by Katzen and Kudlicki (2006), produced extracts containing endoplasmic reticulum (ER) vesicles that retained glycosylation and signal sequence processing activity. ICE has since been used to produce glycoproteins and membrane proteins (Katzen and Kudlicki 2006; Kubick et al. 2009; Sachse et al. 2014; Stech et al. 2014).

More recently, CFPS platforms were developed from mammalian cell lines. Mammalian cells are currently the preferred platform for large-scale in vivo production of human proteins bearing PTMs. The Kubick laboratory has described CFPS technologies using extracts from Chinese hamster ovary (CHO) cells and human K562 cells, a myelogenous leukemia cell line (Brodel et al. 2015). In addition, Mikami et al. (2006) developed a CFPS platform using lysate from HeLa cells, a cervical cancer cell line. Mammalian crude extract CPFS platforms provide a potentially attractive alternative for the synthesis and high-throughput screening of recombinant human proteins.

Parasitic Organism Crude Extract Systems

The *Leishmania tarentolae* extract (LTE) platform, developed by Mureev and colleagues, is the only platform that uses extracts from a parasitic organism. Parasites are attractive candidates for functional proteomic studies because, in some organisms, >50% of proteins have no identifiable homology or predictable function. However, in vivo expression of genes from parasitic or infectious organisms can be challenging because of differences in codon usage and protein synthesis machinery (Mureev et al. 2009). Parasitic organism crude extract systems are potentially advantageous for generating expressed parasite proteomes. In one instance, a parasitic crude extract CFPS system outperformed ECE in bioactive yields of proteins from *Plasmodium falciparum*, a malaria parasite (Kovtun et al. 2010).

L. tarentolae, a single-celled lizard parasite, was chosen for CFPS technology development for two main reasons: (1) it is easily fermentable, and (2) all endogenous mRNAs share the same leader sequence. The latter feature enables complete repression of background translation via inclusion of an antileader oligonucleotide (Mureev et al. 2009). LTE offers a rapid and scalable extract preparation process (Kovtun et al. 2011; Johnston and Alexandrov 2014), and is one of the highest yielding eukaryotic systems in batch; yields of up to 300 µg/mL GFP have been reported (Mureev et al. 2009). Importantly, Mureev and colleagues also developed species-independent translation sequences (SITS) to bypass translational regulation in LTE. The inclusion of SITS allowed cell-free synthesis of GFP in prokaryotic and eukaryotic extracts, including LTE, ECE, WGE, SCE, and ICE, using the same DNA template (Mureev et al. 2009; Kovtun et al. 2011; Johnston and Alexandrov 2014). LTE has been used to synthesize mammalian, *L. tarentolae*, and *P. falciparum* proteins.

Despite recent technological advances, there remain challenges facing CFPS technology that need to be addressed. Yields of bioactive human proteins from CFPS are sometimes lower than those achievable in vivo. In particular, no eukaryotic CFPS platform is currently capable of producing mg/mL quantities of protein in a batch. Increasing eukaryotic CFPS batch yields to 0.5 mg/mL or greater is an important challenge for two reasons: (1) These platforms have the potential to achieve higher soluble, bioactive yields of recombinant human proteins than bacterial platforms, and (2) eukaryotic CFPS systems could be useful as prototyping tools for genetic circuit characterization or production of recombinant proteins in eukaryotic cells. Furthermore, most crude extract CFPS platforms are derived from model organisms, and only the *L. tarentolae* platform is derived from a parasitic organism. Crude extract platforms from nonmodel organisms could have usage for functional genomics studies. Finally, development and optimization of species-independent translation initiation strategies, such as SITS, or novel cap-independent translation initiation sequences (Gan and Jewett 2016) could further increase protein yields and enable expression of proteins from diverse organisms. Addressing these challenges will be important to encourage broad adoption of CFPS technology for fundamental discovery and synthetic biology applications.

COMPLEX PROTEINS AND PROTEIN ASSEMBLIES

Cell-Free Production of Complex Proteins

Most CFPS platforms have been applied to the production of complex proteins. CFPS systems offer distinct advantages over in vivo protein production for applications that require more precise control over the protein synthesis reaction conditions. Such applications include bispecific antibodies, antibody–drug conjugates (ADCs), vaccines, and membrane proteins.

Therapeutics and Vaccines

ECE has been used extensively for production of bioactive recombinant therapeutic proteins, including granulocyte macrophage colony-stimulating factor (GM-CSF), onconase, antibodies, bispecific antibodies, and antibody–drug conjugates (Zawada et al. 2011; Yin et al. 2012; Groff et al. 2014; Stafford et al. 2014; Xu et al. 2014; Zimmerman et al. 2014; Cai et al. 2015; Salehi et al. 2016). Recently, a method for in vitro display of antibody fragments was developed in ECE, providing an exciting new technology for high-throughput production and screening of antibody candidates (Stafford et al. 2014; Cai et al. 2015). Excitingly, recent work has resulted in the development of microfluidic systems to enable CFPS reactions to be performed at the point-of-care (Sullivan et al. 2016; Timm et al. 2016). These systems would enable rapid production of medicines on demand and opens the door to bedside synthesis of personalized protein therapeutics.

 Cite this article as *Cold Spring Harb Perspect Biol* doi: 10.1101/cshperspect.a023853

The eukaryotic WGE, ICE, CHO extract, and K562 extract systems have also been used to produce therapeutic proteins. WGE and ICE have been used to synthesize bioactive antibody fragments, and ICE has been used to produce bioactive tissue-type plasminogen activator (Kawasaki et al. 2003; Stech et al. 2012, 2014). Furthermore, ICE, CHO extract, and K562 extract have been used to synthesize glycosylated erythropoietin (Stech et al. 2013; Brodel et al. 2015).

An especially promising application for CFPS systems is the high-throughput screening and in vitro production of vaccine antigens and virus-like particle (VLP) antigens. The WGE platform was applied to the synthesis of 124 *P. falciparum* genes of potential interest as vaccine antigens, and 75% were successfully synthesized without codon optimization (Tsuboi et al. 2008). ECE has also been used to synthesize vaccines and VLPs, including a B-cell lymphoma vaccine (Ng et al. 2012), anti-influenza VLPs (Lu et al. 2014), and anti-hepatitis B VLPs (Bundy et al. 2008), as well as a virus-like nanoparticle vaccine scaffold (Lu et al. 2015). Notably, the VLPs were produced at significantly higher yields than those possible in vivo, with yields improved up to 15-fold (Bundy et al. 2008). These studies show the potential of CFPS systems as novel vaccine discovery tools and platforms for large-scale vaccine production. Looking forward, because CFPS systems are abiotic, they are potentially well suited for portable and on-demand production of vaccines in resource-poor areas. This is an exciting future application area for CFPS technology.

Membrane Proteins

CFPS systems offer advantages for the production of membrane proteins. Newly synthesized membrane proteins can be stabilized in vitro by including membrane mimics (e.g., surfactants, liposomes, nanodiscs) to the cell-free reaction either posttranslationally, to solubilize proteins, or cotranslationally, to prevent aggregation (Junge et al. 2011; Sachse et al. 2014; Zieleniecki et al. 2016). A recent study even showed that supplying an oil–water interface by encapsulating CFPS reactions in emulsions was sufficient to synthesize functional single-span membrane proteins (Yunker et al. 2016). The ECE, WGE, ICE, CHO extract, and K562 extract systems as well as the PURE system have all been used to synthesize membrane proteins including ATP synthase (Matthies et al. 2011), G-protein-coupled receptors (Kaiser et al. 2008; Wang et al. 2011), membrane proteins from hepatitis C virus (Fogeron et al. 2015), and epidermal growth factor receptor (Stech et al. 2013). The synthesis of membrane proteins is an excellent application for CFPS technology, and has been useful for the structural biology community (Bernhard and Tozawa 2013). Overall, enhanced control over the protein synthesis reaction in vitro enables significantly higher yields of soluble, active protein compared with in vivo expression.

Taken together, these studies show the ability of CFPS technology to produce a variety of recombinant proteins with diverse complexity. ECE and WGE in particular show the potential of CFPS technologies for high-throughput production of many different kinds of proteins of interest for therapeutic development, fundamental discovery, or use as synthetic biological parts.

Macromolecule Production and Assembly in Cell-Free Systems

Much progress has been made toward in vivo synthesis of protein assemblies (Selkrig et al. 2014; Du et al. 2015); however, cellular viability constrains the complexity of the assemblies that can be overproduced. Moreover, cellular complexity may obscure fundamental properties of intricate assembly processes. Recently, CFPS systems have been extended for the production of complex protein assemblies. This advance offers interesting opportunities for studying macromolecule self-assembly and developing synthetic biological devices. CFPS platforms circumvent in vivo macromolecule production and assembly limitations in three key ways. First, the cell-free environment accelerates protein engineering efforts by enabling high-throughput experimentation and simplifying the purification of individual subunits. Second, CFPS platforms are flexible, ranging from crude extract to fully purified transcription–translation systems making it possible to study assembly processes in a controlled reaction environment. Third, CFPS platforms provide an additional level of complexity by coupling encoded protein expression to assembly of protein ensembles, which may elucidate the role of genetic regulation on subunit stoichiometries and assembly mechanisms (Fig. 2A).

Synthesis and Assembly of Naturally Occurring Macromolecules

Early demonstrations of protein assembly in CFPS systems produced macromolecules such as a five-subunit *E. coli* RNA polymerase (Asa-

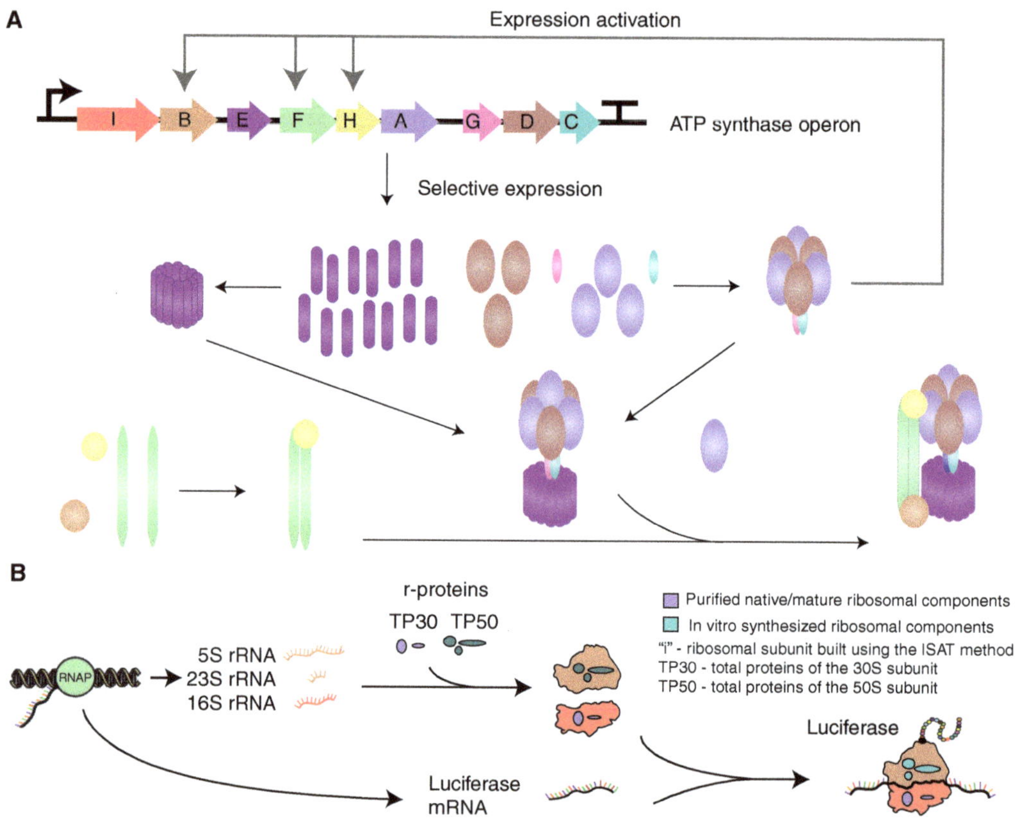

Figure 2. Assembly of macromolecules in cell-free protein synthesis (CFPS) reactions. (*A*) Complete in vitro assembly of adenosine triphosphate (ATP) ATP synthase with hypothesized genetic regulation. The ATP operon is added to a crude *Escherichia coli* extract and transcribed into a single messenger RNA (mRNA). Proteins are expressed at various levels determined by operon regulation. Matthies et al. (2011) suggest synthesis of correctly assembled ATP synthase complexes is dependent on specific expression levels of the subunits, correlated to the subunit stoichiometry in the complex. It is hypothesized that intermediate assemblies may also activate the expression of other subunits in the operon, allowing for sequential assembly processes (Kucharczyk et al. 2009). Thus, combined expression and assembly in CFPS systems allows for an additional level of assembly complexity for analysis. (*B*) In vitro integrated synthesis, assembly, and translation (iSAT) method of constructing ribosomes enables synthesis of active firefly luciferase in a one-pot reaction. iSAT enables one-step coactivation of ribosomal RNA (rRNA) transcription, assembly of ribosomal subunits, and synthesis of active protein by these ribosomes in same compartment. This process begins with T7 RNAP polymerase transcribing rRNA and luciferase mRNA. Ribosomal subunits are reconstituted from mature rRNA and ribosomal components previously purified or synthesized in vitro. Newly assembled ribosomes translate mRNA encoding the reporter protein luciferase to assess its activity.

hara and Chong 2010) and a two subunit hepatitis B core antigen virus-like particle (Bundy et al. 2008). Even these simple assemblies have many applications. For instance, VLPs have been produced in cell-free systems at higher yields than those possible in vivo. This advance can be further applied to the discovery of novel vaccines, and provides straightforward production scale-up processes (Bundy et al. 2008).

As protein production capacity and reaction lifetime increased in CFPS systems, more proteins could be synthesized at once (Zemella et al. 2015). This allowed for the construction of complex structures, such as the pioneering work of Matthies et al. (2011) that performed the synthesis and assembly of *Caldalkalibacillus thermarum* ATP synthase, a 542-kDa membrane protein consisting of 25 individual proteins arranged into eight subunits. Cell-free production and assembly of ATP synthase was made possible by supplementing detergents to crude *E. coli* extract (Fig. 2A). This demonstration opens the way for the construction of novel energy generating biodevices that can be rapidly prototyped in vitro. Multistep coordination of protein ensembles has also been successfully shown in vitro. For instance, Fujiwara et al. (2013) simulated chromosomal DNA replication in a cell-free system by expressing 13 essential genes. In another study, Shin et al. (2012) showed complete in vitro DNA replication, synthesis, and assembly of bacteriophage T7 particles from expression of 60 genes. This is remarkable considering only 35 of the ~60 proteins encoded by the T7 genome have known function (Chan et al. 2005). Full synthesis and assembly of other phages and VLPs have since been shown at higher yields (Garamella et al. 2016) and showcase the use of CFPS in screening efforts (Lu et al. 2015). Overall, this work reaffirms the value of CFPS platforms for interrogating native self-assembly processes and synthesizing self-replicating biological machines in vitro.

Beyond protein machines, the ability to assemble macromolecules with RNA and protein has also significantly advanced. For instance, the *E. coli* ribosome, a large protein/RNA assembly, has been assembled in vitro using a prokaryotic CFPS system (Fig. 2B) (Jewett et al. 2013). By adding DNA encoding the ribosomal RNA (rRNA) and purified ribosomal proteins to an *E. coli* extract lacking native ribosomes, in vitro assembled ribosomes are capable of translation under physiological conditions. Tuning transcription (Fritz and Jewett 2014) and using a continuous-exchange platform to replenish substrates and remove toxic byproducts (Liu et al. 2014) has resulted in superfolder GFP (sfGFP) yields of up to 7.5 μmol/L. This new ribosome construction platform, termed iSAT for integrated synthesis, assembly, and translation of ribosomes, makes possible new ways to probe, dissect, and understand ribosome biogenesis. In addition, it contributes meaningfully toward efforts to build minimal cells and construct engineered ribosomes with novel and useful properties.

Assembly and Evolution of Biological Devices (Biodevices)

To enable in vitro engineering of protein assemblies to create biodevices, cell-free systems can be combined with detection hardware. This is crucial because high molecular structure resolution is needed to follow assembly processes at the nanometer length scale. Interestingly, Heyman and colleagues developed a device that simultaneously synthesizes, assembles, and images nanotubes. This is achieved by patterning anti-hemagglutinin (HA) antibodies onto a TEM grid, trapping CFPS-synthesized nanotube precursor proteins tagged with HA (Heyman et al. 2012). This showed that coupling synthesis to assembly in the presence of an anchoring site could enable on-chip visualization of protein assembly. Microfluidic affinity assays have also been paired with CFPS platforms, enabling 14,792 simultaneous, on-chip experiments to exhaustively measure protein–protein interactions of 43 *Streptococcus pneumoniae* proteins in quadruplicate (Gerber et al. 2009). This study found several physical interactions between proteins that had been previously unreported. In sum, cell-free systems provide a flexible environment for expanding understanding of native assembly processes and the repertoire of synthetic protein assemblies.

SYNTHETIC PATHWAYS IN CFPS SYSTEMS ENABLE SITE-SPECIFIC MODIFICATIONS OF PROTEINS FOR NOVEL FUNCTIONALITIES

The development of highly active CFPS platforms has enabled the adaptation of CFPS technology toward synthesizing proteins with synthetic modifications and novel functionalities. Synthetic pathways have been added to confer the ability of cotranslational and posttranslational modification of recombinant proteins. The ability to site-specifically modify proteins could have many uses in fundamental and applied science. For example, the cotranslational incorporation of acetyllysine (AcK) in vivo has produced recombinant histone proteins that revealed new understanding of epigenetic regulation (Neumann et al. 2009). Site-specific modification will also have usage in the development of therapeutics bearing (a)biological modifications (i.e., phosphorylation, glycosylation, PEGylation, drug conjugates).

Cell-free production of modified proteins complements in vivo efforts and offers many advantages. In particular, modulation of substrate concentrations permits direct monitoring of their effects on protein modifications. This is especially important because synthetic protein modification pathways involve multiple proteins that can show cross talk with the native system. Additionally, an open and abiotic environment allows for the addition of substrates and production of byproducts that cannot be used in vivo because of cell-membrane permeability or cellular toxicity limitations. CFPS platforms offer greater control over the modification process and expand the types of chemistries that can be added to proteins.

Cotranslational Incorporation of Nonstandard Amino Acids

The cotranslational incorporation of nsAAs into proteins expands the chemistry of life. Such an expansion has enabled the incorporation of (a)biological groups into proteins such as biophysical probes (Cornish et al. 1994) (spin-labeled, fluorescent molecules, and photoactivatable cross linkers), redox active groups (Alfonta et al. 2003), and natural PTMs (Neumann et al. 2009; Nguyen et al. 2009; Virdee et al. 2011; Oza et al. 2015). To date, more than 100 nsAAs have been cotranslationally incorporated into proteins (O'Donoghue et al. 2013). This has been possible with the introduction of an orthogonal translation system (OTS), which is capable of genetically encoding nsAAs. Traditionally, OTSs consist of an engineered, orthogonal tRNA/aminoacyl tRNA synthetase (o-tRNA/o-aaRS) pair derived from a phylogenetically distant organism, often *Methanocaldococcus jannaschii*. The o-tRNA has a modified anticodon specific to the UAG stop codon (Fig. 3), whereas the o-aaRS is evolved to bind and aminoacylate the nsAA of interest to the o-tRNA. This process, which was initially pioneered by Schultz and colleagues (Wang et al. 2006), is called amber suppression as it allows the nsAA to be encoded at the UAG (amber) stop codon. An additional, engineered EF-Tu may be required when the nsAA of interest is too bulky to fit into the native EF-Tu amino acid binding pocket or its charge causes inefficient binding (Park et al. 2011). Although it is worth noting that both the WGE (Kiga et al. 2002) and the ICE (Ezure et al. 2014) systems have been used to incorporate nsAAs, for the purposes of this review, we will focus on the extensive work that has been performed in the ECE CFPS platform.

The Swartz laboratory has made many contributions to the development of CFPS platforms for nsAA incorporation, especially in the area of substrate optimization (Goerke and Swartz 2009; Bundy and Swartz 2010). Albayrak and Swartz produced ~1.5 mg/mL of sfGFP protein containing *p*-azido-phenylalnine (pAzF; azide, "click" chemistry, photocrosslinker) (Chin et al. 2002) and *p*-acetylphenylalanine (*p*AcF; keto, orthogonal reactivity) (Lemke 2011) with a 50%–88% amber suppression efficiency (Albayrak and Swartz 2013a). This success was possible by optimizing the concentration of o-tRNA, which was identified as a limiting factor of nsAA incorporation in CFPS reactions (Albayrak and Swartz 2013b). To develop a practical, cost-effective method for supplying more o-tRNA to the reaction, the

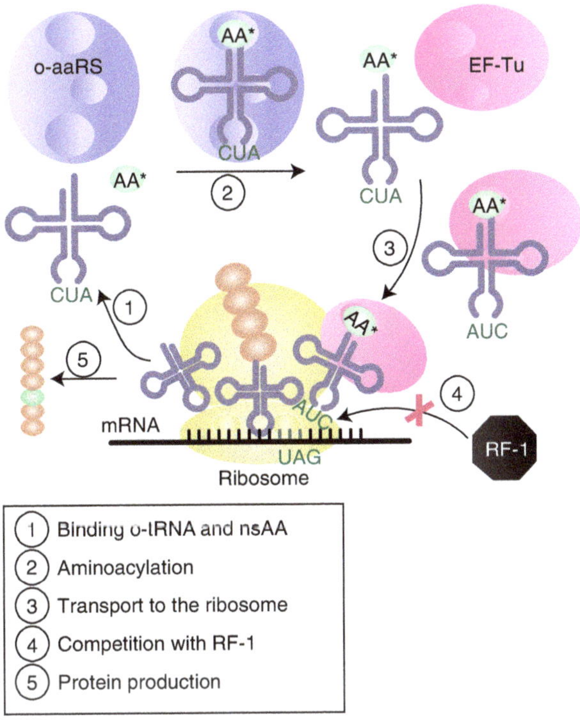

Figure 3. Schematic representation of cotranslational incorporation of a nonstandard amino acid (nsAA) using an orthogonal translation system and amber suppression. The orthogonal aminoacyl-tRNA synthetase (o-aaRS) first binds its cognate nsAA and cognate o-tRNA. The o-aaRS then catalyzes the aminoacylation of the o-tRNA. The aminoacyl-tRNA (aa-tRNA) is then released from the o-aaRS and transported to the ribosome by the elongation factor-Tu (EF-Tu). The aa-tRNA associates with the A-site of the ribosome and its anticodon binds the complementary triplet codon of the messenger RNA (mRNA). The ribosome then ligates the nsAA to the growing peptide chain. When release factor 1 (RF-1) outcompetes the aa-tRNA for binding at the UAG amber stop codon, the protein is truncated, which results in a decrease of nsAA incorporation efficiency. This problem has been overcome by recoding all TAG codons to the synonymous TAA codon, permitting the deletion of RF-1 (Johnson et al. 2011; Loscha et al. 2012b; Ohtake et al. 2012; Lajoie et al. 2013).

o-tRNA was coexpressed in the CFPS reaction with the modified protein. The same method was validated by Hong et al. (2014b) and proved to reduce cost and setup time of nsAA CFPS reactions. It has also been found that increasing o-aaRS concentrations can improve the production of nsAA-containing proteins in CFPS reactions. This not surprising considering that engineered o-aaRS have 500- to 7000-fold lower catalytic efficiencies, as compared with their native counterparts (O'Donoghue et al. 2013). High concentrations of o-aaRS is only possible in vitro as o-aaRSs are known to be toxic in vivo at medium to high concentrations (Hong et al. 2014a). Despite this, Chemla and colleagues have shown that endogenously expressed pyrrolysyl OTS can allow for maximal suppression efficiency for CFPS nsAA incorporation (Chemla et al. 2015).

The efficiency of nsAA incorporation is greatly limited by nsAA aminoacylation efficiency (O'Donoghue et al. 2013). Further evolution of o-aaRS/o-tRNA pairs could increase catalytic efficiencies. Using in vitro compartmentalized directed evolution strategies (Nishikawa et al. 2012), CFPS systems can, in principle, assay larger libraries sizes compared with in vivo evolution methods, which are limited by transformation efficiencies. Additionally, o-aaRS evolved in vitro have the potential to reach lower K_m values by limiting the concentration of nsAA. By comparison, intracellular concentrations of nsAAs are often high because of limited nsAA export and catabolism mechanisms, limiting the K_m values attainable via in vivo evolution strategies.

Major advances have also removed a significant obstacle of amber suppression: nsAA-o-tRNA competition with release factor 1 (RF-1) (Johnson et al. 2011; Loscha et al. 2012b; Ohtake et al. 2012; Lajoie et al. 2013). Endogenous RF-1 recognizes amber codons and subsequently activates hydrolysis of peptidyl-tRNA to release the peptide chain. During nsAA incorporation using amber suppression technology, RF-1 competes with the nsAA-o-tRNA, resulting in a significant amount of truncated product and reduced nsAA incorporation efficiencies. By recoding TAG codons from seven essential genes to the synonymous TAA ochre codons, the RF-1 gene was deleted without dramatically effecting cellular growth. The RF-1 deletion strain increased *O*-phosphoserine incorporation by 120-fold in vivo (Heinemann et al. 2012). A strain with RF-1 knocked out and 13 essential gene TAG codons recoded to TAA was recently developed as a chassis for CFPS. As a result, a 250% improvement was observed compared with the parent stain with RF-1 present (Hong et al. 2014b). More recently, researchers developed an RF-1 deletion strain in which all 321 amber codons were recoded to TAA, completely freeing the amber codon to encode nsAAs (Lajoie et al. 2013). Although successful insertion of multiple nsAAs is possible without these modifications (Loscha et al. 2012a; Ozawa et al. 2012), a total recoded CFPS chassis has allowed for the incorporation of five *p*AcF residues into sfGFP (Hong et al. 2015) without additional extract processing. Further development of this chassis can potentially allow tens of nsAAs to be incorporated into a single protein.

Despite the many advances of nsAA incorporation technology, there is room for improvement and development for novel applications. At present, it is not possible to incorporate more than one kind of nsAA into a single protein in vitro without significant extract processing (Oh et al. 2014). Enabling incorporation of multiple types of nsAAs at multiple locations in recombinant proteins are important technological challenges for cell-free systems that need to be tackled in the near future. The development of further evolved OTSs for improved nsAA incorporation is another important research direction that could be performed in vitro, in which the design space is not constrained by cell viability. In addition, genome-engineering efforts to remove negative effectors of in vitro protein synthesis (e.g., nucleases, proteases) will increase yields of nsAA-containing proteins in recoded chassis strains (Hong et al. 2015). Beyond extract-based systems, it should be noted that purified approaches offer much more freedom of design and control of the translation apparatus for genetic reprogramming (Forster et al. 2003; Murakami et al. 2006; Goto et al. 2011; Passioura and Suga 2013).

Posttranslational Incorporation of Glycans and Metal Centers

PTMs can be performed in CFPS systems by coactivating enzyme pathways and protein synthesis in the cell lysate. The synthetic pathways can be synthesized in vivo in the chassis strain before extract preparation, in vitro via CFPS, or purified and added to the CFPS reaction. These efforts have enabled the in vitro production of proteins bearing glycans (sugars) and accessorized with metal centers.

Glycosylation is particularly important for the production of recombinant protein therapeutics, as improper glycosylation can adversely affect the therapeutic activity or circulation half-life of a therapeutic (Li and d'Anjou 2009). Glycans are synthesized by enzymes called glycosyltransferases (GTs) and attached to proteins by oligosaccharyltransferases (OSTs), which together comprise the glycosylation machinery of the cell (Fig. 4).

Glycosylation is possible in some eukaryotic CFPS systems, including ICE, CHO extract, and K562 extract (Kubick et al. 2009; Stech et al. 2013, 2014; Brodel et al. 2015). However, these platforms harness the endogeneous machinery to carry out glycosylation. As a result, this approach restricts the possible glycan structures to those naturally synthesized by the host cells,

 Cite this article as *Cold Spring Harb Perspect Biol* doi: 10.1101/cshperspect.a023853

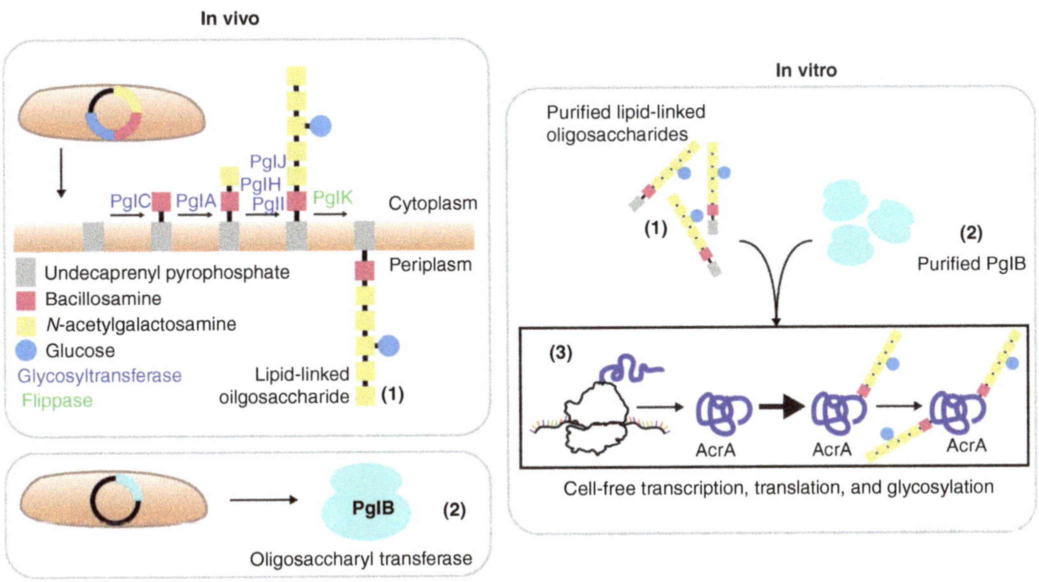

Figure 4. Glycoprotein production in cell-free systems. The *Campylobacter jejuni N*-glycan biosynthesis pathway, which contains glycosyltransferases and a flippase, is expressed in vivo in *Escherichia coli*. These enzymes assemble sugar monomers (bacillosamine, *N*-acetylglucosamine, glucose) onto a lipid anchor (undecaprenyl pyrophosphate) in the membrane to form lipid-linked oligosaccharides (LLOs) (*top*). Similarly, the *C. jejuni* oligosaccharyltransferase (OST), PglB, is expressed in vivo in *E. coli* (*bottom*). LLOs and PglB are purified and added to the in vitro glycoprotein synthesis reaction. Addition of purified LLOs and PglB to the cell-free protein synthesis (CFPS) reaction results in the synthesis of fully glycosylated AcrA, a *C. jejuni* glycoprotein (Guarino and DeLisa 2012).

which may not precisely resemble those found on the recombinant protein of interest. In addition, one would need to learn how to recapitulate protein trafficking to achieve some glycosylation patterns. Recent in vivo glycoengineering efforts have shown that cellular glycosylation machinery can be engineered to synthesize desired glycan structures (Hamilton et al. 2006; Fisher et al. 2011; Valderrama-Rincon et al. 2012; Meuris et al. 2014). More recent efforts are exploiting the open cell-free reaction environment for glycoengineering to enable synthetic glycosylation pathways. In a pioneering study, Guarino and DeLisa (2012) showed that glycosylated proteins can be synthesized in vitro by adding purified bacterial lipid-linked oligosaccharides and the PglB OST to a CPFS reaction. Yields of between 50–100 µg/mL of AcrA, a *Campylobacter jejuni* glycoprotein, were achieved. This result provides a potential path for the incorporation of synthetic, human glycosylation in cell-free systems through the transfer of synthetic enzyme pathways to bacterial cell-free systems. However, significant development of this technology is still needed.

Metalloproteins are important catalysts in biology (Kunamneni et al. 2008). Typically, in vivo production of metalloproteins involves apoprotein production and purification, followed by in vitro metal loading (Lawton and Rosenzweig 2011). CFPS systems potentially offer one-pot synthesis of metal-loaded metalloproteins by addition of free metal atoms and protein chaperones to in vitro reaction mixture. The first example of this approach was the in vitro synthesis of the FeFe hydrogenase metalloprotein (Boyer et al. 2006; Kuchenreuther et al. 2014). Three FeS cluster-containing proteins, HydE, HydF, and HydG, are required to assemble the FeFe hydrogenase. This enzyme pathway was reconstituted in vitro using *E. coli* lysates containing HydE, HydF, and HydG. When the purified apo-form of the FeFe hydrogenase was added to the reaction, the active hydrogenase was synthesized at yields of 100–200 μg/mL (Kuchenreuther et al. 2014). This study provided evidence that enzymatic pathways can be reconstituted in vitro, but the integration of in vitro protein synthesis into this system was not shown. More recently, Li and colleagues showed that two bacterial and two archaeal multicopper oxidase (MCO) enzymes could be synthesized in vitro, with three of the four enzymes produced at yields of greater than 1 mg/mL. They further showed that active MCO could be synthesized in a one-pot in vitro reaction containing 100 μM $CuSO_4$, or by posttranslational addition of $CuSO_4$ to the reaction without the need for purification. MCO synthesis was also scaled from 15 μL to 100 μL without impacting productivity or protein solubility (Li et al. 2016). This study shows the potential of CFPS as a novel method for production of metalloproteins at high titers.

Synthetic PTM mechanisms and enzymatic pathways in cell-free systems have great potential for expanding the types of controllable protein modifications that are possible. Further work in this area will expand opportunities for producing modified proteins with a variety of (a)biological modifications in vitro.

BUILDING UP TO GENETIC NETWORKS IN CELL-FREE SYSTEMS

In addition to using cell-free synthetic biology for the production of individual proteins or assembly of macromolecules, CFPS systems are now used for the construction of sophisticated genetic networks, often referred to as genetic circuits (Brophy and Voigt 2014). Initial efforts to design and implement synthetic genetic networks dominated early in vivo synthetic biology efforts (Gardner et al. 2000), serving as a forward engineering approach for studying natural gene regulation and controlling cellular behavior. Much progress has been made toward developing foundational genetic network modules, such as genetic switches (Bayer and Smolke 2005; Ham et al. 2006, 2008), logic gates (Rackham and Chin 2005; Anderson et al. 2007; Win and Smolke 2008), and memory modules (Friedland et al. 2009), as well as engineering cells with practical applications in the areas of bioremediation (Gilbert et al. 2003), biosensing (Rajendran and Ellington 2008), biofuel production (Inokuma et al. 2010), and therapeutic applications (Ro et al. 2006). Despite the aforementioned progress, translational in vivo synthetic biology is often limited by cellular viability for three key reasons. First, in vivo genetic networks are constrained to relatively small networks owing to restrictions on cellular metabolic load and a limited number of genetic parts. Second, limited high-throughput methods for optimizing regulatory networks slow in vivo genetic network design cycles. Third, interference between host regulation and synthetic networks (cross talk) results in poor computational modeling, reducing prediction capabilities as the network grows in size. Fortunately, many of these challenges can be addressed by implementing genetic networks in vitro using CFPS platforms.

For instance, synthetic gene networks can be studied in isolation as all of the endogenous DNA and mRNA are removed during lysate preparation. This eliminates the issue of cross talk between endogenous regulation and the synthetic network. In addition, the open nature of cell-free systems allows integration with hardware for high-throughput assays and onboard reaction monitoring. Consequently, automation will reduce current design–build–test cycle time, and will provide new biological modules for in vivo characterization. Furthermore, compartmentalizing CFPS reactions within synthetic vesicles provides an opportunity to study artificial cells (Caschera and Noireaux 2016) and the fundamental properties of gene regulation in confined environments.

Implementing Genetic Networks In Vitro

Simple regulatory elements, such as inducible promoters (Noireaux et al. 2003), transcriptional switches (Kim et al. 2006), and multistage cascades (Noireaux et al. 2003), were the first modules to be implemented in vitro. However, because current CFPS systems are optimized for mRNA stability and protein overproduction, in vitro genetic networks are inadequate when constructing oscillatory circuits (Elowitz and Leibler 2000) or larger networks. Complex genetic circuitry requires resources to be efficiently directed toward downstream networks, as well as component half-lives to mimic those found in cells. This is made clear when considering that in vivo global mRNA half-life is ~6.8 min (Selinger et al. 2003), whereas, in vitro mRNA half-life is ~13 min (Shin and Noireaux 2010). This discrepancy causes overall limitations of translational machinery for CFPS reactions as mRNA is accumulated in the reaction (Selinger et al. 2003). To overcome these issues, larger genetic circuits have been implemented in vitro by using purified MazF, a sequence-specific endoribonuclease, which has been shown to incrementally inactivate mRNA (Shin and Noireaux 2010). In particular, AND gates and negative feedback loops were developed in an *E. coli* cell-free expression system using this method (Shin and Noireaux 2012). Karig and colleagues also showed negative feedback loops but without mRNA activation mechanisms, and consequently saw unwanted accumulation of mRNA that ultimately affected the dynamics of the network (Karig et al. 2012). CFPS protein half-life has also been found to be much longer than in in vivo systems because of the reduced proteolytic activity of the CFPS system. This last aspect is important for constructing oscillatory circuits and needs to be further investigated.

Additionally, characterization and optimization of regulatory elements for cell-free systems have been explored (Shin and Noireaux 2012; Garamella et al. 2016). This provides more transcriptional handles to design higher-order genetic networks. Moreover, T7 promoter variants have also been developed in a cell-free context for expanding regulation options and improved switch-like activity (Karig et al. 2012; Iyer et al. 2013). To limit the number of synthesized proteins required for a genetic network, alternative transcriptional control mechanisms, such as ligand-sensitive transcriptional controls, have also been shown (Iyer and Doktycz 2014). Finally, RNA-based genetic circuits represent another solution to limit energy loss from protein production and provide the benefit of quicker response times (Takahashi et al. 2014; Chappell et al. 2015).

The open nature of CFPS also allows for in situ monitoring of genetic networks (Norred et al. 2015). In particular, binary FRET probes have been used to monitor real-time mRNA dynamics (Niederholtmeyer et al. 2013). Fluorescent RNA aptamers have also been used as an alternative simple strategy for monitoring RNA in real-time (Chizzolini et al. 2014). High-throughput analysis of biomolecular interactions can also be performed by encapsulating CFPS reactions into nanoliter droplets and examining interactions via an immunoassay (Rendl et al. 2013). Additionally, HPLC analyses for monitoring metabolite concentrations are straightforward to use because CFPS reactions can be run directly on an HLPC or MS machine.

Collectively, recent efforts hold promise for using more complex circuitry and real-world applications (Pardee et al. 2014). Because of the open environment of cell-free systems and short implementation times, CFPS systems can also be used as a complementary route to prototyping genetic networks in an in vitro environment before in vivo implementation.

Opportunities in Prototyping

In vivo synthetic biology applies molecular biology tools to forward-engineer cellular behavior. This is performed first through in silico design of genetic networks, construction of those networks, followed by testing in vivo (Litcofsky et al. 2012; Sleight and Sauro 2013). However, in vivo engineering cycles require a significant amount of time and financial resources (Ro et al. 2006). In contrast, in vitro biomolecular prototyping promises to improve the overall efficiency of the design–build–test cycle (Fig. 5A) (Sun et al. 2014). There are several demonstrations of prototyping circuits in vitro for in vivo implementation, such as negative feedback loops (Karig et al. 2012), multi-input regulated T7 promoters (Iyer et al. 2013), riboswitches (Martini and Mansy 2014), and oscillators (Niederholtmeyer et al. 2015). However, it is difficult to predict correlations

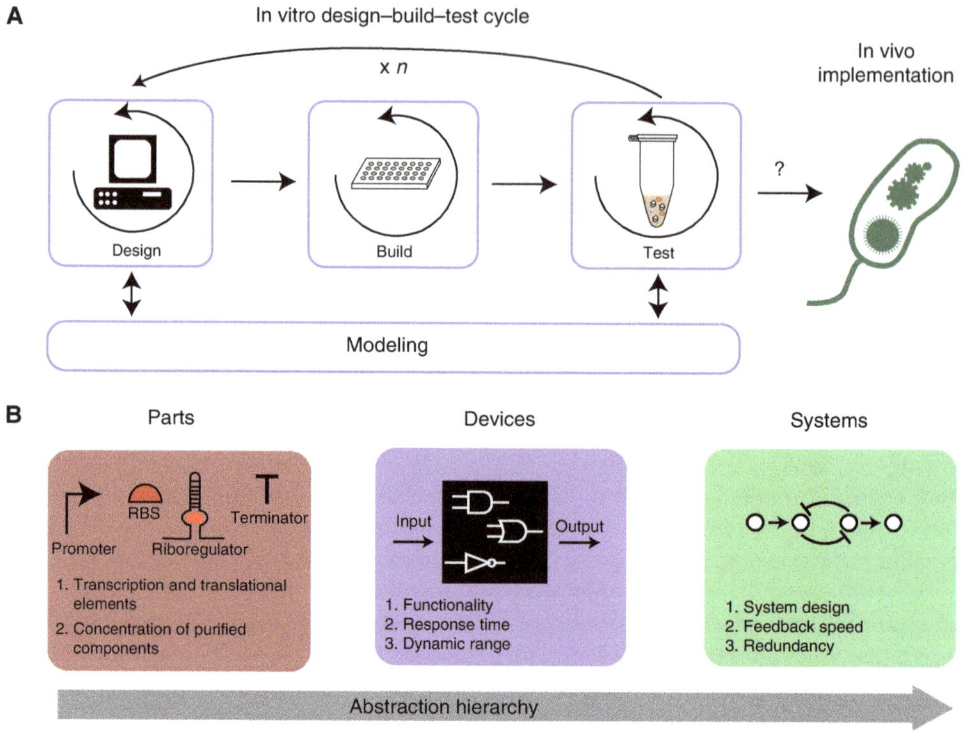

Figure 5. In vitro prototyping of genetic networks. (*A*) Overview of in vitro prototyping for speeding up in vivo design–build–test cycles. In vitro prototyping allows for a genetic part or network to be quickly screened for specific characteristics in vitro before implementation in vivo. First, in the "Design" stage, the genetic parts or networks are designed, informed by computational models or literature. Next, in the "Build" stage, several designs are built. In the "Test" phase, built designs are assayed in vitro. If the Test stage does not yield the desired behavior, one reinitiates the cycle *n* number of times until the desired characteristic is achieved. The Test stage can also inform the modeling and allow for better models for the Design stage. Once the desired characteristics are found, top candidates that behave as expected can be implemented in vivo, with an increased likeliness of being functional. (*B*) Variables for characterization and optimization. In vitro prototyping can occur by characterizing or optimizing various levels on the genetic network "abstraction hierarchy." At the most basic level, transcriptional and translational parts or purified components of a network can be analyzed for functionality. In vitro prototyping can also be applied at the device level to assess input–output relationships. Finally, an in vitro systems level analysis allows for an isolated study of how multiple genetic devices feed into each other and a preview of the overall network behavior.

between parameters (e.g., promoters/RBS strengths) found in vitro and predicted values in vivo. For instance, it was found that characterizing promoter strength in vitro must be performed with circularized DNA for there is a correlation to in vivo promoter strengths (Chappell et al. 2013); however, this is limited to low to medium strength promoters. Additional work is needed to find unknown parameters involved with the transition from in vitro to in vivo implementation.

Despite these issues, in vitro prototyping techniques are compelling and provide a potentially rapid alternative to in vivo systems for circuit design and testing (Niederholtmeyer et al. 2015; de Los Santos et al. 2016). For instance, in vitro prototyping can be accelerated through microscale reactions in micro-well plates and microfluidic systems. Indeed, these methodologies decrease prototyping time with real-time simultaneous or sequential screening of many system parameters (Lee et al. 2007).

Cite this article as *Cold Spring Harb Perspect Biol* doi: 10.1101/cshperspect.a023853

Additionally, complementary, predictive computational models are being developed to better understand CFPS' limitations in prototyping (Karzbrun et al. 2011). This work has confirmed experimental results (Noireaux et al. 2003) that, in CFPS, potential design limitations arise from improperly balanced energy usage. Using a continuous exchange set-up can partially address this issue, and has been proven to be a reliable solution for prototyping (Noireaux et al. 2003). Prototyping efforts have even moved beyond a laboratory environment with development of a CFPS system that is both portable and stable because of its ability to react on paper (Pardee et al. 2014). This allows for a low-cost prototyping environment and provides opportunities in using CFPS for portable, inexpensive diagnostic applications.

Improvements in CFPS systems have opened new opportunities for cell-free synthetic biology, including studying genetic networks in isolation, characterizing new regulatory elements and speeding up design–build–test cycles for in vivo synthetic biology. The size of genetic networks achievable in CFPS systems will only increase as CFPS systems are further developed to meet the specific needs of in vitro genetic networks (e.g., downstream resourcing and matching in vivo component half-lives). Additionally, computational modeling of cell-free reactions will further improve in silico design of in vitro networks and assist in understanding the genetic regulation in isolation. In sum, the development of in vitro genetic networks holds much potential for the synthetic biology field as it complements many in vivo efforts.

CONCLUSIONS AND FUTURE OUTLOOK

Overall, the renewed scientific interest in CFPS in the past two decades has resulted in drastically increased batch yields, active reaction durations, and reaction volumes (Carlson et al. 2012). The variety of different CFPS technology platforms has further enabled the in vitro production of proteins with diverse complexity and species origin. CFPS technology will be invaluable in the near future as an "organism protein factory." In this way, CFPS technology will aid efforts to determine the gene products for the many organisms whose genomes have been sequenced or will be sequenced in the near future, owing to the rapidly decreasing cost of DNA sequencing technology. For example, the Sargasso Seas expedition identified >1.2 million genes, many of which have unknown function and unknown potential as synthetic parts (Venter et al. 2004).

CFPS technology will also have an increasing role in complementing in vivo synthetic biology efforts. Because of its flexibility and high-throughput potential, cell-free systems are perfectly suited to synthesize and assay large libraries, not only for genetic networks as surveyed herein, but also for evolutionary or proteomics studies (Takemori et al. 2015). This will provide in vivo efforts with enriched mutant libraries and accelerated methods for preparing large-scale protein libraries. In sum, CFPS holds tremendous potential to transform our ability to rapidly synthesize and engineer recombinant proteins and its applications in the synthetic biology field will only grow in the future.

ACKNOWLEDGMENTS

The authors would like to acknowledge Jennifer A. Schoborg, Erik D. Carlson, and Filippo Caschera for assistance in editing this manuscript. Select figures in this text were adapted from Jennifer A. Schoborg and Erik D. Carlson. We gratefully acknowledge the National Science Foundation (NSF) (MCB-0943393), the Office of Naval Research (N00014-11-1-0363), the Defense Advanced Research Projects Agency Young Faculty Award (DARPA YFA) Program (N66001-11-1-4137), the Army Research Office (W911NF-11-1-0445), the NSF Materials Network Grant (DMR-1108350), the DARPA Living Foundries Program (N66001-12-C-4211), the David and Lucile Packard Foundation (2011-37152), Advanced Research Projects Agency-Energy (ARPA-E) (DE-AR0000435), and the Chicago Biomedical Consortium with support from the Searle Funds at the Chicago Community Trust for support. J.G.P. is funded by the NSF Graduate Research Fellowship

(DGE-1144469). J.C.S. is funded, in part, by the Northwestern Biotechnology Training Program supported by the National Institutes of Health (NIH) (T32GM008449).

REFERENCES

Albayrak C, Swartz JR. 2013a. Cell-free co-production of an orthogonal transfer RNA activates efficient site-specific non-natural amino acid incorporation. *Nucleic Acids Res* **41**: 5949–5963.

Albayrak C, Swartz JR. 2013b. Using *E. coli*-based cell-free protein synthesis to evaluate the kinetic performance of an orthogonal tRNA and aminoacyl-tRNA synthetase pair. *Biochem Biophys Res Commun* **431**: 291–295.

Alfonta L, Zhang Z, Uryu S, Loo JA, Schultz PG. 2003. Site-specific incorporation of a redox-active amino acid into proteins. *J Am Chem Soc* **125**: 14662–14663.

Anderson JC, Voigt CA, Arkin AP. 2007. Environmental signal integration by a modular AND gate. *Mol Syst Biol* **3**: 133.

Anderson MJ, Stark JC, Eric Hodgman C, Jewett MC. 2015. Energizing eukaryotic cell-free protein synthesis with glucose metabolism. *FEBS Lett* **589**: 1723–1727.

Asahara H, Chong S. 2010. In vitro genetic reconstruction of bacterial transcription initiation by coupled synthesis and detection of RNA polymerase holoenzyme. *Nucleic Acids Res* **38**: e141.

Bayer TS, Smolke CD. 2005. Programmable ligand-controlled riboregulators of eukaryotic gene expression. *Nat Biotechnol* **23**: 337–343.

Bernhard F, Tozawa Y. 2013. Cell-free expression—Making a mark. *Curr Opin Struct Biol* **23**: 374–380.

Boyer ME, Wang CW, Swartz JR. 2006. Simultaneous expression and maturation of the iron-sulfur protein ferredoxin in a cell-free system. *Biotechnol Bioeng* **94**: 128–138.

Brodel AK, Wustenhagen DA, Kubick S. 2015. Cell-free protein synthesis systems derived from cultured mammalian cells. *Methods Mol Biol* **1261**: 129–140.

Brophy JAN, Voigt CA. 2014. Principles of genetic circuit design. *Nat Meth* **11**: 508–520.

Bundy BC, Swartz JR. 2010. Site-specific incorporation of *p*-propargyloxyphenylalanine in a cell-free environment for direct protein–protein click conjugation. *Bioconjug Chem* **21**: 255–263.

Bundy BC, Franciszkowicz MJ, Swartz JR. 2008. *Escherichia coli*–based cell-free synthesis of virus-like particles. *Biotechnol Bioeng* **100**: 28–37.

Buntru M, Vogel S, Spiegel H, Schillberg S. 2014a. Tobacco BY-2 cell-free lysate: An alternative and highly-productive plant-based in vitro translation system. *BMC Biotechnol* **14**: 37.

Buntru M, Vogel S, Stoff K, Spiegel H, Schillberg S. 2014b. A versatile coupled cell-free transcription–translation system based on tobacco BY-2 cell lysates. *Biotechnol Bioeng* **112**: 867–878.

Cai Q, Hanson JA, Steiner AR, Tran C, Masikat MR, Chen R, Zawada JF, Sato AK, Hallam TJ, Yin G. 2015. A simplified and robust protocol for immunoglobulin expression in *Escherichia coli* cell-free protein synthesis systems. *Biotechnol Prog* **31**: 823–831.

Calhoun KA, Swartz JR. 2005a. An economical method for cell-free protein synthesis using glucose and nucleoside monophosphates. *Biotechnol Prog* **21**: 1146–1153.

Calhoun KA, Swartz JR. 2005b. Energizing cell-free protein synthesis with glucose metabolism. *Biotechnol Bioeng* **90**: 606–613.

Carlson ED, Gan R, Hodgman CE, Jewett MC. 2012. Cell-free protein synthesis: Applications come of age. *Biotechnol Adv* **30**: 1185–1194.

Caschera F, Noireaux V. 2014a. Integration of biological parts toward the synthesis of a minimal cell. *Curr Opin Chem Biol* **22**: 85–91.

Caschera F, Noireaux V. 2014b. Synthesis of 2.3 mg/ml of protein with an all *Escherichia coli* cell-free transcription-translation system. *Biochimie* **99**: 162–168.

Caschera F, Noireaux V. 2016. Compartmentalization of an all–*E. coli* cell-free expression system for the construction of a minimal cell. *Artif Life* **2**: 1–11.

Caschera F, Bedau MA, Buchanan A, Cawse J, de Lucrezia D, Gazzola G, Hanczyc MM, Packard NH. 2011. Coping with complexity: Machine learning optimization of cell-free protein synthesis. *Biotechnol Bioeng* **108**: 2218–2228.

Chambers DA, Zubay G. 1969. The stimulatory effect of cyclic adenosine 3′5′-monophosphate on DNA-directed synthesis of β-galactosidase in a cell-free system. *Proc Natl Acad Sci* **63**: 118–122.

Chan LY, Kosuri S, Endy D. 2005. Refactoring bacteriophage T7. *Mol Syst Biol* **1**: 2005.0018.

Chappell J, Jensen K, Freemont PS. 2013. Validation of an entirely in vitro approach for rapid prototyping of DNA regulatory elements for synthetic biology. *Nucleic Acids Res* **41**: 3471–3481.

Chappell J, Takahashi MK, Lucks JB. 2015. Creating small transcription activating RNAs. *Nat Chem Biol* **11**: 214–220.

Chemla Y, Ozer E, Schlesinger O, Noireaux V, Alfonta L. 2015. Genetically expanded cell-free protein synthesis using endogenous pyrrolysyl orthogonal translation system. *Biotechnol Bioeng* **112**: 1663–1672.

Chin JW, Santoro SW, Martin AB, King DS, Wang L, Schultz PG. 2002. Addition of *p*-azido-L-phenylalanine to the genetic code of *Escherichia coli*. *J Am Chem Soc* **124**: 9026–9027.

Chizzolini F, Forlin M, Cecchi D, Mansy SS. 2014. Gene position more strongly influences cell-free protein expression from operons than T7 transcriptional promoter strength. *ACS Synth Biol* **3**: 363–371.

Chong S. 2014. Overview of cell-free protein synthesis: Historic landmarks, commercial systems, and expanding applications. *Curr Protoc Mol Biol* **108**: 16.30.1–16.30.11.

Choudhury A, Hodgman CE, Anderson MJ, Jewett MC. 2014. Evaluating fermentation effects on cell growth and crude extract metabolic activity for improved yeast cell-free protein synthesis. *Biochem Eng J* **91**: 140–148.

Cornish VW, Benson DR, Altenbach CA, Hideg K, Hubbell WL, Schultz PG. 1994. Site-specific incorporation of biophysical probes into proteins. *Proc Natl Acad Sci* **91**: 2910–2914.

de Los Santos EL, Meyerowitz JT, Mayo SL, Murray RM. 2016. Engineering transcriptional regulator effector specificity using computational design and in vitro rapid prototyping: Developing a vanillin sensor. *ACS Synth Biol* **5**: 287–295.

Du D, van Veen HW, Luisi BF. 2015. Assembly and operation of bacterial tripartite multidrug efflux pumps. *Trends Microbiol* **23**: 311–319.

Elowitz MB, Leibler S. 2000. A synthetic oscillatory network of transcriptional regulators. *Nature* **403**: 335–338.

Endo Y, Sawasaki T. 2006. Cell-free expression systems for eukaryotic protein production. *Curr Opin Biotechnol* **17**: 373–380.

Ezure T, Suzuki T, Higashide S, Shintani E, Endo K, Kobayashi S, Shikata M, Ito M, Tanimizu K, Nishimura O. 2006. Cell-free protein synthesis system prepared from insect cells by freeze-thawing. *Biotechnol Prog* **22**: 1570–1577.

Ezure T, Nanatani K, Sato Y, Suzuki S, Aizawa K, Souma S, Ito M, Hohsaka T, von Heijine G, Utsumi T, et al. 2014. A cell-free translocation system using extracts of cultured insect cells to yield functional membrane proteins. *PLoS ONE* **9**: e112874.

Fenz SF, Sachse R, Schmidt T, Kubick S. 2014. Cell-free synthesis of membrane proteins: Tailored cell models out of microsomes. *Biochim Biophys Acta* **1838**: 1382–1388.

Fisher AC, Haitjema CH, Guarino C, Celik E, Endicott CE, Reading CA, Merritt JH, Ptak AC, Zhang S, DeLisa MP. 2011. Production of secretory and extracellular *N*-linked glycoproteins in *Escherichia coli*. *Appl Environ Microbiol* **77**: 871–881.

Fogeron ML, Badillo A, Jirasko V, Gouttenoire J, Paul D, Lancien L, Moradpour D, Bartenschlager R, Meier BH, Penin F, et al. 2015. Wheat germ cell-free expression: Two detergents with a low critical micelle concentration allow for production of soluble HCV membrane proteins. *Protein Expr Purif* **105**: 39–46.

Forster AC, Tan Z, Nalam MN, Lin H, Qu H, Cornish VW, Blacklow SC. 2003. Programming peptidomimetic syntheses by translating genetic codes designed de novo. *Proc Natl Acad Sci* **100**: 6353–6357.

Friedland AE, Lu TK, Wang X, Shi D, Church G, Collins JJ. 2009. Synthetic gene networks that count. *Science* **324**: 1199–1202.

Fritz BR, Jewett MC. 2014. The impact of transcriptional tuning on in vitro integrated rRNA transcription and ribosome construction. *Nucleic Acids Res* **42**: 6774–6785.

Fujiwara K, Katayama T, Nomura SM. 2013. Cooperative working of bacterial chromosome replication proteins generated by a reconstituted protein expression system. *Nucleic Acids Res* **41**: 7176–7183.

Gan R, Jewett MC. 2014. A combined cell-free transcription-translation system from *Saccharomyces cerevisiae* for rapid and robust protein synthe. *Biotechnol J* **9**: 641–651.

Gan R, Jewett MC. 2016. Evolution of translation initiation sequences using in vitro yeast ribosome display. *Biotechnol Bioeng* doi: 10.1002/bit.25933.

Garamella J, Marshall R, Rustad M, Noireaux V. 2016. The all *E. coli* TX-TL toolbox 2.0: A platform for cell-free synthetic biology. *ACS Synth Biol* **5**: 344–355.

Gardner TS, Cantor CR, Collins JJ. 2000. Construction of a genetic toggle switch in *Escherichia coli*. *Nature* **403**: 339–342.

Gerber D, Maerkl SJ, Quake SR. 2009. An in vitro microfluidic approach to generating protein-interaction networks. *Nat Methods* **6**: 71–74.

Gilbert ES, Walker AW, Keasling JD. 2003. A constructed microbial consortium for biodegradation of the organophosphorus insecticide parathion. *Appl Microbiol Biotechnol* **61**: 77–81.

Goerke AR, Swartz JR. 2009. High-level cell-free synthesis yields of proteins containing site-specific non-natural amino acids. *Biotechnol Bioeng* **102**: 400–416.

Goshima N, Kawamura Y, Fukumoto A, Miura A, Honma R, Satoh R, Wakamatsu A, Yamamoto J, Kimura K, Nishikawa T, et al. 2008. Human protein factory for converting the transcriptome into an in vitro-expressed proteome. *Nat Methods* **5**: 1011–1017.

Goto Y, Katoh T, Suga H. 2011. Flexizymes for genetic code reprogramming. *Nat Protoc* **6**: 779–790.

Groff D, Armstrong S, Rivers PJ, Zhang J, Yang J, Green E, Rozzelle J, Liang S, Kittle JD Jr, Steiner AR, et al. 2014. Engineering toward a bacterial "endoplasmic reticulum" for the rapid expression of immunoglobulin proteins. *mAbs* **6**: 671–678.

Guarino C, DeLisa MP. 2012. A prokaryote-based cell-free translation system that efficiently synthesizes glycoproteins. *Glycobiology* **22**: 596–601.

Ham TS, Lee SK, Keasling JD, Arkin AP. 2006. A tightly regulated inducible expression system utilizing the *fim* inversion recombination switch. *Biotechnol Bioeng* **94**: 1–4.

Ham TS, Lee SK, Keasling JD, Arkin AP. 2008. Design and construction of a double inversion recombination switch for heritable sequential genetic memory. *PLoS ONE* **3**: e2815.

Hamilton SR, Davidson RC, Sethuraman N, Nett JH, Jiang Y, Rios S, Bobrowicz P, Stadheim TA, Li H, Choi BK, et al. 2006. Humanization of yeast to produce complex terminally sialylated glycoproteins. *Science* **313**: 1441–1443.

Harbers M. 2014. Wheat germ systems for cell-free protein expression. *FEBS Lett* **588**: 2762–2773.

Heinemann IU, Rovner AJ, Aerni HR, Rogulina S, Cheng L, Olds W, Fischer JT, Soll D, Isaacs FJ, Rinehart J. 2012. Enhanced phosphoserine insertion during *Escherichia coli* protein synthesis via partial UAG codon reassignment and release factor 1 deletion. *FEBS Lett* **586**: 3716–3722.

Heyman Y, Buxboim A, Wolf SG, Daube SS, Bar-Ziv RH. 2012. Cell-free protein synthesis and assembly on a biochip. *Nat Nanotechnol* **7**: 374–378.

Hodgman CE, Jewett MC. 2013. Optimized extract preparation methods and reaction conditions for improved yeast cell-free protein synthesis. *Biotechnol Bioeng* **110**: 2643–2654.

Hodgman CE, Jewett MC. 2014. Characterizing IGR IRES-mediated translation initiation for use in yeast cell-free protein synthesis. *N Biotechnol* **31**: 499–505.

Hong SH, Kwon YC, Jewett MC. 2014a. Non-standard amino acid incorporation into proteins using *Escherichia coli* cell-free protein synthesis. *Front Chem* **2**: 34.

Hong SH, Ntai I, Haimovich AD, Kelleher NL, Isaacs FJ, Jewett MC. 2014b. Cell-free protein synthesis from a release factor 1 deficient *Escherichia coli* activates efficient and multiple site-specific nonstandard amino acid incorporation. *ACS Synth Biol* **3**: 398–409.

Hong SH, Kwon YC, Martin RW, Des Soye BJ, de Paz AM, Swonger KN, Ntai I, Kelleher NL, Jewett MC. 2015. Improving cell-free protein synthesis through genome engineering of *Escherichia coli* lacking release factor 1. *Chembiochem* **16**: 844–853.

Iizuka N, Najita L, Franzusoff A, Sarnow P. 1994. Cap-dependent and cap-independent translation by internal initiation of mRNAs in cell extracts prepared from *Saccharomyces cerevisiae*. *Mol Cell Biol* **14**: 7322–7330.

Inokuma K, Liao JC, Okamoto M, Hanai T. 2010. Improvement of isopropanol production by metabolically engineered Escherichia coli using gas stripping. *J Biosci Bioeng* **110**: 696–701.

Iyer S, Doktycz MJ. 2014. Thrombin-mediated transcriptional regulation using DNA aptamers in DNA-based cell-free protein synthesis. *ACS Synth Biol* **3**: 340–346.

Iyer S, Karig DK, Norred SE, Simpson ML, Doktycz MJ. 2013. Multi-input regulation and logic with T7 promoters in cells and cell-free systems. *PLoS ONE* **8**: e78442.

Jackson K, Kanamori T, Ueda T, Fan ZH. 2014a. Protein synthesis yield increased 72 times in the cell-free PURE system. *Integr Biol (Camb)* **6**: 781–788.

Jackson K, Khnouf R, Fan ZH. 2014b. Cell-free protein synthesis in microfluidic 96-well plates. *Methods Mol Biol* **1118**: 157–168.

Jackson K, Jin S, Fan ZH. 2015. Optimization of a miniaturized fluid array device for cell-free protein synthesis. *Biotechnol Bioeng* **112**: 2459–2467.

Jewett MC, Calhoun KA, Voloshin A, Wuu JJ, Swartz JR. 2008. An integrated cell-free metabolic platform for protein production and synthetic biology. *Mol Syst Biol* **4**: 220.

Jewett MC, Fritz BR, Timmerman LE, Church GM. 2013. In vitro integration of ribosomal RNA synthesis, ribosome assembly, and translation. *Mol Syst Biol* **9**: 678.

Johnson DB, Xu J, Shen Z, Takimoto JK, Schultz MD, Schmitz RJ, Xiang Z, Ecker JR, Briggs SP, Wang L. 2011. RF1 knockout allows ribosomal incorporation of unnatural amino acids at multiple sites. *Nat Chem Biol* **7**: 779–786.

Johnston WA, Alexandrov K. 2014. Production of eukaryotic cell-free lysate from *Leishmania tarentolae*. *Methods Mol Biol* **1118**: 1–15.

Junge F, Haberstock S, Roos C, Stefer S, Proverbio D, Dotsch V, Bernhard F. 2011. Advances in cell-free protein synthesis for the functional and structural analysis of membrane proteins. *N Biotechnol* **28**: 262–271.

Kaiser L, Graveland-Bikker J, Steuerwald D, Vanberghem M, Herlihy K, Zhang S. 2008. Efficient cell-free production of olfactory receptors: Detergent optimization, structure, and ligand binding analyses. *Proc Natl Acad Sci* **105**: 15726–15731.

Karig DK, Iyer S, Simpson ML, Doktycz MJ. 2012. Expression optimization and synthetic gene networks in cell-free systems. *Nucleic Acids Res* **40**: 3763–3774.

Karzbrun E, Shin J, Bar-Ziv RH, Noireaux V. 2011. Coarse-grained dynamics of protein synthesis in a cell-free system. *Phys Rev Lett* **106**: 048104.

Katzen F, Kudlicki W. 2006. Efficient generation of insect-based cell-free translation extracts active in glycosylation and signal sequence processing. *J Biotechnol* **125**: 194–197.

Katzen F, Chang G, Kudlicki W. 2005. The past, present and future of cell-free protein synthesis. *Trends Biotechnol* **23**: 150–156.

Kawasaki T, Gouda MD, Sawasaki T, Takai K, Endo Y. 2003. Efficient synthesis of a disulfide-containing protein through a batch cell-free system from wheat germ. *Eur J Biochem* **270**: 4780–4786.

Kazuta Y, Matsuura T, Ichihashi N, Yomo T. 2014. Synthesis of milligram quantities of proteins using a reconstituted in vitro protein synthesis system. *J Biosci Bioeng* **118**: 554–557.

Kiga D, Sakamoto K, Kodama K, Kigawa T, Matsuda T, Yabuki T, Shirouzu M, Harada Y, Nakayama H, Takio K, et al. 2002. An engineered Escherichia coli tyrosyl-tRNA synthetase for site-specific incorporation of an unnatural amino acid into proteins in eukaryotic translation and its application in a wheat germ cell-free system. *Proc Natl Acad Sci* **99**: 9715–9720.

Kim J, White KS, Winfree E. 2006. Construction of an in vitro bistable circuit from synthetic transcriptional switches. *Mol Syst Biol* **2**: 68.

Kovtun O, Mureev S, Johnston W, Alexandrov K. 2010. Towards the construction of expressed proteomes using a *Leishmania tarentolae* based cell-free expression system. *PloS ONE* **5**: e14388.

Kovtun O, Mureev S, Jung W, Kubala MH, Johnston W, Alexandrov K. 2011. *Leishmania* cell-free protein expression system. *Methods* **55**: 58–64.

Kubick S, Gerrits M, Merk H, Stiege W, Erdmann VA, Larry D. 2009. In vitro synthesis of posttranslationally modified membrane proteins. In *Current topics in membranes* (ed. DeLucas L), Vol. 63, pp. 25–49. Elsevier, Amsterdam.

Kucharczyk R, Zick M, Bietenhader M, Rak M, Couplan E, Blondel M, Caubet SD, di Rago JP. 2009. Mitochondrial ATP synthase disorders: Molecular mechanisms and the quest for curative therapeutic approaches. *Biochim Biophys Acta* **1793**: 186–199.

Kuchenreuther JM, Shiigi SA, Swartz JR. 2014. Cell-free synthesis of the H-cluster: A model for the in vitro assembly of metalloprotein metal centers. *Methods Mol Biol* **1122**: 49–72.

Kunamneni A, Plou FJ, Ballesteros A, Alcalde M. 2008. Laccases and their applications: A patent review. *Recent Pat Biotechnol* **2**: 10–24.

Kuruma Y, Suzuki T, Ueda T. 2010. Production of multi-subunit complexes on liposome through an *E. coli* cell-free expression system. *Methods Mol Biol* **607**: 161–171.

Kuruma Y, Suzuki T, Ono S, Yoshida M, Ueda T. 2012. Functional analysis of membranous F_o-*a* subunit of F_1F_o-ATP synthase by in vitro protein synthesis. *Biochem J* **442**: 631–638.

Kwon YC, Jewett MC. 2015. High-throughput preparation methods of crude extract for robust cell-free protein synthesis. *Sci Rep* **5**: 8663.

Lajoie MJ, Rovner AJ, Goodman DB, Aerni HR, Haimovich AD, Kuznetsov G, Mercer JA, Wang HH, Carr PA, Mosberg JA, et al. 2013. Genomically recoded organisms expand biological functions. *Science* **342**: 357–360.

Lawton TJ, Rosenzweig AC. 2011. Detection and characterization of a multicopper oxidase from *Nitrosomonas europaea*. *Methods Enzymol* **496**: 423–433.

Lee KH, Joung HA, Ahn JH, Kim KO, Oh IS, Shin YB, Kim MG, Kim DM. 2007. Real-time monitoring of cell-free protein synthesis on a surface plasmon resonance chip. *Anal Biochem* **366**: 170–174.

Lemke EA. 2011. Site-specific labeling of proteins for single-molecule FRET measurements using genetically encoded ketone functionalities. *Methods Mol Biol* **751**: 3–15.

Li H, d'Anjou M. 2009. Pharmacological significance of glycosylation in therapeutic proteins. *Curr Opin Biotechnol* **20**: 678–684.

Li J, Lawton TJ, Kostecki JS, Nisthal A, Fang J, Mayo SL, Rosenzweig AC, Jewett MC. 2016. Cell-free protein synthesis enables high yielding synthesis of an active multicopper oxidase. *Biotechnol J* **11**: 212–218.

Lian Q, Cao H, Wang F. 2014. The cost-efficiency realization in the *Escherichia coli*-based cell-free protein synthesis systems. *Appl Biochem Biotechnol* **174**: 2351–2367.

Litcofsky KD, Afeyan RB, Krom RJ, Khalil AS, Collins JJ. 2012. Iterative plug-and-play methodology for constructing and modifying synthetic gene networks. *Nat Methods* **9**: 1077–1080.

Liu Y, Fritz BR, Anderson MJ, Schoborg JA, Jewett MC. 2014. Characterizing and alleviating substrate limitations for improved in vitro ribosome construction. *ACS Synth Biol* **4**: 454–462.

Loscha KV, Herlt AJ, Qi R, Huber T, Ozawa K, Otting G. 2012a. Multiple-site labeling of proteins with unnatural amino acids. *Angew Chem Int Ed Engl* **51**: 2243–2246.

Loscha KV, Herlt AJ, Qi R, Huber T, Ozawa K, Otting G. 2012b. Multiple-site labeling of proteins with unnatural amino acids. *Angew Chem Int Ed Engl* **51**: 2243–2246.

Lu Y, Welsh JP, Swartz JR. 2014. Production and stabilization of the trimeric influenza hemagglutinin stem domain for potentially broadly protective influenza vaccines. *Proc Natl Acad Sci* **111**: 125–130.

Lu Y, Chan W, Ko BY, VanLang CC, Swartz JR. 2015. Assessing sequence plasticity of a virus-like nanoparticle by evolution toward a versatile scaffold for vaccines and drug delivery. *Proc Natl Acad Sci* **112**: 12360–12365.

Madin K, Sawasaki T, Ogasawara T, Endo Y. 2000. A highly efficient and robust cell-free protein synthesis system prepared from wheat embryos: Plants apparently contain a suicide system directed at ribosomes. *Proc Natl Acad Sci* **97**: 559–564.

Martini L, Mansy SS. 2014. Measuring riboswitch activity in vitro and in artificial cells with purified transcription-translation machinery. *Methods Mol Biol* **1111**: 153–164.

Matsubayashi H, Ueda T. 2014. Purified cell-free systems as standard parts for synthetic biology. *Curr Opin Chem Biol* **22**: 158–162.

Matsubayashi H, Kuruma Y, Ueda T. 2014. In vitro synthesis of the *E. coli* Sec translocon from DNA. *Angew Chem Int Ed Engl* **53**: 7535–7538.

Matthies D, Haberstock S, Joos F, Dotsch V, Vonck J, Bernhard F, Meier T. 2011. Cell-free expression and assembly of ATP synthase. *J Mol Biol* **413**: 593–603.

Meuris L, Santens F, Elson G, Festjens N, Boone M, Dos Santos A, Devos S, Rousseau F, Plets E, Houthuys E, et al. 2014. GlycoDelete engineering of mammalian cells simplifies *N*-glycosylation of recombinant proteins. *Nat Biotechnol* **32**: 485–489.

Mikami S, Masutani M, Sonenberg N, Yokoyama S, Imataka H. 2006. An efficient mammalian cell-free translation system supplemented with translation factors. *Protein Expr Purif* **46**: 348–357.

Murakami H, Ohta A, Ashigai H, Suga H. 2006. A highly flexible tRNA acylation method for non-natural polypeptide synthesis. *Nat Methods* **3**: 357–359.

Mureev S, Kovtun O, Nguyen UT, Alexandrov K. 2009. Species-independent translational leaders facilitate cell-free expression. *Nat Biotechnol* **27**: 747–752.

Neumann H, Hancock SM, Buning R, Routh A, Chapman L, Somers J, Owen-Hughes T, van Noort J, Rhodes D, Chin JW. 2009. A method for genetically installing site-specific acetylation in recombinant histones defines the effects of H3 K56 acetylation. *Mol Cell* **36**: 153–163.

Ng PP, Jia M, Patel KG, Brody JD, Swartz JR, Levy S, Levy R. 2012. A vaccine directed to B cells and produced by cell-free protein synthesis generates potent antilymphoma immunity. *Proc Natl Acad Sci* **109**: 14526–14531.

Nguyen DP, Garcia Alai MM, Kapadnis PB, Neumann H, Chin JW. 2009. Genetically encoding N^{ε}-methyl-L-lysine in recombinant histones. *J Am Chem Soc* **131**: 14194–14195.

Niederholtmeyer H, Xu L, Maerkl SJ. 2013. Real-time mRNA measurement during an in vitro transcription and translation reaction using binary probes. *ACS Synth Biol* **2**: 411–417.

Niederholtmeyer H, Sun ZZ, Hori Y, Yeung E, Verpoorte A, Murray RM, Maerkl SJ. 2015. Rapid cell-free forward engineering of novel genetic ring oscillators. *eLife* **4**: e09771.

Nirenberg MW, Matthaei JH. 1961. The dependence of cell-free protein synthesis in *E. coli* upon naturally occurring or synthetic polyribonucleotides. *Proc Natl Acad Sci* **47**: 1588–1602.

Nishikawa T, Sunami T, Matsuura T, Yomo T. 2012. Directed evolution of proteins through in vitro protein synthesis in liposomes. *J Nucleic Acids* **2012**: 923214.

Nishimura K, Matsuura T, Nishimura K, Sunami T, Suzuki H, Yomo T. 2012. Cell-free protein synthesis inside giant unilamellar vesicles analyzed by flow cytometry. *Langmuir* **28**: 8426–8432.

Nishimura K, Tsuru S, Suzuki H, Yomo T. 2014. Stochasticity in gene expression in a cell-sized compartment. *ACS Synth Biol* **4**: 566–576.

Niwa T, Sasaki Y, Uemura E, Nakamura S, Akiyama M, Ando M, Sawada S, Mukai SA, Ueda T, Taguchi H, et al. 2015. Comprehensive study of liposome-assisted synthesis of membrane proteins using a reconstituted cell-free translation system. *Sci Rep* **5**: 18025.

Noireaux V, Bar-Ziv R, Libchaber A. 2003. Principles of cell-free genetic circuit assembly. *Proc Natl Acad Sci* **100**: 12672–12677.

Norred SE, Caveney PM, Retterer ST, Boreyko JB, Fowlkes JD, Collier CP, Simpson ML. 2015. Sealable femtoliter chamber arrays for cell-free biology. *J Vis Exp* doi: 10.3791/52616.

Nozawa A, Tozawa Y. 2014. Modifications of wheat germ cell-free system for functional proteomics of plant membrane proteins. *Methods Mol Biol* **1072**: 259–272.

Nozawa A, Matsubara Y, Tanaka Y, Takahashi H, Akagi T, Seki M, Shinozaki K, Endo Y, Sawasaki T. 2009. Construction of a protein library of Arabidopsis transcription factors using a wheat cell-free protein production system and its application for DNA binding analysis. *Biosci Biotechnol Biochem* **73**: 1661–1664.

O'Donoghue P, Ling J, Wang YS, Soll D. 2013. Upgrading protein synthesis for synthetic biology. *Nat Chem Biol* **9**: 594–598.

Oh SJ, Lee KH, Kim HC, Catherine C, Yun H, Kim DM. 2014. Translational incorporation of multiple unnatural amino acids in a cell-free protein synthesis system. *Biotechnol Bioproc Eng* **19**: 426–432.

Ohtake K, Sato A, Mukai T, Hino N, Yokoyama S, Sakamoto K. 2012. Efficient decoding of the UAG triplet as a full-fledged sense codon enhances the growth of a *prfA*-deficient strain of *Escherichia coli*. *J Bacteriol* **194**: 2606–2613.

Orth JH, Schorch B, Boundy S, Ffrench-Constant R, Kubick S, Aktories K. 2011. Cell-free synthesis and characterization of a novel cytotoxic pierisin-like protein from the cabbage butterfly *Pieris rapae*. *Toxicon* **57**: 199–207.

Oza JP, Aerni HR, Pirman NL, Barber KW, ter Haar CM, Rogulina S, Amrofell MB, Isaacs FJ, Rinehart J, Jewett MC. 2015. Robust production of recombinant phosphoproteins using cell-free protein synthesis. *Nat Commun* **6**: 8168.

Ozawa K, Loscha KV, Kuppan KV, Loh CT, Dixon NE, Otting G. 2012. High-yield cell-free protein synthesis for site-specific incorporation of unnatural amino acids at two sites. *Biochem Biophys Res Commun* **418**: 652–656.

Pardee K, Green AA, Ferrante T, Cameron DE, DaleyKeyser A, Yin P, Collins JJ. 2014. Paper-based synthetic gene networks. *Cell* **159**: 940–954.

Park HS, Hohn MJ, Umehara T, Guo LT, Osborne EM, Benner J, Noren CJ, Rinehart J, Soll D. 2011. Expanding the genetic code of *Escherichia coli* with phosphoserine. *Science* **333**: 1151–1154.

Passioura T, Suga H. 2013. Flexizyme-mediated genetic reprogramming as a tool for noncanonical peptide synthesis and drug discovery. *Chemistry* **19**: 6530–6536.

Pratt J. 1984. Coupled transcription-translation in prokaryotic cell-free systems. In *Transcription and translation: A practical approach* (ed. Hames BD, Higgins SJ), pp. 179–209. Oxford University Press, Cambridge.

Rackham O, Chin JW. 2005. Cellular logic with orthogonal ribosomes. *J Am Chem Soc* **127**: 17584–17585.

Rajendran M, Ellington AD. 2008. Selection of fluorescent aptamer beacons that light up in the presence of zinc. *Anal Bioanal Chem* **390**: 1067–1075.

Rendl M, Brandstetter T, Ruhe J. 2013. Time-resolved analysis of biological reactions based on heterogeneous assays in liquid plugs of nanoliter volume. *Anal Chem* **85**: 9469–9477.

Ro DK, Paradise EM, Ouellet M, Fisher KJ, Newman KL, Ndungu JM, Ho KA, Eachus RA, Ham TS, Kirby J, et al. 2006. Production of the antimalarial drug precursor artemisinic acid in engineered yeast. *Nature* **440**: 940–943.

Sachse R, Dondapati SK, Fenz SF, Schmidt T, Kubick S. 2014. Membrane protein synthesis in cell-free systems: From bio-mimetic systems to bio-membranes. *FEBS Lett* **588**: 2774–2781.

Salehi AS, Smith MT, Bennett AM, Williams JB, Pitt WG, Bundy BC. 2016. Cell-free protein synthesis of a cytotoxic cancer therapeutic: Onconase production and a just-add-water cell-free system. *Biotechnol J* **11**: 274–281.

Sawasaki T, Ogasawara T, Morishita R, Endo Y. 2002. A cell-free protein synthesis system for high-throughput proteomics. *Proc Natl Acad Sci* **99**: 14652–14657.

Schoborg JA, Hodgman CE, Anderson MJ, Jewett MC. 2014. Substrate replenishment and byproduct removal improve yeast cell-free protein synthesis. *Biotechnol J* **9**: 630–640.

Selinger DW, Saxena RM, Cheung KJ, Church GM, Rosenow C. 2003. Global RNA half-life analysis in *Escherichia coli* reveals positional patterns of transcript degradation. *Genome Res* **13**: 216–223.

Selkrig J, Leyton DL, Webb CT, Lithgow T. 2014. Assembly of β-barrel proteins into bacterial outer membranes. *Biochim Biophys Acta* **1843**: 1542–1550.

Shimizu Y, Inoue A, Tomari Y, Suzuki T, Yokogawa T, Nishikawa K, Ueda T. 2001. Cell-free translation reconstituted with purified components. *Nat Biotechnol* **19**: 751–755.

Shimizu Y, Kanamori T, Ueda T. 2005. Protein synthesis by pure translation systems. *Methods* **36**: 299–304.

Shimizu Y, Kuruma Y, Kanamori T, Ueda T. 2014. The PURE system for protein production. *Methods Mol Biol* **1118**: 275–284.

Shin J, Noireaux V. 2010. Study of messenger RNA inactivation and protein degradation in an *Escherichia coli* cell-free expression system. *J Biol Eng* **4**: 9.

Shin J, Noireaux V. 2012. An *E. coli* cell-free expression toolbox: Application to synthetic gene circuits and artificial cells. *ACS Synth Biol* **1**: 29–41.

Shin J, Jardine P, Noireaux V. 2012. Genome replication, synthesis, and assembly of the bacteriophage T7 in a single cell-free reaction. *ACS Synth Biol* **1**: 408–413.

Shrestha P, Holland TM, Bundy BC. 2012. Streamlined extract preparation for *Escherichia coli*-based cell-free protein synthesis by sonication or bead vortex mixing. *Biotechniques* **53**: 163–174.

Siegal-Gaskins D, Tuza ZA, Kim J, Noireaux V, Murray RM. 2014. Gene circuit performance characterization and resource usage in a cell-free "breadboard." *ACS Synth Biol* **3**: 416–425.

Sleight SC, Sauro HM. 2013. Randomized BioBrick assembly: A novel DNA assembly method for randomizing and optimizing genetic circuits and metabolic pathways. *ACS Synth Biol* **2**: 506–518.

Stafford RL, Matsumoto ML, Yin G, Cai Q, Fung JJ, Stephenson H, Gill A, You M, Lin SH, Wang WD, et al. 2014. In vitro Fab display: A cell-free system for IgG discovery. *Protein Eng Des Sel* **27**: 97–109.

Stano P, Luisi PL. 2013. Semi-synthetic minimal cells: Origin and recent developments. *Curr Opin Biotechnol* **24:** 633–638.

Stech M, Merk H, Schenk JA, Stocklein WF, Wustenhagen DA, Micheel B, Duschl C, Bier FF, Kubick S. 2012. Production of functional antibody fragments in a vesicle-based eukaryotic cell-free translation system. *J Biotechnol* **164:** 220–231.

Stech M, Brodel AK, Quast RB, Sachse R, Kubick S. 2013. Cell-free systems: Functional modules for synthetic and chemical biology. *Adv Biochem Eng Biotechnol* **137:** 67–102.

Stech M, Quast RB, Sachse R, Schulze C, Wustenhagen DA, Kubick S. 2014. A continuous-exchange cell-free protein synthesis system based on extracts from cultured insect cells. *PLoS ONE* **9:** e96635.

Sullivan CJ, Pendleton ED, Sasmor HH, Hicks WL, Farnum JB, Muto M, Amendt EM, Schoborg JA, Martin RW, Clark LG, et al. 2016. A cell-free expression and purification process for rapid production of protein biologics. *Biotechnol J* **11:** 238–248.

Sun ZZ, Yeung E, Hayes CA, Noireaux V, Murray RM. 2014. Linear DNA for rapid prototyping of synthetic biological circuits in an *Escherichia coli* based TX-TL cell-free system. *ACS Synth Biol* **3:** 387–397.

Suzuki T, Ito M, Ezure T, Shikata M, Ando E, Utsumi T, Tsunasawa S, Nishimura O. 2007. Preparation of *N*-acylated proteins modified with fatty acids having a specific chain length using an insect cell-free protein synthesis system. *Biosci Biotechnol Biochem* **71:** 261–264.

Swartz J. 2006. Developing cell-free biology for industrial applications. *J Ind Microbiol Biotechnol* **33:** 476–485.

Takahashi MK, Chappell J, Hayes CA, Sun ZZ, Kim J, Singhal V, Spring KJ, Al-Khabouri S, Fall CP, Noireaux V, et al. 2014. Rapidly characterizing the fast dynamics of RNA genetic circuitry with cell-free transcription-translation (TX-TL) systems. *ACS Synth Biol* **4:** 503–515.

Takemori N, Takemori A, Matsuoka K, Morishita R, Matsushita N, Aoshima M, Takeda H, Sawasaki T, Endo Y, Higashiyama S. 2015. High-throughput synthesis of stable isotope-labeled transmembrane proteins for targeted transmembrane proteomics using a wheat germ cell-free protein synthesis system. *Mol Biosyst* **11:** 361–365.

Timm AC, Shankles PG, Foster CM, Doktycz MJ, Retterer ST. 2016. Toward microfluidic reactors for cell-free protein synthesis at the point-of-care. *Small* **12:** 810–817.

Tsuboi T, Takeo S, Iriko H, Jin L, Tsuchimochi M, Matsuda S, Han ET, Otsuki H, Kaneko O, Sattabongkot J, et al. 2008. Wheat germ cell-free system-based production of malaria proteins for discovery of novel vaccine candidates. *Infect Immun* **76:** 1702–1708.

Valderrama-Rincon JD, Fisher AC, Merritt JH, Fan YY, Reading CA, Chhiba K, Heiss C, Azadi P, Aebi M, DeLisa MP. 2012. An engineered eukaryotic protein glycosylation pathway in *Escherichia coli*. *Nat Chem Biol* **8:** 434–436.

Venter JC, Remington K, Heidelberg JF, Halpern AL, Rusch D, Eisen JA, Wu D, Paulsen I, Nelson KE, Nelson W. 2004. Environmental genome shotgun sequencing of the Sargasso Sea. *Science* **304:** 66–74.

Virdee S, Kapadnis PB, Elliott T, Lang K, Madrzak J, Nguyen DP, Riechmann L, Chin JW. 2011. Traceless and site-specific ubiquitination of recombinant proteins. *J Am Chem Soc* **133:** 10708–10711.

Wang L, Xie J, Schultz PG. 2006. Expanding the genetic code. *Annu Rev Biophys Biomol Struct* **35:** 225–249.

Wang X, Liu J, Zheng Y, Li J, Wang H, Zhou Y, Qi M, Yu H, Tang W, Zhao WM. 2008. An optimized yeast cell-free system: Sufficient for translation of human papillomavirus 58 L1 mRNA and assembly of virus-like particles. *J Biosci Bioeng* **106:** 8–15.

Wang X, Corin K, Baaske P, Wienken CJ, Jerabek-Willemsen M, Duhr S, Braun D, Zhang S. 2011. Peptide surfactants for cell-free production of functional G protein–coupled receptors. *Proc Natl Acad Sci* **108:** 9049–9054.

Win MN, Smolke CD. 2008. Higher-order cellular information processing with synthetic RNA devices. *Science* **322:** 456–460.

Xu Y, Lee J, Tran C, Heibeck TH, Wang WD, Yang J, Stafford RL, Steiner AR, Sato AK, Hallam TJ. 2014. Production of bispecific antibodies in "Knobs-into-Holes" using a cell-free expression system. *mAbs* 7: 231–242.

Yin G, Garces ED, Yang J, Zhang J, Tran C, Steiner AR, Roos C, Bajad S, Hudak S, Penta K, et al. 2012. Aglycosylated antibodies and antibody fragments produced in a scalable in vitro transcription-translation system. *mAbs* **4:** 217–225.

Yunker PJ, Asahara H, Hung KC, Landry C, Arriaga LR, Akartuna I, Heyman J, Chong S, Weitz DA. 2016. One-pot system for synthesis, assembly, and display of functional single-span membrane proteins on oil–water interfaces. *Proc Natl Acad Sci* **113:** 608–613.

Zalkin H, Yanofsky C, Squires CL. 1974. Regulated in vitro synthesis of *Escherichia coli* tryptophan operon messenger ribonucleic acid and enzymes. *J Biol Chem* **249:** 465–475.

Zawada JF, Yin G, Steiner AR, Yang J, Naresh A, Roy SM, Gold DS, Heinsohn HG, Murray CJ. 2011. Microscale to manufacturing scale-up of cell-free cytokine production—A new approach for shortening protein production development timelines. *Biotechnol Bioeng* **108:** 1570–1578.

Zemella A, Thoring L, Hoffmeister C, Kubick S. 2015. Cell-free protein synthesis: Pros and cons of prokaryotic and eukaryotic systems. *Chembiochem* **16:** 2420–2431.

Zieleniecki JL, Nagarajan Y, Waters S, Rongala J, Thompson V, Hrmova M, Koper I. 2016. Cell-free synthesis of a functional membrane transporter into a tethered bilayer lipid membrane. *Langmuir* **32:** 2445–2449.

Zimmerman ES, Heibeck TH, Gill A, Li X, Murray CJ, Madlansacay MR, Tran C, Uter NT, Yin G, Rivers PJ, et al. 2014. Production of site-specific antibody-drug conjugates using optimized non-natural amino acids in a cell-free expression system. *Bioconjug Chem* **25:** 351–361.

Building Spatial Synthetic Biology with Compartments, Scaffolds, and Communities

Jessica K. Polka,[1,2] Stephanie G. Hays,[1,2] and Pamela A. Silver[1,2]

[1]Department of Systems Biology, Harvard Medical School, Boston, Massachusetts 02115

[2]Wyss Institute for Biologically Inspired Engineering, Boston, Massachusetts 02115

Correspondence: pamela_silver@hms.harvard.edu

Traditional views of synthetic biology often treat the cell as an unstructured container in which biological reactions proceed uniformly. In reality, the organization of biological molecules has profound effects on cellular function: not only metabolic, but also physical and mechanical. Here, we discuss a variety of perturbations available to biologists in controlling protein, nucleotide, and membrane localization. These range from simple tags, fusions, and scaffolds to heterologous expression of compartments and other structures that confer unique physical properties to cells. Next, we relate these principles to those guiding the spatial environments outside of cells such as the extracellular matrix. Finally, we discuss new directions in building intercellular organizations to create novel symbioses.

Spatial organization of biological molecules is foundational to life. The localization of proteins to distinct regions of the cell—be they as simple as the cytosol, membrane, and nucleoid—restricts and directs enzymatic interactions. The division of cellular contents into compartments further separates otherwise incompatible chemical and enzymatic microenvironments. And finally, the organization of proteins into polymeric structures endows cells with the ability to control their mechanical properties and exert forces in their environments.

Synthetic biology is typically understood to mean endowing cells with novel functions based on genetic manipulations. Of course, modifying transcription necessarily modifies the protein environment in the cell as well. However, simply expressing proteins that are assumed to act independently of their environment severely limits the potential of this discipline to generate new cellular behaviors. Here, we describe a range of engineered modifications to cellular organization. Some of these have reached mainstream use in studying and perturbing cellular functions, whereas others have unrealized potential.

DIRECTING PROTEIN LOCALIZATION

A fundamental source of enzymatic organization within cells is the direction of proteins to specific compartments or locations within the cell (Blobel and Sabatini 1971). These can include the nucleus (Silver et al. 1984), mitochondria (Rapaport 2003), chloroplasts (Tian and Okita 2014), vacuoles (Washida et al. 2009), or, in the case of Gram-negative bacteria, the

Cite this article as *Cold Spring Harb Perspect Biol* doi: 10.1101/cshperspect.a024018

periplasm (Danese and Silhavy 1998). As the responsible elements are amino-terminal sequences on the proteins, this localization mechanism can be readily exploited to direct proteins to novel locations. For example, this technology has recently been combined with variable splicing to create reporters with differential localization to plant cell chloroplasts, peroxisomes, and the cytosol (Voges et al. 2013).

In addition to directing proteins to specific localizations in the cell, their oligomerization state can be modulated. Glutathione *S*-transferase tags typically used for affinity purification form dimers (Parker et al. 1990), and this property alters protein function such as by increasing the binding affinity of a DNA-binding protein (Niedziela-Majka et al. 1998). This property has been exploited as a test of the functional role of the oligomeric state (Lee et al. 2003).

Although induced hetero- or homodimerization is a common technique in molecular and cell biology, these tools and approaches have, nevertheless, been expanded in interesting ways. In addition to localizing individual proteins, they have been used to move whole organelles. A synthetic heterodimerization "crutch" linking the endoplasmic reticulum (ER) to mitochondria was used in a screen to identify factors that naturally colocalize these organelles (Kornmann et al. 2009). Dimerization can also be inducible, as well as constitutive. The FK506-binding protein (FKBP) can be used as a tag that dimerizes on withdrawal of its chemical ligand; mutants can reversibly convert between monomers and dimers (Rollins et al. 2000). Furthermore, engineered phytochrome interaction factor (PIF) pairs are induced to dimerize with light; this permits optogenetic control of protein spatial localization, even in regions of single cells (Levskaya et al. 2009). And, finally, dimerizing tags are available in a variety of affinities, including those that form covalent bonds. The SpyCatcher and SpyTag system is a derivative of a bacterial fibronectin-binding protein that irreversibly forms an isopeptide bond on interaction (Zakeri et al. 2012).

In addition to directing the localization of single proteins, tags can be used to make larger aggregates, which can, in turn, be used to probe cellular function. For example, FM repeats (tandem repeats of FKBP domains) have been fused to enzymes in the Golgi apparatus, causing them to polymerize into aggregates on withdrawal of the ligand. The accumulation of these polymerized enzymes in the *cis*- or *trans*-Golgi have been used to investigate models of cisternal maturation (Rivera et al. 2000; Lavieu et al. 2013; Rizzo et al. 2013).

SCAFFOLDS

Directing the localization of proteins can be used to not only probe cellular function, but also enhance enzymatic activity, or to achieve a metabolic or signaling effect. Multiple proteins can be scaffolded, or coordinated with specified geometry, to enhance their activity. The exact nature of enzymatic enhancement on scaffolds is likely to be a combination of several factors, including a high local concentration of substrates and their "channeling" within a shared hydration shell. The latter has been proposed after introducing an inert bridge protein between glucose oxidase and horseradish peroxidase on a DNA scaffold in vitro, thereby increasing the enzymatic output compared with similar scaffolds lacking the bridge (Fu et al. 2012).

Scaffolding to enhance enzymatic activity occurs naturally in cells. For example, fatty acid synthetase is a single protein with seven enzymatic centers; their direct translational fusion is likely to confer metabolic benefits (Smith 1994). Multiple polypeptides can also be brought together by protein–protein interactions; in *Bacillus subtilis*, the nonribosomal peptide synthetases and polyketide synthetase enzymes assemble to form an ~2.5 MD complex, and these complexes are further localized together in one focus in the cell (Straight et al. 2007).

From a synthetic biology perspective, the simplest engineered enzymatic enhancement is by direct fusion of one protein to another. This approach was used to achieve 15-fold enhancement of resveratrol production in yeast and human cells (Zhang et al. 2006b). Scaffolds, however, offer the additional flexibility of linking components by protein–protein interactions. An early example is a modification of the bacterial cellulosome, which is an assembly that organizes cellulose-degrading enzymes. This work enhanced the natural cellulolytic activity of *Clostridium* by combining enzymes and binding domains from different species (Fierobe et al. 2001). A synthetic protein scaffold with more distantly related components was created in 2009 by recruiting yeast metabolic enzymes to metazoan proteins that naturally interact to function in signaling. This improved melvalonate production between threefold and more than 70-fold in *Escherichia coli*, depending on the concentration of enzymes expressed—at lower concentrations of enzymes, scaffolds were more beneficial, presumably because of a greater relative increase in effective concentration (Dueber et al. 2009). These protein scaffolds are limited in the potential spatial organization of the interacting domains, and by the fact that their assembly is not, by nature, polymeric.

In contrast, the modular properties of nucleic acids can be used to create structures that offer both theoretically infinite size and fine control of enzymatic orientation. RNA can be constructed scaffolds in varying dimensionalities—zero-, one-, and two-dimensional scaffolds can be constructed by varying the base pairing between modular segments (Delebecque et al. 2012). The number of interacting proteins is limited only by the number of aptamers (and aptamer-binding proteins fused to the enzymes of interest). Much like protein scaffolds, RNA scaffold can achieve >50-fold enzymatic enhancements in *E. coli*, although the limited stability of RNA may affect scaffold performance over long timescales (Delebecque et al. 2012).

The potential problems with RNA stability may be alleviated with the use of DNA scaffolds instead, although this platform presents its own inherent limitations. For example, DNA topology is limited to a single dimension, and supercoiling is also a potential problem. Nevertheless, DNA scaffolds have been used to generate up to fivefold enhancements in resveratrol, 1,2-propanediol, and mevalonate production in *E. coli* (Conrado et al. 2012). Each enzyme was fused to a zinc-finger protein; the diversity of this protein family suggests that many enzymes could be arranged in a specific array.

MEMBRANE BARRIERS

Although scaffolds may organize high concentrations of enzymes in specific orientations, the surrounding environment remains essentially identical to the cellular compartment in which they form. This could be disadvantageous for several reasons (Fig. 1). First, reactions may

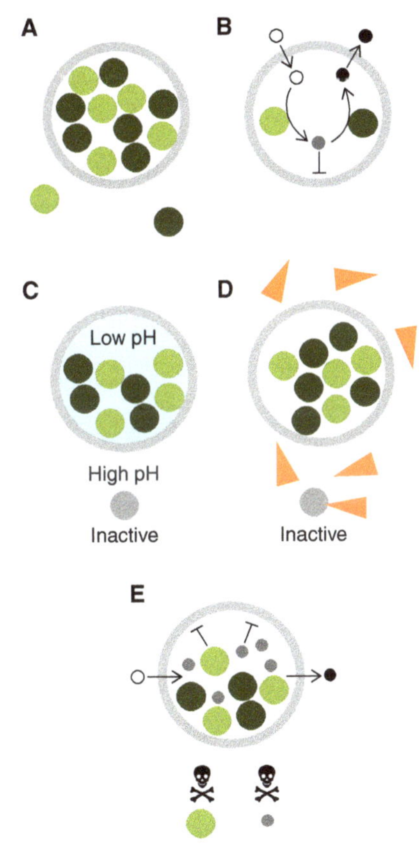

Figure 1. Potential benefits of compartmentalization (with gray circles representing intracellular compartments). (*A*) Concentration of enzymes. (*B*) Increased concentration of intermediates through a selectively permeable diffusion barrier. (*C*) Maintenance of a chemical microenvironment essential for enzymatic function. (*D*) Protection of enzymes from molecules from deactivators or competitors. (*E*) Isolation of toxic enzymes of intermediates from the cell.

produce toxic products, byproducts, or intermediates, such as the formaldehyde formed as an intermediate of methanol metabolism (Fassel et al. 1992). Second, enzymes may need to be insulated from certain substrates to prevent nonproductive side reactions—for example, RuBisCO can catabolize a reaction with molecular oxygen rather than carbon dioxide, drastically reducing enzymatic efficiency. Third, enzymes may require a specific chemical environment in terms of pH, redox potential, or ionic strength that is distinct from the rest of the cytosol, such as is created in mitochondria.

Because most compartments in eukaryotes are membrane based, it is no surprise that many engineering attempts have focused on membrane encapsulation. The most readily implemented form of this technology is the use of targeting sequences to localize enzymes to existing organelles. For example, up to five enzymes in an isobutanol synthesis pathway have been directed to yeast mitochondria; there, they provide modest (approximately threefold) enhancements overexpression in the cytoplasm (Avalos et al. 2013).

In addition, new membrane topologies can also be created from the ER, often through the overexpression of foreign membrane-associated proteins. In tobacco plants, viral proteins form whorled membrane structures termed Z-membranes that appear to derive from the ER (Gong et al. 1996). By making fusion to these viral proteins, various enzymes can be localized on the cytoplasmic surface of these membranes. In mammalian tissue culture cells, ER derivatives having a cuboid or crystalline morphology (termed organized smooth ER) can be generated by expression of certain proteins (Yamamoto et al. 1996; Snapp et al. 2003). This morphology can be produced with chimeric proteins that inducibly dimerize, enabling the topology to be dynamically controlled (Lingwood et al. 2009).

Vesicles have also been generated from other membrane sources, for example, from the cell membrane of *E. coli* (Fig. 2). This was accomplished by overexpressing foreign bacterial glucosyltransferases (Eriksson et al. 2009) or by expressing eukaryotic calveolin (Walser et al. 2012). The latter strategy produces morpholog-

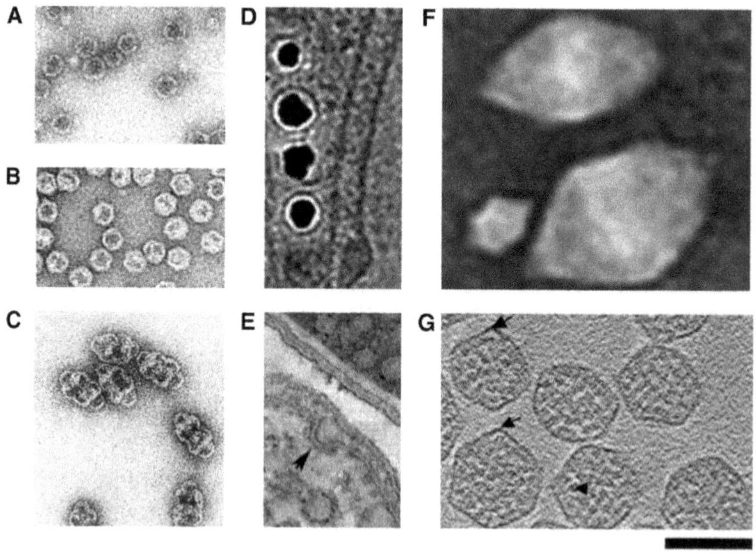

Figure 2. Transmission electron micrographs (TEMs) of cellular compartments, scaled to their relative size. (*A*) Lumazine synthase (Zhang et al. 2006a). (*B*) Encapsulin (Sutter et al. 2008). (*C*) Vaults (Kedersha and Rome 1986). (*D*) Magnetosomes near the cell membrane (Komeili et al. 2006). (*E*) Calveolae: budding from the cell membrane is indicated with an arrow (Walser et al. 2012). (*F*) Gas vesicles (Pfeifer 2012). (*G*) α carboxysomes: arrows indicate vertexes, triangle indicates RuBisCO (Schmid et al. 2006). Scale bar, 100 μm.

 Cite this article as *Cold Spring Harb Perspect Biol* doi: 10.1101/cshperspect.a024018

ically normal calveolae 50 nm in diameter that encapsulate contents derived from the periplasm. Furthermore, by making fusions to the calveolin, the vesicles can be functionalized with proteins that allow them to be targeted to specific cell types after purification.

In addition to directing encapsulation within membrane-bound compartments, we are also able to control release of contents from membrane-bound compartments, such as the cell itself. This takes the shape of inducible bacterial lysis, reviewed in Rice and Bayles (2008). These strategies can involve the expression of lysozyme and pore-forming proteins in response to varied triggers (Pasotti et al. 2011). However, bacteria can also be made to kill specific foreign strains of bacteria with the use of specialized protein machines. For example, pyocins are contractile apparati resembling phage tails that kill target bacteria by depolarizing bacterial membranes (Uratani and Hoshino 1984). Pyocins can be engineered to kill varied bacteria by substitution of their tail fiber proteins from strain-specific phage (Williams et al. 2008). R bodies, proteinaceous toxin delivery devices resembling needles that can extend up to 20-μm long, function to mix the cytoplasms of the expressing bacteria with that of a predatory eukaryote in response to the low-pH environment of a phagosome (reviewed in Pond et al. 1989). As R-bodies can break foreign membranes to release encapsulated protein cargo (Polka and Silver 2016), they could be used to release or mix contents from multiple cells, synthetic vesicles, or compartments.

MICROCOMPARTMENTS

Although further afield from the traditional conception of eukaryotic organelles, proteinaceous compartments may offer a more flexible, robust, and practical platform for engineering compartmentalization. Perhaps the most versatile are microcompartments: icosahedral protein structures, typically between 100 and 200 nm in diameter, found in diverse bacteria. They naturally encapsulate and insulate enzymes from the cellular environment with faceted, selectively permeable shells composed of pentameric and hexameric proteins. Three of the best-studied microcompartment classes include Pdu (propanediol use) microcompartments; Eut (ethanolamine use) microcompartments; and carboxysomes, which contain carbonic anhydrase and RuBisCO. Carboxysomes form two major families: α and β carboxysomes, which differ in the type of RuBisCO incorporated and also by the importance of the matrix in organizing the shell (Schmid et al. 2006; Rae et al. 2013). Microcompartments are now known to be more widespread than originally thought, with 17% of sequenced genomes bearing sequences thought to encode microcompartments of these and other types (Jorda et al. 2013).

Microcompartments are of particular interest because they can be assembled from a small number of protein components and are, therefore, more portable to heterologous expression systems. For example, Pdu compartments from *Citrobacter freundii* have been made in *E. coli* by heterologous expression (Parsons et al. 2008); the empty shell can also be made by expressing only five proteins (Parsons et al. 2010). Because the lumen components were not transferred in the latter study, the compartments are hollow, leaving the opportunity for them to be filled with alternate cargo, as discussed below. Pdu compartments from *Salmonella entericum* have also been made in *E. coli* (Sargent et al. 2013), and empty Eut shells have been reported to form in *E. coli* as well (Choudhary et al. 2012). Finally, α carboxysomes from *Halothiobacillus neapolitanis* have been expressed in *E. coli* from a single operon (Bonacci et al. 2012). Complete β carboxysomes have, so far, resisted heterologous expression, perhaps because of a more complicated genetic organization spanning several operons. However, expression of β carboxysome shell proteins in *E. coli* does produce capsid-like structures (albeit smaller than carboxysomes) that associate with heterologous cargoes tagged with targeting sequences (Cai et al. 2016).

Heterologous expression of carboxysomes, even with their native contents, is of particular interest in light of the goal of improving carbon fixation in plants. Carbon fixation is a metabolic bottleneck for many plants, and the introduction of carboxysomes into the cells of agriculturally important organisms could theoretically increase crop yields by 36%–60% (McGrath and Long 2014). Although the cyanobacterial RuBisCO is faster than plant RuBisCO, it is also more sensitive to molecular oxygen, and thus is detrimental to the growth of its host without encapsulation. Thus, encapsulating this enzyme will be necessary to reap any metabolic benefits of cyanobacterial RuBisCO. Progress has been made toward this end in the expression of carboxysome components in tobacco, with shell and RuBisCO both being successfully expressed (Lin et al. 2014a,b). However, a complete carboxysome in plants has yet to be achieved.

For microcompartments to have novel functions, heterologous cargoes must be targeted to them. In general, this is accomplished by exploiting natural mechanisms of targeting cargoes to the microcompartment lumen. Across many families of microcompartments, short regions at the termini of cargo proteins are thought to form small α helices that bind to the interior surfaces of shell proteins (Fan et al. 2012). The best-developed platform for heterologous targeting is the Pdu microcompartment system. The terminus of PduP from *Salmonella enterica* is necessary and sufficient to drive cargoes to the microcompartment. Green fluorescent protein (GFP) tagged with Pdu^{1-18} cannot only localize to the compartments with fluorescence microscopy, but it is also protected from immunoprecipitation from cell lysates after disruption; this suggests it is located inside the compartment rather than simply associated with the exterior. In addition to PduP, a second enzyme, PduD, also contains an 18-amino-acid sequence sufficient for targeting heterologous cargoes to the *S. enterica* Pdu compartment. PduD is also required for recruiting the other two members of its natural complex, PduCDE, suggesting that not only direct fusions, but also protein–protein interactions, could be used to localize enzymes to the *S. enterica* compartment (Fan and Bobik 2011).

Targeting of engineered cargoes should ideally be functional in a heterologous system. PduC, D, and V can all direct GFP to colocalize with heterologously expressed Pdu compartments in *E. coli* (Parsons et al. 2010). Also in this system, $PduP^{1-18}$ and $PduD^{1-18}$ can be used to target two enzymes involved in ethanol production to the compartment, thereby increasing ethanol yield (Lawrence et al. 2014). It is possible that the enzymes are located on the outside, rather than the inside of the compartment, but nonetheless the Pdu structure succeeds as a scaffold. Finally, a targeting system exists in β carboxysomes. The carboxyl terminus of CcmN is responsible for directing it to the *Synechococcus elongatus* carboxysome (Kinney et al. 2012). The peptide is also necessary for the formation of the β carboxysome shell, although it does not exist in α carboxysomes. Although the CcmN peptide has not yet been shown to localize heterologous cargo, proteins can be targeted to the lumen of the carboxysome by fusion to full-length, native cargoes, like RuBisCO. These fusions, at least in the case of fluorescent proteins, do not impact carboxysome structure, function (Savage et al. 2010), or the establishment of an internal oxidizing microenvironment (Chen et al. 2013), implying that carboxysomes could accommodate small enzymatic cargoes and retain their function.

To optimize enzymatic activity in any engineered compartment, the chemical microenvironments may need to be engineered. The concentration of small molecules inside microcompartments is thought to be controlled by the size and charge of the pores formed in the middle of shell protein hexamers; by exchanging residues in the pores, the selectivity of the permeability barrier could, theoretically, be changed. As a step toward this goal, chimeric carboxysome shells have been made. In these experiments, a shell protein (CsoS1) from *Prochlorococcus marinus* colocalized with *S. elongatus* carboxysomes, and even structurally complemented a deletion of their major shell protein (Cai et al. 2014). Although CsoS1 shares <50% similarity with *S. elongatus* shell proteins, this study shows that components of α (*P. marinus*) and β (*S. elongatus*) carboxysomes can be interchangeable, expanding future engineering possibilities.

In addition to the permeability of the shell, the functionality of an engineered microcompartment would be influenced by its size: not only through the number of molecules that could be encapsulated, but also in the surface to volume ratio, which may influence the compartment's microenvironment. There is a large natural range in microcompartment diameters—from 40-nm *Haliangium ochraceum* compartments to 500-nm β carboxysomes in *Spirulina platensis* (Rae et al. 2013; Lassila et al. 2014). In addition, the size of at least one species of β carboxysome can be modulated genetically. In *S. elongatus*, the protein CcmM is localized to the lumen of carboxysome, where it organizes RuBisCO into a solid matrix. CcmM is present in two isoforms; deleting the short isoform drives the formation of small carboxysomes (Long et al. 2010), whereas fusing a tag to CcmM drives the formation of larger carboxysomes (Long et al. 2007).

ENCAPSULINS

Although microcompartments have shells made up of multiple proteins, a second type of bacteria compartment, encapsulins, have only one shell protein and are characterized by smaller diameters, ~30 nm. They typically encapsulate proteins involved in oxidative stress responses, such as ferritin-like proteins or peroxidases.

The first encapsulin was discovered in *Pyrococcus furiosus*. This compartment is made out of only one protein that functions both as its shell and enzymatic cargo (Namba et al. 2005). Related compartments were later found in other organisms, including *Thermatoga maritima*; this encapsulin is composed of 60 monomers of a single subunit that form a cage 24 nm in diameter (Sutter et al. 2008). The crystal structure of the shell protein shows structural similarity to viral capsids, but like microcompartments, there is no homology on the sequence level. These investigators also expressed the two-component encapsulin operon from *Brevibacterium linens* in *E. coli* and found that the cargo, a single oligomeric complex, is localized within the lumen of the compartments, supporting the hypothesis that a carboxy-terminal sequence targets cargoes to encapsulin lumens.

The purified encapsulin from *Rhodococcus jostii* can be assembled and disassembled in vitro in a pH-dependent fashion, and the assembly of the encapsulin enhances the enzymatic activity of the peroxidase cargo (Rahmanpour and Bugg 2013). The *Rhodococcus erythropolis* encapsulin is capable of both being expressed heterologously in *E. coli* and of packaging heterologous cargo. Proteins tagged with the carboxy-terminal 37 amino acids of its natural cargo are protected from trypsin proteolysis in the presence of the shell (Tamura et al. 2015).

Myxococcus xanthus also produces an encapsulin compartment; the native compartment copurifies with not one, but three cargo proteins. This suggests that these structures could enclose multiple proteins and perhaps support encapsulation of whole enzymatic pathways (McHugh et al. 2014). This encapsulin can be expressed in *E. coli*, although the heterologously expressed protein forms some 18-nm compartments, as well as those of the native 32-nm size. Furthermore, encapsulins are also being explored as a platform for targeted drug delivery (Moon et al. 2014).

LUMAZINE SYNTHASE

A third type of proteinaceous compartment found in bacteria is composed of lumazine synthase, an enzyme involved in riboflavin biosynthesis. In some bacteria, yeast, and plants, this protein forms pentamers (Baldi et al. 2000; Meining et al. 2000), but in other species these pentamers are further assembled into capsids, namely, in *B. subtilis* (Ladenstein et al. 1994) and *Aquifex aeolicus* (Zhang et al. 2001). The capsids from the latter system, termed AaLS, can be expressed and purified from *E. coli*; they assemble, in vitro, into compartments either 16 or 30 nm, depending on pH and buffer conditions (Zhang et al. 2006a). They have no native cargo.

To facilitate encapsulation of heterologous cargoes, the internal surface of lumazine synthase has been mutated to increase electrostatic interactions by charge complementarity. This strategy has been useful for encapsulating proteins bearing a positively charged stretch of 10 arginine residues on the carboxyl terminus coexpressed with these modified AaLS compartments in *E. coli* (Seebeck et al. 2006). Furthermore, this approach can raise the tolerated expression level of human immunodeficiency virus (HIV) protease, which is toxic to *E. coli* (Wörsdörfer et al. 2011b). This selection formed the basis of a screen for mutants of AaLS that could better protect its host, presumably through improving encapsulation owing to increased negative charges on its interior surface. These improved capsids, called AaLS-13, can be isolated and loaded in vitro with monomers of GFP that have been mutated to bear 29 additional positively charged residues (Wörsdörfer et al. 2011a), or even with similarly engineered human ferritin (Beck et al. 2014). Ferritin is itself a small microcompartment, ~12 nm in diameter, that binds and stores iron. Thus, the encapsulation of human ferritin into *A. aeolicus* lumazine synthase generates a nested synthetic organization. It is also reminiscent of natural encapsulins that can bear ferritin cargoes, but here the interaction is dependent on engineered properties of the compartments, which come from different domains of life.

Lumazine synthase capsids have been engineered to produce vaccines because they can act as scaffolds for antigen presentation and are themselves immunogenic (Alfano et al. 2015; Du and Wang 2015). They have also been fused to a chelating agent for use as a contrasting agent for magnetic resonance imaging (MRI) (Song et al. 2015).

VAULTS

Not all protein compartments have a bacterial origin. Vaults, lemon-shaped barrels 70-nm long and 40 nm in diameter, are found in the cytoplasm of eukaryotic cells, but their function is unclear. They are naturally composed of three proteins and an untranslated RNA (Kedersha and Rome 1986; Kickhoefer et al. 1993), but can be purified from insect cells by expression of the major structural protein alone (Stephen et al. 2001). A domain called INT, found on one of the nonessential vault proteins (Kickhoefer et al. 1999), is sufficient to target heterologous cargoes to vaults expressed in insect cells (Kickhoefer et al. 2005). These INT-tagged proteins can be visualized in cryo-EM reconstructions as two rings of additional density encircling each half of the barrel. INT tagging also has biochemical effects. Targeting a fluorescent protein with vaults with INT reduces the speed with which fluorescence decreases in response to high ionic strength, suggesting that they are partially protected from solution. Furthermore, INT-tagged luciferase is functional, yet displays reduced kinetics, suggesting that enzymes can function inside vaults (Kickhoefer et al. 2005). From an engineering perspective, this approach has been used to package a *Chlamydia trachomatis* protein into vaults for vaccine development (Zhu et al. 2014). Manganese peroxidase packaged within vaults is more stable, suggesting potential benefits for remediation of pollutants (Wang et al. 2015).

Because vaults are dynamic structures, cargoes can be loaded in vitro (Poderycki et al. 2006); the two halves of the vault can even completely exchange (Yang et al. 2010). These approaches can be used to package inorganic compounds, like Ni-NTA-nanogold, into the vaults (Goldsmith et al. 2009).

Vaults proteins can also be engineered to gain new functions. For example, an amphipathic helix was fused to the amino terminus of the major vault protein, causing lipids to be encapsulated in the vault (Buehler et al. 2014). The lipid-containing vaults bind with greater avidity to fat-soluble drugs. Purified vaults are readily taken up by cells (Kickhoefer et al. 2005), but fusing a viral endosomolytic peptide enhances the delivery of DNA and other molecules to the cytoplasm (Han et al. 2011).

MAGNETOSOMES

In nature, the function of complex structures is not limited to isolating enzymatic activity. Packing of specialized materials—metals and gas—can serve important roles in natural cellular migration, and have been exploited for diverse practical purposes as well.

 Cite this article as *Cold Spring Harb Perspect Biol* doi: 10.1101/cshperspect.a024018

Magnetosomes are intracellular structures that enable some bacteria to position themselves using the Earth's magnetic field. Morphologically, they are invaginations of the cell membrane that enclose magnetite crystals 50 nm in diameter; these are spatially distributed through the cell by an actin homolog (Komeili et al. 2006). Although they are encoded by large islands of genes, magnetosomes from *Magnetospirillum gryphiswaldense* have been heterologously expressed in *Rhodospirillum rubrum*, with their order and morphology preserved (Kolinko et al. 2014).

MamC, a protein that localizes to the membrane, has been tagged with cameloid antibody fragments (i.e., "nanobodies") in *M. gryphiswaldense*. Because this fusion enables interactions between magenetosomes and the nanobody target protein both in vitro and in vivo, it can be used to generate an intracellular "nanotrap" that localizes natural proteins to the magnetosome in living cells (Pollithy et al. 2011).

Magnetosomes also have many possible applications after purification, for example, their patterning on a surface could be used to create a magnetic memory device (Galloway et al. 2012). Furthermore, a phosphohydrolase has also been fused to MamC in *Magnetospirillum magneticum* to create magnetosomes that could be purified and used in vitro to degrade a pesticide (Ginet et al. 2011). Magnetosomes from this organism can also be functionalized with multiple enzymes with defined stoichiometry for cellulose hydrolysis, among many other fusion partners (Honda et al. 2015).

Magnetic particles in yeast have also been produced, although these inclusions lack the structure of magnetosomes (Nishida and Silver 2012). The particles will form naturally in *Saccharomyces cerevisiae* in the presence of ferric citrate, but the yield and size of the particles can be increased by knocking out a yeast iron transporter and heterologously expressing human ferritin.

GAS VESICLES

Gas vesicles constitute another type of genetically encoded, specialized compartment. Used to regulate bacterial buoyancy during migration to appropriate parts of the water column, gas vesicles are protein-enclosed structures that exclude water and so contain only gas. They have a variety of physical morphologies, including spindles, cylinders, and bicones, and they can grow to sizes ranging from 100 nm to 2 μm in length, depending on the species (Pfeifer 2012).

Bacillus megaterium gas vesicles have been heterologously expressed in *E. coli* (Li and Cannon 1998). Because gas vesicles make cells float, they have been proposed as a tool to facilitate harvests of bacteria in an industrial setting through the formation of a pellicle. Purified gas vesicles can also be used to display antigenic peptides (Stuart et al. 2001), to oxygenate tissue culture media (Sundararajan and Ju 2006), and to serve as contrast agents for ultrasound imaging (Shapiro et al. 2014).

CELL MORPHOLOGY AND THE CYTOSKELETON

In eukaryotic cells, long-range order is achieved through dynamic cytoskeletal proteins and reaction-diffusion systems. If these systems could be controlled, cell shape, migration, and cell size could all be potentially regulated. Currently, the eukaryotic actin and microtubule cytoskeletons can be exploited to direct cargoes to specific positions within cells (reviewed in Goodman et al. 2012). For example, a DNA-binding domain fused to a portion of a microtubule motor, a dynein light chain, can enhance transfection affinity, presumably because the reagent enables the microtubule cytoskeleton to move the DNA toward the nucleus (Toledo et al. 2012). Furthermore, targeting peroxisomes to motors can direct these organelles toward either the nucleus or the periphery of the cell, depending on the polarity of the motors (Kapitein et al. 2010a,b).

Although bacteria and archaea lack ubiquitous, persistent load-bearing cytoskeletons in the traditional sense, they do have protein polymers and reaction-diffusion systems that regulate division, cell-wall deposition, and DNA segregation. In addition, these proteins also control the position of compartments like carboxysomes (Savage et al. 2010) and magnetosomes (Komeili et al. 2006). Therefore, these proteins could play an important role in specifying the intracellular location of synthetic structures in the future.

THE EXTRACELLULAR ENVIRONMENT

Biological organization does not stop at the plasma membrane. In bacteria, the extracellular environment provides a means for protection against toxins and mechanical forces. This matrix can also be modified for scaffolding applications, and to create novel structural and mechanical properties. For example, curli protein forms amyloid fibrils and is a major component of the extracellular matrix made by *E. coli* and related bacteria. By placing curli protein under an inducible promoter, biofilm production can be induced with an external signal (Chen et al. 2014). Curli fibrils have also been tagged with metal-binding peptides to enable them to bind gold nanoparticles and quantum dots to make "nanowires," and the proportion of tagged curli subunits can be regulated temporally with different inducers (Chen et al. 2014). Additionally, curli fibrils can be covalently decorated, via SpyCatcher/SpyTag, with any protein of interest (Nguyen et al. 2014).

The extracellular matrix of eukaryotic cells can also be modified to influence behavior. For example, a SpyCatcher/SpyTag cross-linked elastin network that incorporates leukemia-inhibitory factor has been used as a growth medium for mouse embryonic stem cells; this medium maintains stem-cell pleuripotency (Sun et al. 2014). In theory, this matrix could be genetically encoded.

In addition to matrixes, bacteria make other extracellular structures with biological relevance that are amenable to engineering. For example, *Geobacter sulfurreducens* makes extracellular magnetite particles, the size and composition of which can be altered by changing the composition of available metals (Byrne et al. 2013). A second example is exosomes, extracellular vesicles up to 200 nm in diameter, that naturally package messenger RNA and proteins. Exosomes are used to communicate between cells, and have been engineered to incorporate heterologous cargoes (Marcus and Leonard 2013). Similarly, in Gram-negative bacteria, outer membrane vesicles are produced naturally for many purposes, including carrying toxins and protecting bacteria from antibiotics and phage (reviewed in Kulp and Kuehn 2010; Avila-Calderón et al. 2015). They are also the focus of active engineering efforts, for example, in the production of vaccines (Chen et al. 2010).

SYMBIOSIS

Finally, compartments within cells can be cells themselves. For example, diatoms such as *Hemiaulus hauckii* internalize cyanobacteria like *Richelia intracellularis*; these contain dedicated nitrogen-fixing cells called heterocysts. The specialized compartments allow the cyanobacteria, and thus the diatom, to reap the benefits of both nitrogen fixation and oxygenic photosynthesis (Carpenter et al. 1999; Schouten et al. 2013).

Recapitulating endosymbiotic interactions is a goal of particular interest because it could lead to the development of synthetic organelles. For example, cyanobacteria have been engineered to escape from lysosomes; they have been shown to survive inside zebrafish and mammalian macrophages (Agapakis et al. 2011). Notably, similarly treated *E. coli* were severely cytotoxic in these experiments, suggesting that *S. elongatus* may be inherently better suited to engineering stable endosymbiotic interactions in the future.

Cells do not need to engulf—or be engulfed by—their symbiont to take advantage of inherently dissociated cellular components and capacities. Organization into a consortium allows for the division of labor among different cells, which effectively act as individual compartments. In the example of *Chlorochromatium aggregatum*, the consortium is made up of a single β-proteobacterium surrounded by green sulfur bacteria (Overmann 2010). Although the nonmotile green sulfur bacteria can live alone, they benefit from the specialized motility of the flagellated β-proteobacterium that shows chemotaxis toward light, despite the bacterium not being photosynthetic (Frostl and Over-

 Cite this article as *Cold Spring Harb Perspect Biol* doi: 10.1101/cshperspect.a024018

mann 1998). The motility acquired through symbiosis gives the green sulfur bacteria an advantage over other nonmotile competition. Presumably, the β-proteobacteria also benefits through the provision of photosynthetically fixed carbon.

Often this division of labor endows consortia with capacities far beyond those of individual species. Visually stunning examples of this are microbial mats in which colored striations represent different metabolic capacities: gold, green, pink, peach, or dim green layers correspond to organisms growing together capitalizing on different photosynthetic machinery with metabolisms that differ in ability to withstand exposure to oxygen, reduce sulfate, and fix nitrogen (Stal et al. 1985).

Symbiotic consortia appeal to engineers for two reasons: (1) division of labor lends itself well to modularity, and (2) it is widely believed that consortia are more stable than monocultures. In this vein, synthetic biologists have engineered consortia made up of different strains of the same organism, as well as symbioses involving multiple species. There have been many applied examples of the utility of symbiosis, especially in isobutanol and lipid production from sunlight or cellulosic feedstocks (Ortiz-Marquez et al. 2012; Do Nascimento et al. 2013; Minty et al. 2013). Although at their simplest these artificial consortia are merely novel combinations of existing wild-type strains, there are many engineering interventions that can improve their stability. For example, engineered auxotrophies force interdependence (Shou et al. 2007; Wintermute and Silver 2010; Müller et al. 2014), and quorum-sensing systems can be used to facilitate communication between consortium members (Brenner et al. 2007; Weber et al. 2007; Chen et al. 2015).

It is worth noting that symbiosis can also involve metazoan partners. Consider the bioluminescence provided by *Vibrio fischeri* in the bobtail squid (Naughton and Mandel 2012), amino acids provided by bacterial endosymbionts in aphids (Hansen and Moran 2011), organic carbon and oxygen provided by photosynthetic algae in corals (Roth 2014), and our own microbiomes. These and other examples of natural symbiosis will provide fertile ground for future engineering efforts.

CONCLUDING REMARKS

The ultimate goal of these engineering efforts is to control not only the contents of cells, but also the spatial relationship of these contents to one another. Given that the activity of proteins is dependent on both their potential interacting partners and their local chemical environment, cellular organization must be manipulated to explore the full range of potential biological activity. To gain this mastery, synthetic biologists must adopt into their toolkit some microscopy and biochemical methods associated with traditional cell biology.

ACKNOWLEDGMENTS

We are grateful to Arash Komeili and Ertan Ozyamak (UC Berkeley) for helpful discussions regarding magnetosomes. We thank the National Science Foundation (MCB-1409586, DGE1144152), Jane Coffin Childs Fund, the Wyss Institute for Biologically Inspired Engineering, Department of Energy (DE-SC0012658), and Defense Advanced Research Projects Agency Living Foundries (HR0011-14-C-0072).

REFERENCES

Agapakis CM, Niederholtmeyer H, Noche RR, Lieberman TD, Megason SG, Way JC, Silver PA. 2011. Towards a synthetic chloroplast. *PLoS ONE* **6:** e18877.

Alfano EF, Lentz EM, Bellido D, Dus Santos MJ, Goldbaum FA, Wigdorovitz A, Bravo-Almonacid FF. 2015. Expression of the multimeric and highly immunogenic *Brucella* spp. lumazine synthase fused to bovine rotavirus VP8d as a scaffold for antigen production in tobacco chloroplasts. *Front Plant Sci* **6:** 1170.

Avalos JL, Fink GR, Stephanopoulos G. 2013. Compartmentalization of metabolic pathways in yeast mitochondria improves the production of branched-chain alcohols. *Nat Biotechnol* **31:** 335–341.

Avila-Calderón ED, Araiza-Villanueva MG, Cancino-Diaz JC, López-Villegas EO, Sriranganathan N, Boyle SM, Contreras-Rodríguez A. 2015. Roles of bacterial membrane vesicles. *Arch Microbiol* **197:** 1–10.

Baldi PC, Velikovsky CA, Braden BC, Giambartolomei GH, Fossati CA, Goldbaum FA. 2000. Structural, functional and immunological studies on a polymeric bacterial protein. *Braz J Med Biol Res* **33:** 741–747.

Beck T, Tetter S, Künzle M, Hilvert D. 2014. Construction of Matryoshka-type structures from supercharged protein nanocages. *Angew Chem Int Ed Engl* 54: 937–940.

Blobel G, Sabatini DD. 1971. Ribosome-membrane interaction in eukaryotic cells. In *Biomembranes* (ed. Manson LA), pp. 193–195. Springer, Berlin.

Bonacci W, Teng PK, Afonso B, Niederholtmeyer H, Grob P, Silver PA, Savage DF. 2012. Modularity of a carbon-fixing protein organelle. *Proc Natl Acad Sci* **109:** 478–483.

Brenner K, Karig DK, Weiss R, Arnold FH. 2007. Engineered bidirectional communication mediates a consensus in a microbial biofilm consortium. *Proc Natl Acad Sci* **104:** 17300–17304.

Buehler DC, Marsden MD, Shen S, Toso DB, Wu X, Loo JA, Zhou ZH, Kickhoefer VA, Wender PA, Zack JA, et al. 2014. Bioengineered vaults: Self-assembling protein shell-lipophilic core nanoparticles for drug delivery. *ACS Nano* **8:** 7723–7732.

Byrne JM, Coker VS, Moise S, Wincott PL, Vaughan DJ, Tuna F, Arenholz E, van der Laan G, Pattrick RAD, Lloyd JR, et al. 2013. Controlled cobalt doping in biogenic magnetite nanoparticles. *J R Soc Interface R Soc* **10:** 20130134.

Cai F, Sutter M, Bernstein SL, Kinney JN, Kerfeld CA. 2014. Engineering bacterial microcompartment shells: Chimeric shell proteins and chimeric carboxysome shells. *ACS Synth Biol* **4:** 444–453.

Cai F, Bernstein SL, Wilson SC, Kerfeld CA. 2016. Production and characterization of synthetic carboxysome shells with incorporated luminal proteins. *Plant Physiol* **170:** 1868–1877.

Carpenter EJ, Montoya JP, Burns J, Mulholland MR, Subramaniam A, Capone DG. 1999. Extensive bloom of a N2-fixing diatom/cyanobacterial association in the tropical Atlantic Ocean. *Mar Ecol Prog Ser* **185:** 273–283.

Chen DJ, Osterrieder N, Metzger SM, Buckles E, Doody AM, DeLisa MP, Putnam D. 2010. Delivery of foreign antigens by engineered outer membrane vesicle vaccines. *Proc Natl Acad Sci* **107:** 3099–3104.

Chen AH, Robinson-Mosher A, Savage DF, Silver PA, Polka JK. 2013. The bacterial carbon-fixing organelle is formed by shell envelopment of preassembled cargo. *PloS ONE* **8:** e76127.

Chen AY, Deng Z, Billings AN, Seker UOS, Lu MY, Citorik RJ, Zakeri B, Lu TK. 2014. Synthesis and patterning of tunable multiscale materials with engineered cells. *Nat Mater* **13:** 515–523.

Chen Y, Kim JK, Hirning AJ, Josić K, Bennett MR. 2015. Emergent genetic oscillations in a synthetic microbial consortium. *Science* **349:** 986–989.

Choudhary S, Quin MB, Sanders MA, Johnson ET, Schmidt-Dannert C. 2012. Engineered protein nanocompartments for targeted enzyme localization. *PLoS ONE* **7:** e33342.

Conrado RJ, Wu GC, Boock JT, Xu H, Chen SY, Lebar T, Turnšek J, Tomšič N, Avbelj M, Gaber R, et al. 2012. DNA-guided assembly of biosynthetic pathways promotes improved catalytic efficiency. *Nucleic Acids Res* **40:** 1879–1889.

Danese PN, Silhavy TJ. 1998. Targeting and assembly of periplasmic and outer-membrane proteins in *Escherichia coli*. *Annu Rev Genet* **32:** 59–94.

Delebecque CJ, Silver PA, Lindner AB. 2012. Designing and using RNA scaffolds to assemble proteins in vivo. *Nat Protoc* **7:** 1797–1807.

Do Nascimento M, Dublan Mde L, Ortiz-Marquez JC, Curatti L. 2013. High lipid productivity of an Ankistrodesmus-Rhizobium artificial consortium. *Bioresour Technol* **146:** 400–407.

Du ZQ, Wang JY. 2015. A novel lumazine synthase molecule from *Brucella* significantly promotes the immune-stimulation effects of antigenic protein. *Genet Mol Res* **14:** 13084–13095.

Dueber JE, Wu GC, Malmirchegini GR, Moon TS, Petzold CJ, Ullal AV, Prather KLJ, Keasling JD. 2009. Synthetic protein scaffolds provide modular control over metabolic flux. *Nat Biotechnol* **27:** 753–759.

Eriksson HM, Wessman P, Ge C, Edwards K, Wieslander A. 2009. Massive formation of intracellular membrane vesicles in *Escherichia coli* by a monotopic membrane-bound lipid glycosyltransferase. *J Biol Chem* **284:** 33904–33914.

Fan C, Bobik TA. 2011. The N-terminal region of the medium subunit (PduD) packages adenosylcobalamin-dependent diol dehydratase (PduCDE) into the Pdu microcompartment. *J Bacteriol* **193:** 5623–5628.

Fan C, Cheng S, Sinha S, Bobik TA. 2012. Interactions between the termini of lumen enzymes and shell proteins mediate enzyme encapsulation into bacterial microcompartments. *Proc Natl Acad Sci* **109:** 14995–15000.

Fassel TA, Buchholz LA, Collins ML, Remsen CC. 1992. Localization of methanol dehydrogenase in two strains of methylotrophic bacteria detected by immunogold labeling. *Appl Environ Microbiol* **58:** 2302–2307.

Fierobe HP, Mechaly A, Tardif C, Belaich A, Lamed R, Shoham Y, Belaich JP, Bayer EA. 2001. Design and production of active cellulosome chimeras selective incorporation of dockerin-containing enzymes into defined functional complexes. *J Biol Chem* **276:** 21257–21261.

Frostl J, Overmann J. 1998. Physiology and tactic response of the phototrophic consortium "*Chlorochromatium aggregatum*." *Arch Microbiol* **169:** 129–135.

Fu J, Liu M, Liu Y, Woodbury NW, Yan H. 2012. Interenzyme substrate diffusion for an enzyme cascade organized on spatially addressable DNA nanostructures. *J Am Chem Soc* **134:** 5516–5519.

Galloway JM, Bramble JP, Rawlings AE, Burnell G, Evans SD, Staniland SS. 2012. Biotemplated magnetic nanoparticle arrays. *Small* **8:** 204–208.

Ginet N, Pardoux R, Adryanczyk G, Garcia D, Brutesco C, Pignol D. 2011. Single-step production of a recyclable nanobiocatalyst for organophosphate pesticides biodegradation using functionalized bacterial magnetosomes. *PLoS ONE* **6:** e21442.

Goldsmith LE, Pupols M, Kickhoefer VA, Rome LH, Monbouquette HG. 2009. Utilization of a protein "shuttle" to load vault nanocapsules with gold probes and proteins. *ACS Nano* **3:** 3175–3183.

Gong FC, Giddings TH, Meehl JB, Staehelin LA, Galbraith DW. 1996. Z-membranes: Artificial organelles for overexpressing recombinant integral membrane proteins. *Proc Natl Acad Sci* **93:** 2219–2223.

Goodman BS, Derr ND, Reck-Peterson SL. 2012. Engineered, harnessed, and hijacked: Synthetic uses for cytoskeletal systems. *Trends Cell Biol* **22:** 644–652.

Han M, Kickhoefer VA, Nemerow GR, Rome LH. 2011. Targeted vault nanoparticles engineered with an endosomolytic peptide deliver biomolecules to the cytoplasm. *ACS Nano* **5:** 6128–6137.

Hansen AK, Moran NA. 2011. Aphid genome expression reveals host–symbiont cooperation in the production of amino acids. *Proc Natl Acad Sci* **108:** 2849–2854.

Honda T, Tanaka T, Yoshino T. 2015. Stoichiometrically controlled immobilization of multiple enzymes on magnetic nanoparticles by the magnetosome display system for efficient cellulose hydrolysis. *Biomacromolecules* **16:** 3863–3868.

Jorda J, Lopez D, Wheatley NM, Yeates TO. 2013. Using comparative genomics to uncover new kinds of protein-based metabolic organelles in bacteria. *Protein Sci Publ Protein Soc* **22:** 179–195.

Kapitein LC, Schlager MA, Kuijpers M, Wulf PS, van Spronsen M, MacKintosh FC, Hoogenraad CC. 2010a. Mixed microtubules steer dynein-driven cargo transport into dendrites. *Curr Biol* **20:** 290–299.

Kapitein LC, Schlager MA, van der Zwan WA, Wulf PS, Keijzer N, Hoogenraad CC. 2010b. Probing intracellular motor protein activity using an inducible cargo trafficking assay. *Biophys J* **99:** 2143–2152.

Kedersha NL, Rome LH. 1986. Isolation and characterization of a novel ribonucleoprotein particle: Large structures contain a single species of small RNA. *J Cell Biol* **103:** 699–709.

Kickhoefer VA, Searles RP, Kedersha NL, Garber ME, Johnson DL, Rome LH. 1993. Vault ribonucleoprotein particles from rat and bullfrog contain a related small RNA that is transcribed by RNA polymerase III. *J Biol Chem* **268:** 7868–7873.

Kickhoefer VA, Siva AC, Kedersha NL, Inman EM, Ruland C, Streuli M, Rome LH. 1999. The 193-Kd vault protein, Vparp, is a novel poly(Adp-ribose) polymerase. *J Cell Biol* **146:** 917.

Kickhoefer VA, Garcia Y, Mikyas Y, Johansson E, Zhou JC, Raval-Fernandes S, Minoofar P, Zink JI, Dunn B, Stewart PL, et al. 2005. Engineering of vault nanocapsules with enzymatic and fluorescent properties. *Proc Natl Acad Sci* **102:** 4348–4352.

Kinney JN, Salmeen A, Cai F, Kerfeld CA. 2012. Elucidating essential role of conserved carboxysomal protein CcmN reveals common feature of bacterial microcompartment assembly. *J Biol Chem* **287:** 17729–17736.

Kolinko I, Lohße A, Borg S, Raschdorf O, Jogler C, Tu Q, Pósfai M, Tompa É, Plitzko JM, Brachmann A, et al. 2014. Biosynthesis of magnetic nanostructures in a foreign organism by transfer of bacterial magnetosome gene clusters. *Nat Nanotechnol* **9:** 193–197.

Komeili A, Li Z, Newman DK, Jensen GJ. 2006. Magnetosomes are cell membrane invaginations organized by the actin-like protein MamK. *Science* **311:** 242–245.

Kornmann B, Currie E, Collins SR, Schuldiner M, Nunnari J, Weissman JS, Walter P. 2009. An ER-mitochondria tethering complex revealed by a synthetic biology screen. *Science* **325:** 477–481.

Kulp A, Kuehn MJ. 2010. Biological functions and biogenesis of secreted bacterial outer membrane vesicles. *Annu Rev Microbiol* **64:** 163–184.

Ladenstein R, Ritsert K, Huber R, Richter G, Bacher A. 1994. The lumazine synthase/riboflavin synthase complex of *Bacillus subtilis. Eur J Biochem* **223:** 1007–1017.

Lassila JK, Bernstein SL, Kinney JN, Axen SD, Kerfeld CA. 2014. Assembly of robust bacterial microcompartment shells using building blocks from an organelle of unknown function. *J Mol Biol* **426:** 2217–2228.

Lavieu G, Zheng H, Rothman JE. 2013. Stapled Golgi cisternae remain in place as cargo passes through the stack. *eLife* **2:** e00558.

Lawrence AD, Frank S, Newnham S, Lee MJ, Brown IR, Xue WF, Rowe ML, Mulvihill DP, Prentice MB, Howard MJ, et al. 2014. Solution structure of a bacterial microcompartment targeting peptide and its application in the construction of an ethanol bioreactor. *ACS Synth Biol* **3:** 454–465.

Lee LA, Hoewyk DV, Orr-Weaver TL. 2003. The *Drosophila* cell cycle kinase PAN GU forms an active complex with PLUTONIUM and GNU to regulate embryonic divisions. *Genes Dev* **17:** 2979–2991.

Levskaya A, Weiner OD, Lim WA, Voigt CA. 2009. Spatiotemporal control of cell signalling using a light-switchable protein interaction. *Nature* **461:** 997–1001.

Li N, Cannon MC. 1998. Gas vesicle genes identified in *Bacillus megaterium* and functional expression in *Escherichia coli. J Bacteriol* **180:** 2450–2458.

Lin MT, Occhialini A, Andralojc PJ, Devonshire J, Hines KM, Parry MAJ, Hanson MR. 2014a. β-Carboxysomal proteins assemble into highly organized structures in *Nicotiana* chloroplasts. *Plant J Cell Mol Biol* **79:** 1–12.

Lin MT, Occhialini A, Andralojc PJ, Parry MAJ, Hanson MR. 2014b. A faster Rubisco with potential to increase photosynthesis in crops. *Nature* **513:** 547–550.

Lingwood D, Schuck S, Ferguson C, Gerl MJ, Simons K. 2009. Generation of cubic membranes by controlled homotypic interaction of membrane proteins in the endoplasmic reticulum. *J Biol Chem* **284:** 12041–12048.

Long BM, Badger MR, Whitney SM, Price GD. 2007. Analysis of carboxysomes from *Synechococcus* PCC7942 reveals multiple Rubisco complexes with carboxysomal proteins CcmM and CcaA. *J Biol Chem* **282:** 29323–29335.

Long BM, Tucker L, Badger MR, Price GD. 2010. Functional cyanobacterial β-carboxysomes have an absolute requirement for both long and short forms of the CcmM protein. *Plant Physiol* **153:** 285–293.

Marcus ME, Leonard JN. 2013. FedExosomes: Engineering therapeutic biological nanoparticles that truly deliver. *Pharm Basel Switz* **6:** 659–680.

McGrath JM, Long SP. 2014. Can the cyanobacterial carbon-concentrating mechanism increase photosynthesis in crop species? A theoretical analysis. *Plant Physiol* **164:** 2247–2261.

McHugh CA, Fontana J, Nemecek D, Cheng N, Aksyuk AA, Heymann JB, Winkler DC, Lam AS, Wall JS, Steven AC, et al. 2014. A virus capsid-like nanocompartment that stores iron and protects bacteria from oxidative stress. *EMBO J* **33:** 1896–1911.

Meining W, Mörtl S, Fischer M, Cushman M, Bacher A, Ladenstein R. 2000. The atomic structure of pentameric lumazine synthase from *Saccharomyces cerevisiae* at 1.85 Å resolution reveals the binding mode of a phosphonate intermediate analogue. *J Mol Biol* **299:** 181–197.

Minty JJ, Singer ME, Scholz SA, Bae CH, Ahn JH, Foster CE, Liao JC, Lin XN. 2013. Design and characterization of synthetic fungal–bacterial consortia for direct production of isobutanol from cellulosic biomass. *Proc Natl Acad Sci* **110:** 14592–14597.

Moon H, Lee J, Min J, Kang S. 2014. Developing genetically engineered encapsulin protein cage nanoparticles as a targeted delivery nanoplatform. *Biomacromolecules* **15:** 3794–3801.

Müller MJI, Neugeboren BI, Nelson DR, Murray AW. 2014. Genetic drift opposes mutualism during spatial population expansion. *Proc Natl Acad Sci* **111:** 1037–1042.

Namba K, Hagiwara K, Tanaka H, Nakaishi Y, Chong KT, Yamashita E, Armah GE, Ono Y, Ishino Y, Omura T, et al. 2005. Expression and molecular characterization of spherical particles derived from the genome of the hyperthermophilic euryarchaeote *Pyrococcus furiosus*. *J Biochem (Tokyo)* **138:** 193–199.

Naughton LM, Mandel MJ. 2012. Colonization of *Euprymna scolopes* squid by *Vibrio fischeri*. *J Vis Exp* 2012: e3758.

Nguyen PQ, Botyanszki Z, Tay PKR, Joshi NS. 2014. Programmable biofilm-based materials from engineered curli nanofibres. *Nat Commun* **5:** 4945.

Niedziela-Majka A, Rymarczyk G, Kochman M, Ozyhar A. 1998. GST-Induced dimerization of DNA-binding domains alters characteristics of their interaction with DNA. *Protein Expr Purif* **14:** 208–220.

Nishida K, Silver PA. 2012. Induction of biogenic magnetization and redox control by a component of the target of rapamycin complex 1 signaling pathway. *PLoS Biol* **10:** e1001269.

Ortiz-Marquez JCF, Do Nascimento M, Dublan Mde A, Curatti L. 2012. Association with an ammonium-excreting bacterium allows diazotrophic culture of oil-rich eukaryotic microalgae. *Appl Environ Microbiol* **78:** 2345–2352.

Overmann J. 2010. The phototrophic consortium "*Chlorochromatium aggregatum*"—A model for bacterial heterologous multicellularity. *Adv Exp Med Biol* **675:** 15–29.

Parker MW, Bello ML, Federici G. 1990. Crystallization of glutathione *S*-transferase from human placenta. *J Mol Biol* **213:** 221–222.

Parsons JB, Dinesh SD, Deery E, Leech HK, Brindley AA, Heldt D, Frank S, Smales CM, Lünsdorf H, Rambach A, et al. 2008. Biochemical and structural insights into bacterial organelle form and biogenesis. *J Biol Chem* **283:** 14366–14375.

Parsons JB, Frank S, Bhella D, Liang M, Prentice MB, Mulvihill DP, Warren MJ. 2010. Synthesis of empty bacterial microcompartments, directed organelle protein incorporation, and evidence of filament-associated organelle movement. *Mol Cell* **38:** 305–315.

Pasotti L, Zucca S, Lupotto M, Cusella De Angelis MG, Magni P. 2011. Characterization of a synthetic bacterial self-destruction device for programmed cell death and for recombinant proteins release. *J Biol Eng* **5:** 8.

Pfeifer F. 2012. Distribution, formation and regulation of gas vesicles. *Nat Rev Microbiol* **10:** 705–715.

Poderycki MJ, Kickhoefer VA, Kaddis CS, Raval-Fernandes S, Johansson E, Zink JI, Loo JA, Rome LH. 2006. The vault exterior shell is a dynamic structure that allows incorporation of vault-associated proteins into its interior. *Biochemistry (Mosc)* **45:** 12184–12193.

Polka JK, Silver PA. 2016. A tunable protein piston that breaks membranes to release encapsulated cargo. *ACS Synth Biol* doi: 10.1021/acssynbio.5b00237.

Pollithy A, Romer T, Lang C, Müller FD, Helma J, Leonhardt H, Rothbauer U, Schüler D. 2011. Magnetosome expression of functional camelid antibody fragments (nanobodies) in *Magnetospirillum gryphiswaldense*. *Appl Environ Microbiol* **77:** 6165–6171.

Pond FR, Gibson I, Lalucat J, Quackenbush RL. 1989. R-body-producing bacteria. *Microbiol Rev* **53:** 25–67.

Rae BD, Long BM, Badger MR, Price GD. 2013. Functions, compositions, and evolution of the two types of carboxysomes: Polyhedral microcompartments that facilitate CO_2 fixation in cyanobacteria and some proteobacteria. *Microbiol Mol Biol Rev* **77:** 357–379.

Rahmanpour R, Bugg TDH. 2013. Assembly in vitro of *Rhodococcus jostii* RHA1 encapsulin and peroxidase DypB to form a nanocompartment. *FEBS J* **280:** 2097–2104.

Rapaport D. 2003. Finding the right organelle. Targeting signals in mitochondrial outer-membrane proteins. *EMBO Rep* **4:** 948–952.

Rice KC, Bayles KW. 2008. Molecular control of bacterial death and lysis. *Microbiol Mol Biol Rev* **72:** 85–109.

Rivera VM, Wang X, Wardwell S, Courage NL, Volchuk A, Keenan T, Holt DA, Gilman M, Orci L, Cerasoli F, et al. 2000. Regulation of protein secretion through controlled aggregation in the endoplasmic reticulum. *Science* **287:** 826–830.

Rizzo R, Parashuraman S, Mirabelli P, Puri C, Lucocq J, Luini A. 2013. The dynamics of engineered resident proteins in the mammalian Golgi complex relies on cisternal maturation. *J Cell Biol* **201:** 1027–1036.

Rollins CT, Rivera VM, Woolfson DN, Keenan T, Hatada M, Adams SE, Andrade LJ, Yaeger D, van Schravendijk MR, Holt DA, et al. 2000. A ligand-reversible dimerization system for controlling protein–protein interactions. *Proc Natl Acad Sci* **97:** 7096–7101.

Roth MS. 2014. The engine of the reef: Photobiology of the coral-algal symbiosis. *Front Microbiol* **5:** 422.

Sargent F, Davidson FA, Kelly CL, Binny R, Christodoulides N, Gibson D, Johansson E, Kozyrska K, Lado LL, MacCallum J, et al. 2013. A synthetic system for expression of components of a bacterial microcompartment. *Microbiology* **159:** 2427–2436.

Savage DF, Afonso B, Chen AH, Silver PA. 2010. Spatially ordered dynamics of the bacterial carbon fixation machinery. *Science* **327:** 1258–1261.

Schmid MF, Paredes AM, Khant HA, Soyer F, Aldrich HC, Chiu W, Shively JM. 2006. Structure of *Halothiobacillus neapolitanus* carboxysomes by cryo-electron tomography. *J Mol Biol* **364:** 526–535.

Schouten S, Villareal TA, Hopmans EC, Mets A, Swanson KM, Sinninghe Damsté JS. 2013. Endosymbiotic heterocystous cyanobacteria synthesize different heterocyst glycolipids than free-living heterocystous cyanobacteria. *Phytochemistry* **85:** 115–121.

Seebeck FP, Woycechowsky KJ, Zhuang W, Rabe JP, Hilvert D. 2006. A simple tagging system for protein encapsulation. *J Am Chem Soc* **128:** 4516–4517.

Shapiro MG, Goodwill PW, Neogy A, Yin M, Foster FS, Schaffer DV, Conolly SM. 2014. Biogenic gas nanostructures as ultrasonic molecular reporters. *Nat Nanotechnol* **9:** 311–316.

Shou W, Ram S, Vilar JMG. 2007. Synthetic cooperation in engineered yeast populations. *Proc Natl Acad Sci* **104:** 1877–1882.

Silver PA, Keegan LP, Ptashne M. 1984. Amino terminus of the yeast *GAL4* gene product is sufficient for nuclear localization. *Proc Natl Acad Sci* **81:** 5951–5955.

Smith S. 1994. The animal fatty acid synthase: One gene, one polypeptide, seven enzymes. *FASEB J* **8:** 1248–1259.

Snapp EL, Hegde RS, Francolini M, Lombardo F, Colombo S, Pedrazzini E, Borgese N, Lippincott-Schwartz J. 2003. Formation of stacked ER cisternae by low affinity protein interactions. *J Cell Biol* **163:** 257–269.

Song Y, Kang YJ, Jung H, Kim H, Kang S, Cho H. 2015. Lumazine synthase protein nanoparticle-Gd(III)-DOTA conjugate as a T1 contrast agent for high-field MRI. *Sci Rep* **5:** 15656.

Stal LJ, van Gemerden H, Krumbein WE. 1985. Structure and development of a benthic marine microbial mat. *FEMS Microbiol Lett* **31:** 111–125.

Stephen AG, Raval-Fernandes S, Huynh T, Torres M, Kickhoefer VA, Rome LH. 2001. Assembly of vault-like particles in insect cells expressing only the major vault protein. *J Biol Chem* **276:** 23217–23220.

Straight PD, Fischbach MA, Walsh CT, Rudner DZ, Kolter R. 2007. A singular enzymatic megacomplex from *Bacillus subtilis*. *Proc Natl Acad Sci* **104:** 305–310.

Stuart ES, Morshed F, Sremac M, DasSarma S. 2001. Antigen presentation using novel particulate organelles from halophilic archaea. *J Biotechnol* **88:** 119–128.

Sun F, Zhang WB, Mahdavi A, Arnold FH, Tirrell DA. 2014. Synthesis of bioactive protein hydrogels by genetically encoded SpyTag-SpyCatcher chemistry. *Proc Natl Acad Sci* **111:** 11269–11274.

Sundararajan A, Ju LK. 2006. Use of cyanobacterial gas vesicles as oxygen carriers in cell culture. *Cytotechnology* **52:** 139–149.

Sutter M, Boehringer D, Gutmann S, Günther S, Prangishvili D, Loessner MJ, Stetter KO, Weber-Ban E, Ban N. 2008. Structural basis of enzyme encapsulation into a bacterial nanocompartment. *Nat Struct Mol Biol* **15:** 939–947.

Tamura A, Fukutani Y, Takami T, Fujii M, Nakaguchi Y, Murakami Y, Noguchi K, Yohda M, Odaka M. 2015. Packaging guest proteins into the encapsulin nanocompartment from *Rhodococcus erythropolis* N771. *Biotechnol Bioeng* **112:** 13–20.

Tian L, Okita TW. 2014. mRNA-based protein targeting to the endoplasmic reticulum and chloroplasts in plant cells. *Curr Opin Plant Biol* **22C:** 77–85.

Toledo MAS, Janissen R, Favaro MTP, Cotta MA, Monteiro GA, Prazeres DMF, Souza AP, Azzoni AR. 2012. Development of a recombinant fusion protein based on the dynein light chain LC8 for non-viral gene delivery. *J Control Release* **159:** 222–231.

Uratani Y, Hoshino T. 1984. Pyocin R1 inhibits active transport in *Pseudomonas aeruginosa* and depolarizes membrane potential. *J Bacteriol* **157:** 632–636.

Voges MJ, Silver PA, Way JC, Mattozzi MD. 2013. Targeting a heterologous protein to multiple plant organelles via rationally designed 5′ mRNA tags. *J Biol Eng* **7:** 20.

Walser PJ, Ariotti N, Howes M, Ferguson C, Webb R, Schwudke D, Leneva N, Cho KJ, Cooper L, Rae J, et al. 2012. Constitutive formation of caveolae in a bacterium. *Cell* **150:** 752–763.

Wang M, Abad D, Kickhoefer VA, Rome LH, Mahendra S. 2015. Vault nanoparticles packaged with enzymes as an efficient pollutant biodegradation technology. *ACS Nano* **9:** 10931–10940.

Washida H, Sugino A, Kaneko S, Crofts N, Sakulsingharoj C, Kim D, Choi SB, Hamada S, Ogawa M, Wang C, et al. 2009. Identification of cis-localization elements of the maize 10-kDa δ-zein and their use in targeting RNAs to specific cortical endoplasmic reticulum subdomains. *Plant J Cell Mol Biol* **60:** 146–155.

Weber W, Baba MDE, Fussenegger M. 2007. Synthetic ecosystems based on airborne inter- and intrakingdom communication. *Proc Natl Acad Sci* **104:** 10435–10440.

Williams SR, Gebhart D, Martin DW, Scholl D. 2008. Retargeting R-type pyocins to generate novel bactericidal protein complexes. *Appl Environ Microbiol* **74:** 3868–3876.

Wintermute EH, Silver PA. 2010. Emergent cooperation in microbial metabolism. *Mol Syst Biol* **6:** 407.

Wörsdörfer B, Pianowski Z, Hilvert D. 2011a. Efficient in vitro encapsulation of protein cargo by an engineered protein container. *J Am Chem Soc* **134:** 909–911.

Wörsdörfer B, Woycechowsky KJ, Hilvert D. 2011b. Directed evolution of a protein container. *Science* **331:** 589–592.

Yamamoto A, Masaki R, Tashiro Y. 1996. Formation of crystalloid endoplasmic reticulum in COS cells upon overexpression of microsomal aldehyde dehydrogenase by cDNA transfection. *J Cell Sci* **109:** 1727–1738.

Yang J, Kickhoefer VA, Ng BC, Gopal A, Bentolila LA, John S, Tolbert SH, Rome LH. 2010. Vaults are dynamically unconstrained cytoplasmic nanoparticles capable of half vault exchange. *ACS Nano* **4:** 7229–7240.

Zakeri B, Fierer JO, Celik E, Chittock EC, Schwarz-Linek U, Moy VT, Howarth M. 2012. Peptide tag forming a rapid covalent bond to a protein, through engineering a bacterial adhesin. *Proc Natl Acad Sci* **109:** E690–E697.

Zhang X, Meining W, Fischer M, Bacher A, Ladenstein R. 2001. X-ray structure analysis and crystallographic refinement of lumazine synthase from the hyperthermophile *Aquifex aeolicus* at 1.6 Å resolution: Determinants of thermostability revealed from structural comparisons. *J Mol Biol* **306:** 1099–1114.

Zhang X, Konarev PV, Petoukhov MV, Svergun DI, Xing L, Cheng RH, Haase I, Fischer M, Bacher A, Ladenstein R, et al. 2006a. Multiple assembly states of lumazine synthase: A model relating catalytic function and molecular assembly. *J Mol Biol* **362:** 753–770.

Zhang Y, Li SZ, Li J, Pan X, Cahoon RE, Jaworski JG, Wang X, Jez JM, Chen F, Yu O. 2006b. Using unnatural protein fusions to engineer resveratrol biosynthesis in yeast and mammalian cells. *J Am Chem Soc* **128:** 13030–13031.

Zhu Y, Jiang J, Said-Sadier N, Boxx G, Champion C, Tetlow A, Kickhoefer VA, Rome LH, Ojcius DM, Kelly KA. 2014. Activation of the NLRP3 inflammasome by vault nanoparticles expressing a chlamydial epitope. *Vaccine* **33:** 298–306.

Design Automation in Synthetic Biology

Evan Appleton,[1] Curtis Madsen,[2,3] Nicholas Roehner,[2,3] and Douglas Densmore[2,3]

[1]Department of Genetics, Harvard Medical School, Harvard University, Boston, Massachusetts 02115

[2]Biological Design Center, Boston University, Boston, Massachusetts 02215

[3]Department of Electrical and Computer Engineering, Boston University, Boston, Massachusetts 02215

Correspondence: dougd@bu.edu

Design automation refers to a category of software tools for designing systems that work together in a workflow for designing, building, testing, and analyzing systems with a target behavior. In synthetic biology, these tools are called bio-design automation (BDA) tools. In this review, we discuss the BDA tools areas—specify, design, build, test, and learn—and introduce the existing software tools designed to solve problems in these areas. We then detail the functionality of some of these tools and show how they can be used together to create the desired behavior of two types of modern synthetic genetic regulatory networks.

Synthetic biology as a discipline has concerned itself with "forward engineering" living systems. This is a purposefully grand goal and is faced with numerous scientific, engineering, political, and ethical challenges. These engineering challenges include the storage of biological information, the design of complex, interacting biological systems using that information, the physical creation of these systems, and the dissemination of foundational principles as abstractions on which the next technical advances can be made. What should be clear to even the most casual observer is that computers, and the computational capabilities they bring with them, are going to be required if the field is going to succeed going forward. Debates remain regarding the framing of the computational approaches (Andrianantoandro et al. 2006; Lux et al. 2012; Densmore and Bhatia 2014), but there are few who debate the important and increasingly prominent role of computation in the field.

Synthetic biology and its associated promise have attracted a wide variety of participants. In particular, computer engineers and computer scientists began developing computational tools for engineering these genetic systems using techniques from their disciplines. Electronic design automation (EDA) showed a clear process by which design formalisms could be built, software tools created, and a larger, separate industry developed (see latticeautomation.com; teselagen.com; deskgen.com; benchling.com). This pursuit in the synthetic biology field has recently been coined bio-design automation (BDA) (Densmore 2012). BDA often uses a "divide and conquer" approach for solving small parts of a larger problem one piece at a time and allows for specialists to tackle specific narrow problems in which they have considerable ex-

Cite this article as *Cold Spring Harb Perspect Biol* doi: 10.1101/cshperspect.a023978

pertise. After all necessary pieces are defined and solved, solutions for each subproblem can be automated, connected, and reused to solve larger problems. The central conjecture of this approach to synthetic biology is that the application of these core engineering concepts will accelerate synthetic biology workflows, increase abstraction and reuse, and greatly scale the size and complexity of systems that can be successfully created.

A key contemporary challenge in BDA is that there are few standards and little documentation for repeatedly engineering these systems. In response, multiple efforts have begun working toward community standards (Endy 2005; Arkin 2008; Canton et al. 2008; Kelly et al. 2009; Smolke 2009; Galdzicki et al. 2011) for data models and data exchange. This effort has recently gained more momentum (Galdzicki et al. 2014) and seen support and involvement from larger government organizations (Hayden 2015). Currently, some of the core requirements for representing sequence information have been discussed at length and interest has shifted to establishing standards for additional related subproblems.

The current BDA landscape can be divided into five main areas: specification, design, build, test, and learn. "Specification" concerns a precise, formal definition of the desired function and design of a target genetic system. "Design" concerns the set of decisions needed to determine the DNA constructs to be used and any needed modification to those constructions. "Build" refers to the processes and decisions involved in creating a plan for composing the DNA constructs from their elements using specific cloning methodologies and physically implementing a DNA-assembly plan (e.g., many approaches integrate robotics). "Test" concerns the design and physical implementation of experiments for characterizing engineered systems and the accompanying analysis and interpretation of acquired data. "Learn" includes machine-learning approaches to allow for the automated revision of designs based on processed experimental outcomes.

This review provides a survey of BDA approaches in these areas; it reflects the investigators' experience in the field since 2007 and attempts to complement other similar surveys (Kahl and Endy 2013) by both updating these offerings to reflect the state-of-the-art in 2016 as well as providing examples of some core workflows.

SPECIFICATION AND DESIGN OF GENETIC SYSTEMS

The first step in engineering a genetic system is to specify its intended structure and/or function. Such a specification may take many different forms, ranging from a mathematical description of system behavior to a rule-based description of system architecture. Once a specification has been made, it must be mapped to a set of genetic parts and/or larger devices with defined genetic sequences that can conceivably satisfy its requirements. This mapping process constitutes the problem of genetic design, which can be solved via many alternate approaches. In this section, we will discuss BDA tools for specification and design and classify them as supporting a rational or combinatorial approach to genetic design. We will also highlight the use of several exemplar tools to specify and design a genetic AND gate and dual-feedback oscillator similar to those originally designed by Nielsen et al. (2016) and Stricker et al. (2008), respectively.

Rational Specification and Design

In general, rational approaches to genetic design are those in which design decisions are based on conclusions drawn from a biophysical or biochemical model. For example, a designer might decide to use one genetic part over another based on its impact on gene expression as predicted by a model. Roughly, two-thirds of all the specification and design tools listed in Table 1 can be classified as supporting a rational approach to genetic design. The specification tools belonging to this classification can be further divided into two non-mutually exclusive categories: languages and tools for directly building biophysical/biochemical models of genetic systems (Antimony [Smith et al. 2009], iBioSim [Myers et al. 2009], and ProMot [Mirschel et al.

 Cite this article as *Cold Spring Harb Perspect Biol* doi: 10.1101/cshperspect.a023978

Table 1. A comprehensive list of software tools available for solving bio-design automation tasks

Realm	Tool name	Description
Specification	Eugene (Oberortner and Densmore 2014)	A rule-based specification language, which is able to incorporate abstraction layers and automatically generate combinatorial biological devices from components (eugenecad.org)
	Pigeon (Bhatia and Densmore 2013)	Translates textual representations of nucleic acids into standardized visual representations (pigeoncad.org)
	GEC (Pedersen and Phillips 2009)	A language for modular specification of proteins and genes and their interactions; such specifications can then be compiled into standard biological sequences (biology.azurewebsites.net/gec)
	Phoenix	A tool that guides users through an automated, iterative design–build–test–learn cycle; with an ability to learn from empirical data, it can complete the closed-loop automation
	Proto/Biocompiler (Beal et al. 2011)	Generates optimized genetic regulatory network designs from specifications written in a high-level programming language (synbiotools.bbn.com)
	GSL (Wilson et al. 2016)	A language that facilitates design of large and complex DNA constructs used to engineer genomes; it also incorporates a set of low-level DNA manipulation primitives
	Kera (Dhar 2010)	An object-oriented programming language for synthetic biology, which is tempered by the biopart rule library, Samhita; Kera captures the knowledge regarding the interaction of genome components and catalytic molecules
	Antimony (Smith et al. 2009)	Text-based modular human-readable/writable model definition language with capabilities of easy translation into SBML
	ProMot (Mirschel et al. 2009)	A tool for construction and manipulation of biological dynamic models based on differential-algebraic equations ProMot is matched with the simulation environment Diana (www2.mpi-magdeburg.mpg.de/projects/promot)
	Molecula Maxima	Contains a programming IDE called Cytostudio for a synthetic biology programming language called Synthetic; Synthetic is a bio-hardware description language (moleculamaxima.com)
	MoSeC (Misirli et al. 2011)	Java application capable of generating DNA sequences. MoSeC has the ability to generate FASTA, GenBank, EMBL, and SBOL formats from SBML and CellML models (ico2s.org/software/mosec.html)
	Asmparts (Rodrigo et al. 2007)	A deprecated tool for modular specification of biological devices; Asmparts takes advantage of standardization by considering a model for each part in the design
Design	Cello (Nielsen et al. 2016)	Compiles Verilog code specifying a logic circuit to a DNA sequence, with prediction of the design correctness; outputs are Eugene and SBOL compatible (cellocad.org)
	Merlin (Quintin et al. 2016)	Genomic reprogramming tool exploiting Multiplex Automated Genome Engineering (MAGE), resulting in simultaneous multiregion genome modification (merlincad.org)
	Double Dutch (Roehner et al. 2016a)	Web application for designing libraries of biosynthetic pathway variants based on design of experiments (DOE) matrices (clothocad.org/doubleDutch)
	RBS Calculator (Salis et al. 2009; Borujeni et al. 2013)	Allows for predicting and controlling translation initiation and protein expression in bacteria; can also be used for optimizing synthetic RBS sequences to achieve a targeted translation initiation rate (denovodna.com/software)

Continued

Table 1. *Continued*

Realm	Tool name	Description
	GenoCAD (Czar et al. 2009)	A gene design and simulation tool with its own repository of biological parts; it incorporates various grammars/libraries and supports GenBank format (genocad.com)
	TeselaGen	A cloud-based SBOL compliant solution for sequence editing, combinatorial gene design, and assembly automation (teselagen.com)
	BioJADE (Goler 2004)	A java-based design and simulation tool for synthetic biological systems with interactive use of BioBrick repositories (web.mit.edu/jagoler/www/biojade)
	DeviceEditor	Maps the DNA sequence and feature annotations underlying each part to a graphic icon; icons can then be drag-and-dropped about a visual design canvas (j5.jbei.org/bin/deviceeditor.pl)
	SBROME (Huynh et al. 2013)	Given genetic circuit topology and dynamics, designs circuit with parts that minimize cross talk and incompatibility (tagkopouloslab.ucdavis.edu/software.html)
	Salis Lab calculators (Borujeni and Salis 2016)	A number of mathematical tools for synthetic biology design process (salislab.net/software)
	DeskGen	A genome-editing platform that combines the use of a CRISPR-Cas9 tool with the convenience of a genome browser (deskgen.com)
	Genome Compiler	A software platform incorporating advanced design tools, a variety of plasmids and parts repositories, direct connections to an array of DNA synthesis providers and laboratory automation services (genomecompiler.com)
	MatchMaker (Yaman et al. 2012)	Automates three steps: feature matching for regulatory relationship satisfaction, signal matching for expression level compatibility, and part matching to find physical DNA sequence in a database
	Mfold (Zuker 2003)	Analysis of DNA/RNA folding as well as free energy determination and structure display (unafold.rna .albany.edu)
	Rosetta (Leaver-Fay et al. 2011)	Computes the expected structure of natural proteins and design of new proteins (boinc.bakerlab.org/ rosetta)
	GLAMM (Bates et al. 2011)	Provides maps for metabolic pathways, biosynthesis of secondary metabolites, and macrobial metabolism (glamm.lbl.gov)
	OptCircuit (Dasika and Maranas 2008)	An optimization-based framework that automatically identifies the circuit components from a list and connectivity that brings about the desired functionality.
	AutoBioCAD (Rodrigo and Jaramillo 2013)	Fully automated computational design of regulatory circuits in *Escherichia coli*; is able to compile the design into a nucleotide sequence
	Parts & Pools (Marchisio 2014)	A framework for modular design of bacterial synthetic genetic circuits from standard biological parts and pools of molecules as signal carriers

Build/assembly	Puppeteer (Vasilev et al. 2011)	Liquid-handling automation tool capable of handling various protocols and laboratory resources; output is Common Human Robot Instruction Set (CHRIS); a compiler then translates CHRIS into a human/machine executable language
	Raven (Appleton et al. 2014a)	DNA assembly planning tool for high throughput cloning; supports modern assembly methods such as BioBrick, MoClo, and Gibson Assembly; can output computer-readable or human-readable assembly instructions for liquid-handling automation (ravencad.org)
	DNALD (Blakes et al. 2014)	DNA assembly planning tool for binary assembly; improves on algorithm performance of prior binary algorithms (dnald.org/planner/index.html)
	Fluigi (Huang and Densmore 2014)	Automates the design of microfluidic devices used in synthetic biology by optimizing their layout; specifications are in the text-based MINT language
	3DuF	Visual CAD tool for the design of microfluidic devices; outputs STL for CNC milling and SVG for lithography, as well as a textual JSON representation (cidarlab.github.io/3DuF)
	j5 (Hillson et al. 2011)	Web-based DNA assembly automation tool with ability to design assemblies and oligonucleotides for SLIC/Gibson/CPEC/SLiCE and Golden Gate; also performs cost analysis for choices of full synthesis, PCR/SOE, or oligo embedding (j5.jbei.org/bin/j5_entry_form.pl)
	PR-PR (Linshiz et al. 2014)	A cross-platform high-level biologist-friendly standard language used for management of liquid-handling robotics and microfluidic automated platforms.
	Aquarium	An assembly design tool for generating scalable, reproducible, and transferable molecular biology workflows (aquarium.bio)
	Primer3 (Koressaar and Remm 2007; Untergasser et al. 2012)	A tool used to design and analyze primers for PCR reactions with the ability to select primers for sequencing reactions and hybridization probes (bioinfo.ut.ee/primer3)
Test/analysis	TASBE tools	A variety of synthetic biology automation software tool for tasks from high-level specification to part assignment and assembly (synthetic-biology.bbn.com/tasbe.html)
	Bioconductor (Huber et al. 2015)	Analysis and comprehension tool that uses the open-source statistical language, R; AMI and Docker images are available (bioconductor.org)
	FlowCal (Castillo-Hair et al. 2016)	A library for reading, analyzing, and calibrating flow cytometry data; it accepts FCS files as input and is compatible with different calibration particles, fluorescent probes, and cell types
	FlowJo	A software application with an integrated environment for viewing and analyzing flow cytometry data; the user interface (UI) is presented as the workspace of experimental data, statistics, gates, as well as tabular and graphical layouts (flowjo.com)
Data	ICE (Ham et al. 2012)	A registry platform for robust data storage for DNA components, integrated tools for part characterization, and secure access and information sharing (public-registry.jbei.org)

Continued

Table 1. *Continued*

Realm	Tool name	Description
	SBOL Stack (Madsen et al. 2016)	An RDF-based repository of SBOL represented synthetic biology parts and devices with an ability to execute search queries (sbolstack.org)
	GenBank (Bilofsky and Christian 1988)	A genetic sequence database containing a complete repository of public DNA sequences with corresponding annotations; GenBank also has a widely used file format for representing such sequences (ncbi.nlm.nih.gov/genbank)
	Registry of Standard Biological Parts	The iGEM repository including parts that are used in BioBrick assembly; this repository has more than 20,000 documented parts (parts.igem.org)
	Clotho (Densmore et al. 2009)	A database for storing synthetic biological parts data as well as a framework for applications to engineer them (clothocad.org)
	Owl (Appleton et al. 2014b)	A tool for creating datasheets for biological entities with an ability to perform data analytics services (cidar.bu.edu/owl)
	Phagebook	A Clotho 3.0 app that serves as a social network that also integrates laboratory inventory management; publication announcements are also supported
	Viz-a-Brick	A search and navigation tool for the iGEM registry of standard biological parts with quick and easy access to global and local relationships between parts in the registry (gcat.davidson.edu/vizabrick)
	BioModels database (Chelliah et al. 2013)	A reference repository hosting mathematical models that describe the dynamic interactions of biological components at various scales; most model components are cross-linked with external resources to facilitate interoperability (ebi.ac.uk/biomodels-main)
Simulation	iBioSim (Myers et al. 2009)	A simulation and design tool focused on genetic circuits specified in SBML or SBOL format; has the ability to analyze metabolic networks and cell-signaling pathways, as well as modeling and visualization of multicellular and spatial models (async.ece.utah.edu/iBioSim)
	COPASI (Hoops et al. 2006)	Simulation and analysis tool for biochemical networks and their dynamics; accepts models in SBML format and analyzes them using ODE as well as stochastic algorithms (copasi.org)
	SynBioSS (Hill et al. 2008)	A suite of software for the modeling and simulation of synthetic genetic constructs; it uses the registry of standard biological parts, a database of kinetic parameters, and both graphical and command-line interfaces to multiscale simulation algorithms (synbioss.sourceforge.net)
	TinkerCell (Chandran et al. 2009)	Incorporates a diagram that is detailed enough so that it can be mapped to models or experimental results; various mathematical analyses can then be run on those models (tinkercell.com)
	Gro (Jang et al. 2012)	A language for programming, modeling, specifying, and simulating the behavior of cells in growing microcolonies of microorganisms; Gro can simulate cell growth, division, intrinsic and extrinsic noise, as well as diffusing molecular signals (depts.washington.edu/soslab/gro)

	Morpheus (Starruß et al. 2014)	A modeling environment for simulating cell-based models with ordinary differential equations and reaction–diffusion systems; multiscale biological models can be defined in biological terms and mathematical expressions—it supports model construction, simulation, visualization, and batch processing (imc.zih.tu-dresden.de/wiki/morpheus/doku.php)
	PySCeS (Olivier et al. 2004)	Provides a variety of tools for the analysis of cellular systems; these include a human readable model description language, a structural, and a bifurcation analysis module—PySCeS supports SBML and SED-ML (pysces.sourceforge.net)
	CellDesigner (Funahashi et al. 2008)	A structured diagram editor for drawing gene-regulatory and biochemical networks, with integrated SBML ODE solver, SBML simulation core, and COPASI (celldesigner.org)
	Jarnac (Sauro 2000)	A Systems Biology Workbench (SBW) reaction simulator that includes both deterministic and stochastic time course simulation, steady-state analysis, basic structural properties of networks, and more (sbw.kgi.edu/software/jarnac.htm)
	JDesigner	An SBW network design tool with ability to export SBML file format; JDesigner is capable of connecting to Jarnac to perform simulation tasks (sbw.kgi.edu/software/jdesigner.htm)
	Mathematica	Incorporates an exclusive solution for synthetic biology; it is capable of performing the complete workflow from data import to analysis
	RoVerGeNe	A tool for analysis of dynamical properties regarding genetic regulatory networks; RoVerGeNe is capable of robustness analysis and parameter constraint synthesis (sites.bu.edu/hyness/rovergene)
	BioPSy (Madsen et al. 2015)	Performs guaranteed parameter set synthesis for ODE biological models expressed in SBML given a desired behavior expressed by time-series data (github.com/dreal/biology)
	D-VASim	A genetic network simulator that is used to perform the timing and propagation delay analysis on genetic logic circuits; D-VASim supports SBML format (bda.compute.dtu.dk/tools-2/d-vasim)
	RoadRunner	A portable simulation engine for systems and synthetic biology models in SBML format; RoadRunner is written in C# and incorporates C, C++, and Python APIs (libroadrunner.org)
Sequence editing	Benchling	Sequence repository capable of storing and showing annotations and searching; Benchling is also a cloud-based framework for life sciences applications (benchling.com)
	ApE	A sequence editor supporting FASTA, GenBank, DNA Strider, EMBL, and ABI formats; it also supports direct connection to BLAST (biologylabs.utah.edu/jorgensen/wayned/ape)
	VectorEditor	Web-based DNA sequence-editing and analysis tool with restriction enzyme manager, amino acid translation, and gel digest prediction (public-registry.jbei.org/vectoreditor)
	GeneDesigner	Encompasses a set of in silico sequence design tools and algorithms, such as cloning, codon optimization, back translation and primer design; supports GenBank and FASTA formats (dna20.com/resources/genedesigner)

Continued

Table 1. *Continued*

Realm	Tool name	Description
	VectorNTI/Express	Sequence analysis and design tool with an integrated graphical environment that provides the ability to manage, view, analyze, transform, share, and publish molecular biology data
	Geneious (Drummond et al. 2010)	Sequence editing and alignment tool with support for annotations, incorporating multiple alignment algorithms and easy switch between them (geneious.com/features/genome-alignment)
	GeneDesign	Web-based synthetic gene designer including various modules such as reverse translator, codon juggler, etc. for sequence manipulation (54.235.254.95/gd)
	Sequence REFINER (Chakrabarti et al. 2006)	Sequence alignment tool that refines a multiple sequence alignment by iterative realignment of its individual sequences with the predetermined conserved core model of a protein family (ftp.ncbi.nih.gov/pub/REFINER)
	SnapGene Editor	A sequence-editing software capable of creating, browsing, and sharing richly annotated DNA sequence files up to 1 Gb in length (snapgene.com)
	Synthetic Gene Designer (Wu et al. 2006)	Generates optimal gene sequences taking codon bias, like GC content, into consideration (userpages.umbc.edu/wug1/codon/sgd)
	SBOL Designer	CAD tool for creating and manipulating genetic construct sequences using the SBOL data model (async.ece.utah.edu/SBOLDesigner)

Cite this article as *Cold Spring Harb Perspect Biol* doi: 10.1101/cshperspect.a023978

2009]) and languages and tools for composing less-detailed descriptions of system behavior and/or architecture that are compiled into more detailed models (Cello [Nielsen et al. 2016], GEC [Pedersen and Phillips 2009], GenoCAD [Czar et al. 2009], iBioSim [Myers et al. 2009], Kera [Dhar 2010], the Proto Biocompiler [Beal et al. 2011], and TinkerCell [Chandran et al. 2009]). Specification tools belonging to the first class tend to give the user greater control over the form of model built with them, whereas specification tools belonging to the second class tend to remove this control from the user to automate specification and reduce the expert knowledge required during this process.

While rational specification tools are used to specify the intended function of a genetic system, rational design tools are used to map from such a specification to an implementing set of genetic parts and/or devices with defined sequences. A majority of the rational design tools in Table 1 (AutoBioCAD [Rodrigo and Jaramillo 2013], BioJade [Goler 2004], Cello [Nielsen et al. 2016], GEC [Pedersen and Phillips 2009], GenoCAD [Czar et al. 2009], iBioSim [Myers et al. 2009], Kera [Dhar 2010], Matchmaker [Yaman et al. 2012], MoSeC [Misirli et al. 2011], OptCircuit [Dasika and Maranas 2008], Parts&Pools [Marchisio 2014], and SBROME [Huynh et al. 2013; Huynh and Tagkopoulos 2014]) feature dedicated support for designing genetic regulatory networks/circuits, but none feature the same level of support for designing biosynthetic/metabolic pathways. This omission is partly a reflection of the current membership of the BDA community, many of whom work at the intersection between synthetic biology, systems biology, and electrical and computer engineering, thus favoring the development of tools for genetic circuit design. However, this omission also highlights that existing tools for rational pathway design, such as Cobra 2.0 (Schellenberger et al. 2011) and OptFlux (Rocha et al. 2010), have been developed based on traditional approaches to pathway engineering as opposed to newer approaches from synthetic biology. The former approaches tend to focus on knocking out genes from a known pathway to optimize its performance (Kim et al. 2015), although the latter approaches emphasize the optimal design of a synthetic pathway under the control of standardized parts (Smanski et al. 2014). Of the remaining rational design tools in Table 1, Rosetta (Leaver-Fay et al. 2011) and the RBS Calculator (Salis et al. 2009) have perhaps the broadest application in the sense that they are based on principles from biophysics and can be used to design some of the primitive genetic parts (proteins and RBSs, respectively) found in most genetic systems.

The example at the end of this section features Cello, GEC, and the RBS Calculator as effective BDA tools for rational specification and design. Cello was selected as it represents one of the most dedicated and experimentally vetted tools for genetic circuit design. Cello is also notable for basing its designs on an empirical model of circuit behavior in which the transfer curves for the individual logic gates that make up a circuit are multiplied to obtain a steady-state relationship between its input and output. This is in stark contrast with the mechanistic models of genetic regulatory networks leveraged by most other circuit design tools. In this way, Cello obviates the need to directly measure or estimate chemical kinetic parameters, but at the cost of being unable to predict the transient, non-steady-state behavior of its designed circuits. Consequently, for our example, Cello is best suited to the design of the AND gate as opposed to the dual-feedback oscillator. For the design of the latter circuit, GEC was selected by reason of its highly developed language for abstractly specifying genetic function and its capability to export biochemical models written in the Systems Biology Markup Language (SBML) (Hucka et al. 2003), a standard leveraged by many other BDA tools, including simulation tools featured in a later section (see Data Analysis and Simulation) (iBioSim [Myers et al. 2009] and COPASI [Hoops et al. 2006]). Last, the RBS calculator was chosen for its immediate accessibility compared with other biophysics-based tools, its high potential for complementing other BDA tools, and its widespread use among synthetic biologists outside of the BDA community.

Combinatorial Specification and Design

Combinatorial approaches to genetic design typically involve designing many different variants of a system, such that these variants can be efficiently screened and/or undergo selection to obtain those that optimize system function. For example, an engineer might design a library of biosynthetic pathway variants in which the genetic parts controlling expression are varied to enable screening for the set of parts that optimize pathway yield (Smanski et al. 2014). Alternatively, one might design a library of oligonucleotides to make a large number of targeted edits to the genome of a host organism and then select for the most productive strain (Wang et al. 2009). One-third of the specification and design tools listed in Table 1 can be classified as supporting a combinatorial approach to genetic design. At least four of these tools are dedicated to combinatorial specification, with three tools providing languages for specifying library architectures and/or target sequences for genome engineering (Eugene [Bilitchenko et al. 2011], GSL [Wilson et al. 2016], and Molecula Maxima) and one tool providing the means to generate experimental design matrices that can be interpreted as specifications of library function (Double Dutch [Roehner et al. 2016a]).

While combinatorial specification tools are used to specify the intended structure and/or function of a genetic library, combinatorial design tools are used to map from such a specification to an implementing set of genetic parts and/or devices with defined sequences. The combinatorial design tools in Table 1 are a superset of the combinatorial specification tools in this table and can be broadly divided into three groups: tools focused on genome engineering (MAGE [Wang et al. 2009] with Merlin [Quintin et al. 2016] and CRISPR [Cong et al. 2013] with DeskGen [see deskgen.com]), tools for general purpose design and/or assembly of combinatorial libraries (Device Editor [Hillson 2014], Eugene [Oberortner et al. 2014], Genome Compiler, GSL [Wilson et al. 2016], j5 [Hillson et al. 2011], and Teselagen [see teselagen.com]), and tools focused on designing libraries that adhere to experimental designs (Double Dutch [Roehner et al. 2016a]). These three groups can be viewed as a spectrum in which tools from the first and second groups are well suited to combinatorial approaches that rely exclusively on screening and/or selection, whereas tools from the second and third groups are well suited to semirational approaches that supplement screening with statistical methods such as regression analysis to guide the search for optimal variants. The majority of the combinatorial design tools in Table 1 belong to the first half of this spectrum rather than the second, which suggests that the BDA community has thus far focused its attention on combinatorial design and assembly as opposed to semirational design. This gap in tooling may also reflect a difference in research focus between academia and industry, given that the academic tools in Table 1 are overrepresented in the category of rational design, whereas the industrial tools are overrepresented in the category of combinatorial design but underrepresented in the subcategory of semirational design.

The example at the end of this section features Eugene and Double Dutch as effective BDA tools for combinatorial specification and design. Eugene was chosen for its flexibility as a domain-specific programming language, its incorporation into both academic and industrial tools (Cello [Nielsen et al. 2016], Device Editor [Hillson 2014], j5 [Hillson et al. 2011], and Teselagen [see teselagen.com]), and its capacity to export the Synthetic Biology Open Language (SBOL) (Galdzicki et al. 2011, 2014). SBOL is a prominent design standard that has been developed by a subset of the BDA community and has growing support among BDA tools (see section on Data Exchange and Standardization). Last, Double Dutch was chosen as the sole combinatorial design tool in Table 1 with explicit support for semirational design of biological systems.

Example

In this example, we will discuss how Cello (Nielsen et al. 2016) and Eugene (Oberortner et al. 2014) can be used to design a genetic AND gate, and how GEC (Pedersen and Phillips 2009), the

 Cite this article as *Cold Spring Harb Perspect Biol* doi: 10.1101/cshperspect.a023978

RBS Calculator (Salis et al. 2009), Eugene, and Double Dutch (Roehner et al. 2016a) can be used to design a dual-feedback oscillator (Fig. 1). The first step in the design of the genetic AND gate is to submit a Verilog (Thomas and Moorby 1995) case statement for the appropriate Boolean logic as input to Cello. Cello then compiles this Verilog specification to a gate netlist and optimizes the assignment of genetic logic gates to this netlist (colors the netlist) based on predicted circuit behavior. Because the correct behavior of a genetic circuit in vivo is the result of many factors besides just gate choice, oftentimes an additional combinatorial design step is performed to increase the likelihood of success. Among the outputs of Cello is a Eugene file that encodes the theoretical best circuit design. This file can be modified through the addition or subtraction of Eugene rules to customize the design and potentially specify a library of design variants. In this case, the rules specifying gene order are removed, whereas the rules specifying scar positions are kept. When the modified Eugene file is executed, the result is a library of six AND gates, one for every possible gene order. Following testing, we can attempt to analyze the effect of gene order on circuit performance and use the circuit variant that best satisfies the original Verilog specification made using Cello.

Next, the design of the dual-feedback oscillator begins with writing a specification in the GEC language for its genetic architecture (number and types of genetic parts) and regulatory relationships. GEC then compiles this specification to all genetic circuit designs that satisfy the specified constraints. In this example, only one design is generated because the default part database provided with GEC is relatively small and must be extended with a dual-regulatory promoter from the iGEM (Smolke 2009; Vilanova and Porcar 2014) Registry (BioBrick K094120) to obtain any solutions. To predict which design will have optimal performance, one must use GEC to compile each design to a biochemical model and simulate each model using GEC or another tool. In practice, however, it can be difficult to obtain predictive biochemical models for this step, as many part databases lack the chemical kinetic parameters necessary to inform such models. Hence, in this example, the RBS calculator is used to design a library of nine RBSs with translation initiation rates spanning three orders of magnitude. This set of parts is used later as part of a semirational design strategy with Double Dutch. Unlike Cello, GEC does not produce a Eugene file that can be used for combinatorial design. Instead, this file must be written by the user. In this case, the Eugene file for the oscillator varies not only gene order, but also gene orientation, constraining the latter with a rule that requires genes to have alternating orientations (e.g., forward, reverse, forward). This rule can be useful for reducing read-through transcription of genes. Once the most promising of the six rule-conformant architectures has been identified through testing, a library of oscillators that share this architecture but have different RBSs is designed using Double Dutch. Double Dutch bases this library design on a design of experiments (DOE) matrix so that a user can perform regression analysis following testing and correlate RBS strength with oscillator performance. Consequently, further design decisions can be based on a regression model, and the oscillator can be fine-tuned when a biochemical model like those used in GEC is unavailable or not predictive.

BUILDING AND EDITING DNA SEQUENCES

Once a specification has been mapped to a set of target DNA constructs, they must be assembled. This problem, called the DNA assembly problem, has both design optimization and physical implementation aspects and can be broken into at least three subproblems: (1) vector design; (2) assembly planning; and (3) liquid handling. Traditionally, many of these problems have been solved manually and simultaneously in a process that became known as molecular cloning. In the early days of molecular cloning, these methods were all ad hoc; for each target clone, a human would have to study the fragment and vector of interest and select an intelligent set of restriction enzymes to cut the vector and the amplified template and then perform a separate

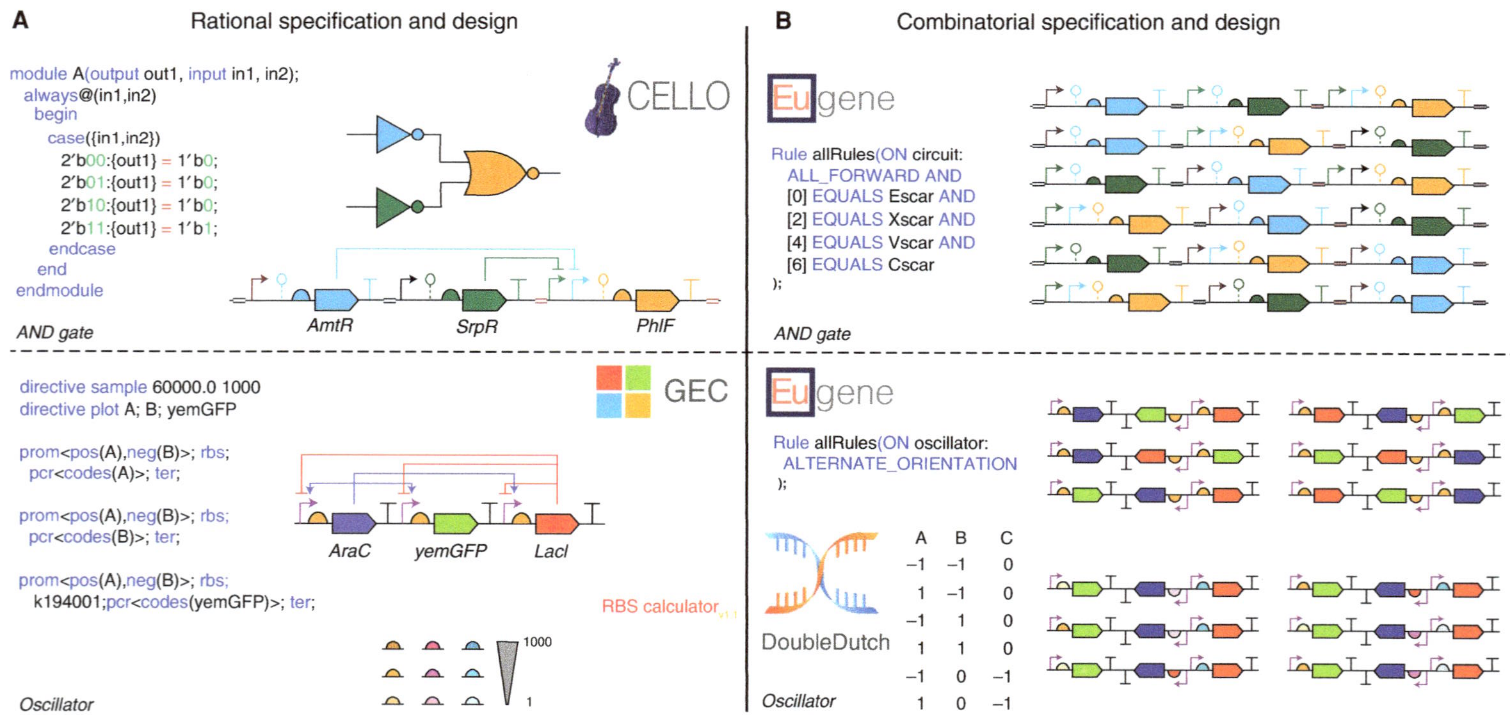

Figure 1. Specification and design of a genetic AND gate and dual-feedback oscillator. All genetic part symbols are defined by the SBOL Visual standard (Quinn et al. 2015). Parts of the same symbol but different colors have different sequences, whereas parts of different symbols but the same color belong to same genetic logic gate. (*A*) Cello (Nielsen et al. 2016) takes a Verilog (Thomas and Moorby 1995) specification for an AND gate as input, compiles and colors a gate netlist, and outputs a genetic circuit design (*top*). GEC (Pedersen and Phillips 2009) takes a GEC specification for a dual-feedback oscillator as input and outputs a genetic circuit design (*bottom*). The RBS calculator (Salis et al. 2009) outputs a library of RBSs with a range of translation initiation rates (*bottom*). (*B*) Eugene (Oberortner et al. 2014) takes Eugene specifications as input and outputs a library of AND gate designs that vary in gene order (*top*) and a library of oscillator designs that vary in gene order and orientation (*bottom*). Double Dutch (Roehner et al. 2016a) takes a design of experiments (DOE) matrix as input and outputs a commensurate library of oscillator designs with varying RBSs designed by the RBS calculator (*bottom*).

ligation reaction to create the final clone. Some of the challenges of this approach include selecting compatible sets of restriction enzymes and compatible sets of sticky ends that were likely to result in a successful ligation. This method of cloning is called "traditional cloning."

As our knowledge of cloning enzymes and DNA synthesis technologies improved, some groups started to gain interest in high-throughput DNA assembly where many fragments could be assembled at once in parallel. For these technologies, we could not require humans to make each assembly decision in such a large set and so some groups invented standard cloning frameworks called "DNA assembly methods," which asserted standard conventions for the design of flanking ends of DNA fragments and the enzymes used in cloning reactions. In recent years, numerous methods have been developed with variations in these categories (Li and Elledge 2007; Engler et al. 2008; Shetty et al. 2008; Gibson et al. 2009; Quan and Tian 2009; Weber et al. 2010; Sarrion-Perdigones et al. 2011; Iverson et al. 2016), but all have the goal of scaling and modularizing DNA assembly capabilities. Some methods place a heavy emphasis on modularity (Weber et al. 2010; Iverson et al. 2016) and standardization (Shetty et al. 2008), whereas others place a heavier emphasis on the scale of fragments that can be reliably assembled (Gibson et al. 2009). In addition to these biochemical technology developments, software was developed to assist problem solving in each of the three DNA assembly problem areas.

Vector Design

The first category of problems to solve in DNA assembly is the conversion of designs into replicating vectors or sets of replicating vectors. This problem, called vector design, is a commonly overlooked subproblem in the current software landscape. Presently, there are no known algorithmic software tools dedicated exclusively to this task and this problem is generally solved manually within sequence-editing software. To perform this task, a user copy–pastes their sequences into the editor and highlights important features and candidate restriction sites (Fig. 2A). There are a variety of popular proprietary (VectorNTI, Benchling, Geneious, Vector Express, SnapGene) and open-source (ApE, GeneDesign, Sequence Refiner, Vector Editor, Synthetic Gene Designer) sequence-editing tools available for solving this problem. These tools often include many additional features for curating sequence libraries and data management and have become the standard technology for storing sequence information in most laboratories.

Assembly Planning

Hierarchical assembly planning refers to the selection of which cloning reactions to perform in parallel and in which series to ultimately create all final clones. For large assemblies or large sets of assemblies, it is desirable to maximize fragment reuse and minimize cloning steps to minimize time and material expense. Some algorithms have been developed to automate this design process (Densmore et al. 2010; Appleton et al. 2014a; Blakes et al. 2014) and have resulted in two web-based open-source software tools: DNALD (see dnald.org/planner) and Raven (see cidar.bu.edu/ravencad). DNALD inputs sets of target parts made of combinations of "basic part" primitives and determines an optimized hierarchical assembly plan for building all target parts assuming that only two fragments can ever be joined in one cloning step (called "binary assembly"). Raven also calculates hierarchical assembly plans for sets of target parts composed of "basic part" primitives, but offers several additional features. First, it partially addresses the vector design problem by allowing for the association of parts with the vectors they originate from and the vectors the final clone must be in. Second, it has algorithms for "n-ary" assembly in which more than two fragments may be connected in one cloning reaction, which is a common trait of modern "one-pot" cloning techniques categories (Fig. 2B) (Li and Elledge 2007; Engler et al. 2008; Shetty et al. 2008; Gibson et al. 2009; Quan and Tian 2009; Weber et al. 2010; Sarrion-Perdigones et al. 2011; Iverson et al. 2016).

After the hierarchical assembly plan is determined, flanking sequences for each of these

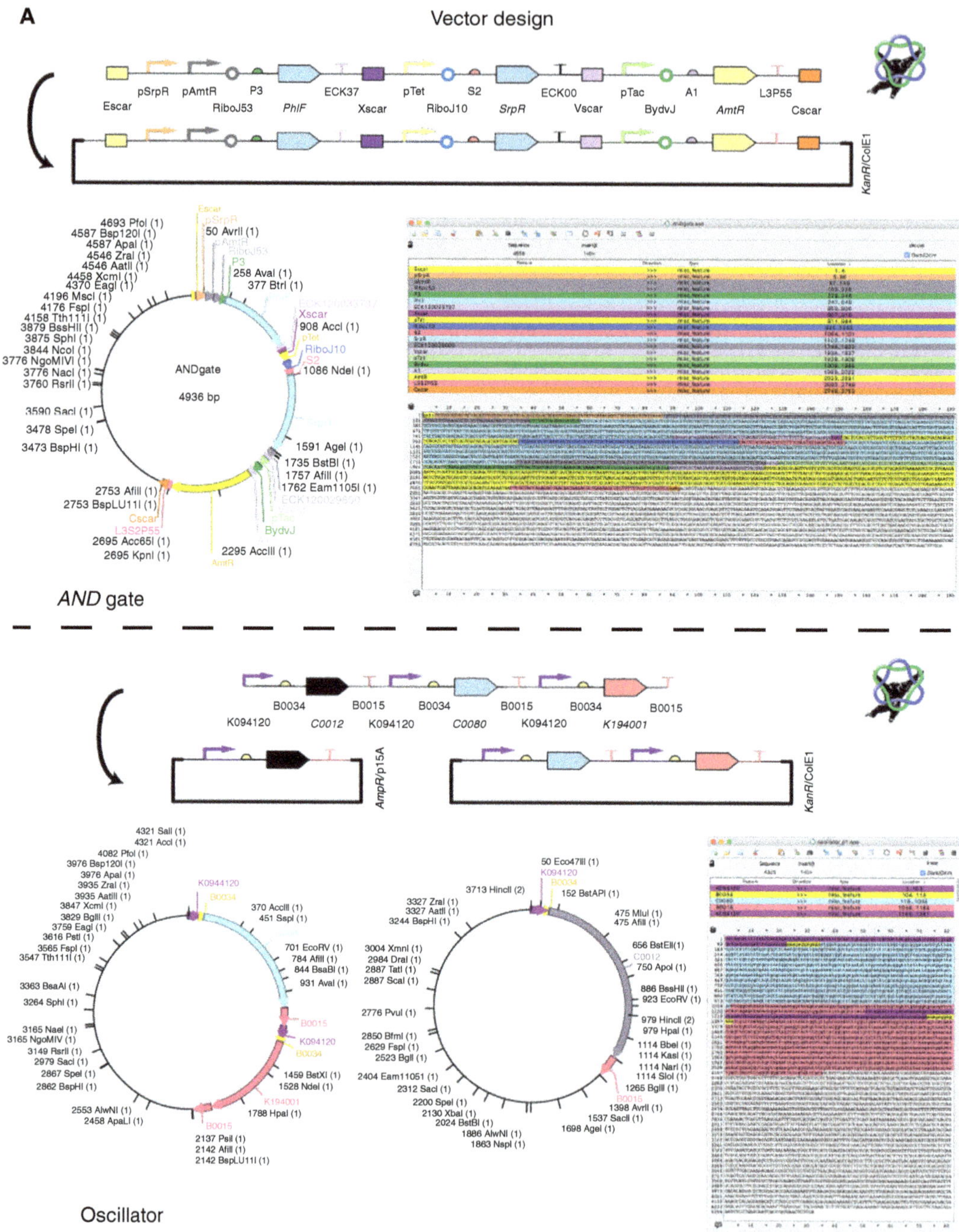

Figure 2. Construction of a complex logic gate and oscillator. (*A*) Using ApE, one can take a sequence of parts from a design, assign to one or more vectors, and edit any parts or junctions at the DNA sequence level. (*Continued on following page.*)

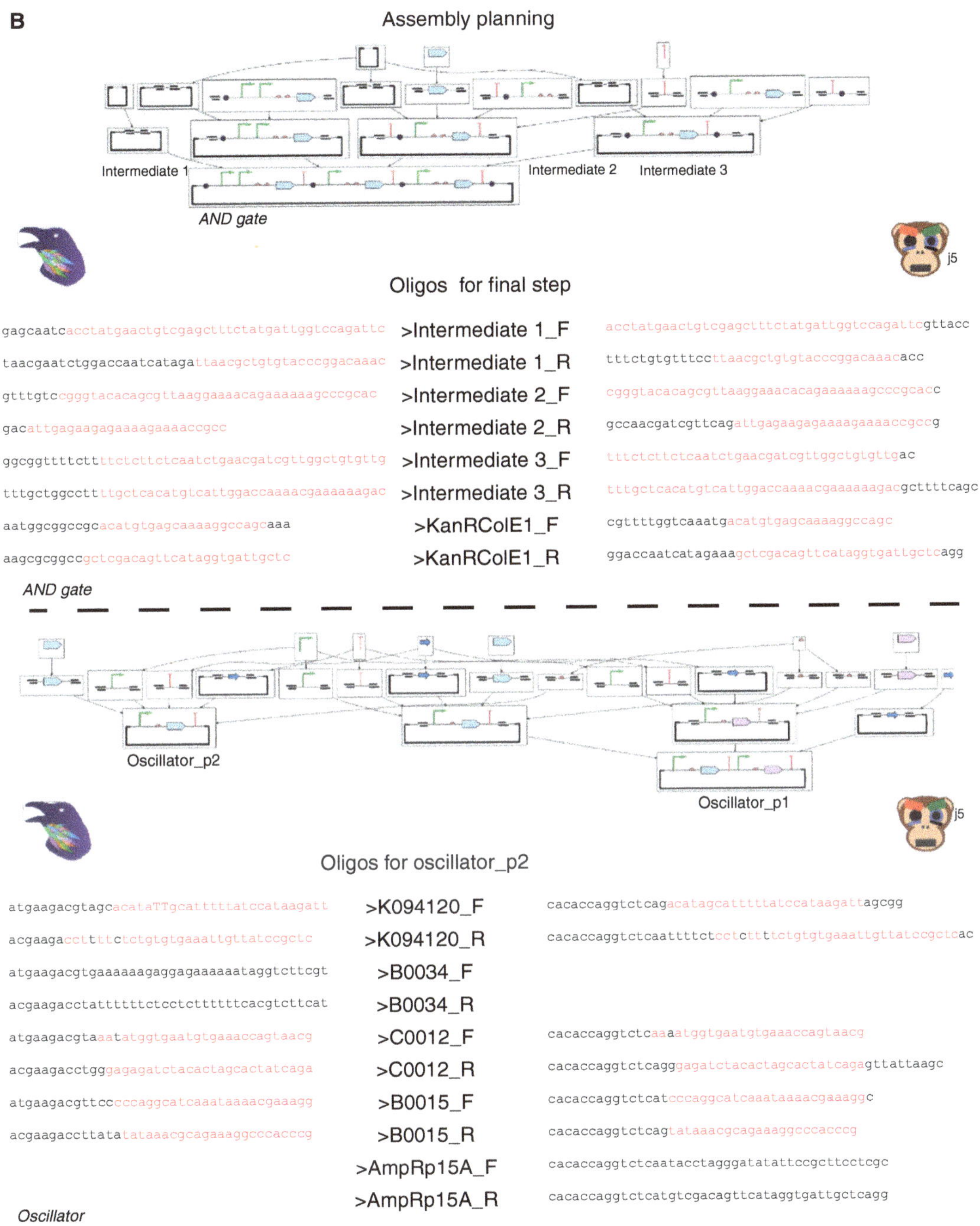

Figure 2. (*Continued.*) (*B*) Raven is used to determine a hierarchical assembly plan for each construct using a specific assembly method. Assembling an AND gate with Gibson requires two stages, four steps, four synthesis steps, and six polymerase chain reaction (PCR) steps, whereas assembling the oscillator constructs with GoldenGate requires two stages, four steps, and 16 PCR steps. Raven generates oligonucleotides to perform these assemblies, but j5 can be used for more refined primer designs for specific cloning steps. The common sequences are highlighted in red. (*Continued on following page.*)

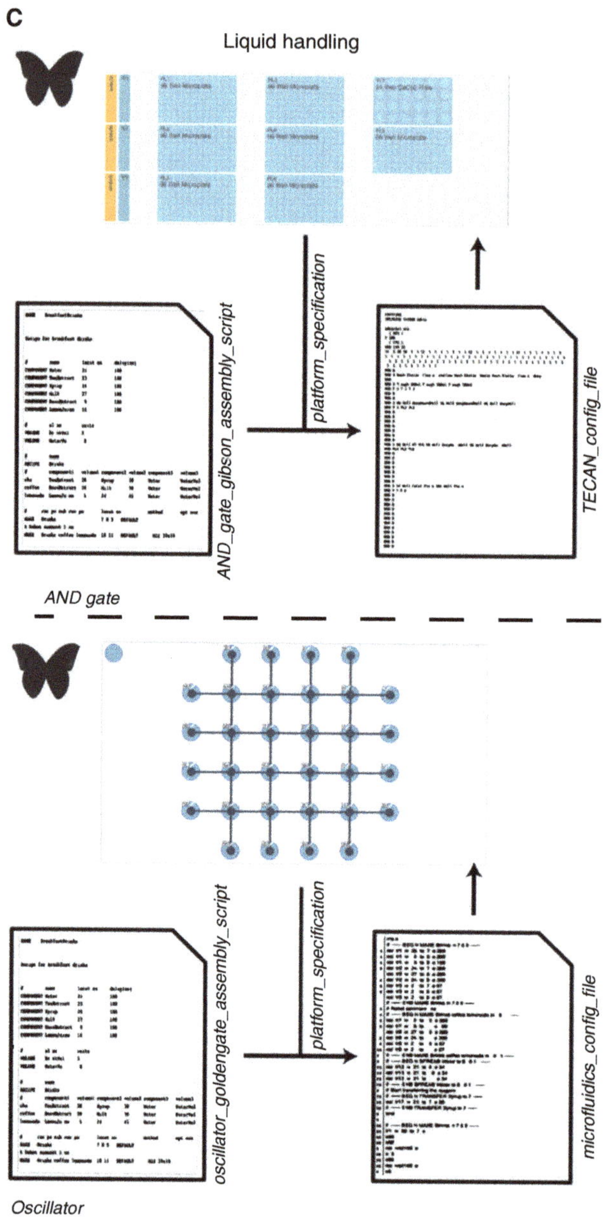

Figure 2. (*Continued.*) (*C*) Once a complete assembly plan is determined, PR-PR can be used to create instruction files for liquid-handling robotics or microfluidics platforms to distribute liquids.

fragments must be determined and primers must be designed for polymerase chain reaction (PCR) such that the correct flanking regions are added to each fragment. Raven provides algorithms for optimizing assembly junction sequence reuse, which is important when cloning libraries for combinatorial assembly (Engler et al. 2008; Weber et al. 2010; Sarrion-Perdigones et al. 2011; Iverson et al. 2016), and provides basic primer designs but does optimize them for assembly-method-specific cloning reaction chemistry concerns. However, the j5 tool addresses this problem in great detail for many contemporary cloning methods (Hillson et al. 2011) and produces highly optimized primers with a high likelihood of improving the success of each individual cloning reaction (Fig. 2B). Most sequence-editing software also has integrated tools for informing these decisions, but do not optimize primers as well as j5. Other tools like Primer3 can optimize primers for sequencing or other assembly-related tasks such as genome editing.

Liquid Handling

After assembly planning, the selection of chemical protocols and plans for dispensing liquid to carry out these protocols must be determined. Some of these protocols are standardized by the chemical manufacturer, but many cloning reaction reagents are not available in the form of a standard master mix, so they must be designed by a biologist. After a set of protocols is determined, the problem of how to execute all necessary protocols efficiently is called the liquid-handling problem. Some open-source tools (PR-PR [Linshiz et al. 2014] and Aquarium) have been created to solve these problems and subsequently produce step-by-step instructions for executing these tasks in an effort to reduce human cost and improve protocol efficiency plus reproducibility. One such tool for this task, PR-PR, inputs a formatted file to describe all of the reactions that must be performed and a file to describe the configuration of a supported platform before generating liquid-handling instructions for a liquid-handling robot, microfluidic platform, or human user (Fig. 2C).

Example

To provide an example of how to use these tools, we will examine the construction of the AND gate and oscillator designs from the previous section (Fig. 2). To solve the vector design problem, one can import the sequences from each part of the overall design into ApE and paste them into a single (as per the AND gate) or multiple (as per the oscillator) replicating vector(s) (Fig. 2A). Next, we can use Raven to determine hierarchical assembly plans using Gibson assembly for the AND gate and GoldenGate for oscillator. Raven generates primers and synthesis fragments for these plans, but j5 can offer more refined primers for each specific cloning step that can improve the likelihood that the cloning step will work on the first try (Fig. 2B) (in this case, we determine primers for the last cloning step of the AND gate and for the smaller plasmid from the oscillator). Finally, this assembly plan can be written up in a standard format and converted to protocol planning files that can be compiled by the PR-PR software to produce assembly instructions that can be interpreted by specific liquid-handling robotics platforms, microfluidic rigs, or a human with a pipettor (Fig. 2C).

DATA ANALYSIS AND SIMULATION

Data generated in the previous steps can be simulated and analyzed to evaluate and validate the biological system. The first step in this analysis process typically involves the application of a computational model. This model can either be simulated to produce traces representing the behavior of the system or analyzed to validate properties of the system. Depending on what types of properties a researcher wants to check, there are different modeling techniques as well as a variety of simulation and analysis methods that can be used.

Modeling

To reason about biological data, researchers must construct models. There are a plethora of different modeling techniques that can be used to accomplish this goal, which mainly fall into two categories: qualitative and quantitative (Jong 2002). Qualitative approaches typically abstract many details of the system, reducing it to a collection of states and transitions. An example of one of these types of models is a Boolean network that views each species in the system as being either in the "on" or "off" state and each reaction as a logical transition of species from "on" to "off" and vice versa. If more "levels" are desired (e.g., "high," "low," and "medium"), the model can be extended into a qualitative network. These types of models are often used in multicellular simulators such as Gro or Morpheus, in which simplifying the logic for each cell aids in reducing the complexity of the entire multicellular system.

Another type of qualitative model for describing chemical processes is a Petri net (Petri 1966). Petri nets are useful for modeling systems with a lot of concurrency and are comprised of bipartite graphs made up of a collection of transitions (i.e., events or reactions) and places (i.e., states or conditions). Each place can have a number of tokens representing anything such as the concentration of a species or the number of times a reaction has occurred. Tokens are moved from one place to another by firing a transition between them. This firing can represent many things such as an interaction of two species or the advancement of time. Among its modeling options, iBioSim supports the creation of Petri net models of biological systems.

Although qualitative models have been used to successfully predict different behaviors in biological systems, they often fall short in producing quantitative results. Quantitative models, on the other hand, strive to be as real-to-life as possible and often include very complex interactions that are described using reaction rate equations. In these types of models, the modeler will often define the quantity of each chemical species, a set of chemical reactions between species, and a collection of parameters used in the rate equations. These models can then be either translated into a collection of ordinary differential equations (ODEs) using classical chemical kinetics (Waage and Guldberg 1864) or into a collection of propensity functions using stochastic chemical kinetics (Gillespie 1992), depending on which types of simulation or analysis a user wishes to perform. Quantitative models are the most widely used in synthetic biology as they can be built and tuned to give the most predictive power. Nearly all of the simulation tools presented in this review are capable of producing and/or processing quantitative models.

Simulation

Simulation largely falls into two categories: deterministic and stochastic (or probabilistic) (Lecca et al. 2013). Deterministic simulation is often performed after applying classical chemical kinetics to a model to produce a set of ODEs. These ODEs are then solved using a standard method such as the Euler or Runge-Kutta methods to produce time series data for each species in the system. ODE simulation is deterministic because the solution to the ODEs is always the same given the same initial conditions on the model. Hence, the system only needs to be simulated once to produce a trace representing its behavior.

Classical chemical kinetics assume that concentrations of species in a system are large and that changes in these concentrations occur continuously and deterministically. This assumption does not always hold for biological systems, particularly genetic regulatory circuits where molecule counts are small and reaction firings cause large fluctuations in species counts (McAdams and Arkin 1999). For these systems, stochastic simulation is necessary to produce accurate results.

To prepare the model for stochastic simulation, the propensity of each reaction in the system must be calculated. Each propensity represents the probability that its associated reaction will fire within the next infinitesimal time interval, and they are computed by multiplying the reactants in the reaction by the reaction rate constants. With these propensities, the simulator can use a method such as Gillespie's stochastic simulation algorithm (SSA) (Gillespie 1976, 1977). This algorithm roughly works by determining the next reaction to fire based on propensities, probabilistically determining a time for the reaction to fire, stepping over all of the uneventful time to this firing time, firing the reaction to update the state, and then repeating this process until the desired time limit has been reached. Traces produced by this algorithm will vary each time it is run because the next reaction and the next firing times are always selected probabilistically. This fact means that a system may need to be simulated many times to determine its average behavior or to observe a rare event. Additionally, some systems have many fast reactions (reactions with large propensities), which cause the simulator to spend an exorbitant amount of simulation time firing these reactions while not advancing time very much. For these systems, improvements to the SSA such as tau-leaping (Gillespie and Petzold 2003) (selecting a larger time step and firing several reactions instead of one) and slow-scale SSA (Cao et al. 2005) (partitioning the system into fast and slow reactions and simulating them on different time scales with the help of steady-state approximations) have aided in their simulation.

Almost all of the tools mentioned in this review with simulation capabilities support both deterministic and stochastic simulation. Some of them (e.g., TinkerCell, CellDesigner, JDesigner), however, connect to another simulation tool (e.g., COPASI, Jarnac) for this support.

Analysis

Beyond simulation, there exist many types of analysis that can be performed on a model. These include model checking over steady-state or transient properties of the system, flux balance analysis, parameter estimation, and sensitivity analysis. Model checking approaches, especially those with a probabilistic component, can either be analyzed with statistical or numerical approaches (Younes et al. 2006; Kwiatkowska et al. 2007). Statistical approaches typically involve stochastically simulating the system a large number of times and then computing statistics over the resulting traces, whereas numerical analysis usually involves finding the state space of the model, often represented with a matrix, and performing computations directly on it. Without translating the model into the appropriate format and exporting it to an external model-checking tool like PRISM (Hinton et al. 2006), the only biological simulation tool capable of performing model checking is iBioSim.

Flux balance analysis is used to calculate the steady-state metabolic fluxes of biological systems by assuming that the input to the system roughly equals the output plus or minus one or a few factors (Raman and Chandra 2009). It is fairly easy to implement and is included as a feature in iBioSim, COPASI, PySCeS, and Jarnac.

Parameter estimation methods attempt to find values for parameters based on some experimental data (Lynch and Walsh 1998; Thompson et al. 2005). Most of these methods involve choosing some values for the parameters and then simulating the system to see how closely the resulting trace matches the data. If the trace matches well enough, the parameters are returned as acceptable; otherwise, the parameters are perturbed and the process repeats. This perturbation can be performed using a variety of methods, from genetic algorithms to particle swarms to hill-climbing techniques. iBioSim has newly added support for parameter estimation, but COPASI has the largest number of implemented parameter estimation methods. For those looking to take parameter estimation a step further and instead try to solve the parameter synthesis problem to determine ranges of acceptable parameters, BioPSy can be used.

Sensitivity analysis involves determining how much different parameters affect the dynamics of the system (Nestorov 1999). Like model checking, sensitivity analysis can either be performed through simulations of the model or by analyzing the model numerically. Many tools including COPASI, PySCeS, and Jarnac are capable of performing this type of analysis.

Example

In this example, we apply modeling and simulation to the genetic AND gate and the dual-feedback genetic oscillator using iBioSim and COPASI (Fig. 3). The first step is generating a mathematical model for each of these systems

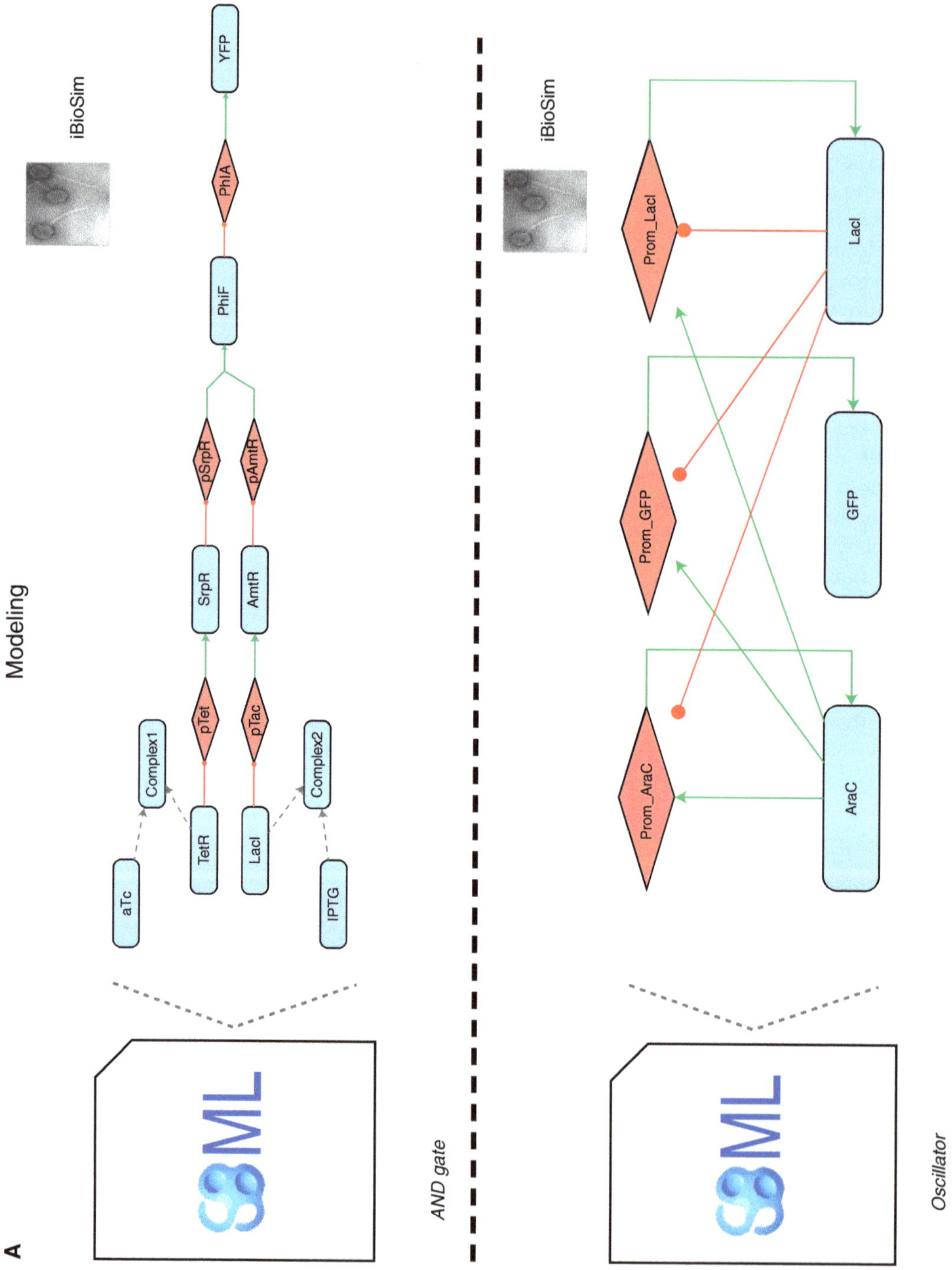

Figure 3. *See legend on p. 158.*

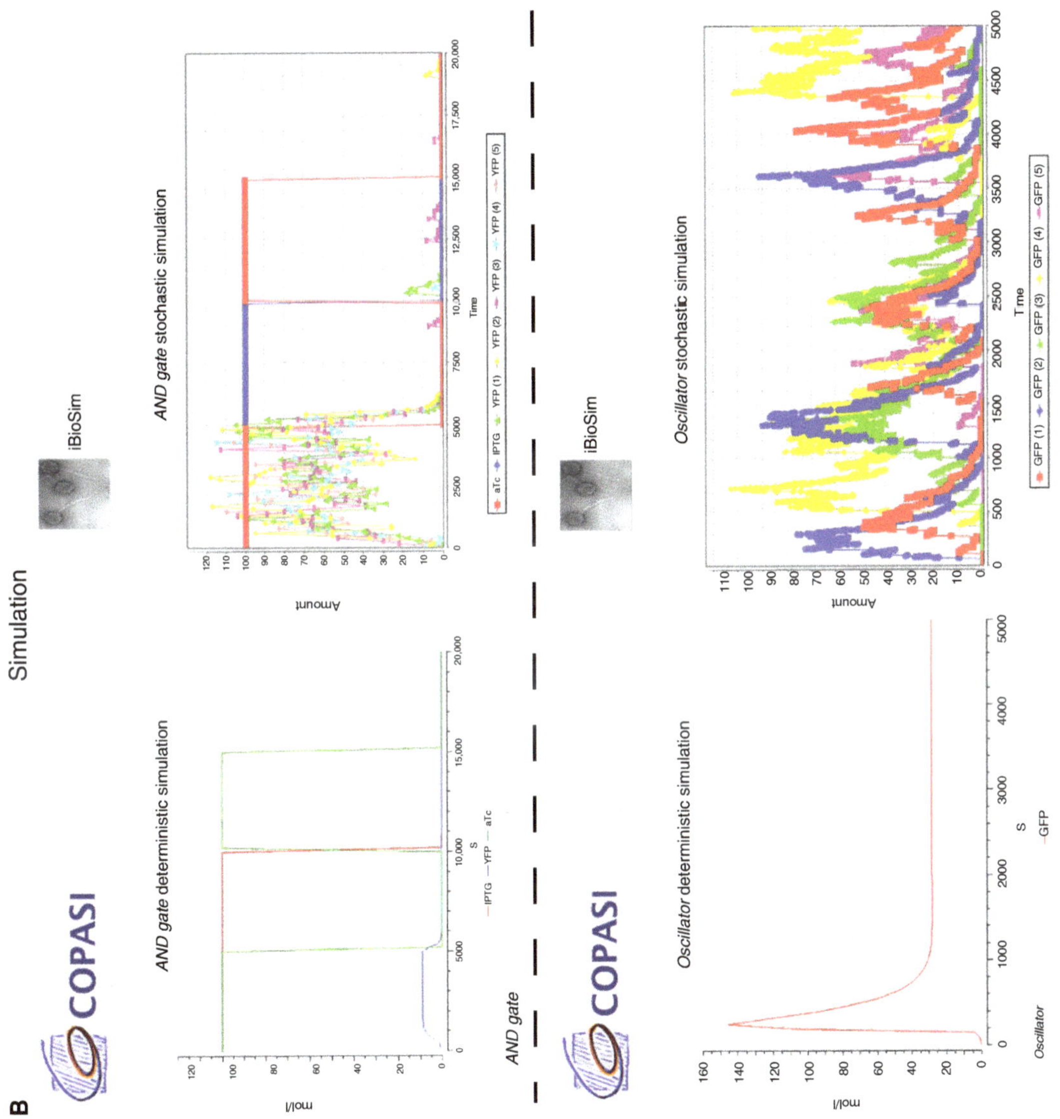

Figure 3. (*Continued.*) *See legend on p. 158.*

that we can simulate. For the genetic AND gate, the SBOL file generated by Cello is imported into iBioSim and converted into SBML. Although the converted model can be simulated outright, it is missing tunable inputs (i.e., the small molecules aTc and IPTG) and an output (i.e., the florescent protein YFP). Using the model editor in iBioSim, aTc, and IPTG are added to the model in addition to their complex formation reactions with TetR and LacI, respectively. Additionally, YFP is added, a repression arc is drawn between PhlF and YFP, and the final model is shown (Fig. 3A). This model is then exported as an SBML model that is passed to COPASI and simulated using ODEs while varying the values of aTc and IPTG (Fig. 3B shows simulation plots for this model). The ODE simulation shows that the value of YFP is highest when both aTc and IPTG are present and lowest when only one or neither are present. However, the high value of YFP is still fairly low as the ODE simulation treats reactions as continuous, deterministic events. Using iBioSim, we can perform a few stochastic simulations while still varying the values of aTc and IPTG. These simulations show that again the value of YFP is highest when both aTc and IPTG are present and lowest when only one or neither are present. Because of stochastic noise, this simulation reveals that there is occasionally sporadic production of YFP, but YFP quickly degrades away because of the lack of one or both inputs.

The SBML model for the dual-feedback genetic oscillator produced by GEC was not able to be simulated directly using iBioSim because of some differences in support of SBML packages; however, this model is recreated in iBioSim (Fig. 3A). As with the genetic AND gate, the oscillator's model is exported to SBML that is fed into COPASI and simulated using deterministic methods (Fig. 3B). In this simulation, the value of GFP initially increases to a peak value and then declines until it levels off at a steady-state value. Needless to say, this is not the behavior we would expect from an oscillator, and it is obvious that the ODE simulation is failing to capture some of the noisy, stochastic events that lead to oscillations. Using iBioSim, on the other hand, several stochastic simulations are performed, which show this oscillatory behavior.

DATA EXCHANGE AND STANDARDIZATION

At all stages of the design cycle described in the introductory section, there exist opportunities to store and exchange data on genetic designs, including their structure (genetic sequences, X-ray crystal structures, etc.), function (regulatory relationships, chemical kinetic models, etc.), and measurements (flow cytometry histograms, microscope images, etc.), among other types of data. Table 1 contains a list of BDA databases and tools for storing and exchanging these data types. In this section, we will briefly describe two of the standards most widely used by these and other BDA tools, and discuss how these standards facilitate data exchange in the examples presented in this review.

The Synthetic Biology Open Language (SBOL)

SBOL (Galdzicki et al. 2014; Roehner et al. 2016b) is an emerging data exchange standard for synthetic biology with growing support among BDA tools, including database tools such as ICE (Ham et al. 2012) and the SBOL Stack (Madsen et al. 2016). SBOL has been de-

Figure 3. Modeling and simulation results of a genetic AND gate and dual-feedback oscillator. (*A*) SBML models are created in iBioSim for the AND gate and oscillator. The AND gate's model is automatically generated by converting the SBOL file produced by Cello to SBML. This model is then extended to include the regulation by the proteins LacI and TetR, the complex formation of LacI and TetR with the small molecules IPTG and aTc, respectively, and the regulation of the florescent protein YFP. The oscillator model is reconstructed using information (e.g., parameter values, rate constants) from the SBML generated by GEC. (*B*) Simulations are performed using both COPASI and iBioSim. Deterministic simulations are performed using the LSODA ODE solver in COPASI. Stochastic simulations are performed in iBioSim using an implementation of Gillespie's SSA.

veloped by members of the BDA and greater synthetic biology communities to document genetic components (DNA, RNA, protein, etc.) for the purpose of engineering design. Unlike existing standards that were originally conceived for documenting naturally occurring genetic sequences, such as FASTA (Pearson and Lipman 1988) and GenBank (Bilofsky and Christian 1988), SBOL can be used to document partial genetic designs and recursively annotate the sequences of genetic components with other components in a hierarchical fashion. These capabilities of SBOL address the iterative, modular character of engineering designs in a way that current standards for genetic sequences neglect. Furthermore, SBOL is an extensible standard that can be adapted to meet the evolving needs of the synthetic biology community, such as the need to document the context-dependence of functioning genetic systems.

In the examples presented in this review, SBOL facilitates the exchange of design data for the genetic AND gate. In particular, Cello exports an SBOL file for the AND gate that is imported and compiled by iBioSim into an SBML model (Roehner and Myers 2014). This model is then extended and simulated as described above (see sections Data Analysis and Simulation and Example). In this way, the SBOL-documented structure and function of the AND gate directly inform the creation of a quantitative model of its chemical kinetics.

The Systems Biology Markup Language (SBML)

SBML (Hucka et al. 2003) is a biological modeling standard that has been developed by the systems biology community and is currently supported by more than 250 different software tools. There also exists a public database for storing and accessing SBML models known as the BioModels database (Chelliah et al. 2013). The primary goal of SBML is to enable intersoftware exchange of most essential features of models for biological systems, especially biochemical models for describing cell signaling, metabolism, and genetic regulation, among other phenomena. Accordingly, SBML is a machine-readable Extensible Markup Language (XML) that is independent of any proprietary software language or particular analytical framework (e.g., ODE or stochastic analysis).

In the examples presented in this review, SBML enables models of the genetic AND gate and dual-feedback oscillator to be simulated across tools. In particular, models of the AND gate and oscillator exported from iBioSim (Myers et al. 2009) can be imported and simulated by COPASI (Hoops et al. 2006). This allows users to take advantage of the analysis methods that are unique to each tool without having to translate their models between formats and risk losing information. Even in the case when tools adhering to a standard are unable to directly exchange models because of implementation errors or differences in support, the standard can facilitate the recreation of models across tools. This is currently the case when exchanging SBML models between GEC (Pedersen and Phillips 2009) and iBioSim. Even though the oscillator model exported by GEC cannot be directly imported into iBioSim, its SBML encoding permits a researcher to more easily interpret and recreate this model in iBioSim.

CONCLUSION

As shown in this overview, great progress has been made in design automation in synthetic biology in a relatively short period of time (<9 years at the time of writing as measured by the first iGEM software competition). These efforts have validated many approaches (Beal et al. 2012), created interwoven design flows (Galdzicki et al. 2014), and began to foster a self-sustaining industrial development ecosystem and associated academic discipline. However, much more work needs to be performed. Examples include:

1. Specification—the specification of the desired performance (e.g., timing, yield, expression level) of a biological system separated from the behavior of that system remains largely unaddressed. Formal languages for describing the diverse domains in synthetic biology (e.g., metabolic net-

works, genetic circuits, engineered organisms) will need to be developed and related to one another when appropriate. Abstraction terminology should also work toward standardization.

2. Design—rules and constraints need to be formally captured and shared more broadly. These constraints need to be validated experimentally and parameterized for specific contexts (e.g., host organisms, environmental conditions).

3. Build—microfluidics and liquid-handling robot platforms need to be more fully integrated across the entire assembly process (e.g., device design to organism growth). Interfaces for these devices need to be standardized and widely understood. Support for "cloud labs" (emeraldcloudlab.com; transcriptic.com) should increase and be part of the software toolchain.

4. Test—standardized assays amendable to BDA tools should be encouraged and expected of experimentalists in the field. Measurement standards and metrics (programmingbiology.org; Beal 2015) should be developed and promoted.

5. Learn—large-scale data mining and storage need to be available to communities outside of synthetic biology so that experts in these fields can apply the latest analytical methods.

Recently SBOL has been adopted by the *ACS Synthetic Biology* journal (Hillson et al. 2016) as an acceptable format to submit along with research manuscripts. This represents an important milestone in the promotion of BDA principles.

In addition, curated open-source software efforts related to BDA may prove valuable for academics as a means to sustain their software over long periods of time and unpredictable funding climates. These may also be useful for industry looking to benefit from early technical advances with little upfront development costs. The Nona Research Foundation (see nonasoftware.org) is a U.S.-based nonprofit that looks to provide open-source software to the entire community (academic and industrial). This organization and others like the BioBricks Foundation and BioBuilder can also serve as educational hubs to communicate the importance of BDA.

REFERENCES

Andrianantoandro E, Basu S, Karig DK, Weiss R. 2006. Synthetic biology: New engineering rules for an emerging discipline. *Mol Syst Biol* **2:** 2006.0028.

Appleton E, Tao J, Haddock T, Densmore D. 2014a. Interactive assembly algorithms for molecular cloning. *Nat Methods* **11:** 657–662.

Appleton E, Tao J, Wheatley FC, Desai DH, Lozanoski TM, Shah PD, Awtry JA, Jin SS, Haddock TL, Densmore DM. 2014b. Owl: Electronic datasheet generator. *ACS Synth Biol* **3:** 966–968.

Arkin A. 2008. Setting the standard in synthetic biology. *Nat Biotechnol* **26:** 771–774.

Bates JT, Chivian D, Arkin AP. 2011. GLAMM: Genome-linked application for metabolic maps. *Nucleic Acids Res* **39:** W400–W405.

Beal J. 2015. Signal-to-noise ratio measures efficacy of biological computing devices and circuits. *Front Bioeng Biotechnol* **3:** 93.

Beal J, Lu T, Weiss R. 2011. Automatic compilation from high-level biologically oriented programming language to genetic regulatory networks. *PLoS ONE* **6:** e22490.

Beal J, Weiss R, Densmore D, Adler A, Appleton E, Babb J, Bhatia S, Davidsohn N, Haddock T, Loyall J, et al. 2012. An end-to-end workflow for engineering of biological networks from high-level specifications. *ACS Synth Biol* **1:** 317–331.

Bhatia S, Densmore D. 2013. Pigeon: A design visualizer for synthetic biology. *ACS Synth Biol* **2:** 348–350.

Bilitchenko L, Liu A, Cheung S, Weeding E, Xia B, Leguia M, Anderson JC, Densmore D. 2011. Eugene: A domain specific language for specifying and constraining synthetic biological parts, devices, and systems. *PLoS ONE* **6:** e18882.

Bilofsky HS, Christian B. 1988. The GenBank genetic sequence data bank. *Nucleic Acids Res* **16:** 1861–1863.

Blakes J, Raz O, Feige U, Bacardit J, Widera P, Ben-Yehezkel T, Shapiro E, Krasnogor N. 2014. Heuristic for maximizing DNA reuse in synthetic DNA library assembly. *ACS Synth Biol* **3:** 529–542.

Borujeni AE, Salis HM. 2016. Translation initiation is controlled by RNA folding kinetics via a ribosome drafting mechanism. *J Am Chem Soc* **138:** 7016–7023.

Borujeni AE, Channarasappa AS, Salis HM. 2013. Translation rate is controlled by coupled trade-offs between site accessibility, selective RNA unfolding and sliding at upstream standby sites. *Nucleic Acids Res* **42:** 2646–2659.

Canton B, Labno A, Endy D. 2008. Refinement and standardization of synthetic biological parts and devices. *Nat Biotechnol* **26:** 787–793.

Cao Y, Gillespie DT, Petzold LR. 2005. The slow-scale stochastic simulation algorithm. *J Chem Phys* doi: 10.1063/1.1824902.

Castillo-Hair SM, Sexton JT, Landry BP, Olson EJ, Igoshin OA, Tabor JJ. 2016. FlowCal: A user-friendly, open source software tool for automatically converting flow cytometry data from arbitrary to calibrated units. *ACS Synth Biol* **5:** 774–780.

Chakrabarti S, Lanczycki CJ, Panchenko AR, Przytycka TM, Thiessen PA, Bryant SH. 2006. Refining multiple sequence alignments with conserved core regions. *Nucleic Acids Res* **34:** 2598–2606.

Chandran D, Bergmann F, Sauro H. 2009. TinkerCell: Modular CAD tool for synthetic biology. *J Biol Eng* **3:** 19.

Chelliah V, Laibe C, Novere NL. 2013. BioModels database: A repository of mathematical models of biological processes. *Methods Mol Biol* **1021:** 189–199.

Cong L, Ran FA, Cox D, Lin S, Barretto R, Habib N, Hsu PD, Wu X, Jiang W, Marraffini LA, et al. 2013. Multiplex genome engineering using CRISPR/Cas systems. *Science* **339:** 819–823.

Czar MJ, Cai Y, Peccoud J. 2009. Writing DNA with GenoCAD. *Nucleic Acids Res* **37:** W40–W47.

Dasika MS, Maranas CD. 2008. OptCircuit: An optimization based method for computational design of genetic circuits. *BMC Syst Biol* **2:** 24.

Densmore D. 2012. Bio-design automation: Nobody said it would be easy. *ACS Synth Biol* **1:** 296–296.

Densmore DM, Bhatia S. 2014. Bio-design automation: Software + biology + robots. *Trends Biotechnol* **32:** 111–113.

Densmore D, Van Devender A, Johnson M, Sritanyaratana N. 2009. A platform-based design environment for synthetic biological systems. In *The Fifth Richard Tapia Celebration of Diversity in Computing Conference: Intellect, Initiatives, Insight, and Innovations*, pp. 24–29. Portland, OR.

Densmore D, Hsiau TH, Kittleson JT, DeLoache W, Batten C, Anderson JC. 2010. Algorithms for automated DNA assembly. *Nucleic Acids Res* **38:** 2607–2616.

Dhar PK. 2010. The language of life. *Nat India* doi: 10.1038/nindia.2010.145.

Drummond A, et al. 2010. Geneious v5.5. http://www.geneious.com.

Endy D. 2005. Foundations for engineering biology. *Nature* **438:** 449–453.

Engler C, Kandzia R, Marillonnet S. 2008. A one pot, one step, precision cloning method with high throughput capability. *PLoS ONE* **3:** e3647.

Funahashi A, Matsuoka Y, Jouraku A, Morohashi M, Kikuchi N, Kitano H. 2008. CellDesigner 3.5: A versatile modeling tool for biochemical networks. *Proc IEEE* **96:** 1254–1265.

Galdzicki M, Wilson ML, Rodriguez CA, Adam L, Adler A, Anderson JC, Beal J, Chandran D, Densmore D, Drory OA, et al. 2011. Synthetic Biology Open Language (SBOL) Version 1.0.0. RFC 85. doi: 1721.1/66172.

Galdzicki M, Clancy KP, Oberortner E, Pocock M, Quinn JY, Rodriguez CA, Roehner N, Wilson ML, Adam L, Anderson JC, et al. 2014. The Synthetic Biology Open Language (SBOL) provides a community standard for communicating designs in synthetic biology. *Nat Biotechnol* **32:** 545–550.

Gibson DG, Young L, Chuang RY, Venter JC, Hutchison CA III, Smith HO. 2009. Enzymatic assembly of DNA molecules up to several hundred kilobases. *Nat Methods* **6:** 343–345.

Gillespie DT. 1976. A general method for numerically simulating the stochastic time evolution of coupled chemical reactions. *J Comp Phys* **22:** 403–434.

Gillespie DT. 1977. Exact stochastic simulation of coupled chemical reactions. *J Phys Chem* **81:** 2340–2361.

Gillespie DT. 1992. *Markov processes: An introduction for physical scientists*. Academic, New York.

Gillespie DT, Petzold LR. 2003. Tau leaping. *J Chem Phys* **119:** 8229–8234.

Goler J. 2004. "BioJADE: A design and simulation tool for synthetic biological systems." Master's thesis, MIT.

Ham TS, Dmytriv Z, Plahar H, Chen J, Hillson NJ, Keasling JD. 2012. Design, implementation and practice of JBEI-ICE: An open source biological part registry platform and tools. *Nucleic Acids Res* **40:** e141.

Hayden EC. 2015. Synthetic biologists seek standards for nascent field. *Nature* **520:** 141–142.

Hill AD, Tomshine JR, Weeding EMB, Sotiropoulos V, Kaznessis YN. 2008. SynBioSS: The synthetic biology modeling suite. *Bioinformatics* **24:** 2551–2553.

Hillson NJ. 2014. *j5 DNA assembly design automation*, pp. 245–269. Humana, Totowa, NJ.

Hillson N, Rosengarten RD, Keasling J. 2011. j5 DNA assembly design automation software. *ACS Synth Biol* **1:** 14–21.

Hillson NJ, Plahar HA, Beal J, Prithviraj R. 2016. Improving synthetic biology communication: Recommended practices for visual depiction and digital submission of genetic designs. *ACS Synth Biol* **5:** 449–451.

Hinton A, Kwiatkowska M, Norman G, Parker D. 2006. *12th International Conference on Tools and Algorithms for the Construction and Analysis of Systems (TACAS'06)* (ed. Hermanns H, Palsberg J), pp. 441–444. Springer, Berlin.

Hoops S, Sahle S, Gauges R, Lee C, Pahle J, Simus N, Singhal M, Xu L, Mendes P, Kummer U. 2006. COPASI—A COmplex PAthway SImulator. *Bioinformatics* **22:** 3067–3074.

Huang H, Densmore D. 2014. Fluigi: Microfluidic device synthesis for synthetic biology. *ACM J Emerg Technol Comput Syst* **11:** 1–19.

Huber W, Carey VJ, Gentleman R, Anders S, Carlson M, Carvalho BS, Bravo HC, Davis S, Gatto L, Girke T, et al. 2015. Orchestrating high-throughput genomic analysis with Bioconductor. *Nat Methods* **12:** 115–121.

Hucka M, Finney A, Sauro HM, Bolouri H, Doyle JC, Kitano H, Arkin AP, Bornstein BJ, Bray D, Cornish-Bowden A, et al. 2003. The systems biology markup language (SBML): A medium for representation and exchange of biochemical network models. *Bioinformatics* **19:** 524–531.

Huynh L, Tagkopoulos I. 2014. Optimal part and module selection for synthetic gene circuit design automation. *ACS Synth Biol* **3:** 556–564.

Huynh L, Tsoukalas A, Köppe M, Tagkopoulos I. 2013. SBROME: A scalable optimization and module matching framework for automated biosystems design. *ACS Synth Biol* **2:** 263–273.

Iverson SV, Haddock TL, Beal J, Densmore DM. 2016. CIDAR MoClo: Improved MoClo assembly standard and new *E. coli* part library enable rapid combinatorial design for synthetic and traditional biology. *ACS Synth Biol* **5:** 99–103.

Jang SS, Oishi KT, Egbert RG, Klavins E. 2012. Specification and simulation of synthetic multicelled behaviors. *ACS Synth Biol* **1:** 365–374.

Jong HD. 2002. Modeling and simulation of genetic regulatory systems: A literature review. *J Comp Biol* **9:** 67–103.

Kahl LJ, Endy D. 2013. A survey of enabling technologies in synthetic biology. *J Biol Eng* **7:** 1–19.

Kelly JR, Rubin AJ, Davis JH, Ajo-Franklin CM, Cumbers J, Czar MJ, de Mora K, Glieberman AL, Monie DD, Endy D. 2009. Measuring the activity of BioBrick promoters using an in vivo reference standard. *J Biol Eng* **3:** 4.

Kim TY, Park JM, Kim HU, Cho KM, Lee SY. 2015. Design of homo-organic acid producing strains using multi-objective optimization. *Metab Eng* **28:** 63–73.

Koressaar T, Remm M. 2007. Enhancements and modifications of primer design program Primer3. *Bioinformatics* **23:** 1289–1291.

Kwiatkowska M, Norman G, Parker D. 2007. Stochastic model checking. In *Formal Methods for the Design of Computer, Communication and Software Systems: Performance Evaluation (SFM'07)*, Vol. 4486 (ed. Bernardo M, Hillston J), pp. 220–270. Springer, Berlin.

Leaver-Fay A, Tyka M, Lewis SM, Lange OF, Thompson J, Jacak R, Kaufman K, Renfrew PD, Smith CA, Sheffler W, et al. 2011. ROSETTA3: An object-oriented software suite for the simulation and design of macromolecules. *Methods Enzymol* **487:** 545–574.

Lecca P, Laurenzi I, Jordan F. 2013. *Deterministic versus stochastic modelling in biochemistry and systems biology.* Woodhead, Oxford.

Li MZ, Elledge SJ. 2007. Harnessing homologous recombination in vitro to generate recombinant DNA via SLIC. *Nat Methods* **4:** 251–256.

Linshiz G, Stawski N, Goyal G, Bi C, Poust S, Sharma M, Mutalik V, Keasling JD, Hillson NJ. 2014. PR-PR: Cross-platform laboratory automation system. *ACS Synth Biol* **3:** 515–524.

Lux M, Bramlett BW, Ball DA, Peccoud J. 2012. Genetic design automation: Engineering fantasy or scientific renewal? *Trends Biotechnol* **30:** 120–126.

Lynch M, Walsh B. 1998. *Genetics and analysis of quantitative traits.* Sinauer Associates, Sunderland, MA.

Madsen C, Shmarov F, Zuliani P. 2015. *Computational methods in systems biology*, pp. 182–194. Springer, New York.

Madsen C, McLaughlin JA, Misirli G, Pocock M, Flanagan K, Hallinan J, Wipat A. 2016. The SBOL Stack: A platform for storing, publishing, and sharing synthetic biology designs *ACS Synth Biol* **5:** 487–497.

Marchisio MA. 2014. Parts & pools: A framework for modular design of synthetic gene circuits. *Front Bioeng Biotechnol* **2:** 42.

McAdams HH, Arkin A. 1999. Genetic regulation at the nanomolar scale: It's a noisy business. *Trends Genet* **15:** 65–69.

Mirschel S, Steinmetz K, Rempel M, Ginkel M, Gilles ED. 2009. PROMOT: Modular modeling for systems biology. *Bioinformatics* **25:** 687–689.

Misirli G, Hallinan JS, Yu T, Lawson JR, Wimalaratne SM, Cooling MT, Wipat A. 2011. Model annotation for synthetic biology: Automating model to nucleotide sequence conversion. *Bioinformatics* **27:** 973–979.

Myers CJ, Barker N, Jones K, Kuwahara H, Madsen C, Nguyen NP. 2009. iBioSim: A tool for the analysis and design of genetic circuits. *Bioinformatics* **25:** 2848–2849.

Nestorov IA. 1999. Sensitivity analysis of pharmacokinetic and pharmacodynamic systems. I: A structural approach to sensitivity analysis of physiologically based pharmacokinetic models. *J Pharmacokinet Biopharm* **27:** 577–596.

Nielsen AA, Der BS, Shin J, Vaidyanathan P, Paralanov V, Strychalski EA, Ross D, Densmore D, Voigt CA. 2016. Genetic circuit design automation. *Science* **352:** aac7341.

Oberortner E, Densmore D. 2014. Web-based software tool for constraint-based design specification of synthetic biological systems. *ACS Synth Biol* **4:** 757–760.

Oberortner E, Bhatia S, Lindgren E, Densmore D. 2014. A rule-based design specification language for synthetic biology. *ACM J Emerg Technol Comput Syst* **11z:** 25.

Olivier BG, Rohwer JM, Hofmeyr JHS. 2004. Modelling cellular systems with PySCeS. *Bioinformatics* **21:** 560–561.

Pearson WR, Lipman DJ. 1988. Improved tools for biological sequence comparison. *Proc Natl Acad Sci* **85:** 2444–2448.

Pedersen M, Phillips A. 2009. Towards programming languages for genetic engineering of living cells. *J R Soc Interface* **6:** S437–S450.

Petri CA. 1966. "Communication with automata." PhD thesis, Universität Hamburg, Hamburg, Germany.

Quan J, Tian J. 2009. Circular polymerase extension cloning of complex gene libraries and pathways. *PLoS ONE* **4:** e6441.

Quinn JY, Cox RS III, Adler A, Beal J, Bhatia S, Cai Y, Chen J, Clancy K, Galdzicki M, Hillson NJ, et al. 2015. SBOL Visual: A graphical language for genetic designs. *PLoS Biol* **13:** 1–9.

Quintin M, Ma NJ, Ahmed S, Bhatia S, Lewis A, Isaacs FJ, Densmore D. 2016. Merlin: Computer-aided oligonucleotide design for large scale genome engineering with MAGE. *ACS Synth Biol* **5:** 452–458.

Raman K, Chandra N. 2009. Flux balance analysis of biological systems: Applications and challenges. *Brief Bioinform* **10:** 435–449.

Rocha I, Maia P, Evangelista P, Vilaça P, Soares S, Pinto JP, Nielsen J, Patil KR, Ferreira EC, Rocha M. 2010. OptFlux: An open-source software platform for in silico metabolic engineering. *BMC Syst Biol* **4:** 1–12.

Rodrigo G, Jaramillo A. 2013. AutoBioCAD: Full biodesign automation of genetic circuits. *ACS Synth Biol* **2:** 230–236.

Rodrigo G, Carrera J, Jaramillo A. 2007. Asmparts: Assembly of biological model parts. *Syst Synth Biol* **1:** 167–170.

Roehner N, Myers CJ. 2014. A methodology to annotate systems biology markup language models with the synthetic biology open language. *ACS Synth Biol* **3:** 57–66.

Roehner N, Young EM, Voigt CA, Gordon DB, Densmore D. 2016a. Double Dutch: A tool for designing combinatorial libraries of biological systems. *ACS Synth Biol* **5:** 507–517.

Roehner N, Beal J, Clancy K, Bartley B, Misirli G, Grünberg R, Oberortner E, Pocock M, Bissell M, Madsen C, et al. 2016b. Sharing structure and function in biological design with SBOL 2.0. *ACS Synth Biol* **5:** 498–506.

Salis HM, Mirsky EA, Voigt CA. 2009. Automated design of synthetic ribosome binding sites to control protein expression. *Nat Biotechnol* **27:** 946–950.

Sarrion-Perdigones A, Falconi EE, Zandalinas SI, Juárez P, Fernández-del-Carmen A, Granell A, Orzaez D. 2011. GoldenBraid: An iterative cloning system for standardized assembly of reusable genetic modules. *PLoS ONE* **6:** e21622.

Sauro H. 2000. Jarnac: An interactive metabolic systems language in computation in cells. In *Proceedings of an EPSRC Emerging Computing Paradigms Workshop* (ed. Bolouri H, Paton R). University of Hertfordshire, Hertfordshire, UK.

Schellenberger J, Que R, Fleming RM, Thiele I, Orth JD, Feist AM, Zielinski DC, Bordbar A, Lewis NE, Rahmanian S, et al. 2011. Quantitative prediction of cellular metabolism with constraint-based models: The COBRA Toolbox v2.0. *Nat Protoc* **1:** 1290–1307.

Shetty RP, Endy D, Knight TF. 2008. Engineering BioBrick vectors from BioBrick parts. *J Biol Eng* **2:** 5.

Smanski MJ, Bhatia S, Zhao D, Park Y, B A Woodruff L, Giannoukos G, Ciulla D, Busby M, Calderon J, Nicol R, et al. 2014. Functional optimization of gene clusters by combinatorial design and assembly. *Nat Biotechnol* **32:** 1241–1249.

Smith LP, Bergmann FT, Chandran D, Sauro HM. 2009. Antimony: A modular model definition language. *Bioinformatics* **25:** 2452–2454.

Smolke CD. 2009. Building outside of the box: iGEM and the BioBricks Foundation. *Nat Biotechnol* **27:** 1099–1102.

Starruß Jo, de Back W, Brusch L, Deutsch A. 2014. Morpheus: A user-friendly modeling environment for multiscale and multicellular systems biology. *Bioinformatics* **30:** 1331–1332.

Stricker J, Cookson S, Bennett MR, Mather WH, Tsimring LS, Hasty J. 2008. A fast, robust and tunable synthetic gene oscillator. *Nature* **456:** 516–519.

Thomas DE, Moorby P. 1995. *The Verilog hardware description language* (2nd ed.). Kluwer, New York.

Thompson R, Brotherstone S, White IMS. 2005. Estimation of quantitative genetic parameters. *Philos Trans R Soc B Biol Sci* **360:** 1469–1477.

Untergasser A, Cutcutache I, Koressaar T, Ye J, Faircloth BC, Remm M, Rozen SG. 2012. Primer3—New capabilities and interfaces. *Nucleic Acids Res* **40:** e115.

Vasilev V, Liu C, Haddock T, Bhatia S, Adler A, Yaman F, Beal J, Babb J, Weiss R, Densmore D. et al. 2011. A software stack for specification and robotic execution of protocols for synthetic biological engineering. In *3rd International Workshop on Bio-Design Automation (IWDBDA)*, pp. 24–25. San Diego, CA.

Vilanova C, Porcar M. 2014. iGEM 2.0—Refoundations for engineering biology. *Nat Biotechnol* **32:** 420–424.

Waage P, Guldberg CM. 1864. Studies concerning affinity. *J Chem Educ* **63:** 1044.

Wang HH, Isaacs FJ, Carr PA, Sun ZZ, Xu G, Forest CR, Church GM. 2009. Programming cells by multiplex genome engineering and accelerated evolution. *Nature* **460:** 894–898.

Weber E, Engler C, Gruetzner R, Werner S, Marillonet S. 2010. A modular cloning system for standardized assembly of multigene constructs. *PLoS ONE* doi: 10.1371/journal.pone.0016765.

Wilson EH, Sagawa S, Weis JW, Schubert MG, Bissell M, Hawthorne B, Reeves CD, Dean J, Platt D. 2016. Genotype specification language. *ACS Synth Biol* **5:** 471–478.

Wu G, Bashir-Bello N, Freeland SJ. 2006. The synthetic gene designer: A flexible web platform to explore sequence manipulation for heterologous expression. *Prot Expres Purif* **47:** 441–445.

Yaman F, Bhatia S, Adler A, Densmore D, Beal J. 2012. Automated selection of parts for genetic regulatory networks. *ACS Synth Biol* **1:** 332–344.

Younes H, Kwiatkowska M, Norman G, Parker D. 2006. Numerical vs. statistical probabilistic model checking. *Int J Softw Tools Technol Transf* **8:** 216–228.

Zuker M. 2003. Mfold web server for nucleic acid folding and hybridization prediction. *Nucleic Acids Res* **31:** 3406–3415.

The Need for Integrated Approaches in Metabolic Engineering

Anna Lechner,[1,2] Elizabeth Brunk,[3] and Jay D. Keasling[1,2,4]

[1]Joint Bioenergy Institute (JBEI), Emeryville, California 94608

[2]Department of Chemical & Biomolecular Engineering, Department of Bioengineering, University of California, Berkeley, California 94720

[3]Department of Bioengineering, University of California, San Diego, California 92093

[4]Physical Biosciences Division, Lawrence Berkeley National Laboratory, Berkeley, California 94720

Correspondence: jdkeasling@lbl.gov

This review highlights state-of-the-art procedures for heterologous small-molecule biosynthesis, the associated bottlenecks, and new strategies that have the potential to accelerate future accomplishments in metabolic engineering. We emphasize that a combination of different approaches over multiple time and size scales must be considered for successful pathway engineering in a heterologous host. We have classified these optimization procedures based on the "system" that is being manipulated: transcriptome, translatome, proteome, or reactome. By bridging multiple disciplines, including molecular biology, biochemistry, biophysics, and computational sciences, we can create an integral framework for the discovery and implementation of novel biosynthetic production routes.

CURRENT APPROACHES IN METABOLIC ENGINEERING AND THEIR LIMITATIONS

Microbial organisms are able to package a set of biocatalysts into "molecular factories" for the energy-efficient generation of value-added compounds derived from simple sugars (Keasling 2010). Making use of these molecular factories is an attractive alternative to organic syntheses that rely on petrochemical feedstocks, finite resources, or environmentally unfriendly production processes. However, microbial organisms have not evolved to meet the demands of a scaled-up production process at the industrial scale and tend to have a poor ratio of achieved versus theoretical yield. Thus, one of the main goals of metabolic engineering is to transform organisms into efficient systems for the production of active pharmaceutical ingredients, commodity chemicals, and energy. Metabolic engineering has already provided sustainable access to a number of chemical classes. A recent milestone of bio-based industrial production is the engineered microbial biosynthesis of artemisinic acid, a plant-derived precursor to the antimalarial drug artemisinin (Paddon et al. 2013). Additional high-value compounds produced via sustainable processes include antibiotics, such as erythromycin A, daptomycin, and penicillin, as well as precursors to pharmaceuticals like taxadien-5a-acetoxy-10b-ol, which are reviewed elsewhere (Lee et al. 2009). Industrial

Cite this article as *Cold Spring Harb Perspect Biol* doi: 10.1101/cshperspect.a023903

fermentation processes have also been established for production of commodity chemicals like 1,3-propanediol, 1,4-butanediol, farnesene, and other terpenoids (for a review on this topic, see Klein-Marcuschamer et al. 2007). Further progress in metabolic engineering could facilitate industrial biosynthesis of the biofuels isobutanol, isopentenol, bisabolene, and pinene, as well as other hydrocarbons (for a review on this topic, see Lennen and Pfleger 2013). However, some important classes of small molecules, such as alkanes, alkenes, alkaloids, polyketides, and peptides, remain accessible mostly from native sources or through organic synthesis.

Why do we succeed in biomanufacturing certain compound classes while others remain recalcitrant? One limitation is the lack of integrated approaches toward pathway optimization. Most strategies deal with bottlenecks on only one of multiple levels in pathway engineering, often addressing either gene expression or enzyme engineering. This semirational approach is popular because quantitative behavior remains difficult to predict for complex biological systems. Inaccurate predictions in turn are partly a result of unknown or unpredictable cellular behavior most likely caused by additional undefined layers of control. Sometimes this is compounded by limitations in genetic accessibility of the host. Of the many microbial species that have potential as efficient factories, only a small number are currently genetically tractable. The recent discovery of the CRISPR-Cas system has the potential to readily engineer previously intractable species, but this technique has yet to be implemented widely among species (Jakočiūnas et al. 2015). In all cases, metabolic engineering is limited by analytical methods, which require specific method optimization for each compound class being produced. Despite these shortcomings, there are clear advancements in the development of new -omics data acquisition and analysis techniques that enable metabolic engineering on various levels. These challenges and perspectives highlight the need to implement a multilayer optimization framework to perfect the "design–build–test–learn" engineering cycle of synthetic biology.

THE NEED FOR INTEGRATED APPROACHES IN METABOLIC ENGINEERING

The ideal design for maximal metabolite production would generate active enzymes with sufficient catalytic turnover in consideration of timing and cellular location. This is a hard goal to achieve because of the complexity of the cell. Overlaid systems interact at different time scales and a single manipulation may have a positive impact at one metabolic or regulatory layer and a negative or neutral impact at another. Modifications at the transcriptome level adjust the timing and strength of gene expression, whereas changes to the translatome (measured by ribosome occupancy) influence protein synthesis rates, local translational speed, as well as protein solubility, activity, and specificity. Changes to the sequence and structure of proteins affect the catalytic efficiency of pathway enzymes and allosteric regulation at the proteome level. At the reactome level, balanced enzyme activity, sufficient transfer of intermediates between enzymes, and cofactor balance are all required for high flux through the engineered pathway.

A multilevel engineering approach can be subdivided into manageable processes (Fig. 1). At the transcriptome level, mRNA amounts are controlled by promoter strength, gene copy number, and mRNA stability. At the translatome level, translational efficiency and protein solubility are tuned by ribosome-binding site (RBS) strength, mRNA secondary structure, and codon usage. At the protein/proteome level, efficient catalysis is engineered by site-specific enzyme modifications and release of feedback inhibition. At the reactome level, enzyme ratio balancing and protein colocalization affect the efficient turnover of pathway intermediates. Robust and optimal calibration of all of these nested and interlocked mechanisms is necessary to achieve maximum flux through an engineered pathway. We propose a multilayer framework for metabolic engineering by integrating design and control elements on multiple levels to accelerate the implementation and optimization of novel biosynthetic production routes.

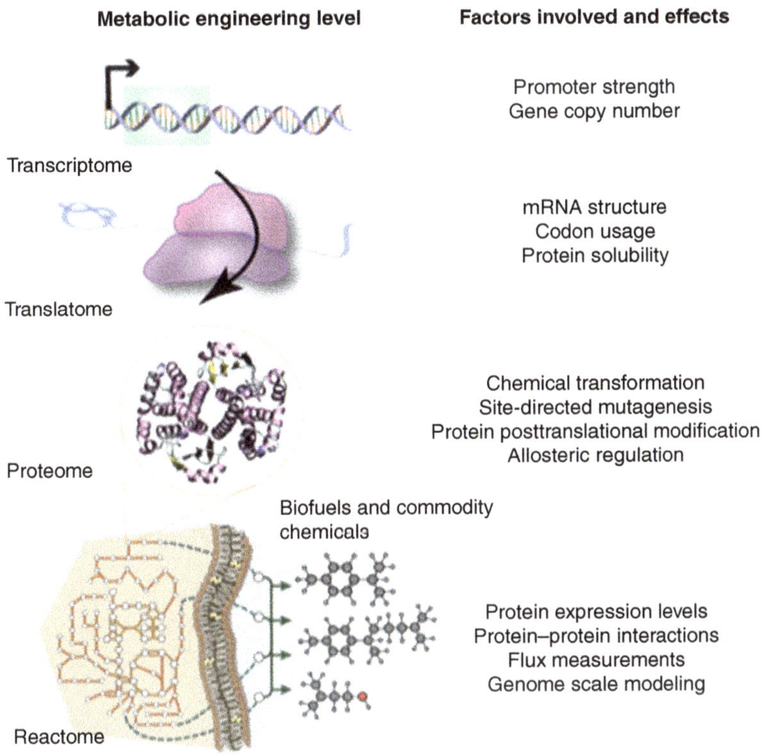

Figure 1. Integrated approaches in metabolic engineering. To achieve high metabolite flux through a biosynthetic pathway, it is essential to consider bottlenecks on the transcriptome, translatome, protein/proteome, and reactome level. Integrated metabolic engineering approaches need to occur both on the molecular-scale and system-scale level.

ENGINEERING AT THE TRANSCRIPTOME AND TRANSLATOME LEVEL

The most straightforward way to achieve a desired gene expression profile is to regulate the transcription and translation processes, which span a wide dynamic range (10^2–10^5-fold) for mRNA and protein amounts (Salis et al. 2009; Blazeck et al. 2012). Metabolic engineers would like to temporally regulate the expression strength of nonnative genes to minimize the metabolic burden of protein synthesis because nonoptimal expression may draw cellular resources away from essential functions and lower the overall fitness of the host (Poelwijk et al. 2011). This, in turn, can lead to decreased pathway productivity, reducing the design space at the reactome and proteome levels. Synthetic biology efforts have provided characterized native and synthetic promoters to control mRNA expression levels (Leavitt and Alper 2015). However, our ability to forward engineer heterologous gene expression with a precise outcome regarding protein amounts and activity is currently lacking (Figs. 2 and 3). To avoid unnecessary rounds of expression optimization, it is advisable to identify common bottlenecks that can be bypassed in advance. We will discuss a few studies that have addressed these issues and suggest an improved workflow.

Predictable and Reliable Gene Expression

It is well known that mRNA and protein levels do not perfectly correlate in native or engineered systems (Jayapal et al. 2008; Vogel and Marcotte 2012; Payne 2015). This variation is especially prevalent in prokaryotic expression, but remained unaccounted for in genetic design

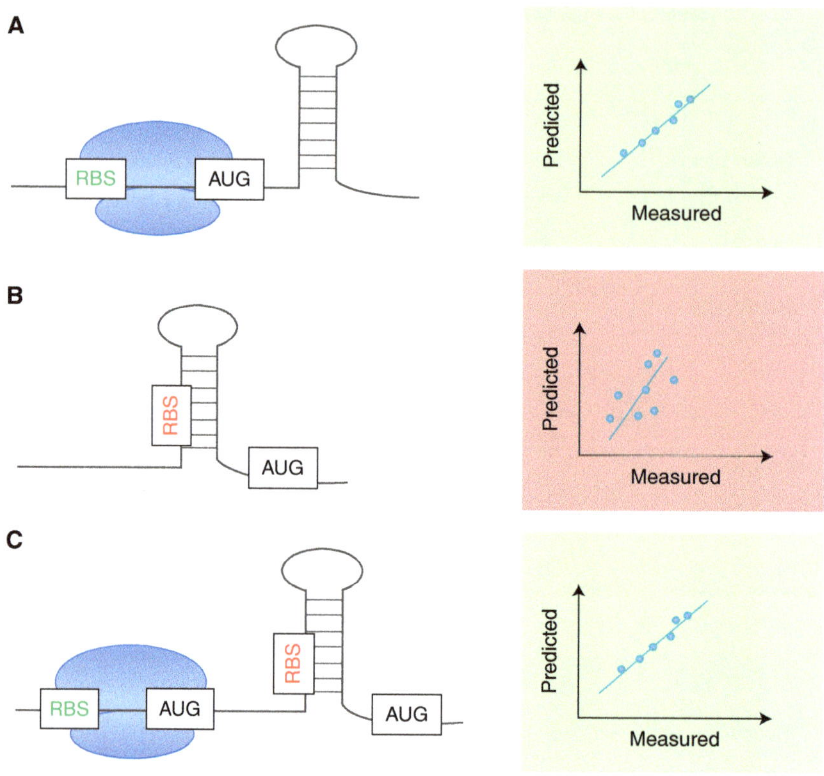

Figure 2. Predictable and reliable gene expression. (*A*) To precisely predict translation initiation rates, the ideal transcript would be depleted in secondary structures close to the ribosome-binding site (RBS). However, mRNA secondary structures are crucial to overall transcript stability and translation speed control. (*B*) Secondary structures present in most nascent mRNA molecules complicate accurate computational predictions of expression strength. Depending on the location of the hairpin, secondary structure formation can slow down translation initiation rates or completely prevent translation initiation in case of an occluded RBS. (*C*) The bicistronic gene design developed by Mutalik and coworkers (Davis et al. 2011; Mutalik et al. 2013) allows for higher predictability, because a constant short open reading frame is placed upstream of the target expression cassette that allows for read-through despite the presence of downstream hairpins.

principles until recently. Synthetic biology is striving to establish a knowledge base for modular expression design based on a comprehensive understanding of the relationship between mRNA sequence and protein abundance. Toward this goal, Kosuri et al. (2013) analyzed a library of >12,500 promoter and RBS combinations in *Escherichia coli*. As a result of this study, a short list of part combinations was created to guide future expression constructs with desired mRNA and protein level profiles. However, when predictions and measurements were compared in this study, the correlation was much lower for protein abundance ($R^2 = 0.82$) than for mRNA abundance ($R^2 = 0.96$). A small fraction of constructs displayed an even greater discrepancy, which indicates the presence of unknown variables in this combinatorial design. One significant contributor to the context-dependent variation is sequence-dependent mRNA secondary structure formation around the translation start site, which greatly influences synthesized protein amounts and complicates accurate predictions (Fig. 2).

To close the gap between designed and tested gene-expression patterns, Mutalik et al. created a new bicistronic expression cassette in combination with a previously reported insulated promoter design (Davis et al. 2011; Muta-

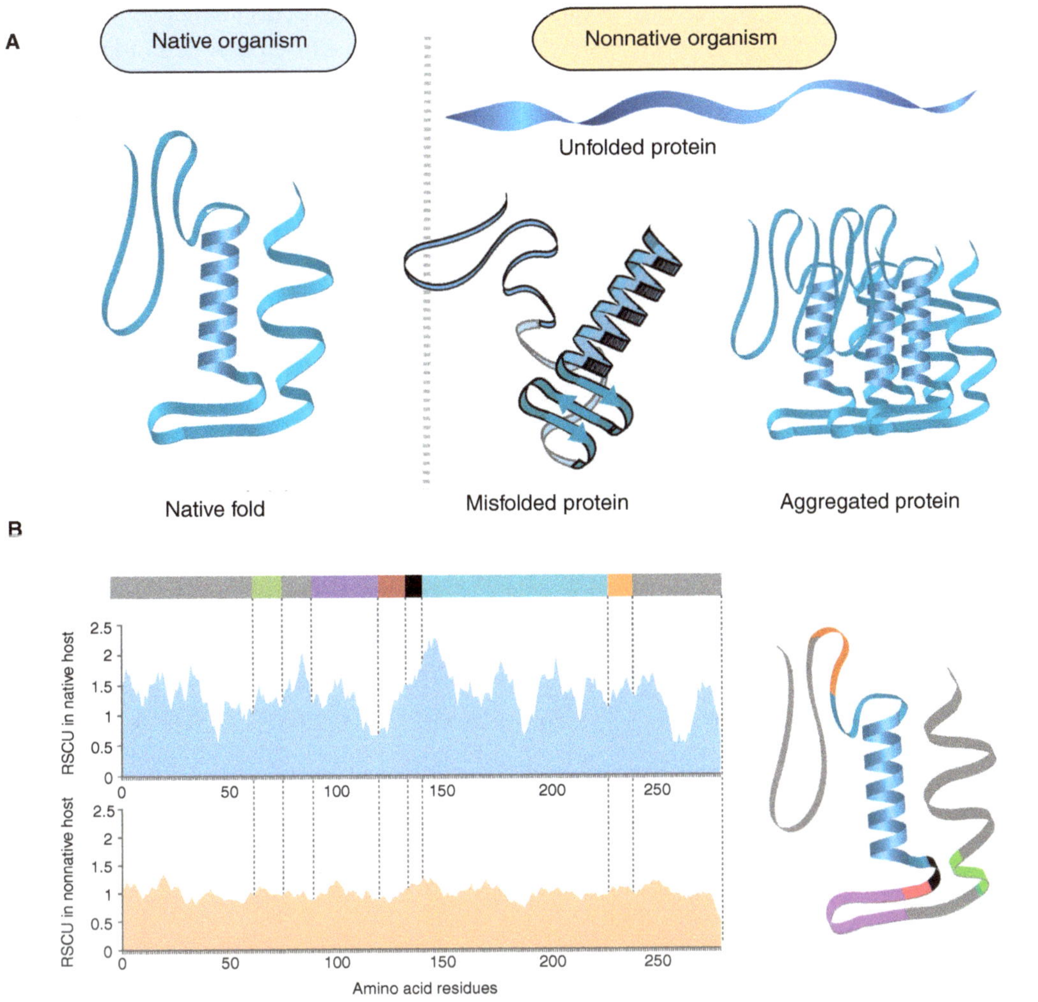

Figure 3. Soluble and active protein biosynthesis. (*A*) Proteins expressed in their native host naturally assume a soluble and catalytically active tertiary structure, whereas protein synthesis in nonnative hosts leads to misfolded or aggregated proteins. (*B*) Relative synonymous codon usage plots (RSCUs) for a representative protein in native and nonnative hosts correlated with structural motifs within the protein.

lik et al. 2013). In this design, the RBS initiating translation of the reporter gene was embedded within an upstream open reading frame for a short leader peptide (Fig. 2). The investigators suggest that the intrinsic helicase activity of the translation machinery is able to unwind any secondary structures present in the target RBS when reading through the upstream gene. Testing a library of promoter and RBS combination arranged as part of the bicistronic design, the investigators were able to significantly reduce the error rate in forward engineering. The bicistronic design had a correlation rate of 93% within a twofold window of expected expression levels compared with 84% for the monocistronic design. This allowed the characterization of >500 composable parts over a wide dynamic transcription and translation range to enable predictable engineering in *E. coli.*

The existence of context-dependency has also been described for eukaryotic gene expression. Dvir et al. (2013) measured a sevenfold change in protein abundance when manipulating a very small nucleotide sequence space surrounding the RBS of a reporter gene in *Saccharomyces cerevisiae*. Based on these measurements, a quantitative model was developed to correctly predict 74% of the protein levels considering a list of factors (i.e., mRNA secondary structure, in-frame and out-of-frame start codons, predicted nucleosome occupancy, sequence composition statistics, and arbitrary k-mer sequences). Translation efficiency has also been correlated with mRNA secondary structure formation in the 5′ coding region (Cannarozzi et al. 2010; Tuller et al. 2010; Goodman et al. 2013). Importantly, these studies imply that protein synthesis rates are affected by noncoding and coding sequences in a context-dependent manner, which may be disentangled using computational approaches. Finally, to test protein-specific influences on translation efficiency and to drive these predictions toward general design guidelines for expression of all types of protein classes, additional studies with proteins other than fluorescence reporters are required. Additional and unknown layers of transcriptional, translational, and posttranslational control remain to be identified.

Soluble and Active Protein Biosynthesis

The successful implementation of a heterologous pathway also requires the production of functional enzymes, whereas suboptimal translation of mRNA into misfolded proteins can lead to low catalytic turnover (reviewed in Li 2015). The use of enzymes from hosts that are distantly related to the expression host is one well-known stumbling block for expression of catalytically active proteins (Fig. 3). Certain bottlenecks in heterologous expression can be attributed to intrinsic host factors, such as translation speed, tRNA abundance, and chaperones. For example, tRNA abundance and codon usage bias is known to vary between and even within organisms depending on the growth stage (Ikemura 1985; Dong et al. 1996; Kanaya et al. 1999). Different temporal factors in recombinant protein expression are important, from the growth stage of the host organism to the initiation and elongation rate of nascent peptide chains (Nissley and O'Brien 2014). Changing expression timing and translation rates can affect not only translational efficiency but also important protein properties, such as solubility, activity, and specificity (Plotkin and Kudla 2011; Hunt et al. 2014). Other bottlenecks arise from the gene coding sequence itself, such as mRNA folding, transcript length, type of encoded amino acids, codon context, and codon usage.

The most influential factor affecting translational efficiency in bacterial expression is believed to be codon usage bias (Lithwick and Margalit 2003). This notion is an essential motivation for the common use of codon-adjusted transgenes (Gustafsson et al. 2004; Gould et al. 2014). However, synonymous codon changes pose the potential risk of disrupting important characteristics of a nucleotide sequence, such as mRNA stability and translation rates. It comes as no surprise that idiosyncratic reports reveal a poor correlation between codon "optimization" and functional protein levels. There are a myriad of variables affecting translation, such as mRNA stability, translation speed and accuracy, cotranslational folding, and cotranslational localization. However, the influence of these factors in distinct organisms or even protein families might vary. For instance, transgenic expression in eukaryotic systems is burdened by gene-silencing issues or low expression levels (Jackson et al. 2014). On the other hand, bacterial host systems are known to commonly produce misfolded proteins (Sander et al. 2014). There is a great need for synthetic codon adaptation strategies that consider multiple prevailing design parameters in a host- and time-dependent manner.

Recently, Lanza et al. (2014) created a condition-specific codon optimization approach, which considers the dynamic character of protein translation. The investigators generated a codon usage table based on a subset of highly expressed genes at the target growth stage and were able to improve heterologous protein levels of a bacterial gene in yeast. Other studies have used combinatorial libraries with synonymous codon variants in the coding region to experimentally disentangle the factors involved in translational efficiency and protein integrity (Goodman et al. 2013; Cheong et al. 2015).

Such studies have established that synonymous codons are translated at different rates affecting cotranslational folding (Ciryam et al. 2013; Chevance et al. 2014) and localization (Fig. 3) (Fluman et al. 2014; Pechmann et al. 2014). In summary, we emphasize that the interrelation between codon usage, cotranslational folding, and protein integrity can greatly affect the catalytic activity of heterologous proteins (Kimchi-Sarfaty et al. 2007; Agashe et al. 2013; Zhou et al. 2013).

ENGINEERING AT THE PROTEIN LEVEL

Increasing yields of desired compounds can further be influenced by the capacity of the proteins themselves to catalyze a desired reaction. Native and nonnative enzymes commonly require further optimization to meet the demands of biomanufacturing (Fig. 4). Recent advances in enzyme engineering bring about significant expansions in the catalytic repertoire of enzymes with respect to increased efficiencies, improved selectivities, and novel substrate and cofactor specificities. In addition, the de novo engineering of enzymes has enabled the design of novel catalysts in which natural enzymes are not available (Bolon et al. 2002; Kaplan and DeGrado 2004; Jiang et al. 2008; Röthlisberger et al. 2008). Numerous engineering strategies have been developed to tailor enzyme activities for specific industrial purposes, which include directed evolution and mutagenesis (Moore and Arnold 1996; Arnold 1998; Brustad and Arnold 2011; Illanes et al. 2012; Wang et al. 2012b), modular enzyme components (Dueber et al. 2009; Good et al. 2011), and computer-aided design techniques (Bornscheuer and Pohl 2001; Mandell and Kortemme 2009; Verma et al. 2012; Hilvert 2013; Damborsky and Brezovsky 2014) (for a recent review of the impact of directed evolution on synthetic biology, see Currin et al. 2015). The enzymatic characteristics that are most often targeted by these techniques include activity, stability, selectivity, solubility, and optimal pH (Fig. 4). For example, engineered fungal endoglucanases operate at optimal temperatures of 17°C higher than the wild-type enzyme and hydrolyze 1.5 times as much cellulose over 60 h (Trudeau et al. 2014). Engineering stable enzymes such as these has benefited the production of isobutanol at elevated temperatures in thermophilic *Geobacillus thermoglucosidasius* (Lin et al. 2014). However, in contrast to the numerous successful accounts of directly evolved single enzyme catalysts, limited examples exist for directly evolved enzymes that exist as catalysts within a biosynthetic pathway. In general, this can be explained by the fact that a coordinated improvement of performance of an entire pathway of enzymes has typically not been discovered through the optimization of a single gene.

The complexity of pathway design oftentimes requires overcoming metabolic bottlenecks, such as accumulation of toxic intermediates, cofactor imbalance, and inefficient enzyme activities, which remains a significant challenge for metabolic engineering and the focus of numerous research studies (Fig. 4). Metabolic engineering has traditionally integrated existing enzymes from different organisms into the production host to modify an existing pathway or create a nonnatural pathway (e.g., a set of enzymes that are not normally found within the host metabolism). This approach has generated significant improvements in product yields. For example, a 120% yield improvement was reported for the isoprenoid-derived sesquiterpene, amorphadiene, which was produced by an engineered strain of *E. coli* using the seven-gene mevalonate pathway from *S. cerevisiae* (Ma et al. 2011). However, the catalytic efficiencies and selectivities of many naturally occurring enzymes may require further optimization to become highly optimized biocatalysts with capacities to achieve higher rates, yields, and product purities. This section reviews the efficiency of the individual enzyme components themselves within existing or nonnatural biosynthetic pathways and poses the following question: "Can engineering individual enzymes transform and improve the performance of biosynthetic routes?" Here, we focus on several successful examples, in which engineered characteristics of individual enzymes did improve the efficiency of entire biosynthetic pathways. The engineered characteristic enzyme

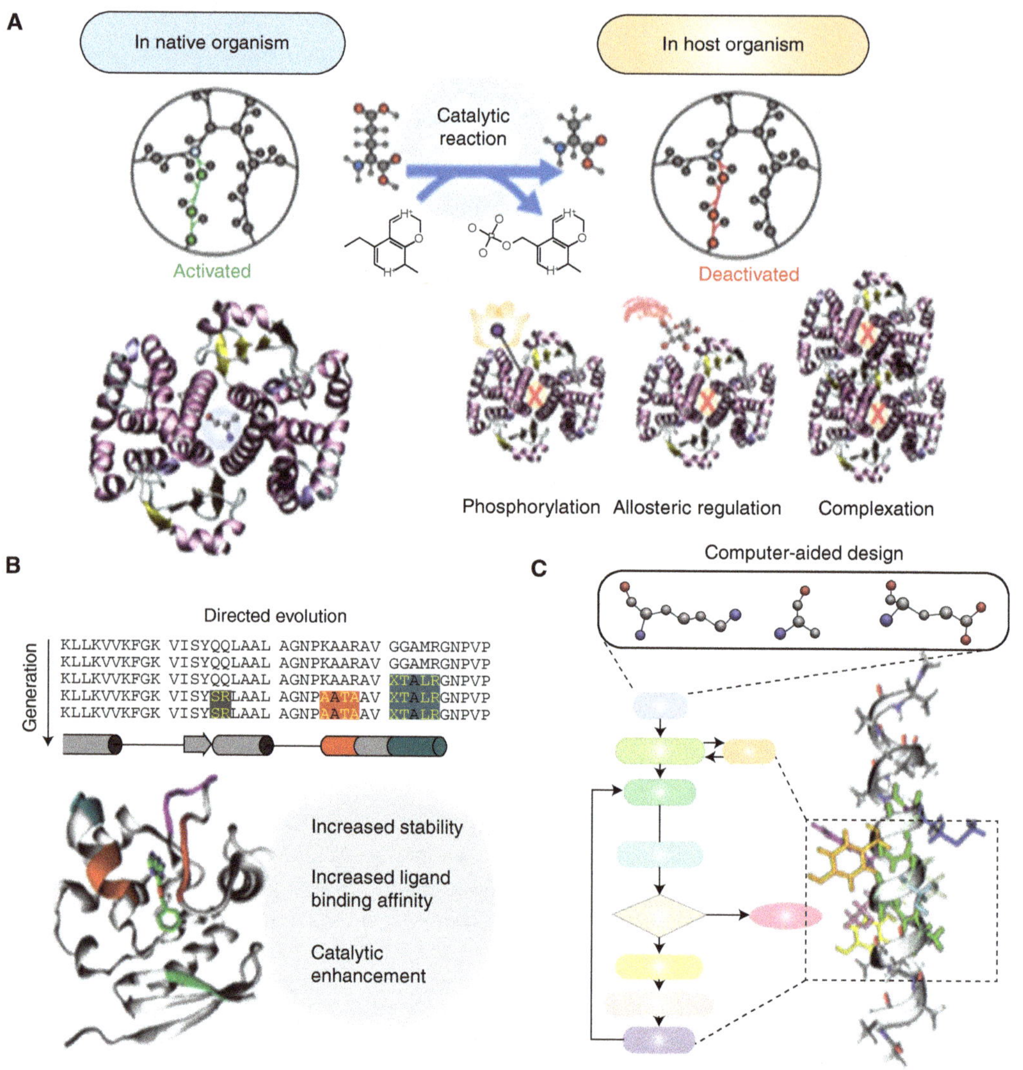

Figure 4. Proteome-level optimization. (*A*) Optimal versus nonoptimal enzyme-level activities in engineered designs. Nonoptimal factors that influence activities may include allosteric regulation, protein posttranslational modifications, incorrect protein–protein interactions, and many others. (*B*) Graphical representation of directed evolution of enzyme that directly influences its stability and reactivity. (Adapted from Mollwitz et al. 2012.) (*C*) An in silico workflow to establish rational engineering through novel complementary computational methods, such as machine learning, genetic algorithms, and molecular modeling.

properties that are highlighted in this section include enzyme regulation and enzyme specificity.

Reduced Feedback Inhibition

Heterologous proteins expressed in host organisms can suffer from a number of design issues, which include being tightly regulated via posttranslational modifications and/or allosteric inhibition (Fig. 4). In cases such as these, directed evolution of key enzymes within metabolic pathways has shown to benefit product yields (Lee et al. 2012; Brinkmann-Chen et al. 2013) (for reviews on these topics, see Johannes and Zhao 2006; Cobb et al. 2013;

Renata et al. 2015). For example, an improved variant of glucosamine synthase (GlmS) was generated via an error-prone polymerase chain reaction (PCR) and its overexpression in an engineered strain of *E. coli* led production of glucosamine at 17 GL^{-1} (Deng et al. 2005). The directed evolution of GlmS enabled increased production, in part, by generating variants that showed significantly reduced sensitivity to product inhibition (Deng et al. 2005). Another example of enzyme engineering that reduced feedback-inhibition mechanisms is that of L-phenylalanine production. By performing DNA shuffling on *pheA*, a chorismate mutase-prephenate dehydratase enzyme, mutations led to activation of enzymatic activities at low phenylalanine concentrations and nearly complete resistance to feedback inhibition of prephenate dehydratase by phenylalanine concentrations up to 200 mM in certain variants (Nelms et al. 1992). These variants were incorporated into an *E. coli* K12 strain to construct a pathway producing *S*-mandelic acid via this modified L-phenylalanine pathway (Sun et al. 2011). In addition to directed evolution techniques, computational design strategies provide further possibilities for configuring novel allosteric modulators (for a review of this subject, see Lu et al. 2014).

Increased Substrate Specificity

Enzyme engineering can also play a key role in shaping enzyme specificity and, as a result, limiting erroneous and potentially deleterious side reactions and products. By using existing promiscuous enzyme activities as a starting point for directed evolution, the creation of new enzymatic functions, such as catalyzing reactions on nonnative substrates, has been shown (Broadwater et al. 2002). The knowledge of the promiscuous activities catalyzed by an enzyme not only provides a starting point for engineering novel functionalities, but it has shown to guide the engineering of native activities with increased specificity and higher activity (Yoshikuni et al. 2006). For example, γ-humulene synthase, a sesquiterpene synthase from *Abies grandis* (Steele et al. 1998; Little and Croteau 2002), produces 52 different sesquiterpenes from a sole substrate, farnesyl diphosphate, through a wide variety of cyclization mechanisms (Lesburg et al. 1997; Starks et al. 1997; Caruthers et al. 2000; Rynkiewicz et al. 2001). Yoshikuni et al. (2006) showed that site-directed mutagenesis of a set of previously identified plasticity residues shifted the relative selectivity of one product for another by 100- to 1000-fold. An additional example of how directed evolution improved enzyme specificity has been reported in xylose isomerase-based pathways in *S. cerevisiae*, which typically suffer from poor ethanol productivity and low xylose consumption rates. Improving the specific activity of a heterologous enzyme, *Piromyces* sp. xylose isomerase, via rounds of random mutagenesis and growth-based screening has led to a variant that showed a 77% increase in activity. When expressed in a minimally engineered yeast host, the strain improved its aerobic growth rate by 61-fold and ethanol and xylose consumption rates by nearly eightfold (Lee et al. 2012). Identifying potentially promiscuous enzyme activities is a major challenge and has been one of the main roles of certain computational pathway prediction algorithms, such as the Biochemical Network Integrated Computational Explorer (BNICE) framework (Hatzimanikatis et al. 2005). BNICE uses existing enzyme chemistry as a groundwork for novel enzyme mechanisms that can be fine-tuned and tailored via minimal enzyme reengineering. Several examples of in silico strategies for the reengineering of naturally occurring enzymes into novel biocatalysts are shown by the work of (1) Cho et al. (2010), who predicted novel enzyme activities on the basis of physicochemical properties of substrate/product pairs, (2) Brunk et al. (2012), who predicted novel enzyme activity within a biosynthetic pathway (Henry et al. 2010) on the basis of 3D protein structural characteristics and molecular dynamics simulations (Fig. 4), and (3) Campodonico et al. (2014), who predicted enzyme similarity on the basis of enzyme commission classification. In certain cases, a comprehensive assessment was performed for 20 predicted heterologous pathways in *E. coli* (Campodonico et al. 2014).

ENGINEERING AT THE REACTOME LEVEL

In the previous sections, we have shown the importance of engineering correctly folded and catalytically active pathway enzymes. Here, we address the significance of balanced metabolite fluxes. Achieving maximum protein abundance does not necessarily give rise to highest product titers (Jones et al. 2000). Simultaneously, an increase in V_{max} ($=k_{cat} \times [E]$) of a heterologous enzyme can create a metabolic sink causing nutrient limitation and growth inhibition (Fig. 5). Excessive catalytic turnover can also lead to intermediate buildup causing feedback inhibition or even toxicity (Ajikumar et al. 2010). Therefore, it is important to balance protein synthesis and activity levels to engineer a pathway with stable steady-state behavior and minimal metabolic burden (Fig. 5). In this section, we review methods to assess the optimal design space of metabolic pathways. Studies are beginning to use multivariate statistical approaches to guide predictable forward engineering efforts. We will also present a putative workflow to guide the integration of a heterologous pathway within the host metabolic and regulatory network.

Balanced Pathway Fluxes

Unlike other areas of engineering, biological systems are extremely nonlinear, which makes predicting how pathway networks function in vivo a challenge. Often, engineered pathways have been optimized through sampling of large design spaces (Pfleger et al. 2006; Lütke-Ever-

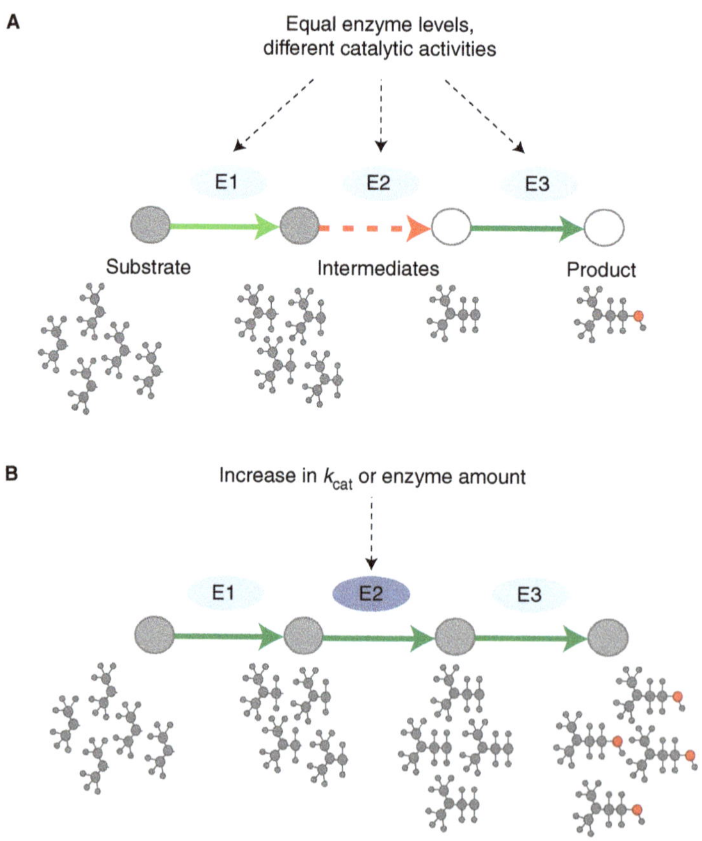

Figure 5. Reactome-level optimization. (*A*) Nonoptimal metabolite flux limited by low turnover of one pathway enzyme. (*B*) Optimal flux through the pathway can be achieved by engineering the bottleneck enzyme leading to either higher enzyme levels or increased catalytic activity.

Cite this article as *Cold Spring Harb Perspect Biol* doi: 10.1101/cshperspect.a023903

sloh and Stephanopoulos 2008; Du et al. 2012; Wang et al. 2012a; Xu et al. 2013; Nowroozi et al. 2014). Such combinatorial approaches are specifically valuable for organisms with less-defined genetic and metabolic networks (Oliver et al. 2014). However, combinatorial libraries are limited by analytical capabilities and require high-throughput screens or selection assays. For a review on high-throughput library selection and screening strategies, the reader is referred to Dietrich et al. (2010).

The main difficulty remains in understanding how interactions among components of a biochemical network within an organism give rise to the function and behavior of the entire system. In lieu of probing large design spaces through traditional metabolite analysis, a few studies have used smaller data sets with multivariate statistical analysis to predict behavior. The resulting models can be used to forward-engineer required enzyme levels for optimal pathway flux. Lee et al. (2013), for example, created a regression model to analyze a subset of a combinatorial promoter library. To assess the capacity of the generated model, they chose to express the highly branched violacein pathway and trained the model on only a small fraction of the measured data (~100 random clones, 3% of total library). The model was able to correctly predict production levels of the violacein congeners under different expression conditions with a correlation coefficient between 0.77 and 0.92, which was sufficient to guide further optimization efforts.

Alonso-Gutierrez et al. (2014) used principal component analysis of quantitative proteomics data (PCAP) to correlate product titers in *E. coli*. Gaining insight into optimal pathway stoichiometry of limonene biosynthesis, the investigators were able to further improve yields by ~40%. Expression timing was also addressed in this study, and there is an indication that different induction times lead to largely varied production levels because of toxic intermediate accumulation. Altogether, this study revealed the delicate balance between enzyme levels that is required for optimal pathway performance. At the same time, the holistic approach of PCAP emphasized that multiple optima exist in the combinatorial expression space of a mevalonate-dependent pathway. These examples show pathway-centric approaches by correlating protein synthesis and product flux. Such statistical approaches can be further complemented with constraint-based flux balance analysis (FBA) to create an integrative approach, which also considers the host metabolism (Brunk et al. 2016).

FUTURE DIRECTIONS TOWARD INTEGRATED APPROACHES IN METABOLIC ENGINEERING

To tackle the existing challenges in metabolic engineering, strategies need to account for all aspects of product synthesis in which control can be exerted: the transcriptome, the translatome, the proteome, and the reactome (Fig. 1). First, engineering the transcriptome involves manipulation of mRNA stability and expression levels, which is commonly mastered through promoter and terminator strength engineering. Second, translatome engineering requires an understanding of the factors involved in translational efficiency and cotranslational processes such as folding and translocation. Third, the design of chemical transformations and protein engineering is controlled at the proteome level. And last, reactome manipulations allow us to address aspects of enzymes ratios and, therefore, metabolite flux (Fig. 1). Inaccuracies at any of these levels can prohibit successful implementation of an engineered pathway. Here, we emphasize the current potentials at the different control levels.

Translatome Manipulations Toward Reliable Enzyme Expression

Despite having made great strides in metabolic engineering and synthetic biology (i.e., orthogonal design, parts characterization), there remains a gap between designed and measured enzyme activity. For example, protein solubility and, therefore, activity is currently optimized through a rather random approach (Batard et al. 2000; Burgess-Brown et al. 2008; Cheong et al. 2015). General optimization strategies for heterologous enzyme synthesis that do not suffer losses of catalytic activity are currently lacking. To narrow this gap between design and biology, a deeper characterization of the factors involved in gene expression is desired (Latif et al. 2014). Currently, computational models only exist for the analysis and prediction of translation initiation rates (for a review on this topic, see Reeve et al. 2014). Beyond controlling translational efficiency, there is a prevailing need to establish design guidelines for robust and specific codon optimization approaches. For this goal, emerging methods to experimentally assess ribosomal speed for different hosts have the potential to provide a reliable foundation for computational codon optimization strategies (Ingolia et al. 2009; Chevance et al. 2014; Hockenberry et al. 2014; Latif et al. 2014).

Protein-Level Manipulations Toward Efficient Enzyme Catalysis

Rational design will not move beyond its current limitations until our understanding of the underlying mechanisms of enzyme catalysis has improved. A great deal of advancement has been made in understanding structure–function–activity relationships in enzyme chemistry (Thornton et al. 2000). This is, in part, a result of the ever-increasing number of high-resolution structures available through X-ray crystallography (Berman et al. 2003) as well as the insights on detailed enzyme catalytic mechanisms from numerous theoretical-based studies (Colombo et al. 2002; Brunk and Rothlisberger 2015). Understanding protein promiscuity and its great potential for the design of novel catalysts has been the subject of many recent reviews (Nobeli et al. 2009; Humble and Berglund 2011). Synergistic efforts to couple computational enzyme design and metabolic pathway engineering with directed evolution brings promise of reengineering biosynthetic pathways and, in particular, in secondary metabolism, which offers a wider range of metabolic and biosynthetic routes, an increased scope of substrates and products, and has been shown to be less shaped by natural selection pressures than that of primary metabolism (Bar-Even and Tawfik 2013).

Reactome Manipulations Toward Optimal Production Fluxes

Typically, designing a new technical system requires starting from scratch and testing various potential prototypes, usually by means of trial and error. Thus, designing a procedure of interpretation or translation from biology to technology is a necessary goal to overcome the engineering bottlenecks. Although databases, such as KEGG (Kanehisa 2002) and BRENDA (Schomburg et al. 2004; Chang et al. 2009), provide information mainly focused on primary pathways in metabolism, categorization of novel biosynthetic pathways, the enzymes involved in biosynthesis, and the small molecules tailored to desired tasks will also be needed. Computational tools that can have access to resources such as these will undoubtedly have the potential to predict pathways, identify variations in enzymes, and model these changes in genome-scale metabolic networks of candidate host systems. Being able to model, understand, and predict complex metabolic networks has been made possible by recent developments of genome-scale metabolic models (GEMs), which have been reconstructed for >60 organisms (Reed and Palsson 2003; Monk et al. 2013; Bordbar et al. 2014). These reconstructions are converted into constraint-based models (Becker et al. 2007), allowing useful calculations like FBA (Orth et al. 2010) to be performed. Methods such as these are complementary and amenable to pathway design and offer great promise for advancing the field of synthetic pathway engineering and the design of microbial biofactories (Monk et al. 2014; Brunk et al. 2016).

CONCLUSIONS

Recent developments in the field of metabolic engineering bring promise to the design of biosynthetic pathways, in which entire metabolic pathways can be (re)designed and expressed in host organisms. We attempt to give the reader a comprehensive view on the aspects pertinent to successful pathway engineering. Additionally, we emphasize that there remains a gap between optimization efforts at the molecular level and those at the systems level. One important consideration is to address the different complications or bottlenecks that may occur on vastly different chemical and biological scales. Feasibility of engineered components can be assessed using molecular-scale analyses, which typically evaluate feasibility in terms of the ability of enzymes to catalyze a desired transformation on a substrate, and systems-level analyses, which generally assess feasibility on the basis of changes in the range of intracellular metabolites and enzyme concentrations (e.g., proteome allocations, pathway ratios). At the same time, these remaining challenges reveal many opportunities to fill knowledge gaps and develop more sophisticated design tools in metabolic engineering. First, we emphasize that computational predictions of pathway performance will only be as accurate as our models that heavily rely on existing omics data sets. Within the context of the design–build–test–learn cycle, we are successfully designing and building production strains; however, there is a delay in the learning process because "testing" is highly limited by the capabilities of existing high-throughput analytical methods. Additional omics data sets of rationally engineered microbes are required to build accurate prediction models and to facilitate the de novo design of constructs. Second, there remains a need to create an open access knowledge database to organize and analyze large data sets. And last, "learning" is informed by our basic understanding of cellular processes. Yet, there remain multiple control levels that are not well understood on a cellular scale such as allosteric regulation exerted through interaction between proteins and metabolites. Posttranslational regulation is an additional control layer that has recently attracted much attention because of novel methods assessing posttranslational modifications on an organism scale. Future milestones in "test and learn" will transform the field of metabolic engineering from a semirational to a fully rational forward engineering discipline.

ACKNOWLEDGMENTS

We are grateful to Dr. Daniel Liu, Dr. Sarah Richardson, Dr. Nathan Hillson, and Dr. Christopher Petzold for valuable comments and discussion on this review. The authors also thank Prof. Vassily Hatzimankatis and Prof. Ursula Rothlisberger for the many useful discussions on computational design strategies over the years. This work is funded by the Department of Energy Joint BioEnergy Institute (jbei.org) supported by the U.S. Department of Energy, Office of Science, Office of Biological and Environmental Research, through contract DE-AC02-05CH11231 between Lawrence Berkeley National Laboratory and the U.S. Department of Energy and by the Synthetic Biology Engineering Research Center (SynBERC) through National Science Foundation Grant NSF EEC 0540879. The authors acknowledge support from the Swiss National Science Foundation (Grant p2elp2_148961 to E.B.).

REFERENCES

Agashe D, Martinez-Gomez NC, Drummond DA, Marx CJ. 2013. Good codons, bad transcript: Large reductions in gene expression and fitness arising from synonymous mutations in a key enzyme. *Mol Biol Evol* **30:** 549–560.

Ajikumar PK, Xiao WH, Tyo KEJ, Wang Y, Simeon F, Leonard E, Mucha O, Phon TH, Pfeifer B, et al. 2010. Isoprenoid pathway optimization for taxol precursor overproduction in *Escherichia coli*. *Science* **330:** 70–74.

Alonso-Gutierrez J, Kim E-M, Batth TS, Cho N, Hu Q, Chan LJG, Petzold CJ, Hillson NJ, Adams PD, Keasling JD, et al. 2014. Principal component analysis of proteomics (PCAP) as a tool to direct metabolic engineering. *Metabol Eng* **28C:** 123–133.

Arnold FH. 1998. Design by directed evolution. *Acc Chem Res* **31:** 125–131.

Bar-Even A, Tawfik DS. 2013. Engineering specialized metabolic pathways—Is there a room for enzyme improvements? *Curr Opin Biotechnol* **24:** 310–319.

Batard Y, Hehn A, Nedelkina S, Schalk M, Pallett K, Schaller H, Werck-Reichhart D. 2000. Increasing expression of P450 and P450-reductase proteins from monocots in heterologous systems. *Arch Biochem Biophys* **379:** 161–169.

Becker SA, Feist AM, Mo ML, Hannum G, Palsson BO, Herrgard MJ. 2007. Quantitative prediction of cellular metabolism with constraint-based models: The COBRA toolbox. *Nat Protoc* **2:** 727–738.

Berman H, Henrick K, Nakamura H. 2003. Announcing the worldwide protein data bank. *Nat Struct Biol* **10:** 980.

Blazeck J, Garg R, Reed B, Alper HS. 2012. Controlling promoter strength and regulation in *Saccharomyces cerevisiae* using synthetic hybrid promoters. *Biotechnol Bioeng* **109:** 2884–2895.

Bolon DN, Voigt CA, Mayo SL. 2002. De novo design of biocatalysts. *Curr Opin Chem Biol* **6:** 125–129.

Bordbar A, Monk JM, King ZA, Palsson BO. 2014. Constraint-based models predict metabolic and associated cellular functions. *Nat Rev Genet* **15:** 107–120.

Bornscheuer UT, Pohl M. 2001. Improved biocatalysts by directed evolution and rational protein design. *Curr Opin Chem Biol* **5:** 137–143.

Brinkmann-Chen S, Flock T, Cahn JKB, Snow CD, Brustad EM, McIntosh JA, Meinhold P, Zhang L, Arnold FH. 2013. General approach to reversing ketol-acid reductoisomerase cofactor dependence from NADPH to NADH. *Proc Natl Acad Sci* **110:** 10946–10951.

Broadwater JA, Whittle E, Shanklin J. 2002. Desaturation and hydroxylation. Residues 148 and 324 of *Arabidopsis* FAD2, in addition to substrate chain length, exert a major influence in partitioning of catalytic specificity. *J Biol Chem* **277:** 15613–15620.

Brunk E, Rothlisberger U. 2015. Mixed quantum mechanical/molecular mechanical molecular dynamics simulations of biological systems in ground and electronically excited states. *Chem Rev* **115:** 6217–6263.

Brunk E, Neri M, Tavernelli I, Hatzimanikatis V, Rothlisberger U. 2012. Integrating computational methods to retrofit enzymes to synthetic pathways. *Biotechnol Bioeng* **109:** 572–582.

Brunk E, George KW, Alonso-Gutierrez J, Thompson M, Baidoo E, Wang G, Petzold CJ, McCloskey D, Monk J, Yang L, et al. 2016. Characterizing strain variation in engineered *E. coli* using a multi-omics-based workflow. *Cell Syst* **2:** 335–346.

Brustad EM, Arnold FH. 2011. Optimizing non-natural protein function with directed evolution. *Curr Opin Chem Biol* **15:** 201–210.

Burgess-Brown NA, Sharma S, Sobott F, Loenarz C, Oppermann U, Gileadi O. 2008. Codon optimization can improve expression of human genes in *Escherichia coli*: A multi-gene study. *Protein Expr Purif* **59:** 94–102.

Campodonico MA, Andrews BA, Asenjo JA, Palsson BO, Feist AM. 2014. Generation of an atlas for commodity chemical production in *Escherichia coli* and a novel pathway prediction algorithm, GEM-Path. *Metab Eng* **25:** 140–148.

Cannarozzi G, Cannarrozzi G, Schraudolph NN, Faty M, von Rohr P, Friberg MT, Roth AC, Gonnet P, Gonnet G, Barral Y. 2010. A role for codon order in translation dynamics. *Cell* **141:** 355–367.

Caruthers JM, Kang I, Rynkiewicz MJ, Cane DE, Christianson DW. 2000. Crystal structure determination of aristolochene synthase from the blue cheese mold, penicillium roqueforti. *J Biol Chem* **275:** 25533–25539.

Chang A, Scheer M, Grote A, Schomburg I, Schomburg D. 2009. BRENDA, AMENDA and FRENDA the enzyme information system: New content and tools in 2009. *Nucleic Acids Res* **37:** D588–D592.

Cheong DE, Ko KC, Han Y, Jeon HG, Sung BH, Kim GJ, Choi JH, Song JJ. 2015. Enhancing functional expression of heterologous proteins through random substitution of genetic codes in the 5′ coding region. *Biotechnol Bioeng* **112:** 822–826.

Chevance FF, Le Guyon S, Hughes KT. 2014. The effects of codon context on in vivo translation speed. *PLoS Genet* **10:** e1004392.

Cho A, Yun H, Park JH, Lee SY, Park S. 2010. Prediction of novel synthetic pathways for the production of desired chemicals. *BMC Syst Biol* **4:** 35.

Ciryam P, Morimoto RI, Vendruscolo M, Dobson CM, O'Brien EP. 2013. In vivo translation rates can substantially delay the cotranslational folding of the *Escherichia coli* cytosolic proteome. *Proc Natl Acad Sci* **110:** E132–E140.

Cobb RE, Sun N, Zhao H. 2013. Directed evolution as a powerful synthetic biology tool. *Methods* **60:** 81–90.

Colombo MC, Guidoni L, Laio A. 2002. Hybrid QM/MM Car–Parrinello simulations of catalytic and enzymatic reactions. *CHIMIA Int J Chem* **56:** 13–19.

Currin A, Swainston N, Day PJ, Kell DB. 2015. Synthetic biology for the directed evolution of protein biocatalysts: Navigating sequence space intelligently. *Chem Soc Rev* **44:** 1172–1239.

Damborsky J, Brezovsky J. 2014. Computational tools for designing and engineering enzymes. *Curr Opin Chem Biol* **19:** 8–16.

Davis JH, Rubin AJ, Sauer RT. 2011. Design, construction and characterization of a set of insulated bacterial promoters. *Nucleic Acids Res* **39:** 1131–1141.

Deng MD, Severson DK, Grund AD, Wassink SL, Burlingame RP, Berry A, Running JA, Kunesh CA, Song L, Jerrell TA, et al. 2005. Metabolic engineering of *Escherichia coli* for industrial production of glucosamine and *N*-acetylglucosamine. *Metab Eng* **7:** 201–214.

Dietrich JA, McKee AE, Keasling JD. 2010. High-throughput metabolic engineering: Advances in small-molecule screening and selection. *Annu Rev Biochem* **79:** 563–590.

Dong H, Nilsson L, Kurland CG. 1996. Co-variation of tRNA abundance and codon usage in *Escherichia coli* at different growth rates. *J Mol Biol* **260:** 649–663.

Du J, Yuan Y, Si T, Lian J, Zhao H. 2012. Customized optimization of metabolic pathways by combinatorial transcriptional engineering. *Nucleic Acids Res* **40:** e142.

Dueber JE, Wu GC, Malmirchegini GR, Moon TS, Petzold CJ, Ullal AV, Prather KLJ, Keasling JD. 2009. Synthetic protein scaffolds provide modular control over metabolic flux. *Nat Biotechnol* **27:** 753–759.

Dvir S, Velten L, Sharon E, Zeevi D, Carey LB, Weinberger A, Segal E. 2013. Deciphering the rules by which 5′-UTR sequences affect protein expression in yeast. *Proc Natl Acad Sci* **110:** E2792–E2801.

Fluman N, Navon S, Bibi E, Pilpel Y. 2014. mRNA-programmed translation pauses in the targeting of *E. coli* membrane proteins. *eLife* doi: 10.7554/eLife.03440.

Good MC, Zalatan JG, Lim WA. 2011. Scaffold proteins: Hubs for controlling the flow of cellular information. *Science* **332:** 680–686.

Goodman DB, Church GM, Kosuri S. 2013. Causes and effects of N-terminal codon bias in bacterial genes. *Science* **342:** 475–479.

Gould N, Hendy O, Papamichail D. 2014. Computational tools and algorithms for designing customized synthetic genes. *Front Bioeng Biotechnol* **2:** 41.

Gustafsson C, Govindarajan S, Minshull J. 2004. Codon bias and heterologous protein expression. *Trends Biotechnol* **22:** 346–353.

Hatzimanikatis V, Li C, Ionita JA, Henry CS, Jankowski MD, Broadbelt LJ. 2005. Exploring the diversity of complex metabolic networks. *Bioinformatics* **21:** 1603–1609.

Henry CS, Broadbelt LJ, Hatzimanikatis V. 2010. Discovery and analysis of novel metabolic pathways for the biosynthesis of industrial chemicals: 3-Hydroxypropanoate. *Biotechnol Bioeng* **106:** 462–473.

Hilvert D. 2013. Design of protein catalysts. *Annu Rev Biochem* **82:** 447–470.

Hockenberry AJ, Sirer MI, Nunes Amaral LA, Jewett MC. 2014. Quantifying position-dependent codon usage bias. *Mol Biol Evol* **31:** 1880–1893.

Humble MS, Berglund P. 2011. Biocatalytic promiscuity. *Eur J Organic Chem* **2011:** 3391–3401.

Hunt RC, Simhadri VL, Iandoli M, Sauna ZE, Kimchi-Sarfaty C. 2014. Exposing synonymous mutations. *Trends Genet* **30:** 308–321.

Ikemura T. 1985. Codon usage and tRNA content in unicellular and multicellular organisms. *Mol Biol Evol* **2:** 13–34.

Illanes A, Cauerhff A, Wilson L, Castro GR. 2012. Recent trends in biocatalysis engineering. *Biores Technol* **115:** 48–57.

Ingolia NT, Ghaemmaghami S, Newman JRS, Weissman JS. 2009. Genome-wide analysis in vivo of translation with nucleotide resolution using ribosome profiling. *Science* **324:** 218–223.

Jackson MA, Sternes PR, Mudge SR, Graham MW, Birch RG. 2014. Design rules for efficient transgene expression in plants. *Plant Biotechnol J* **12:** 925–933.

Jakočiūnas T, Bonde I, Herrgård M, Harrison SJ, Kristensen M, Pedersen LE, Jensen MK, Keasling JD. 2015. Multiplex metabolic pathway engineering using CRISPR/Cas9 in *Saccharomyces cerevisiae*. *Metab Eng* **28:** 213–222.

Jayapal KP, Philp RJ, Kok YJ, Yap MGS, Sherman DH, Griffin TJ, Hu WS. 2008. Uncovering genes with divergent mRNA-protein dynamics in *Streptomyces coelicolor*. *PLoS ONE* **3:** e2097.

Jiang L, Althoff EA, Clemente FR, Doyle L, Röthlisberger D, Zanghellini A, Gallaher JL, Betker JL, Tanaka F, Barbas CF III, et al. 2008. De novo computational design of retro-aldol enzymes. *Science* **319:** 1387–1391.

Johannes TW, Zhao H. 2006. Directed evolution of enzymes and biosynthetic pathways. *Curr Opin Microbiol* **9:** 261–267.

Jones KL, Kim SW, Keasling JD. 2000. Low-copy plasmids can perform as well as or better than high-copy plasmids for metabolic engineering of bacteria. *Metab Eng* **2:** 328–338.

Kanaya S, Yamada Y, Kudo Y, Ikemura T. 1999. Studies of codon usage and tRNA genes of 18 unicellular organisms and quantification of Bacillus subtilis tRNAs: Gene expression level and species-specific diversity of codon usage based on multivariate analysis. *Gene* **238:** 143–155.

Kanehisa M. 2002. The KEGG database. *Novartis Found Symp* **247:** 91–101; discussion 101–103, 119–128, 244–252.

Kaplan J, DeGrado WF. 2004. De novo design of catalytic proteins. *Proc Natl Acad Sci* **101:** 11566–11570.

Keasling JD. 2010. Manufacturing molecules through metabolic engineering. *Science* **330:** 1355–1358.

Kimchi-Sarfaty C, Oh JM, Kim IW, Sauna ZE, Calcagno AM, Ambudkar SV, Gottesman MM. 2007. A "silent" polymorphism in the *MDR1* gene changes substrate specificity. *Science* **315:** 525–528.

Klein-Marcuschamer D, Ajikumar PK, Stephanopoulos G. 2007. Engineering microbial cell factories for biosynthesis of isoprenoid molecules: Beyond lycopene. *Trends Biotechnol* **25:** 417–424.

Kosuri S, Goodman DB, Cambray G, Mutalik VK, Gao Y, Arkin AP, Endy D, Church GM. 2013. Composability of regulatory sequences controlling transcription and translation in *Escherichia coli*. *Proc Natl Acad Sci* **110:** 14024–14029.

Lanza A, Curran K, Rey L, Alper H. 2014. A condition-specific codon optimization approach for improved heterologous gene expression in *Saccharomyces cerevisiae*. *BMC Syst Biol* **8:** 33.

Latif H, Szubin R, Tan J, Brunk E, Lechner A, Zengler K, Palsson BO. 2014. A streamlined ribosome profiling protocol for the characterization of microorganisms. *Biotechniques* **58:** 329–332.

Leavitt JM, Alper HS. 2015. Advances and current limitations in transcript-level control of gene expression. *Curr Opin Biotechnol* **34:** 98–104.

Lee SY, Kim HU, Park JH, Park JM, Kim TY. 2009. Metabolic engineering of microorganisms: General strategies and drug production. *Drug Discov Today* **14:** 78–88.

Lee SM, Jellison T, Alper HS. 2012. Directed evolution of xylose isomerase for improved xylose catabolism and fermentation in the yeast *Saccharomyces cerevisiae*. *Appl Environ Microbiol* **78:** 5708–5716.

Lee ME, Aswani A, Han AS, Tomlin CJ, Dueber LE. 2013. Expression-level optimization of a multi-enzyme pathway in the absence of a high-throughput assay. *Nucl Acids Res* **41:** 10668–10678.

Lennen RM, Pfleger BF. 2013. Microbial production of fatty acid-derived fuels and chemicals. *Curr Opin Biotechnol* **24:** 1044–1053.

Lesburg CA, Zhai G, Cane DE, Christianson DW. 1997. Crystal structure of pentalenene synthase: Mechanistic insights on terpenoid cyclization reactions in biology. *Science* **277:** 1820–1824.

Li GW. 2015. How do bacteria tune translation efficiency? *Curr Opin Microbiol* **24:** 66–71.

Lin PP, Rabe KS, Takasumi JL, Kadisch M, Arnold FH, Liao JC. 2014. Isobutanol production at elevated temperatures in thermophilic *Geobacillus thermoglucosidasius*. *Metab Eng* **24:** 1–8.

Lithwick G, Margalit H. 2003. Hierarchy of sequence-dependent features associated with prokaryotic translation. *Genome Res* **13:** 2665–2673.

Little DB, Croteau RB. 2002. Alteration of product formation by directed mutagenesis and truncation of the multiple-product sesquiterpene synthases δ-selinene synthase and γ-humulene synthase. *Arch Biochem Biophys* **402:** 120–135.

Lu S, Huang W, Zhang J. 2014. Recent computational advances in the identification of allosteric sites in proteins. *Drug Discov Today* **19:** 1595–1600.

Lütke-Eversloh T, Stephanopoulos G. 2008. Combinatorial pathway analysis for improved L-tyrosine production in *Escherichia coli*: Identification of enzymatic bottlenecks by systematic gene overexpression. *Metab Eng* **10:** 69–77.

Ma SM, Garcia DE, Redding-Johanson AM, Friedland GD, Chan R, Batth TS, Haliburton JR, Chivian D, Keasling JD, Petzold CH, et al. 2011. Optimization of a heterologous mevalonate pathway through the use of variant HMG-CoA reductases. *Metab Eng* **13:** 588–597.

Mandell DJ, Kortemme T. 2009. Computer-aided design of functional protein interactions. *Nat Chem Biol* **5:** 797–807.

Mollwitz B, Brunk E, Schmitt S, Pojer F, Bannwarth M, Schiltz M, Rothlisberger U, Johnsson K. 2012. Directed evolution of the suicide protein O^6-alkylguanine-DNA alkyltransferase for increased reactivity results in an alkylated protein with exceptional stability. *Biochemistry* **51:** 986–994.

Monk JM, Charusanti P, Aziz RL, Lerman JA, Premyodhin N, Orth JD, Feist AM, Palsson BO. 2013. Genome-scale metabolic reconstructions of multiple *Escherichia coli* strains highlight strain-specific adaptations to nutritional environments. *Proc Natl Acad Sci* **110:** 20338–20343.

Monk J, Nogales J, Palsson BO. 2014. Optimizing genome-scale network reconstructions. *Nat Biotechnol* **32:** 447–452.

Moore JC, Arnold FH. 1996. Directed evolution of a *para*-nitrobenzyl esterase for aqueous-organic solvents. *Nat Biotechnol* **14:** 458–467.

Mutalik VK, Guimaraes JC, Cambray G, Lam C, Christoffersen MJ, Mai QA, Tran AB, Paull M, Keasling JD, Arkin AP, et al. 2013. Precise and reliable gene expression via standard transcription and translation initiation elements. *Nat Methods* **10:** 354–360.

Nelms J, Edwards RM, Warwick J, Fotheringham I. 1992. Novel mutations in the pheA gene of *Escherichia coli* K-12 which result in highly feedback inhibition-resistant variants of chorismate mutase/prephenate dehydratase. *Appl Environ Microbiol* **58:** 2592–2598.

Nissley DA, O'Brien EP. 2014. Timing is everything: Unifying codon translation rates and nascent proteome behavior. *J Am Chem Soc* **136:** 17892–17898.

Nobeli I, Favia AD, Thornton JM. 2009. Protein promiscuity and its implications for biotechnology. *Nat Biotechnol* **27:** 157–167.

Nowroozi FF, Baidoo EEK, Ermakov S, Redding-Johanson AM, Batth TS, Petzold CJ, Keasling JD. 2014. Metabolic pathway optimization using ribosome binding site variants and combinatorial gene assembly. *Appl Microbiol Biotechnol* **98:** 1567–1581.

Oliver JWK, Machado IMP, Yoneda H, Atsumi S. 2014. Combinatorial optimization of cyanobacterial 2,3-butanediol production. *Metab Eng* **22:** 76–82.

Orth JD, Thiele I, Palsson BO. 2010. What is flux balance analysis? *Nat Biotechnol* **28:** 245–248.

Paddon CJ, Westfall PJ, Pitera DJ, Benjamin K, Fisher K, McPhee D, Leavell MD, Tai A, Main A, Eng D, et al. 2013. High-level semi-synthetic production of the potent antimalarial artemisinin. *Nature* **496:** 528–532.

Payne SH. 2015. The utility of protein and mRNA correlation. *Trends Biochem Sci* **40:** 1–3.

Pechmann S, Chartron JW, Frydman J. 2014. Local slowdown of translation by nonoptimal codons promotes nascent-chain recognition by SRP in vivo. *Nat Struct Mol Biol* **21:** 1100–1105.

Pfleger BF, Pitera DJ, Smolke CD, Keasling JD. 2006. Combinatorial engineering of intergenic regions in operons tunes expression of multiple genes. *Nat Biotechnol* **24:** 1027–1032.

Plotkin JB, Kudla G. 2011. Synonymous but not the same: The causes and consequences of codon bias. *Nat Rev Genet* **12:** 32–42.

Poelwijk FJ, Heyning PD, de Vos MGJ, Kiviet DJ, Tans SJ. 2011. Optimality and evolution of transcriptionally regulated gene expression. *BMC Syst Biol* **5:** 128.

Reed JL, Palsson BO. 2003. Thirteen years of building constraint-based in silico models of *Escherichia coli*. *J Bacteriol* **185:** 2692–2699.

Reeve B, Hargest T, Gilbert C, Ellis T. 2014. Predicting translation initiation rates for designing synthetic biology. *Front Bioeng Biotechnol* **2:** 1.

Renata H, Wang ZJ, Arnold FH. 2015. Expanding the enzyme universe: Accessing non-natural reactions by mechanism-guided directed evolution. *Angew Chem Int Ed Engl* **54:** 3351–3367.

Röthlisberger D, Khersonsky O, Wollacott AM, Jiang L, DeChancie J, Betker J, Gallaher JL, Althoff EA, Zanghellini A, Dym O, et al. 2008. Kemp elimination catalysts by computational enzyme design. *Nature* **453:** 190–195.

Rynkiewicz MJ, Cane DE, Christianson DW. 2001. Structure of trichodiene synthase from fusarium sporotrichioides provides mechanistic inferences on the terpene cyclization cascade. *Proc Natl Acad Sci* **98:** 13543–13548.

Salis HM, Mirsky EA, Voigt CA. 2009. Automated design of synthetic ribosome binding sites to control protein expression. *Nat Biotechnol* **27:** 946–950.

Sander IM, Chaney JL, Clark PL. 2014. Expanding Anfinsen's principle: Contributions of synonymous codon selection to rational protein design. *J Am Chem Soc* **136:** 858–861.

Schomburg I, Chang A, Ebeling C, Gremse M, Heldt C, Huhn G, Schomburg H. 2004. BRENDA, the enzyme database: Updates and major new developments. *Nucleic Acids Res* **32:** D431–D433.

Starks CM, Back K, Chappell J, Noel JP. 1997. Structural basis for cyclic terpene biosynthesis by tobacco 5-epi-aristolochene synthase. *Science* **277:** 1815–1820.

Steele CL, Crock J, Bohlmann J, Croteau R. 1998. Sesquiterpene synthases from Grand Fir (*Abies grandis*). Comparison of constitutive and wound-induced activities, and cDNA isolation, characterization, and bacterial expression of δ-selinene synthase and γ-humulene synthase. *J Biol Chem* **273:** 2078–2089.

Sun Z, Ning Y, Liu L, Liu Y, Sun B, Jiang W, Yang C, Yang S. 2011. Metabolic engineering of the L-phenylalanine pathway in *Escherichia coli* for the production of *S*- or *R*-mandelic acid. *Microb Cell Fact* **10:** 71.

Thornton JM, Todd AE, Milburn D, Borkakoti N, Orengo CA. 2000. From structure to function: Approaches and limitations. *Nat Struct Biol* **7:** 991–994.

Trudeau DL, Lee TM, Arnold FH. 2014. Engineered thermostable fungal cellulases exhibit efficient synergistic cellulose hydrolysis at elevated temperatures. *Biotechnol Bioeng* **111:** 2390–2397.

Tuller T, Carmi A, Vestsigian K, Navon S, Dorfan Y, Zaborske J, Pan T, Dahan O, Furman I, Pilpel Y. 2010. An evolutionarily conserved mechanism for controlling the efficiency of protein translation. *Cell* **141:** 344–354.

Verma R, Schwaneberg U, Roccatano D. 2012. Computer-aided protein directed evolution: A review of web servers, databases and other computational tools for protein engineering. *Comput Struct Biotechnol J* **2:** e201209008.

Vogel C, Marcotte EM. 2012. Insights into the regulation of protein abundance from proteomic and transcriptomic analyses. *Nat Rev Genet* **13:** 227–232.

Wang HH, Kim H, Cong L, Jeong J, Bang D, Church GM. 2012a. Genome-scale promoter engineering by coselection MAGE. *Nat Methods* **9:** 591–593.

Wang M, Si T, Zhao H. 2012b. Biocatalyst development by directed evolution. *Biores Technol* **115:** 117–125.

Xu P, Gu Q, Wang W, Wong L, Bower AGW, Collins CH, Koffas MAG. 2013. Modular optimization of multi-gene pathways for fatty acids production in *E. coli*. *Nat Commun* **4:** 1409.

Yoshikuni Y, Ferrin TE, Keasling JD. 2006. Designed divergent evolution of enzyme function. *Nature* **440:** 1078–1082.

Zhou M, Guo J, Cha J, Chae M, Chen S, Barral JM, Sachs MS, Liu Y. 2013. Non-optimal codon usage affects expression, structure and function of clock protein FRQ. *Nature* **495:** 111–115.

Synthetic Biology of Natural Products

Rainer Breitling and Eriko Takano

Manchester Centre for Synthetic Biology of Fine and Speciality Chemicals (SYNBIOCHEM), Manchester Institute of Biotechnology, Faculty of Life Sciences, University of Manchester, Manchester M1 7DN, United Kingdom

Correspondence: rainer.breitling@manchester.ac.uk; eriko.takano@manchester.ac.uk

The diversity and natural modularity of their biosynthetic pathways has turned natural products into attractive, but challenging, targets for synthetic biology approaches. Here, we discuss the current state of the field, highlighting recent advances and remaining bottlenecks. Global genomic assessments of natural product biosynthetic capacities across large parts of microbial diversity provide a first survey of the available natural parts libraries and identify evolutionary design rules for further engineering. Methods for compound and pathway detection and characterization are developed increasingly on the basis of synthetic biology tools, contributing to an accelerated translation of genomic information into usable building blocks for pathway assembly. A wide range of methods is also becoming available for accessing ever larger parts of chemical space by rational diversification of natural products, guided by rapid progress in our understanding of the underlying biochemistry and enzymatic mechanisms. Enhanced genome assembly and editing tools, adapted to the needs of natural products research, facilitate the realization of ambitious engineering strategies, ranging from combinatorial library generation to high-throughput optimization of product titers. Together, these tools and concepts contribute to the emergence of a new generation of revitalized natural product research.

THE NATURAL PRODUCTS LANDSCAPE

The engineering of microbial systems for the enhanced production of natural products is one of the most obvious and popular applications of synthetic biology methodologies. Numerous excellent reviews, covering all aspects of the field, have been published during the last couple of years (Keasling 2008; Medema et al. 2011b; Cobb et al. 2013; Hranueli et al. 2013; Zakeri and Lu 2013; Cummings et al. 2014; Genilloud 2014; Luo et al. 2014; Porro et al. 2014; Quin and Schmidt-Dannert 2014; Singh 2014; Sleator 2014; Unkles et al. 2014; Wilson et al. 2014; Wright 2014; Breitling and Takano 2015). After a brief overview of the general ambitions and achievements of synthetic biology applications for natural products, we will then focus on selected aspects in which we either see major recent progress or important remaining bottlenecks.

Natural product production using engineered microorganisms is probably the application domain of synthetic biology in which its boundaries with classic genetic modification methods and traditional metabolic engineering

Cite this article as *Cold Spring Harb Perspect Biol* doi: 10.1101/cshperspect.a023994

are most easily blurred. Nonetheless, following a number of highly successful case studies, it is considered one of the prime targets for biotechnological deployment of synthetic biology in the near future. Early commercialized examples range from the semisynthetic malaria drug artemisinin (University of California Berkeley/Amyris, Emeryville, CA/Sanofi, Paris) to the first consumer-market synthetic biology product, "natural" vanillin (Evolva, Reinach Switzerland/International Flavors and Fragrances, New York), both of them produced by fermentation in genetically modified yeast (Paddon and Keasling 2014; Kurita et al. 2015).

These success stories of synthetic biology have not only provided renewed motivation for the exploration of microbes as production systems for high-value compounds, they have also highlighted the challenges faced when moving from proof-of-principle heterologous synthesis of a biomolecule to its economically viable production on an industrial scale. The number and complexity of the necessary genetic modifications and the diversity of bottlenecks as a result of limited understanding of microbial physiology have been unexpected.

According to Porro et al. (2014), the necessary titer, yield, and productivity of microbial cell factories for industrial applications have rarely been achieved so far. Based on a thorough analysis of examples from primary metabolite production, they conclude that thermodynamic feasibility, and stoichiometric, redox, and energy balance, all need to be considered when assessing the economic feasibility of a microbial production system, both in terms of process efficiency and expected yield and purity. Moreover, they identify flux coupling and the introduction of irreversible reactions as important engineering targets to drive product generation. This is particularly important for natural products, for which the natural systems have typically not evolved toward high yields and titers, as minute amounts of bioactive molecules are all that is necessary for their biological function.

Optimizing all of these aspects has long been the aim of metabolic engineering, and synthetic biology offers additional opportunities for achieving some of these goals. For example, Medema et al. (2011b) suggested that spatial engineering of biosynthetic pathways at various scales (from protein scaffolds to bacterial microcompartments to microbial consortia) can potentially be used to optimize product titers and increase metabolic pull by irreversibly trapping the desired end product in a compartment or vesicle, at the same time avoiding the potential toxicity seen for many natural products at high production levels. In particular, various protein-based bacterial microcompartments seem promising. For example, the Pdu microcompartment of *Citrobacter freundii* has already been engineered to encapsulate the pyruvate decarboxylase and alcohol dehydrogenase of *Zymomonas mobilis* to successfully produce ethanol in *Escherichia coli* (Lawrence et al. 2014; Chessher et al. 2015).

Another area of intense research is the discovery, modification, and overproduction of bioactive natural products from a variety of natural sources. In particular, the natural modularity of the biosynthetic pathways for some major classes of these bioactive molecules has attracted the interest of synthetic biologists, who see the potential of a "Lego-ization" of natural product biosynthesis. Two groups of natural products stand out in this respect, nonribosomal peptides and polyketides, both immensely diverse classes of compounds with a wide range of desirable bioactivities (from antibiotics and anticancer activities to cholesterol-lowering effects). Compounds in both of these groups are typically produced by large molecular assembly lines, which show natural modularity at multiple levels: enzymatic complexes responsible for the sequential incorporation of chemical units into a growing core compound are made up of individual catalytic domains in predictable patterns, and are, in turn, linked with different combinations of tailoring enzymes to yield the final product. Swapping of biosynthetic modules and domains and tailoring enzymes for the creation of chemical novelty has long been considered a promising strategy (often referred to as combinatorial biosynthesis), and recent advances in genome-editing capabilities have opened up the prospect of a plug-and-play approach to natural product discovery and production (Medema et al. 2011b; Frasch et al. 2013).

In the case of polyketide synthesis, the easy genome-based discovery (Helfrich et al. 2014) and extreme modularity of the system, and the resulting prospect of complete programmability, has led to predictions that around 20 of the 150 major commodity chemicals currently produced by the petrochemical industry could be synthesized by "mixing and matching" of naturally occurring polyketide synthase domains (Poust et al. 2014). Currently, only a small part of the theoretical chemical space accessible via polyketide synthesis is actually available through natural or engineered synthesis systems, but retrosynthetic approaches to predict and manipulate the specificity of polyketide synthase domains are constantly improving, and learning on a large scale from the successes and failures of repeated iterations of the design–build–test cycle of synthetic biology is expected to yield generalizable rules and principles for accessing ever larger fractions of chemical space. Some of the formidable challenges still faced in this domain, illustrating both the importance of synthetic biology tools and the present limitations of rational engineering of biochemical complexity, have recently been discussed by Cummings et al. (2014).

One additional exciting aspect of the synthetic biology approach is its systematic lowering of phylogenetic barriers. Traditionally, many complex natural products have been discovered in nonmodel organisms, such as plants and actinomycetes. Even today, industrial production processes often rely on native host strains that have been optimized through tedious rounds of mutagenesis and selection. Heterologous production was impeded by practical difficulties in genetically manipulating the huge biosynthetic gene clusters involved, especially when the extreme GC-rich nature of actinomycete genomes interfered with expression in model organisms, such as *E. coli*. Furthermore, extensive engineering of primary metabolism, providing essential precursors in sufficient amounts, were often required to allow production in a heterologous system. Synthetic biology promises to change this. Not only is it becoming easier to refactor entire gene clusters for optimized activity in a nonnative environment, computational metabolic modeling and rapid genome editing make it possible to design production chassis with tailored primary metabolism quickly enough to be economically feasible (e.g., Song and Lee 2013). On the other hand, synthetic biology approaches are now used to streamline the genomes of actinomycete strains to obtain suitable generic hosts for a diverse range of secondary metabolites (Rebets et al. 2014), and computational modeling results indicate that a much broader range of species might have a metabolic network preadapted for the production of natural products (Zakrzewski et al. 2012). Moreover, the toolbox of synthetic biology, ranging from promoter libraries and expression vectors to reporter genes and regulatory circuitry, is rapidly growing, both in actinomycetes (Rebets et al. 2014; Weber et al. 2015b) and in a diverse range of other potential chassis organisms, including model species with a long track record of industrial application, such as yeast (Siddiqui et al. 2012) and *E. coli* (Li and Neubauer 2014).

Seyedsayamdost and Clardy (2014) have recently suggested the conceptual division of synthetic biology of natural products into three major areas. First, work on the "known knowns," such as artemisinin or vanilla, in which the chemistry of the target compound, as well as its biosynthetic pathway, are well known and pathway engineering mainly aims at improved production, often in a more amenable host than the natural one. Second, research into the "known unknowns," compounds whose existence is easily inferred from the analysis of bacterial genomes, but which have not yet been detected; in these cases, the contribution of synthetic biology is mostly foreseen in the awakening of the biosynthesis of cryptic metabolites, facilitating their chemical and functional characterization and, ultimately, their overproduction using the methods established for "known knowns." And, finally, screen for the "unknown unknowns," the large diversity of interesting molecules that are suspected to be still entirely undiscovered, belonging to new chemical classes not covered by current genome-based discovery methods. As the continued discovery of new compounds and the associated biosynthetic enzyme activities forms the basis of the growing synthetic biology toolbox, we will start our discussion by exploring recent developments in the accelerated discovery of new natural products.

DIVERSITY OF NATURAL PRODUCTS

When the first genomic sequences of versatile secondary metabolite producers became available, it was soon discovered that a huge treasure trove of potential natural products had been missed by classic screening-based discovery strategies. Strains that were known to produce perhaps a handful of bioactive compounds were found to encode the machinery required for the biosynthesis of dozens of complex chemicals. Recently, a number of comprehensive analyses of the wealth of microbial genome data currently available have revealed the general patterns underlying this unexpected abundance of chemical diversity.

Cimermancic and colleagues (2014) annotated the genomes of 1154 microbial genomes across the entire prokaryotic section of the tree of life, using a computational algorithm that tries to minimize a bias toward the detection of known classes of biosynthetic gene clusters (BGCs). Using antiSMASH, 33,351 detected putative BGCs were annotated (Medema et al. 2011a; Blin et al. 2013; Weber et al. 2015a), and 10,724 of them were considered to be "high-confidence" clusters, as they either belonged to well-characterized classes or passed a manual inspection. Surprisingly, 40% of the putative BGCs turned out to encode for the biosynthesis of complex saccharides. This result was unexpected because this group of compounds was previously often neglected compared to known hyperdiverse classes, such as polyketides and nonribosomal peptides, despite the fact that bioactive saccharides have been reported before (Kersten et al. 2013). The function of most of the putative saccharides in the global analysis remains unknown, but they are clearly very widespread (93% of species contain at least one saccharide BGC and, in one-third of the species, they constitute the majority of the putative clusters) and surprisingly diverse (they are rarely shared even between closely related species, even less so than the notoriously diverse polyketides).

Even more strikingly, a global map of all high-confidence BGCs revealed a network containing large cliques of widely distributed BGC classes, which did not contain any experimentally characterized clusters. These orphan cliques, sometimes containing hundreds of members spread across a wide range of bacterial diversity, represent the most promising candidates for chemical and functional novelty in the expanded universe of natural product diversity. The most diverse of them, with a total of 1021 members across all subphyla of proteobacteria, as well as a range of other Gram-negative species, was shown experimentally by Cimermancic et al. (2014) to be responsible for the biosynthesis of aryl polyenes. The deep sequence divergence between cluster members and the broad, but discontinuous, range of the phylogenetic distribution indicates that the chemical diversity of this group of compounds, which has putative antioxidant activity, has been considerably underestimated before.

A study of the predicted biosynthetic gene clusters in 830 genome sequences of actinobacteria confirms the general pattern of astonishing diversity, detecting 11,422 putative BGCs in 4122 gene cluster families, each potentially encoding a different natural product scaffold. Extrapolating from this dataset to the complete diversity of actinobacteria, the investigators predict a total biosynthetic capacity of hundreds of thousands of natural products from this phylum alone, despite the fact that their analysis focused only on the "known unknowns," that is, members of a handful of major well-characterized classes of secondary metabolites. The results of Cimermancic et al. (2014) indicate that this may still be an underestimate of the actual chemical potential of microbial metabolism.

The global assessment of biosynthetic capacity not only provides a map toward the identification of new building blocks for synthetic biology, it also may help guide the actual process of cluster engineering. A follow-up analysis of the same data set explored the patterns of gene

 Cite this article as *Cold Spring Harb Perspect Biol* doi: 10.1101/cshperspect.a023994

cluster evolution to identify rules that could help predict the interoperability of biosynthetic modules (Medema et al. 2014a). Observing how natural chemical diversity is created by evolution, they identified three specific constraints that should be taken into account when attempting cluster engineering for chemical novelty. First, their analysis delineates multienzyme subclusters that evolve as independent units in a process they describe as "bricks and mortar" evolution. These subclusters, rather than individual enzymes, could thus be considered as natural "biobrick" equivalents and would be the most promising starting point for a systematic recombination of biosynthetic functions for the assembly of new pathways. Second, they detect patterns of concerted evolution for groups of polyketide and nonribosomal peptide biosynthetic assembly lines, which have resulted in sets of sequence-homogenized catalytic domains; these groups are predicted to be particularly amenable to domain-shuffling strategies for diversity generation. And third, they highlight the fact that different classes of clusters show quite different evolutionary patterns with different levels of subcluster transfers, domain recruitment, concerted evolution, and internal duplication and diversification. Engineering strategies should take the specific evolutionary constraints revealed by this into account for optimized efficiency.

A more targeted genome survey by Yu et al. (2013) explored the diversity of phosphonate biosynthetic pathways using a signature gene (phosphoenolpyruvate mutase) as an indicator. This analysis revealed that phosphonates, which were previously considered biochemical oddities, are in fact widespread in microbes, the biosynthetic capacity being found in at least 5% of sequenced genomes from a diverse range of species. Further study of the genomic neighborhood of the signature genes suggests that the products of the phosphonate pathways are chemically very diverse (on the order of at least several hundred different chemical entities, as indicated by rarefaction analysis) and probably have a number of new and clinically relevant bioactivities. This analysis was later expanded to cover the entire predicted metabolome of actinomycetes, extrapolating to a potential chemical diversity of hundreds of thousands of possible drug leads encoded by members of this phylum alone (Doroghazi et al. 2014). Related approaches using signature genes in combination with high-throughput real-time polymerase chain reaction (PCR) have recently been suggested as an efficient means of strain prioritization for natural product discovery programs (Hindra et al. 2014), and their potential to identify "talented strains" was illustrated for diterpenoid-producing actinomycetes (Xie et al. 2014), albeit without a negative control, so that it remains possible that the successful discovery of numerous diterpenoids in one of the prioritized strains, *Streptomyces griseus* CB00830, is the result of the general giftedness of actinomycetes rather than a validation of the prioritization strategy.

One of the insights from genome-based natural product discovery was the realization that specialized metabolites can be found in a variety of unexpected sources. One of the underexploited sources of bioactive metabolites was recently highlighted by Sharon et al. (2014) who discuss the diverse set of metabolites produced by the human commensal microbiome and their potential effects on human physiology. As these metabolites have evolved in direct interaction with the human host, they are expected to be enriched in interesting drug leads, and metabolites from the microbiome turn out to play an important role in the development of cardiovascular disease, cancer, and other diseases, as well as influencing human behavior. Large-scale analysis of biosynthetic gene cluster complement of the human microbiome has already resulted in the discovery of a new thiopeptide antibiotic, lactocillin, from a member of the vaginal microbiome (Donia et al. 2014).

Another source of metabolic diversity is only recently becoming accessible. Advances in single-cell genome analysis allow, for the first time, the comprehensive exploration of the biosynthetic capacity of talented uncultivated bacteria, a rich source of potentially novel chemistry, as was shown, for example, by the diversity and uniqueness of biosynthetic gene clusters identified in the symbiotic bacteria of the marine sponge *Theonella swinhoei* (Wilson et al. 2014).

DISCOVERY OF NATURAL PRODUCTS

Although the genomic analyses help in the identification of building blocks and generic engineering strategies for the synthetic biology of natural products, chemistry-directed discovery is needed for turning "known unknowns" into "known knowns" as targets for drug development and biotechnology (Bologa et al. 2013). Luo et al. (2014) subdivide the field of natural products discovery into two main groups of approaches: top-down and bottom-up. The first, top-down, they consider to be the traditional primary means of discovery, using various perturbations to induce production of new compounds in their natural hosts without requiring prior knowledge of their biosynthetic machinery. The second class of approaches, in contrast, depends on genomic information and subsequent genetic manipulation to achieve the synthesis of a compound of interest, often in a heterologous host. Both of these strategies can be supported and accelerated by the techniques of synthetic biology, and progress in genomic and metabolomic technologies, as well as in bioinformatics, typically leads to rapid transition from top-down to bottom-up work once a compound of interest has first been detected.

Guo et al. (2015) introduced a strategy that combines a promoter reporter system that allows the color-based identification of strains overexpressing a previously cryptic target gene cluster with genome-scale random mutagenesis to create a library of genetic variants. They successfully applied this approach to the *pga* gene cluster of *Streptomyces* species PGA64, which had been recalcitrant to previous activation attempts, and identified two new anthroquinone aminoglycosides as the products associated with this gene cluster. In a complementary approach, Seyedsayamdost (2014) developed a high-throughput platform for awakening silent gene clusters using a library of small-molecule elicitors, avoiding the random mutagenesis required by the reporter-guided mutant selection strategy of Guo et al. (2015). In Seyedsayamdost's strategy, the activity of a silent gene cluster, for example, one that has been identified by genome analysis, is monitored by a targeted genetic reporter fusion, based on GFP or *lacZ* translationally fused to an essential gene of the cluster. Screening the resulting reporter strain against a library of bioactive compounds can rapidly identify chemical inducers of the silent cluster, as illustrated for the malleilactone and the burkholderac gene cluster. Once an efficient elicitor has been found, the induced product of the cluster can be detected and characterized, for example, by mass spectrometry-based methods. Multiplex strategies in which several clusters are monitored in parallel based on multiple reporters can easily be envisaged.

Computational methods for the detection of biosynthetic gene clusters in microbial genomes are continuously evolving, and the popular antiSMASH pipeline, which handles the majority of annotation tasks in one unified interface, is updated regularly (Medema et al. 2011a; Blin et al. 2013, 2014; Weber et al. 2015a). Like other genome annotation tools, antiSMASH relies on the availability of trustworthy reference annotations. The recently published community standard for the annotation of secondary metabolite biosynthetic gene clusters (MiBIG) provides a long-awaited tool for the rigorous documentation of the necessary information and has already been successfully applied in a large-scale standardized recuration of a vast body of legacy biochemical data on secondary metabolite biosynthesis (Medema et al. 2015). It is expected that, from now on, all major publications on the experimental characterization of biosynthetic gene clusters will be accompanied by MiBIG-compliant annotation data.

Many of the available tools, as well as other computational methods for the synthetic biology of natural products, were recently reviewed by Weber (2014) and Medema et al. (2012). The most interesting advances are provided by those tools that allow the direct integration of computational and experimental approaches. A recent example is Pep2Path (Medema et al. 2014b), a method for the mass spectrometry-guided mining for peptidic natural products. The Pep2Path software automates the process of matching observed novel peptide metabolites to the most likely biosynthetic gene cluster using a rapid Bayesian probabilistic approach with high accuracy and precision. In combination with nanoDESI mass spectrometry (Watrous et al. 2013), this method can also be used for peptide discovery from environmental samples, matching observed compounds to related clusters in already sequenced genomes. Pep2Path is applicable to nonribosomal peptides (NRPs), as well as for ribosomally synthesized and post-translationally modified peptides (RiPPs), the latter being a group of compounds with unexpected prevalence in global genome analyses (Cimermancic et al. 2014). Methods for the rapid structural elucidation of NRPs and RiPPs by mass spectrometry have recently been developed (Mohimani et al. 2014a,b; Zhang et al. 2014), and a large-scale "metabologenomics" strategy matching bacterial genome sequences to mass spectrometry–based profiles of secreted metabolomes led to the discovery of a novel type of chlorinated nonribosomal peptide, tambromycin (Goering et al. 2016).

Although these approaches combine bottom-up and top-down discovery, progress is also made in pure bottom-up strategies. One example of this should suffice to illustrate the contribution of synthetic biology in this respect: Unkles et al. (2014) tackled the challenges of heterologous expression of silent fungal gene clusters by establishing a generic synthetic biology toolbox for this purpose. This includes improved in vivo cloning approaches, based on recombination in *Saccharomyces cerevisiae*; a talented host strain, *Aspergillus nidulans*, which is a native producer of natural products and therefore more likely to contain the necessary auxiliary machinery for successful biosynthesis; and the assembly of polycistronic genes using viral 2A peptide sequences to direct cotranslational cleavage, thus avoiding the requirements for individual marker selection systems and well-characterized promoter sequences for every single gene in a complex biosynthetic pathway. The application of this toolbox for the heterologous expression of the entire penicillin biosynthetic gene cluster of *Penicillium chrysogenum* from a single polycistronic gene showed the convenience of this strategy. A recently developed platform for the rapid exchange of all native promoters in a target gene cluster with constitutively active promoters, based on yeast homologous recombination (Montiel et al. 2015), complements this system and further increases the speed with which silent or cryptic biosynthetic pathways can be awakened.

In related work, Ross et al. (2014) used synthetic biology tools to establish a platform for the capture and expression of biosynthetic gene clusters in *E. coli*. Using transformation-associated recombination in yeast they were able to capture targeted pathways without the usual size limitations associated with the traditional screening of large, randomized genomic libraries. Following the initial capture and heterologous expression of their proof-of-concept cluster, encoding the alterochromide pathway of *Pseudoalteromonas piscicida*, they proceeded to increase production, using λ-red recombination to bring the cluster under the control of a T7 promoter, together with the accessory phosphopantetheinyl transferase required for functionalizing the NRP synthase thiolation (T) domains. The product titers achieved were sufficient for detailed characterization of the biosynthetic pathway without further optimization.

All these approaches for the discovery of novel natural products will benefit from advances in functional screening methods that link chemical compounds to new bioactivities and modes of action, as illustrated by the proof-of-concept work of Kurita et al. (2015) on the functional annotation of natural product extracts by a combination of untargeted metabolomics and image-based phenotypic screening.

DESIGN OF NOVEL NATURAL PRODUCTS

Genome mining for new biosynthetic pathways and the discovery of new natural products from a variety of sources are only the first steps in a much more ambitious synthetic biology approach aiming at the generation of chemical novelties in the form of "nonnatural" natural products.

Beer et al. (2014) provided an important proof-of-concept advance in the engineering of nonribosomal peptide biosynthesis. Targeting a single-module nonribosomal peptide synthase, IndC, which is responsible for the biosynthesis of the blue pigment indigoidin, they show successful exchange of the T domain of the enzyme, replacing it with natural and synthetic domains from different sources. In the process, they identified general "design principles" when considering the importance of the linker region between the T domain and its preceding adenylation domain. These principles and additional rules were incorporated in computer-aided design software that aims at helping users to create nonribosomal peptide synthases with any desired composition. Although the present experiments only aimed at exchanging T domains with the same substrate specificity, it is still an encouraging step forward (Calcott and Ackerley 2014 review some of the remaining challenges of a truly general NRP reengineering strategy). As a side note, it is amazing to realize that the major part of the work by Beer et al. (2014) was performed as part of a 6-month iGEM competition summer project—an achievement that illustrates that the synthetic biology of natural products has come a long way toward its goal of easy and predictable engineering of biological systems. Recent advances in understanding the molecular basis of the modularity of nonribosomal peptide synthesis, both at the level of the peptide synthases (Reimer et al. 2016) and at the level of precursor biosynthesis (Diez et al. 2015), will further expand our ability to create chemical novelty within this class in a predictable manner.

Sardar et al. (2015) advanced our ability to program the biosynthesis of RiPPs by refining our knowledge regarding the recognition sequences that direct processing enzymes to specific peptides. They show that, to a large extent, these recognition sequences are portable, and posttranslational modifications can be exchanged between RiPPs by importing the relevant enzyme and adding the corresponding recognition sequence to the precursor peptide, leading to diverse nonnatural products. As suggested by Medema et al. (2014a), Sardar et al. (2015) emphasize the importance of large libraries of RiPP pathways as a prerequisite for evolutionary analysis of sufficient breadth to allow predictable engineering of the full scope of peptidic natural products.

In the area of polyketide biosynthesis, Sugimoto et al. (2014) managed for the first time to achieve full predictable reprogramming of a modular polyketide synthase to produce a product that was originally isolated in a different species. Using recombination and domain exchanges, they convert the aurethin biosynthetic pathway into a luteoreticulin assembly line, mimicking the natural evolutionary process of gene transfer, recombinations, and mutations. In the process, they generated "evolutionary" intermediate strains producing various luteoreticulin analogs, but also identified a number of unprecedented domain functions and context dependencies that reemphasize our need for deepened biochemical understanding of natural product biosynthetic machineries as a prerequisite for successful pathway engineering.

In another contribution to polyketide diversification, Riva et al. (2014) generated libraries of novel polyketide intermediates by feeding functionalized chemical probes to engineered bacterial strains that attach these probes to stalled biosynthetic intermediates that accumulate because of targeted mutations in the acyl carrier protein (ACP) domains of the native polyketide synthases. In their proof-of-concept study, they used this approach to generate a complex library of unnatural polyethers of potential therapeutic interest and suggest that similar libraries could be created by reinitiation of stalled biosynthesis on any ketosynthase, using diverse mimics of the natural malonyl-ACP extension intermediates. They also emphasize that the electron cryomicroscopy-based three-dimensional structure of a full-length polyketide synthase module (Dutta et al. 2014) will be a helpful guide for designing unnatural malonyl-ACP analogs with the best fit to all relevant active sites of the multienzyme.

Kung et al. (2014) take the structure-based approach one step further and use crystal structure information to guide the engineering of geranylgeranyl reductase for the controlled biosynthesis of diverse isoprenoids. They manage to achieve considerable control over the extent of reduction of the substrate, geranylgeranyl pyrophosphate, GGPP (ranging from dihydro to hexahydro products). As GGPP is a key intermediate in the biosynthesis of an enormous range of isoprenoids (ranging from diterpenes and retinoids to carotenoids), this ability of controlled reduction opens up large new areas of chemical space for synthetic biology.

Another development that might have important implications for the rapid diversification of natural products is the increasing availability of cell-free platforms for synthetic biology (Toogood et al. 2015; Garamella et al. 2016). These systems allow the rapid prototyping, optimization, and combinatorial exploration of biosynthetic pathways, before introduction to a suitable chassis for overproduction.

BUILDING NATURAL PRODUCT PRODUCTION STRAINS

The engineering of natural product production strains obviously benefits from general advances in synthetic biology methodology for genome editing and assembly (Ellis et al. 2011; Esvelt and Wang 2013; Gibson 2014; Si et al. 2014; Selle and Barrangou 2015). However, recent years have also seen a number of advances specific to natural product research. For example, Cobb et al. (2014b) established an engineered CRISPR-Cas9 system for multiplex genome editing in *Streptomyces*, a genus notable for its abundant repertoire of secondary metabolites. The constructed pCRISPomyces plasmids are compatible with Golden Gate or isothermal assembly and are expected to make a major contribution to the large-scale genome editing for the purposes of awakening silent gene clusters, engineering and recombining pathways, as well as optimizing hosts strains. Other examples using the CRISPR system in actinomycetes have also been reported recently, making this groundbreaking technology widely applicable for this important group of natural product production hosts (Huang et al. 2015; Tong et al. 2015).

The necessary high-throughput DNA assembly methods able to deal with the challenges posed by the large and highly repetitive gene clusters for natural products have also advanced continuously, including methods facilitating the domain swapping underlying combinatorial biosynthesis approaches toward creating natural product libraries (Cobb et al. 2014a). Methods based on site-specific recombination have recently been shown to provide further opportunities for the rapid assembly and reshuffling of metabolic pathways (Colloms et al. 2014). These methods have so far been shown on pathways of relatively limited size and complexity, but their potential to construct libraries of combinatorial pathway assemblies, with randomized gene order or ribosome binding site strength, as well as to modify pathways by facile addition or replacement of selected genes will be of considerable interest for natural product research.

The concept of combinatorial assembly has recently been taken to new extremes in the case of the *Klebsiella oxytoca* nitrogen fixation gene cluster, containing a total of 16 genes and 103 individual parts (open reading frames, promoters, ribosome binding sites, terminators) (Smanski et al. 2014). Starting from a comprehensively refactored and modularized version of the pathways and using a hierarchical DNA assembly strategy to construct libraries of variants with rearranged gene orders, orientation, and operon combinations, as well as simultaneous substitutions of the ribosome binding sites of all 16 genes, they achieved substantial improvements of cluster performance through multiple rounds of the design–build–test cycle of synthetic biology. Although the proof-of-concept target was a complex pathway of primary metabolism, the applicability of the same tools for natural product pathway engineering is obvious. The availability of integrating plasmids in actinobacteria, which are particularly talented producers of natural products, enables powerful multiplexed approaches for the combinatorial assembly of biosynthetic pathway variants (Fayed et al. 2015), which, in the future, could be applied not only for combinatorial biosynthesis, but also for more general refactoring strategies in these species.

CONCLUSION

It is obvious from the selected examples that the synthetic biology approaches to natural products research are progressing at a rapid and accelerating pace. Particularly noteworthy is the deepening integration between classical approaches to natural products and the new concepts of synthetic biology. This new, integrated community is quickly moving toward the critical mass that is necessary for translating numerous proof-of-concept studies and technological advances into an economically viable and generalizable strategy for the biotechnological industry, realizing the potential of natural products as drugs and chemicals for a wide variety of applications. It is also obvious from the preceding discussion that we are not quite there yet and that many challenges still remain to be addressed. We need more detailed knowledge of the engineering constraints imposed by enzymatic mechanisms and evolutionary history; we need more generic hosts optimized for rapid scale-up of production levels of natural products of interest; we need improved computational tools for pathway and host design; and we need even faster, more robust, and versatile genome assembly methods for increased throughput and improved predictability of the engineering process. If progress continues at the present rate, the synthetic biology of natural products is moving toward a bright future.

REFERENCES

Beer R, Herbst K, Ignatiadis N, Kats I, Adlung L, Meyer H, Niopek D, Christiansen T, Georgi F, Kurzawa N, et al. 2014. Creating functional engineered variants of the single-module non-ribosomal peptide synthetase IndC by T domain exchange. *Mol Biosyst* **10:** 1709–1718.

Blin K, Medema MH, Kazempour D, Fischbach MA, Breitling R, Takano E, Weber T. 2013. antiSMASH 2.0—A versatile platform for genome mining of secondary metabolite producers. *Nucleic Acids Res* **41:** W204–W212.

Blin K, Kazempour D, Wohlleben W, Weber T. 2014. Improved lanthipeptide detection and prediction for antiSMASH. *PloS ONE* **9:** e89420.

Bologa CG, Ursu O, Oprea TI, Melancon CE III, Tegos GP. 2013. Emerging trends in the discovery of natural product antibacterials. *Curr Opin Pharmacol* **13:** 678–687.

Breitling R, Takano E. 2015. Synthetic biology advances for pharmaceutical production. *Curr Opin Biotechnol* **35C:** 46–51.

Calcott MJ, Ackerley DF. 2014. Genetic manipulation of non-ribosomal peptide synthetases to generate novel bioactive peptide products. *Biotechnol Lett* **36:** 2407–2416.

Chessher A, Breitling R, Takano E. 2015. Bacterial microcompartments: Biomaterials for synthetic biology-based compartmentalization strategies. *ACS Biomater Sci Eng* **1:** 345–351.

Cimermancic P, Medema MH, Claesen J, Kurita K, Wieland Brown LC, Mavrommatis K, Pati A, Godfrey PA, Koehrsen M, Clardy J, et al. 2014. Insights into secondary metabolism from a global analysis of prokaryotic biosynthetic gene clusters. *Cell* **158:** 412–421.

Cobb RE, Luo Y, Freestone T, Zhao H. 2013. Drug discovery and development via synthetic biology. In *Synthetic biology*, pp. 183–206. Elsevier, Amsterdam.

Cobb RE, Ning JC, Zhao H. 2014a. DNA assembly techniques for next-generation combinatorial biosynthesis of natural products. *J Ind Microbiol Biotechnol* **41:** 469–477.

Cobb RE, Wang Y, Zhao H. 2014b. High-efficiency multiplex genome editing of *Streptomyces* species using an engineered CRISPR/Cas system. *ACS Synth Biol* **4:** 723–728.

Colloms SD, Merrick CA, Olorunniji FJ, Stark WM, Smith MCM, Osbourn A, Keasling JD, Rosser SJ. 2014. Rapid metabolic pathway assembly and modification using serine integrase site-specific recombination. *Nucleic Acids Res* **42:** e23.

Cummings M, Breitling R, Takano E. 2014. Steps towards the synthetic biology of polyketide biosynthesis. *FEMS Microbiol Lett* **351:** 116–125.

Diez V, Loznik M, Taylor S, Winn M, Rattray NJ, Podmore H, Micklefield J, Goodacre R, Medema MH, Muller U, et al. 2015. Functional exchangeability of oxidase and dehydrogenase reactions in the biosynthesis of hydroxyphenylglycine, a nonribosomal peptide building block. *ACS Synth Biol* **4:** 796–807.

Donia MS, Cimermancic P, Schulze CJ, Wieland Brown LC, Martin J, Mitreva M, Clardy J, Linington RG, Fischbach MA. 2014. A systematic analysis of biosynthetic gene clusters in the human microbiome reveals a common family of antibiotics. *Cell* **158:** 1402–1414.

Doroghazi JR, Albright JC, Goering AW, Ju KS, Haines RR, Tchalukov KA, Labeda DP, Kelleher NL, Metcalf WW. 2014. A roadmap for natural product discovery based on large-scale genomics and metabolomics. *Nat Chem Biol* **10:** 963–968.

Dutta S, Whicher JR, Hansen DA, Hale WA, Chemler JA, Congdon GR, Narayan AR, Hakansson K, Sherman DH, Smith JL, et al. 2014. Structure of a modular polyketide synthase. *Nature* **510:** 512–517.

Ellis T, Adie T, Baldwin GS. 2011. DNA assembly for synthetic biology: From parts to pathways and beyond. *Integr Biol (Cambr)* **3:** 109–118.

Esvelt KM, Wang HH. 2013. Genome-scale engineering for systems and synthetic biology. *Mol Syst Biol* **9:** 641.

Fayed B, Ashford DA, Hashem AM, Amin MA, El Gazayerly ON, Gregory MA, Smith MC. 2015. Multiplexed integrating plasmids for engineering of the erythromycin gene cluster for expression in *Streptomyces* spp. and combinatorial biosynthesis. *Appl Environ Microbiol* **81:** 8402–8413.

Frasch HJ, Medema MH, Takano E, Breitling R. 2013. Design-based re-engineering of biosynthetic gene clusters:

Plug-and-play in practice. *Curr Opin Biotechnol* **24:** 1144–1150.

Garamella J, Marshall R, Rustad M, Noireaux V. 2016. The all *E. coli* TX-TL toolbox 2.0: A platform for cell-free synthetic biology. *ACS Synth Biol* **5:** 344–355.

Genilloud O. 2014. The re-emerging role of microbial natural products in antibiotic discovery. *Antonie Van Leeuwenhoek* **106:** 173–188.

Gibson DG. 2014. Programming biological operating systems: Genome design, assembly and activation. *Nat Methods* **11:** 521–526.

Goering AW, McClure RA, Doroghazi JR, Albright JC, Haverland NA, Zhang Y, Ju K-S, Thomson RJ, Metcalf WW, Kelleher NL. 2016. Metabologenomics: Correlation of microbial gene clusters with metabolites drives discovery of a nonribosomal peptide with an unusual amino acid monomer. *ACS Cent Sci* **2:** 99–108.

Guo F, Xiang S, Li L, Wang B, Rajasarkka J, Grondahl-Yli-Hannuksela K, Ai G, Metsa-Ketela M, Yang K. 2015. Targeted activation of silent natural product biosynthesis pathways by reporter-guided mutant selection. *Metabolic Eng* **28:** 134–142.

Helfrich EJN, Reiter S, Piel J. 2014. Recent advances in genome-based polyketide discovery. *Curr Opin Biotechnol* **29:** 107–115.

Hindra, Huang T, Yang D, Rudolf JD, Xie P, Xie G, Teng Q, Lohman JR, Zhu X, Huang Y, et al. 2014. Strain prioritization for natural product discovery by a high-throughput real-time PCR method. *J Nat Prod* **77:** 2296–2303.

Hranueli D, Starcevic A, Zucko J, Rojas JD, Diminic J, Baranasic D, Gacesa R, Padilla G, Long PF, Cullum J. 2013. Synthetic biology: A novel approach for the construction of industrial microorganisms. *Food Technol Biotechnol* **51:** 3–11.

Huang H, Zheng G, Jiang W, Hu H, Lu Y. 2015. One-step high-efficiency CRISPR/Cas9-mediated genome editing in *Streptomyces*. *Acta Biochim Biophys Sinica* **47:** 231–243.

Keasling JD. 2008. Synthetic biology for synthetic chemistry. *ACS Chem Biol* **3:** 64–76.

Kersten RD, Ziemert N, Gonzalez DJ, Duggan BM, Nizet V, Dorrestein PC, Moore BS. 2013. Glycogenomics as a mass spectrometry-guided genome-mining method for microbial glycosylated molecules. *Proc Natl Acad Sci* **110:** E4407–E4416.

Kung Y, McAndrew RP, Xie X, Liu CC, Pereira JH, Adams PD, Keasling JD. 2014. Constructing tailored isoprenoid products by structure-guided modification of geranylgeranyl reductase. *Structure* **22:** 1028–1036.

Kurita KL, Glassey E, Linington RG. 2015. Integration of high-content screening and untargeted metabolomics for comprehensive functional annotation of natural product libraries. *Proc Natl Acad Sci* **112:** 11999–12004.

Lawrence AD, Frank S, Newnham S, Lee MJ, Brown IR, Xue WF, Rowe ML, Mulvihill DP, Prentice MB, Howard MJ, et al. 2014. Solution structure of a bacterial microcompartment targeting peptide and its application in the construction of an ethanol bioreactor. *ACS Synth Biol* **3:** 454–465.

Li J, Neubauer P. 2014. *Escherichia coli* as a cell factory for heterologous production of nonribosomal peptides and polyketides. *New Biotechnol* **31:** 579–585.

Luo Y, Cobb RE, Zhao H. 2014. Recent advances in natural product discovery. *Curr Opin Biotechnol* **30:** 230–237.

Medema MH, Blin K, Cimermancic P, de Jager V, Zakrzewski P, Fischbach MA, Weber T, Takano E, Breitling R. 2011a. antiSMASH: Rapid identification, annotation and analysis of secondary metabolite biosynthesis gene clusters in bacterial and fungal genome sequences. *Nucleic Acids Res* **39:** W339–W346.

Medema MH, Breitling R, Bovenberg R, Takano E. 2011b. Exploiting plug-and-play synthetic biology for drug discovery and production in microorganisms. *Nat Rev Microbiol* **9:** 131–137.

Medema MH, van Raaphorst R, Takano E, Breitling R. 2012. Computational tools for the synthetic design of biochemical pathways. *Nat Rev Microbiol* **10:** 191–202.

Medema MH, Cimermancic P, Sali A, Takano E, Fischbach MA. 2014a. A systematic computational analysis of biosynthetic gene cluster evolution: Lessons for engineering biosynthesis. *PLoS Comput Biol* **10:** e1004016.

Medema MH, Paalvast Y, Nguyen DD, Melnik A, Dorrestein PC, Takano E, Breitling R. 2014b. Pep2Path: Automated mass spectrometry-guided genome mining of peptidic natural products. *PLoS Comput Biol* **10:** e1003822.

Medema MH, Kottmann R, Yilmaz P, Cummings M, Biggins JB, Blin K, de Bruijn I, Chooi YH, Claesen J, Coates RC, et al. 2015. Minimum information about a biosynthetic gene cluster. *Nat Chem Biol* **11:** 625–631.

Mohimani H, Kersten RD, Liu WT, Wang M, Purvine SO, Wu S, Brewer HM, Pasa-Tolic L, Bandeira N, Moore BS, et al. 2014a. Automated genome mining of ribosomal peptide natural products. *ACS Chem Biol* **9:** 1545–1551.

Mohimani H, Liu WT, Kersten RD, Moore BS, Dorrestein PC, Pevzner PA. 2014b. NRPquest: Coupling mass spectrometry and genome mining for nonribosomal peptide discovery. *J Nat Prod* **77:** 1902–1909.

Montiel D, Kang HS, Chang FY, Charlop-Powers Z, Brady SF. 2015. Yeast homologous recombination-based promoter engineering for the activation of silent natural product biosynthetic gene clusters. *Proc Natl Acad Sci* **112:** 8953–8958.

Paddon CJ, Keasling JD. 2014. Semi-synthetic artemisinin: A model for the use of synthetic biology in pharmaceutical development. *Nat Rev Microbiol* **12:** 355–367.

Porro D, Branduardi P, Sauer M, Mattanovich D. 2014. Old obstacles and new horizons for microbial chemical production. *Curr Opin Biotechnol* **30:** 101–106.

Poust S, Hagen A, Katz L, Keasling JD. 2014. Narrowing the gap between the promise and reality of polyketide synthases as a synthetic biology platform. *Curr Opin Biotechnol* **30:** 32–39.

Quin MB, Schmidt-Dannert C. 2014. Designer microbes for biosynthesis. *Curr Opin Biotechnol* **29:** 55–61.

Rebets Y, Brötz E, Tokovenko B, Luzhetskyy A. 2014. Actinomycetes biosynthetic potential: How to bridge in silico and in vivo? *J Ind Microbiol Biotechnol* **41:** 387–402.

Reimer JM, Aloise MN, Harrison PM, Schmeing TM. 2016. Synthetic cycle of the initiation module of a formylating nonribosomal peptide synthetase. *Nature* **529:** 239–242.

Riva E, Wilkening I, Gazzola S, Li WM, Smith L, Leadlay PF, Tosin M. 2014. Chemical probes for the functionalization of polyketide intermediates. *Angew Chem Int Ed Engl* **53:** 11944–11949.

Ross AC, Gulland LE, Dorrestein PC, Moore BS. 2014. Targeted capture and heterologous expression of the *Pseudoalteromonas* Alterochromide gene cluster in *Escherichia coli* represents a promising natural product exploratory platform. *ACS Synth Biol* **4:** 414–420.

Sardar D, Pierce E, McIntosh JA, Schmidt EW. 2015. Recognition sequences and substrate evolution in cyanobactin biosynthesis. *ACS Synth Biol* **4:** 167–176.

Selle K, Barrangou R. 2015. Harnessing CRISPR-Cas systems for bacterial genome editing. *Trends Microbiol* **23:** 225–232.

Seyedsayamdost MR. 2014. High-throughput platform for the discovery of elicitors of silent bacterial gene clusters. *Proc Natl Acad Sci* **111:** 7266–7271.

Seyedsayamdost MR, Clardy J. 2014. Natural products and synthetic biology. *ACS Synth Biol* **3:** 745–747.

Sharon G, Garg N, Debelius J, Knight R, Dorrestein PC, Mazmanian SK. 2014. Specialized metabolites from the microbiome in health and disease. *Cell Metab* **20:** 719–730.

Si T, Xiao H, Zhao H. 2014. Rapid prototyping of microbial cell factories via genome-scale engineering. *Biotechnol Adv* **33:** 1420–1432.

Siddiqui MS, Thodey K, Trenchard I, Smolke CD. 2012. Advancing secondary metabolite biosynthesis in yeast with synthetic biology tools. *FEMS Yeast Res* **12:** 144–170.

Singh V. 2014. Recent advancements in synthetic biology: Current status and challenges. *Gene* **535:** 1–11.

Sleator RD. 2014. The synthetic biology future. *Bioengineered* **5:** 69–72.

Smanski MJ, Bhatia S, Zhao D, Park Y, Woodruff L BA, Giannoukos G, Ciulla D, Busby M, Calderon J, Nicol R, et al. 2014. Functional optimization of gene clusters by combinatorial design and assembly. *Nat Biotechnol* **32:** 1241–1249.

Song CW, Lee SY. 2013. Rapid one-step inactivation of single or multiple genes in *Escherichia coli*. *Biotechnol J* **8:** 776–784.

Sugimoto Y, Ding L, Ishida K, Hertweck C. 2014. Rational design of modular polyketide synthases: Morphing the aureothin pathway into a luteoreticulin assembly line. *Angew Chem Int Ed Engl* **53:** 1560–1564.

Tong Y, Charusanti P, Zhang L, Weber T, Lee SY. 2015. CRISPR-Cas9 based engineering of actinomycetal genomes. *ACS Synth Biol* **4:** 1020–1029.

Toogood HS, Ni Cheallaigh A, Tait S, Mansell DJ, Jervis A, Lygidakis A, Humphreys L, Takano E, Gardiner JM, Scrutton NS. 2015. Enzymatic menthol production: One-pot approach using engineered *Escherichia coli*. *ACS Synth Biol* **4:** 1112–1123.

Unkles SE, Valiante V, Mattern DJ, Brakhage AA. 2014. Synthetic biology tools for bioprospecting of natural products in eukaryotes. *Chem Biol* **21:** 502–508.

Watrous J, Roach P, Heath B, Alexandrov T, Laskin J, Dorrestein PC. 2013. Metabolic profiling directly from the Petri dish using nanospray desorption electrospray ionization imaging mass spectrometry. *Anal Chem* **85:** 10385–10391.

Weber T. 2014. In silico tools for the analysis of antibiotic biosynthetic pathways. *Int J Med Microbiol* **304:** 230–235.

Weber T, Blin K, Duddela S, Krug D, Kim HU, Bruccoleri R, Lee SY, Fischbach MA, Muller R, Wohlleben W, et al. 2015a. antiSMASH 3.0—A comprehensive resource for the genome mining of biosynthetic gene clusters. *Nucleic Acids Res* **43:** W237–W243.

Weber T, Charusanti P, Musiol-Kroll EM, Jiang X, Tong Y, Kim HU, Lee SY. 2015b. Metabolic engineering of antibiotic factories: New tools for antibiotic production in actinomycetes. *Trends Biotechnol* **33:** 15–26.

Wilson MC, Mori T, Ruckert C, Uria AR, Helf MJ, Takada K, Gernert C, Steffens UA, Heycke N, Schmitt S, et al. 2014. An environmental bacterial taxon with a large and distinct metabolic repertoire. *Nature* **506:** 58–62.

Wright G. 2014. Perspective: Synthetic biology revives antibiotics. *Nature* **509:** S13.

Xie P, Ma M, Rateb ME, Shaaban KA, Yu Z, Huang SX, Zhao LX, Zhu X, Yan Y, Peterson RM, et al. 2014. Biosynthetic potential-based strain prioritization for natural product discovery: A showcase for diterpenoid-producing actinomycetes. *J Nat Prod* **77:** 377–387.

Yu X, Doroghazi JR, Janga SC, Zhang JK, Circello B, Griffin BM, Labeda DP, Metcalf WW. 2013. Diversity and abundance of phosphonate biosynthetic genes in nature. *Proc Natl Acad Sci* **110:** 20759–20764.

Zakeri B, Lu TK. 2013. Synthetic biology of antimicrobial discovery. *ACS Synth Biol* **2:** 358–372.

Zakrzewski P, Medema MH, Gevorgyan A, Kierzek AM, Breitling R, Takano E. 2012. MultiMetEval: Comparative and multi-objective analysis of genome-scale metabolic models. *PloS ONE* **7:** e51511.

Zhang Q, Ortega M, Shi Y, Wang H, Melby JO, Tang W, Mitchell DA, van der Donk WA. 2014. Structural investigation of ribosomally synthesized natural products by hypothetical structure enumeration and evaluation using tandem MS. *Proc Natl Acad Sci* **111:** 12031–12036.

Engineering Gene Circuits for Mammalian Cell–Based Applications

Simon Ausländer[1] and Martin Fussenegger[1,2]

[1]Department of Biosystems Science and Engineering, ETH Zurich, CH-4058 Basel, Switzerland
[2]Faculty of Science, University of Basel, CH-4058 Basel, Switzerland
Correspondence: fussenegger@bsse.ethz.ch

Synthetic gene switches are basic building blocks for the construction of complex gene circuits that transform mammalian cells into useful cell-based machines for next-generation biotechnological and biomedical applications. Ligand-responsive gene switches are cellular sensors that are able to process specific signals to generate gene product responses. Their involvement in complex gene circuits results in sophisticated circuit topologies that are reminiscent of electronics and that are capable of providing engineered cells with the ability to memorize events, oscillate protein production, and perform complex information-processing tasks. Microencapsulated mammalian cells that are engineered with closed-loop gene networks can be implanted into mice to sense disease-related input signals and to process this information to produce a custom, fine-tuned therapeutic response that rebalances animal metabolism. Progress in gene circuit design, in combination with recent breakthroughs in genome engineering, may result in tailored engineered mammalian cells with great potential for future cell-based therapies.

The engineering of mammalian cells holds great promise for cell-based technologies, such as stem cell reprogramming, biopharmaceutical manufacturing, and for cell-based therapies and diagnostics (Weber and Fussenegger 2012). The manipulation of complex, multicellular processes involving metabolic pathways, protein secretion, and production, and cell differentiation and development allows their precise control, improvement, and adaption and could promote the power of engineered mammalian cells in future cell-based applications. Central to synthetic biology-inspired engineering approaches are tailor-made gene controllers that enable cells to interact and respond to specific signal inputs and that can be rewired into programmable gene circuits to execute defined functions (Benenson 2012; Ausländer and Fussenegger 2013). When integrated into mammalian host cells, programmable gene circuits may enable the temporal control of stem cell reprogramming, advance cell-based therapies by arming autologous cells with closed-loop therapeutic gene circuits, and enhance the cellular production capacity in biopharmaceutical manufacturing.

Synthetic biology exploits the vast diversity of functional components found in the genetic material of all life forms. Because the cellular language of DNA, RNA, and protein is universal

Cite this article as *Cold Spring Harb Perspect Biol* doi: 10.1101/cshperspect.a023895

throughout all kingdoms of life, bioengineers can exchange these biological components between organisms to provide cells with a novel, customized function. For example, ligand-binding sensors are engineered to modulate diverse biological activities (Wieland and Fussenegger 2010; Ausländer and Fussenegger 2013), enzymes from different organisms are coupled to produce new biomolecules (Müller et al. 2012), and programmable DNA-binding proteins are harnessed for gene-editing purposes (Gaj et al. 2013). Moreover, we are beginning to create synthetic components that are not found in nature, expanding the set of appropriate biological components, and equipping engineered cells with special functionalities for useful applications. Synthetic nucleotides and amino acids are already being incorporated into cell-produced genomes (Malyshev et al. 2014) and proteins (Mandell et al. 2015; Rovner et al. 2015), respectively, and synthetic RNA sensors are being used to control gene expression in living mammalian cells (Culler et al. 2010; Saito et al. 2010; Ausländer et al. 2014c). However, the implementation of new biological components often requires precise connection and dosing of endogenous and synthetic components to assemble a functional system. The efficient engineering of mammalian cells remains a challenge because bioengineers often cannot predict the behavior of a cell on component integration. Additionally, technical barriers and an as-yet limited understanding of the entire living mammalian cell must be overcome.

Living cells continuously sense and respond to environmental and endogenous signals and drive cellular programs that adapt cellular behavior, for example, regulate cell death, fine-tune metabolic pathways, or differentiate into specified cell types. Fundamental to this adaptability is the dynamic regulation of gene-expression patterns. Mammalian cells use various complex mechanisms to influence the rates of transcription and translation of individual or multiple genes, enabling global or local alterations in cellular gene-expression patterns. Using an engineering approach, synthetic biology aims to control gene expression in a predictable manner by enabling bioengineers to program engineered cells to perform specific tasks. Central to this concept are programmable gene switches that are able to execute the gene regulatory function of heterologous or endogenous genes in a ligand-dependent manner (Benenson 2012; Ausländer and Fussenegger 2013). In principle, these gene switches can be exploited to execute information-processing tasks in living cells. Sensor units measure input signals and process the input state by rendering a defined output, that is, the gene-expression level. To run cellular programs that are more complex, multiple gene switches are combined to assemble gene circuits with specified network topologies. The design of such complex gene circuits is challenging because they must function precisely and specifically in the complicated cellular environment without influencing essential endogenous processes. For example, gene circuits are designed to improve, tune, or adapt the switching performance of target genes (Deans et al. 2007; Lapique and Benenson 2014; Prochazka et al. 2014) or to perform specific tasks such as biocomputing (Xie et al. 2011; Ausländer et al. 2012a) and oscillatory functions (Tigges et al. 2009), as well as recording environmental and endogenous events in synthetic memory devices (Burrill et al. 2012). Moreover, engineered cell–cell communication devices enable mammalian cells to interact and assemble synthetic multicellular systems (Bacchus et al. 2012). Basic gene switches and complex gene circuits have great potential for biomedical applications because they enable encapsulated, engineered cells to sense specific input signals in the body and to control the expression of specific outputs, that is, proteins, to initiate a therapeutic intervention (Ausländer et al. 2012b).

Here we describe a synthetic biology-inspired approach for engineering mammalian cells. We cover classic and emerging design strategies for synthetic gene switches and their different modes of action. In addition, we focus on different blueprints for connecting basic gene switches to build more complex gene circuits with preset network topologies. We also introduce advanced genome-engineering technologies that have the potential to revolutionize the manipulation and regulation of gene expression via the incorporation of synthetic DNA-encoded gene circuits. Finally, we review recent progress in cell-based applications in biomedicine and biotechnology and provide a perspective regarding future applications for mammalian cell–based technologies.

GENE SWITCHES

Gene expression in mammalian cells is divided into multiple steps and can be briefly described as follows: Transcription factors are recruited to regulatory sites to activate or repress the transcription of target genes, thereby controlling the production of nascent messenger RNA (mRNA) transcripts. In the nucleus, posttranscriptional modifications produce a mature mRNA molecule. A 5′ cap and 3′ poly(A) tail are attached to the nascent mRNA molecule, and introns are removed during the splicing process. Then mRNA is exported into the cytoplasm and subject to translation. In this section, we review synthetic gene switches that can intervene and control mammalian gene expression at different levels and that can serve as basic components for gene circuit design.

Transcriptional Gene Switches

In principle, a transcription factor consists of a DNA-binding domain (DBD) and a transcriptional modulator. When bound to the corresponding DNA target sequence, transcription factors recruit either positive or negative regulatory proteins, depending on the transcriptional modulator, to either induce or repress, respectively, the transcription of the target gene. The activity of transcription factors can be regulated by their availability at the DNA target site. Although transcription factors with a programmable DBD such as transcription activator-like effector (TALE) or zinc-finger (ZF) proteins cannot be controlled by a ligand, others possess ligand-responsive DBDs that modulate their affinity for a cognate DNA target sequence. Another mechanism to link signal sensing to a transcriptional response is based on the rewiring of endogenous signaling pathways that control the expression of target genes. For example, G protein–coupled receptor (GPCR)-dependent signaling pathways control the activity of transcription factors. Ligand binding to GPCRs induces signal transduction based on second messenger molecules and phosphorylation cascades that culminate in transcription factor activity. The number of divergent classes of transcription factors that have different characteristics with potential advantages or disadvantages for custom applications in mammalian cells is increasing.

Prokaryotic regulator protein families are a widespread class of sensor proteins (e.g., >200,000 prokaryotic regulator protein-like sequences have been identified) that enable cells to adapt rapidly to environmental and metabolic changes by controlling the transcription of target genes depending on specific input signals (Ramos et al. 2005; Cuthbertson and Nodwell 2013). Most prokaryotic regulator proteins consist of an amino-terminal DBD that recognizes a unique DNA operator sequence and a carboxy-terminal sensory domain (SD) that specifically binds to a ligand molecule (Fig. 1A) (Cuthbertson and Nodwell 2013). Their diversity and sensory capacity render these proteins attractive for mammalian gene regulation and gene circuit design (Wieland and Fussenegger 2012). In a synthetic biology-inspired approach, prokaryotic regulator-based systems can be transferred easily into mammalian cells when DNA operator and protein sequence information and ligand identity are known. In brief, a transcriptional modulator (e.g., VP16, Krüppel-associated box [KRAB]) is fused to the transcriptional regulator (TR) protein, and DNA operator repeats are attached to suitable minimal or constitutive promoter sequences that drive target gene expression (Fig. 1A). Because the ligand-binding domain is part of the protein, the binding of the ligand to the regulator dimer leads to the dissociation of the operator–dimer complex, thus influencing gene-expression rates. Based on this design strategy, mammalian cells have been engineered with various synthetic systems to sense different ligand inputs (e.g., antibiotics, Gossen and Bujard 1992; Fussenegger et al. 2000; Urlinger et al. 2000; Weber et al. 2002, 2003; and small molecule ligands, Ned-

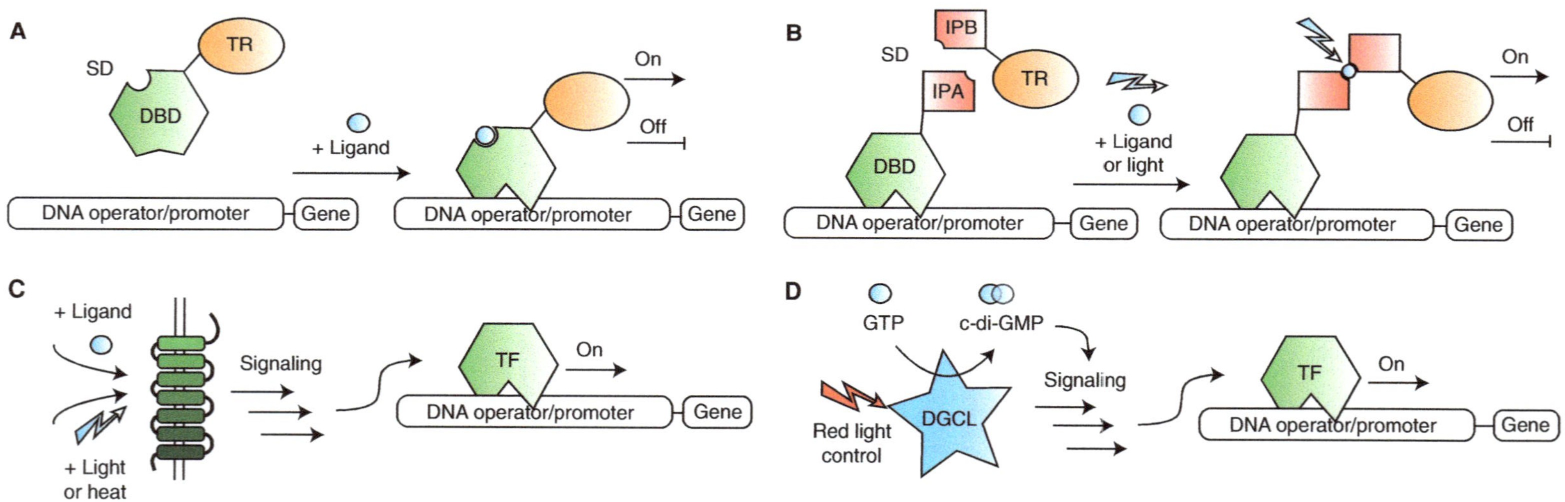

Figure 1. Transcriptional gene switches. (*A*) Prokaryotic regulator proteins are fused to transcriptional regulator proteins and bind to DNA operator sequences to control the transcription of target genes in a ligand-responsive manner. (*B*) Combining prokaryotic regulator proteins with ligand- or light-induced dimerization systems (DSs) enables the signal-dependent recruitment of transcriptional regulator proteins. (*C*) Cell-surface-located G protein–coupled receptors (GPCRs) sense extracellular signals and trigger signal transduction via signaling pathways to control cellular processes such as gene transcription in target cells. (*D*) An engineered diguanylate cyclase (DGCL) synthesizes the second messenger c-di-GMP in a red-light-responsive manner, triggering a downstream signaling pathway and leading to the transcriptional activation of target genes. DBD, DNA-binding domain; SD, sensory domain; TR, transcriptional regulator; IPA, interaction partner A; IPB, interaction partner B; TF, transcription factor; GTP, guanosine triphosphate.

Cite this article as *Cold Spring Harb Perspect Biol* doi: 10.1101/cshperspect.a023895

dermann et al. 2003; Weber et al. 2006, 2009; Hartenbach et al. 2007; Gitzinger et al. 2009, 2012; Kemmer et al. 2010; Bacchus et al. 2013) and proven to be useful tools for diverse biological applications. Moreover, a synthetic system that enables dual-input gene-expression control was developed using a single engineered prokaryotic regulator protein fused to the transrepressor domain KRAB, termed KRAB–CbaR (Xie et al. 2014). The first input, benzoic acid, is able to induce target gene expression, whereas the second input, vanillic acid, leads to an increase in the affinity of KRAB–CbaR to the CbaR-specific DNA operator sequence, further enhancing target gene repression. Furthermore, chemical- or light-inducible dimerization systems (DSs) provide a second layer of gene control, for example, interaction partner A (IPA) is fused to the DBD of a prokaryotic regulator protein, and the interaction partner B (IPB) is connected to a TR, resulting in a split transcription factor (Fig. 1B). For example, the interaction of *Arabidopsis thaliana*–derived protein partners phytochrome-interacting factor 6 (PIF6) and phytochrome B (PhyB) can be triggered and reverted in a red- and far-red-controlled manner, respectively. This interaction has been combined with the prokaryotic repressor protein TetR to generate a synthetic light-dependent gene-regulation system (Müller et al. 2013a). The same approach was also used in the design of ultraviolet (UV)-responsive (Müller et al. 2013b) and fatty acid-inducible (Rössger et al. 2013b) gene-regulation systems. The following are characteristics of prokaryotic regulator-based systems: (1) orthogonality to mammalian cells, (2) high diversity of different prokaryotic regulators that bind to various ligands, (3) a plasticity simplifying gene switch design, (4) high specificity that enables the usage of multiple synthetic systems in single cells, and (5) robust switching performance in mammalian cells.

Another widespread family of ligand-responsive gene switches is based on a family of integral eukaryotic membrane proteins called GPCRs (<800 members in the human genome) (Heng et al. 2013). GPCRs detect, first, messengers (extracellular signal inputs) and activate a signaling cascade that includes the production of second messengers to transduce information into the cell and to modulate cellular processes such as gene transcription (Fig. 1C). Specified cell types in the human body use GPCRs to respond to changes in hormone levels, metabolite concentrations, and other disease-related ligand molecule alterations. Therefore, GPCRs are extraordinarily suited for sensing therapeutic-relevant signal molecules in the appropriate physiological or pathological range and are often used in mammalian cell–based biomedical applications. Many mammalian cell lines express the components required to assemble intracellular signaling pathways, making the rewiring of custom GPCR sensory proteins to synthetic promoters straightforward. A synthetic GPCR-based gene-regulation system consists of a constitutive GPCR-expression unit and a target gene-expression unit. The latter harbors a synthetic promoter sequence with cognate DNA response elements (CRE, SRE, NFAT-RE) that recruit signaling pathway-specific transcription factors. Thus far, synthetic GPCR-based gene control systems have been engineered that respond to metabolites, hormones, and drug ligands, including luteinizing hormone (Kemmer et al. 2011), histamine (Ausländer et al. 2014b), dopamine (Rössger et al. 2013a), and guanabenz (Ye et al. 2013), in addition to other signals, such as pH changes (Ausländer et al. 2014a), heat (Stanley et al. 2012), and blue light (Ye et al. 2011). The orthogonal usage of multiple GPCRs in individual cells is impeded if they use the same signaling pathway. However, GPCRs can also be engineered to rewire their signaling pathway to synthetic or alternative endogenous signaling pathways or to respond to synthetic ligands. For example, a set of GPCR-based designer receptors activated exclusively by designer drugs (DREADDs) have been engineered to activate GPCR signaling cascades selectively using the pharmacologically inert molecule clozapine-*N*-oxide (CNO) (Urban and Roth 2015). This tool has been used for modulating cell type-specific GPCRs in neuroscience (Wess et al. 2013) and for engineering a synthetic cell motility device in mammalian cells (Park et al. 2014). Another class of synthetic

GPCR hybrids is based on so-called OptoXRs, which consist of engineered intracellular loops that are used to rewire GPCR signaling pathways, providing spatiotemporal light-responsive activation control (Airan et al. 2009). The following are characteristics of GPCR systems: (1) high compatibility with mammalian cells, (2) high diversity in sensing various disease-related ligands, (3) a simple gene switch design, (4) low connectivity of multiple GPCRs, and (5) robust switching performance in mammalian cells.

Stimulator of interferon genes (STING) is an endogenous, intracellular sensor protein that detects cytosolic DNA and intracellular pathogens by binding to cyclic dinucleotides (c-di-GMP, cGAMP). The activation of STING initiates a signaling cascade that leads to the nuclear translocation of interferon-regulatory factor 3 (IRF3), a transcription factor that specifically binds IRF3 DNA response elements to induce target gene expression. Although mammalian cells produce the cGAS protein, which synthesizes cGAMP on the detection of cytosolic DNA, c-di-GMP is only produced in specific prokaryotic cells. For example, the phototrophic bacterium *Rhodobacter sphaeroides* possesses a multidomain protein consisting of a diguanylate cyclase (DGCL) that synthesizes c-di-GMP, a phosphodiesterase (PDE) that breaks down c-di-GMP and a red-light sensor that controls the cyclase activity of DGCL. The integration of an engineered PDE-deficient variant of this protein into mammalian cells enables the red-light-responsive synthesis of c-di-GMP, thus rewiring the endogenous STING-dependent signaling pathway to activate synthetic target gene expression (Fig. 1D) (Folcher et al. 2014). This example shows the ability of synthetic biology-inspired concepts to enable the successful transfer and usage of prokaryotic second messenger molecules, such as c-di-GMP, in mammalian cells.

Because the aforementioned transcription factors bind to defined, invariable DNA sequences, correct insertion in proximity to promoter sequences is crucial, thus limiting their usage to regulate endogenous genes without laborious genetic engineering. Recent progress in genome engineering has produced advanced gene-editing systems capable of targeting and modifying any DNA sequence in the genome (Gaj et al. 2013). These systems are based on programmable, sequence-specific DNA-binding modules that can be equipped with nonspecific nucleases that cleave their target in proximity to the DNA-binding site. The three most popular gene-editing systems differ in their composition and characteristics. ZF proteins are commonly found in eukaryotes, and TALEs are derived from plant pathogens, for example, *Xanthomonas* (Gaj et al. 2013). To engineer these proteins and to determine their specific function, effector proteins such as nucleases or transcriptional regulators are fused to the customized DBD (Hsu et al. 2014). ZFs and TALEs consist of modular amino acid repeats that recognize either three (ZFs) or one (TALE) base pair(s) and that can be connected to one another to bind to DNA sequences of increased lengths. This modularity enables the engineering of specific ZFs or TALEs that bind to pre-defined sequences with high specificity within large genomes to execute defined functions. In contrast to ZFs and TALEs, the clustered regulatory interspaced short palindromic repeat (CRISPR)-CRISPR-associated (Cas)-based gene-editing system facilitates DNA recognition and binding using RNA (Doudna and Charpentier 2014). This system is derived from bacteria, in which it is involved in the cellular defense mechanism against viruses and plasmids. This system is based on the DNA endonuclease protein Cas9 and an engineered guide RNA (gRNA) that consists of two essential elements: a 20-nt-long RNA sequence that matches the target site and a structured RNA element that binds to the Cas9 effector protein and that facilitates DNA binding and cleavage. The recognition and binding of the gRNA is based on Watson–Crick base pairing, which allows for simple, predictable, and efficient reengineering of the gRNA to target any DNA sequence of interest. Moreover, the usage of multiple gRNAs enables multiplexed applications of the CRISPR-Cas system to manipulate distinct genomic sites simultaneously (Cong et al. 2013). The replacement of the nuclease domain with a transcriptional activator or repressor protein allows the three gene-editing systems to synthesize gene controllers with the potential ability to target any stretch of DNA sequence in the genome and thus to regulate the expression of any endogenous gene of interest (Doudna and Charpentier 2014). These systems, which have attracted great interest in synthetic biology, have been used to control heterologous gene expression and to design logic gates. Thus far, however, only a few options are available that enable the linkage of these synthetic gene controllers to sensor units to achieve ligand responsiveness. One option is to use chemical- or light-inducible DS, which consist of two proteins that interact in a ligand- or light-controllable manner (Fig. 1B). Coupling one of the DS proteins to a ZF or TALE and the other to a transcriptional regulatory protein results in ligand- or light-regulated recruitment of the transcriptional regulatory protein to the DNA-bound ZF or TALE and thus control of gene expression (Beerli et al. 2000; Polstein and Gersbach 2012; Konermann et al. 2013; Mercer et al. 2014). Based on the same concept, the CRISPR-Cas9 system has been engineered with light- or rapamycin-inducible dimerization domains enabling optogenetic or chemical control of RNA-guided gene regulation, respectively (Polstein and Gersbach 2015; Zetsche et al. 2015). The CRISPR-Cas9 system has been further adapted to allow for RNA-mediated recruiting of transcriptional regulators to specific genomic sites. In this adapted system, natural RNA-protein interactions are exploited by fusing different protein-binding RNA motifs to specific gRNAs that determine the genomic target locus (Zalatan et al. 2015). Because different transcriptional regulators can be fused to different RNA-binding proteins, simultaneous activation and repression of CRISPR-based gene regulation is possible. More recently, it has been shown that the CRISPR-Cas9 system is a highly flexible platform that accepts not only simple protein-binding aptamers but also much more complex RNA structures, such as long noncoding RNAs (lncRNAs) or fluorescent aptamers (Shechner et al. (2015). This powerful technology known as CRISPR-Display enables the expression of multiple gRNAs in single cells where each gRNA shows a different function. This enables, for example, the simultaneous activation and repression of multiple endogenous genes and at the same time the fluorescence-based visualization of genomic loci, thus providing new opportunities for basic research as well as synthetic biology.

Posttranscriptional Gene Switches

Transcriptional gene regulation relies on transcription factors that control the production of mRNAs and thus the amount of transcripts that are subject to translation. An additional layer of gene-expression control is provided by post-transcriptional regulatory modes that fine-tune and balance global gene expression. These control modes act on target mRNA transcripts and influence their stability, splicing, and translation. The versatility, modularity, and programmability of RNA render it attractive for the synthetic biology–inspired engineering of new functionalities (Saito and Inoue 2009; Grabow and Jaeger 2014). Synthetic RNA devices are engineered from basic RNA modules that provide diverse functionality. Natural and synthetic RNA aptamers bind to ligands with high specificity and give RNA the ability to sense small molecules, metabolites, and proteins (Famulok et al. 2007). Moreover, aptamers are often able to transmit the binding state via structural changes within the RNA molecule itself (Wunnicke et al. 2011). This ability can be exploited to manipulate the function of other RNA modules (e.g., short interfering RNAs [siRNAs] or ribozymes) in a ligand-responsive manner (Wieland and Fussenegger 2010). Endogenous gene-regulation pathways, such as RNA interference (RNAi) and mRNA splicing, can also be rewired and placed under the control of synthetic regulators.

The mammalian RNAi pathway enables sequence-specific gene knockdown targeted by 21- to 23-nt RNA species called siRNAs (Mittal 2004). The programming of these siRNAs is simple and enables the targeting of virtually any mRNA sequence. Ligand-responsive ribozymes (aptazymes) have been fused to siRNA precursors to block siRNA maturation to construct ligand-responsive siRNA-based gene switches

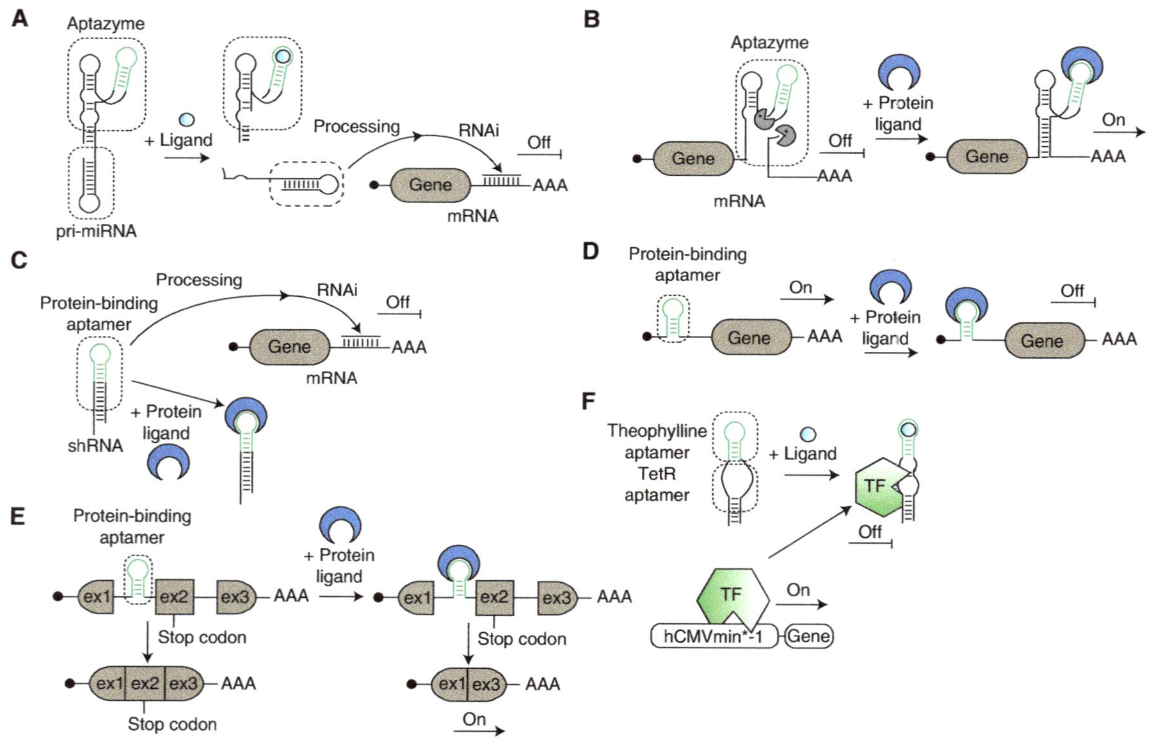

Figure 2. Posttranscriptional gene switches. (*A*) Aptazymes are fused to primary microRNA (pri-miRNA) molecules, enabling the ligand-responsive control of pri-miRNA processing and posttranscriptional target gene control. (*B*) Protein-responsive aptazymes are integrated into messenger RNAs (mRNAs) to regulate their stability, depending on the presence or absence of the protein ligand. (*C*) Protein binding to protein-binding aptamers that are integrated into small hairpin RNAs (shRNAs) inhibits shRNA processing and allows for protein-controlled target gene control. (*D*) Protein-binding aptamers integrated into the 5′ untranslated regions (UTRs) of mRNAs can be used to control translational initiation in a protein-dependent manner. (*E*) Integration of protein-binding aptamers into close proximity of splicing sites allows for protein-responsive alternative splicing regulation. (*F*) A TetR-binding aptamer is combined with a theophylline-responsive aptamer to enable the theophylline-dependent folding of the TetR-binding aptamer. When bound to its cognate aptamer, the TetR protein loses its DNA operator binding ability and influences gene expression at the transcriptional level. TF, Transcription factor; RNAi, RNA interference.

Cite this article as *Cold Spring Harb Perspect Biol* doi: 10.1101/cshperspect.a023895

in mammalian cells (Fig. 2A) (Kumar et al. 2009; Nomura et al. 2012). The addition of the small-molecule ligand theophylline or guanine induces ribozyme self-cleavage and releases the respective siRNA precursor, which further matures into a functional siRNA molecule that can repress target gene expression. Ribozymes are potent building blocks for the engineering of RNA-based devices because of their efficient self-cleavage activity, which can be exploited to control the function of other RNA molecules. To regulate the self-cleavage activity of ribozymes by ligands, aptamers can be assembled into ribozymes, rendering them ligand-responsive molecules (Wieland et al. 2012). The integration of these engineered ribozymes into mRNA transcripts leads to ligand-dependent cleavage of the target mRNA and thus control over gene expression (Win and Smolke 2007; Ausländer et al. 2010; Nomura et al. 2012, 2013). Recently, a new design strategy for protein-responsive ribozymes and their implementation into gene circuits has been described (Fig. 2B) (Ausländer et al. 2014c). In another study, a novel design of aptamer–siRNA fusions enabled siRNA-processing control in mammalian cells (Fig. 2C) (Kashida et al. 2012). Protein-binding RNA aptamers (e.g., U1A protein-U1A RNA motif) have been engineered into siRNA precursors that can replace RNA stem loops that are essential for siRNA processing. Protein binding to the aptamer blocks efficient processing and thus impedes gene knockdown. Protein-responsive aptamers can also be integrated into the 5′ untranslated regions (UTRs) of mRNAs to control translational initiation in a protein-dependent manner (Fig. 2D) (Saito et al. 2010). This simple approach was first shown using the L7Ae protein-C/D box RNA motif, which was engineered into reporter mRNA transcripts. In subsequent studies, further advancements have been made to tune input–output functions and to use multiple protein-responsive aptamers for the robust control of endogenous pathways and gene circuits (Saito et al. 2011; Goldfless et al. 2012; Stapleton et al. 2012; Endo et al. 2013). In a similar approach, protein-binding aptamers (MS2, p65, p50, and β-catenin aptamers) have been introduced into the proximity of splicing recognition sites of an alternative exon element of mRNA precursors (Fig. 2E) (Culler et al. 2010). However, the alternative splicing pattern was influenced by the cognate protein ligands and varied in terms of exon inclusion or exclusion for different protein ligands. Because p65 and β-catenin proteins are downstream modulators from signaling pathways, alternative splicing regulation could be rewired using the cognate inducers tumor necrosis factor α (TNF-α) or LTD_4. Aptamer–aptamer fusions have been developed to control the function of an engineered, bacterial repressor-based transcription factor (Fig. 2F) (Ausländer et al. 2011). The TetR-binding aptamer is able to bind to the DBD of the TetR-VP16 fusion protein, thereby releasing DNA operator-bound TetR and blocking gene expression. By rational design, the TetR-binding aptamer was fused to and placed under the control of a theophylline-binding aptamer, leading to an RNA-based transcriptional gene-regulation system.

ENGINEERING GENE CIRCUITS

Synthetic biology is an engineering discipline that aims to find solutions for biological questions by creating and developing new biological machines and systems that are based on biomolecules and synthetic materials to perform customized tasks. Network abstractions of other engineering disciplines can be used to describe synthetic biology-inspired biological systems. For example, "circuit topology" specifies the interconnection of all components in a circuit and is commonly used in electronics. Circuit topology can also be applied in biology to show synthetic biological networks and to describe input–output functions. Gene circuits with different topologies, such as oscillators (Tigges et al. 2009), logic gates (Kramer et al. 2004; Rinaudo et al. 2007; Win and Smolke 2007, 2008; Ausländer et al. 2012a, 2014a), hysteresis (Kramer and Fussenegger 2005), band-pass (Greber and Fussenegger 2010), time delay (Weber et al. 2007; Lapique and Benenson 2014), bow-tie (Prochazka et al. 2014), and memory devices (Burrill et al. 2012) have already been constructed in living mammalian cells. Because the above-described synthetic gene switches are able to process input information and to produce a specific output, they are well-suited building blocks for the design of programmable gene circuits reminiscent of electronic circuits. In a seminal publication, Kramer et al. (2004) described the design of various gene circuits with two and three inputs that process a single output reporter protein and set the basis for programmable gene circuits. Bacterial repressor-based transcriptional systems were cascaded to imitate the input–output function of logic gates found in electronics. Different combinations of small molecule ligands controlled the DNA binding of synthetic repressor proteins, resulting in a specific output. Based on the concept of programmable gene circuits, many different biological types of engineering decision-making systems have been developed in mammalian cells. These systems differ in their component composition and complexity, input–output identity and performance characteristics. Gene circuits can also be based on the sole application of posttranscriptional gene switches such as engineered ligand-responsive ribozymes (Win and Smolke 2008). However, the combination of both transcriptional and posttranscriptional systems promises to advance gene circuit design because it enables gene control at both the DNA and RNA levels. The leakiness of a gene switch could be significantly improved by combining RNAi methods with common transcriptional control (Deans et al. 2007). The linkage of transcriptional and posttranscriptional control was also applied to design gene circuits that mimic half-adder circuits capable of performing basic arithmetic (Ausländer et al. 2012a) and showing oscillatory gene-expression behavior (Tigges et al. 2009) in mammalian cells. For the design of a three-input AND gate, an engineered protein-responsive hammerhead ribozyme that regulates mRNA stability was combined with a protein-responsive aptamer that blocks translation (Fig. 3A) (Ausländer et al. 2014c). Coupled to three different transcriptional systems, each of the three small molecule ligands controlled the production, the stability, or the translation of the reporter mRNA. This concept simplifies gene circuit design because it enables the combination of multiple gene switches to act on individual transcripts. The eligibility of a biological component for gene circuit design is determined by its performance, connectivity, programmability, and orthogonality. Specifically, synthetic gene switches must function in a complex cellular environment without influencing either endogenous or other synthetic components. In a recent publication, sets of different building blocks were used to resemble logic gates in mammalian cells (Fig. 3B) (Ausländer et al. 2014a). In particular, a GPCR-based CO_2/H^+-inducible gene-regulation system was combined with bacterial repressor- and protein dimerization–based systems to build gas-inducible gene circuits. Gene circuits have also been designed to detect siRNA states in mammalian cells (Rinaudo et al. 2007; Tigges et al. 2009; Leisner et al. 2010; Xie et al. 2011; Lapique and Benenson 2014; Prochazka et al. 2014). siRNA or miRNA target sites are integrated into synthetic components to process siRNA information logically, enabling either the programming of the resulting gene circuit by the exogenous application of siRNA molecules, the intracellular, ligand-inducible production of siRNA ligands or endogenous miRNA species. Gene circuit performance can be improved with the addition of regulatory connections and new circuit designs. Site-specific recombinases, which have the ability to excise, flip, or invert DNA sequences, have been established as powerful building blocks for genome engineering and transcriptional gene control in mammalian cells (Nern et al. 2011). Site-specific recombination sites are engineered to flank genes, promoters, or transcriptional blockers to influence gene expression and to introduce regulatory knots to decouple inputs and outputs (Lapique and Benenson 2014; Prochazka et al. 2014). When integrated into gene circuits in a bow-tie topology, recombinases can act as a "knot" to transmit the circuit's information and thereby improve the performance of gene circuits significantly by reducing component leakage and input and output decoupling, as well as the timely production of individual components induced

 Cite this article as *Cold Spring Harb Perspect Biol* doi: 10.1101/cshperspect.a023895

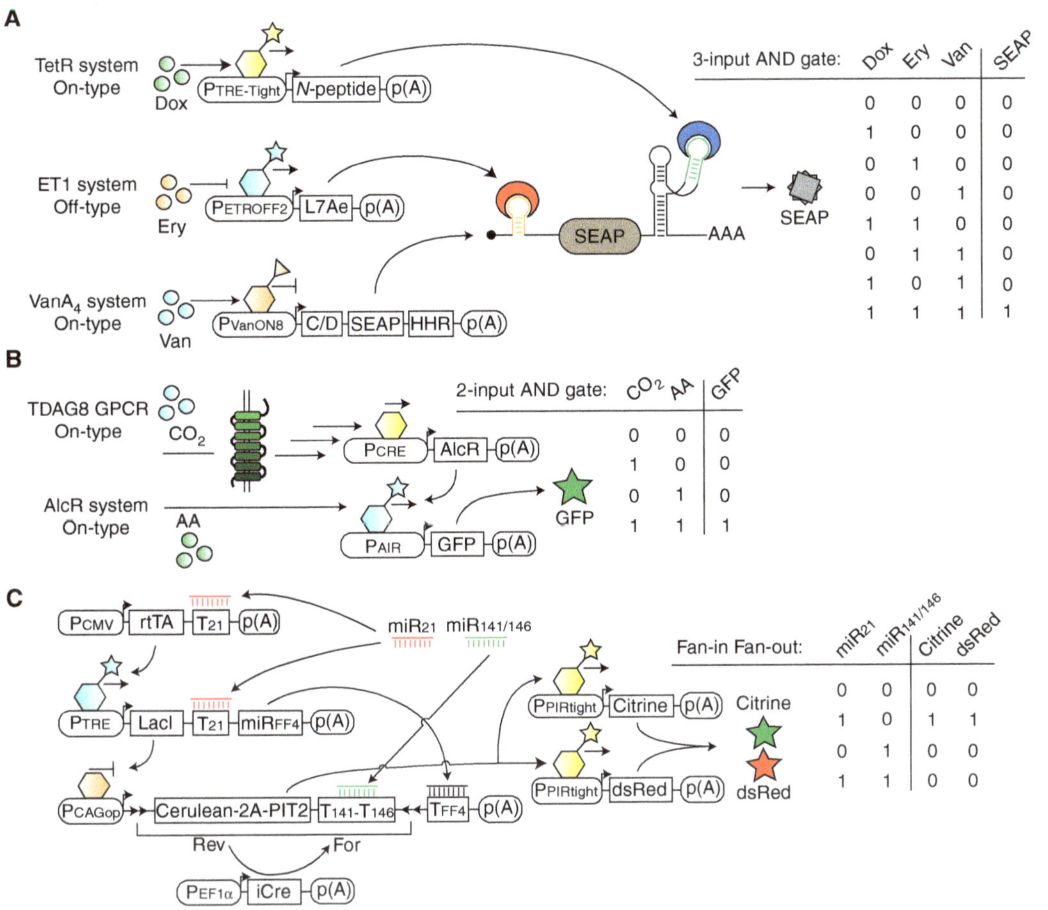

Figure 3. Engineered gene circuits. (*A*) The three ligand-responsive, transcriptional gene-control systems based on the transcriptional regulators TetR, ET1, and VanA4 are combined with two posttranscriptional gene switches that use C/D RNA motif-L7Ae protein and *N*-peptide-responsive ribozyme-*N*-peptide interactions to assemble a three-input AND gate. Mammalian cells engineered with this gene circuit can be programmed with the three ligands erythromycin (Ery), doxycycline (Dox), and vanilic acid (Van) to control output gene expression as shown in the truth table. (*B*) Design of a two-input AND gate that can be controlled by the two gaseous molecules carbon dioxide (CO_2) and acetaldehyde (AA). Activation of the G protein–coupled receptor (GPCR) TDAG8 by low pH triggers a signaling pathway that leads to the expression of AlcR, which binds to its cognate promoter exclusively in the presence of AA, thus producing the output protein GFP. (*C*) Transcriptional activators (rtTA and PIT2) and repressors (LacI), in addition to the recombinase iCre, are connected to build a gene circuit using bow-tie architecture. Two miRNA inputs (miR21 and miR141/146) manipulate the gene circuit to express the two output proteins in the exclusive presence of miR21.

by gene flipping (Fig. 3C) (Prochazka et al. 2014). The production of an output protein within gene circuits is used to wire different gene switches within the circuit or to have a convenient readout protein for output quantification. Engineered mammalian cells harboring different gene circuits can communicate with each other when the sender cell population produces a signaling molecule that can be sensed by the receiver cell population. For example, sender cells can produce the enzyme tryptophan synthase that converts indole into L-tryptophan, which, in turn, can be detected by receiver cells that measure tryptophan levels using a transcriptional controller (Bacchus et al. 2012). The combination of this system with a second synthetic cell–cell communication system based on the communication molecule acetaldehyde resulted in a two-way multicellular communication network that could temporally program endothelial cells to form blood vessels. Although the above-described gene circuits are based on ligand-responsive gene switches, ligand-unresponsive components can also be connected to build gene networks that can be controlled by the presence or absence of their components. For example, TALE- or ZF-based transcription factors have been used to build basic logic gates in which component-encoding plasmids serve as input signals (Lohmueller et al. 2012; Lienert et al. 2013).

ENGINEERING MAMMALIAN CELLS FOR BIOMEDICAL APPLICATIONS

Reporter genes are used as output proteins for the design and characterization of new gene switches and circuits because they provide information regarding switching performance. The replacement of the reporter gene with output proteins that induce cell apoptosis or differentiation or that have a therapeutic effect allows for the use of gene circuit-engineered mammalian cells in biomedical applications. The design of a cell classifier circuit based on the input level detection of different miRNA species allows for the differentiation of different cell types (Xie et al. 2011). TetR- and LacI-based transcriptional regulation systems control the expression of mRNA transcripts that sense specific miRNA inputs and that are logically connected to produce the proapoptotic factor human Bcl-2-associated X protein (hBax). Cell death is only induced in cancer cells that express a specific miRNA pattern. Via a similar strategy, transcription factor patterns that are present only in cancerous cells can be detected by their cognate promoter target sequences (CXCL1, SSX1, and H2A1) to drive the expression of split transcription factor units (Gal4-Coh2 and DocS-VP16 fusion proteins) (Nissim and Bar-Ziv 2010). When both units are expressed, Coh2 and DocS heterodimerization results in upstream activation sequence (UAS)-binding of Gal4 and VP16-dependent activation of the adjacent promoter. The output protein thymidine kinase type 1 derived from the Herpes simplex virus converts the prodrug ganciclovir into a toxic molecule that induces cell death. A therapeutic application of the above-described circuits would require an efficient and complete delivery of all DNA components into a patient's target cells.

In contrast, therapeutic cell implants consist of engineered mammalian cells that integrate into the metabolic processes of an organism such that these cells are able to produce a therapeutic effector (Ausländer et al. 2012b; Weber and Fussenegger 2012). Inert carriers such as alginate- or cellulose-based capsules permit the diffusion of nutrients and output molecules while protecting the engineered cells from host immune system attacks. Moreover, these capsules serve as an interface between bodily fluids and engineered cells and enable the detection of disease-related input signals and the incorporation of cell-produced therapeutic effectors into circulation. Encapsulated cells that contain ligand-responsive gene switches have shown the ability to regulate the production of therapeutic proteins in a ligand-dependent manner in mice. Although small molecules that are not found in an organism (e.g., antibiotics, Weber et al. 2002; and drugs, Ye et al. 2013) they must be invasively administered, gene-control systems based on radio waves (Stanley et al. 2012), light (Ye et al. 2011; Folcher et al. 2014; Kim et al. 2015), food additives (Gitzinger et al. 2012; Rössger et al. 2013b), and skin cream (Gitzinger et al. 2009), providing more convenient, noninvasive options to manipulate gene expression. Accordingly, recent work has shown the use of a brain–computer interface to control the illumination of red-light-emitting diodes (LEDs) to control gene expression in living cells (Folcher et al. 2014). These engineered mammalian cells contain a red-light-responsive gene-expression system that allows for the fine-tuning of target protein production.

Human diseases often originate from an imbalance of metabolites, hormones, or other disease-related factors. Healthy individuals manage this balance via specified closed-loop circuits that are generally dysfunctional in sick

 Cite this article as *Cold Spring Harb Perspect Biol* doi: 10.1101/cshperspect.a023895

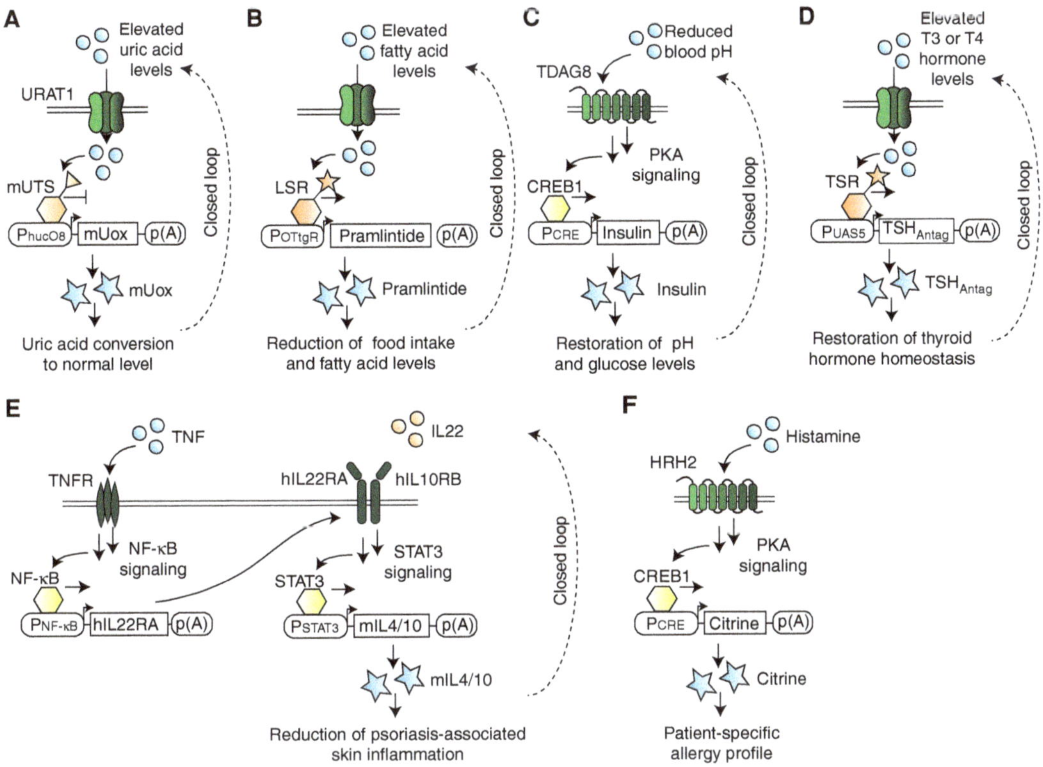

Figure 4. Gene circuits for biomedical applications. (*A*) Closed-loop gene circuit designed to detect elevated uric acid levels. Uric acid is transported by URAT1 into the cell, where it binds to the engineered uric acid sensor mUTS. This interaction induces the transcription of the secreted enzyme mUox, leading to decreased uric acid levels and thereby closing the gene circuit loop. (*B*) Closed-loop gene circuit designed to detect elevated fatty acid levels. Fatty acids bind to the engineered fatty acid lipid-sensing receptor (LSR) and induce transcription of the secreted hormone pramlintide, leading to decreased fatty acid levels and thereby closing the gene circuit loop. (*C*) Closed-loop gene circuit designed to detect decreased blood pH levels. Protons bind to the G protein–coupled receptor (GPCR) TDAG8, triggering a signaling pathway that induces transcription of the secreted hormone insulin. This cascade results in restored blood pH levels, thereby closing the gene circuit loop. (*D*) Closed-loop gene circuit designed to detect elevated thyroid hormone levels in Graves disease. Thyroid hormones T3 and T4 are transported into the cell and bind the synthetic thyroid-sensing receptor (TSR) that induces transcription of the engineered thyroid-stimulating hormone (TSH_{Antag}) receptor. TSH_{Antag} competes with endogenous TSH receptor for thyroid hormone and autoantibody binding, thereby restoring the thyrotrophic feedback control of the hypothalamus–pituitary–thyroid axis. (*E*) Antipsoriatic cytokine converter, a closed-loop gene circuit converting the presence of proinflammatory cytokines tumor necrosis factor (TNF) and interleukin 22 (IL22) levels to production of the antiinflammatory cytokines IL4 and IL10 with AND-gate-expression logic. TNF activates the endogenous TNF receptor on the cell surface, and induces the NF-κB signaling pathway. This leads to the expression of the hIL22RA, which heterodimerizes with endogenous hIL10RB to form a functional IL22 receptor. In the presence of IL22, IL22 receptor activates the STAT3 signaling pathway that drives the expression of the secreted, antiinflammatory cytokines mIL4 and mIL10. These cytokines attenuate psoriasis-associated skin inflammations. (*F*) Gene circuit designed to sense blood-derived histamine. Histamine binds to the GPCR HRH2 and triggers a signaling pathway that induces the transcription of a reporter protein that enables the readout of allergic reactions using engineered mammalian cells. TNFR, Tumor necrosis factor receptor; PKA, protein kinase A.

patients. Prosthetic cell implants consist of engineered mammalian cells that can sense alterations in metabolite levels and, accordingly, that can produce a therapeutic response to counteract the disease (input) state, thus emulating closed-loop circuits (Ausländer et al. 2012b; Weber and Fussenegger 2012; Ausländer and Fussenegger 2013). Thus far, prosthetic cell implants have been developed to target (1) gouty arthritis, scoring elevated uric acid concentrations and producing a secreted version of the urate oxidase that converts uric acid to allantoin, thereby restoring uric acid homeostasis in the blood (Fig. 4A) (Kemmer et al. 2010); (2) obesity, measuring increased blood fatty acid levels and expressing the anorectic peptide hormone pramlintide, thereby reducing caloric food intake and body weight (Fig. 4B) (Rössger et al. 2013b); (3) diabetic ketoacidosis, sensing reduced blood pH levels and producing the hormone insulin, thereby restoring blood glucose and pH homeostasis (Fig. 4C) (Ausländer et al. 2014a); (4) Graves disease, sensing thyroid hormone levels and producing an engineered variant of the human thyroid-stimulating hormone (TSH), thereby restoring the thyrotrophic feedback control of the hypothalamus–pituitary–thyroid axis (Fig. 4D) (Saxena et al. 2016); and (5) psoriasis, sensing proinflammatory cytokines TNF as well as interleukin 22 (IL22) and producing anti-inflammatory cytokines IL4 and IL10 only in the presence of both input signals (AND-gate-expression logic), thereby preventing or attenuating psoriatic flares (Fig. 4E) (Schukur et al. 2015). In response to these disease states, specific therapeutic effectors are produced by the engineered cell implants to treat the respective medical condition in a closed-loop manner.

Engineered cells are useful for both therapeutic and diagnostic applications because they are able to sense disease-relevant molecules in the physiological range. A prime example is the design of histamine-responsive mammalian cells that can be used to profile the histamine levels of allergen-treated immune effector cells in human whole blood (Fig. 4F) (Ausländer et al. 2014b). These mammalian cells have been engineered with the histamine-sensing GPCR HRH2 and with a cognate synthetic response cassette that consists of an HRH2 signaling–induced promoter and a reporter gene for simple readout. By treating human blood with a variety of allergens and by subsequently detecting immune cell–derived histamine using these engineered cells, researchers were able to generate personalized allergy profiles of human patients.

CONCLUDING REMARKS

The design of novel, synthetic gene controllers is of central importance in synthetic biology because they are the basic building blocks for the construction of complex gene circuits. Artificial gene circuits transcend the usage of mere transcriptional regulators and combine RNA- and protein-based switches to increase their complexity and functionality. Using the design of cell–cell communication systems, circuits that are more complex are envisaged to distribute complexity into simpler tasks that, when combined, form designer cell implants capable of autonomously sensing and treating multiple diseases. Engineered mammalian cells have also proven to be useful for diagnostic purposes because of their ability to detect disease-relevant molecules in the physiological range. Recent advancements in genome-editing technologies may soon enable the stable integration of multicomponent gene circuits, resulting in next-generation engineered mammalian cells with the potential to accelerate the transfer of cell-based treatment strategies into clinical settings.

ACKNOWLEDGMENTS

We thank David Ausländer for generous advice and comments on the manuscript.

REFERENCES

Airan RD, Thompson KR, Fenno LE, Bernstein H, Deisseroth K. 2009. Temporally precise in vivo control of intracellular signalling. *Nature* **458:** 1025–1029.

Ausländer S, Fussenegger M. 2013. From gene switches to mammalian designer cells: Present and future prospects. *Trends Biotechnol* **31:** 155–168.

Ausländer S, Ketzer P, Hartig JS. 2010. A ligand-dependent hammerhead ribozyme switch for controlling mammalian gene expression. *Mol Biosyst* **6:** 807–814.

Ausländer D, Wieland M, Ausländer S, Tigges M, Fussenegger M. 2011. Rational design of a small molecule-responsive intramer controlling transgene expression in mammalian cells. *Nucleic Acids Res* **39:** e155.

Ausländer S, Ausländer D, Müller M, Wieland M, Fussenegger M. 2012a. Programmable single-cell mammalian biocomputers. *Nature* **487:** 123–127.

Ausländer S, Wieland M, Fussenegger M. 2012b. Smart medication through combination of synthetic biology and cell microencapsulation. *Metab Eng* **14:** 252–260.

Ausländer D, Ausländer S, Charpin-El Hamri G, Sedlmayer F, Müller M, Frey O, Hierlemann A, Stelling J, Fussenegger M. 2014a. A synthetic multifunctional mammalian pH sensor and CO_2 transgene-control device. *Mol Cell* **55:** 397–408.

Ausländer D, Eggerschwiler B, Kemmer C, Geering B, Ausländer S, Fussenegger M. 2014b. A designer cell-based histamine-specific human allergy profiler. *Nat Commun* **5:** 4408.

Ausländer S, Stucheli P, Rehm C, Ausländer D, Hartig JS, Fussenegger M. 2014c. A general design strategy for protein-responsive riboswitches in mammalian cells. *Nat Method* **11:** 1154–1160.

Bacchus W, Lang M, El-Baba MD, Weber W, Stelling J, Fussenegger M. 2012. Synthetic two-way communication between mammalian cells. *Nat Biotechnol* **30:** 991–996.

Bacchus W, Weber W, Fussenegger M. 2013. Increasing the dynamic control space of mammalian transcription devices by combinatorial assembly of homologous regulatory elements from different bacterial species. *Metab Eng* **15:** 144–150.

Beerli RR, Schopfer U, Dreier B, Barbas CF III. 2000. Chemically regulated zinc finger transcription factors. *J Biol Chem* **275:** 32617–32627.

Benenson Y. 2012. Biomolecular computing systems: Principles, progress and potential. *Nat Rev Genet* **13:** 455–468.

Burrill DR, Inniss MC, Boyle PM, Silver PA. 2012. Synthetic memory circuits for tracking human cell fate. *Genes Dev* **26:** 1486–1497.

Cong L, Ran FA, Cox D, Lin S, Barretto R, Habib N, Hsu PD, Wu X, Jiang W, Marraffini LA, et al. 2013. Multiplex genome engineering using CRISPR/Cas systems. *Science* **339:** 819–823.

Culler SJ, Hoff KG, Smolke CD. 2010. Reprogramming cellular behavior with RNA controllers responsive to endogenous proteins. *Science* **330:** 1251–1255.

Cuthbertson L, Nodwell JR. 2013. The TetR family of regulators. *Microbiol Mol Biol Rev* **77:** 440–475.

Deans TL, Cantor CR, Collins JJ. 2007. A tunable genetic switch based on RNAi and repressor proteins for regulating gene expression in mammalian cells. *Cell* **130:** 363–372.

Doudna JA, Charpentier E. 2014. Genome editing. The new frontier of genome engineering with CRISPR-Cas9. *Science* **346:** 1258096.

Endo K, Hayashi K, Inoue T, Saito H. 2013. A versatile *cis*-acting inverter module for synthetic translational switches. *Nat Commun* **4:** 2393.

Famulok M, Hartig JS, Mayer G. 2007. Functional aptamers and aptazymes in biotechnology, diagnostics, and therapy. *Chem Rev* **107:** 3715–3743.

Folcher M, Oesterle S, Zwicky K, Thekkottil T, Heymoz J, Hohmann M, Christen M, Daoud El-Baba M, Buchmann P, Fussenegger M. 2014. Mind-controlled transgene expression by a wireless-powered optogenetic designer cell implant. *Nat Commun* **5:** 5392.

Fussenegger M, Morris RP, Fux C, Rimann M, von Stockar B, Thompson CJ, Bailey JE. 2000. Streptogramin-based gene regulation systems for mammalian cells. *Nat Biotechnol* **18:** 1203–1208.

Gaj T, Gersbach CA, Barbas CF III. 2013. ZFN, TALEN, and CRISPR/Cas-based methods for genome engineering. *Trends Biotechnol* **31:** 397–405.

Gitzinger M, Kemmer C, El-Baba MD, Weber W, Fussenegger M. 2009. Controlling transgene expression in subcutaneous implants using a skin lotion containing the apple metabolite phloretin. *Proc Natl Acad Sci* **106:** 10638–10643.

Gitzinger M, Kemmer C, Fluri DA, El-Baba MD, Weber W, Fussenegger M. 2012. The food additive vanillic acid controls transgene expression in mammalian cells and mice. *Nucleic Acids Res* **40:** e37.

Goldfless SJ, Belmont BJ, de Paz AM, Liu JF, Niles JC. 2012. Direct and specific chemical control of eukaryotic translation with a synthetic RNA-protein interaction. *Nucleic Acids Res* **40:** e64.

Gossen M, Bujard H. 1992. Tight control of gene expression in mammalian cells by tetracycline-responsive promoters. *Proc Natl Acad Sci* **89:** 5547–5551.

Grabow WW, Jaeger L. 2014. RNA self-assembly and RNA nanotechnology. *Acc Chem Res* **47:** 1871–1880.

Greber D, Fussenegger M. 2010. An engineered mammalian band-pass network. *Nucleic Acids Res* **38:** e174.

Hartenbach S, Daoud-El Baba M, Weber W, Fussenegger M. 2007. An engineered L-arginine sensor of *Chlamydia pneumoniae* enables arginine-adjustable transcription control in mammalian cells and mice. *Nucleic Acids Res* **35:** e136.

Heng BC, Aubel D, Fussenegger M. 2013. An overview of the diverse roles of G-protein coupled receptors (GPCRs) in the pathophysiology of various human diseases. *Biotechnol Adv* **31:** 1676–1694.

Hsu PD, Lander ES, Zhang F. 2014. Development and applications of CRISPR-Cas9 for genome engineering. *Cell* **157:** 1262–1278.

Kashida S, Inoue T, Saito H. 2012. Three-dimensionally designed protein-responsive RNA devices for cell signaling regulation. *Nucleic Acids Res* **40:** 9369–9378.

Kemmer C, Gitzinger M, Daoud-El Baba M, Djonov V, Stelling J, Fussenegger M. 2010. Self-sufficient control of urate homeostasis in mice by a synthetic circuit. *Nat Biotechnol* **28:** 355–360.

Kemmer C, Fluri DA, Witschi U, Passeraub A, Gutzwiller A, Fussenegger M. 2011. A designer network coordinating bovine artificial insemination by ovulation-triggered release of implanted sperms. *J Control Release* **150:** 23–29.

Kim T, Folcher M, Doaud-El Baba M, Fussenegger M. 2015. A synthetic erectile optogenetic stimulator enabling blue-light-inducible penile erection. *Angew Chem Int Ed Engl* **54:** 5933–5938.

Konermann S, Brigham MD, Trevino AE, Hsu PD, Heidenreich M, Cong L, Platt RJ, Scott DA, Church GM, Zhang F. 2013. Optical control of mammalian endogenous transcription and epigenetic states. *Nature* **500:** 472–476.

Kramer BP, Fussenegger M. 2005. Hysteresis in a synthetic mammalian gene network. *Proc Natl Acad Sci* **102:** 9517–9522.

Kramer BP, Fischer C, Fussenegger M. 2004. BioLogic gates enable logical transcription control in mammalian cells. *Biotechnol Bioeng* **87:** 478–484.

Kumar D, An CI, Yokobayashi Y. 2009. Conditional RNA interference mediated by allosteric ribozyme. *J Am Chem Soc* **131:** 13906–13907.

Lapique N, Benenson Y. 2014. Digital switching in a biosensor circuit via programmable timing of gene availability. *Nat Chem Biol* **10:** 1020–1027.

Leisner M, Bleris L, Lohmueller J, Xie Z, Benenson Y. 2010. Rationally designed logic integration of regulatory signals in mammalian cells. *Nat Nanotechnol* **5:** 666–670.

Lienert F, Torella JP, Chen JH, Norsworthy M, Richardson RR, Silver PA. 2013. Two- and three-input TALE-based AND logic computation in embryonic stem cells. *Nucleic Acids Res* **41:** 9967–9975.

Lohmueller JJ, Armel TZ, Silver PA. 2012. A tunable zinc finger-based framework for Boolean logic computation in mammalian cells. *Nucleic Acids Res* **40:** 5180–5187.

Malyshev DA, Dhami K, Lavergne T, Chen T, Dai N, Foster JM, Correa IR Jr, Romesberg FE. 2014. A semi-synthetic organism with an expanded genetic alphabet. *Nature* **509:** 385–388.

Mandell DJ, Lajoie MJ, Mee MT, Takeuchi R, Kuznetsov G, Norville JE, Gregg CJ, Stoddard BL, Church GM. 2015. Biocontainment of genetically modified organisms by synthetic protein design. *Nature* **518:** 55–60.

Mercer AC, Gaj T, Sirk SJ, Lamb BM, Barbas CF 3rd. 2014. Regulation of endogenous human gene expression by ligand-inducible TALE transcription factors. *ACS Synth Biol* **3:** 723–730.

Mittal V. 2004. Improving the efficiency of RNA interference in mammals. *Nat Rev Genet* **5:** 355–365.

Müller M, Ausländer S, Ausländer D, Kemmer C, Fussenegger M. 2012. A novel reporter system for bacterial and mammalian cells based on the non-ribosomal peptide indigoidine. *Met Eng* **14:** 325–335.

Müller K, Engesser R, Metzger S, Schulz S, Kampf MM, Busacker M, Steinberg T, Tomakidi P, Ehrbar M, Nagy F, et al. 2013a. A red/far-red light-responsive bi-stable toggle switch to control gene expression in mammalian cells. *Nucleic Acids Res* **41:** e77.

Müller K, Engesser R, Schulz S, Steinberg T, Tomakidi P, Weber CC, Ulm R, Timmer J, Zurbriggen MD, Weber W. 2013b. Multi-chromatic control of mammalian gene expression and signaling. *Nucleic Acids Res* **41:** e124.

Neddermann P, Gargioli C, Muraglia E, Sambucini S, Bonelli F, De Francesco R, Cortese R. 2003. A novel, inducible, eukaryotic gene expression system based on the quorum-sensing transcription factor TraR. *EMBO Rep* **4:** 159–165.

Nern A, Pfeiffer BD, Svoboda K, Rubin GM. 2011. Multiple new site-specific recombinases for use in manipulating animal genomes. *Proc Natl Acad Sci* **108:** 14198–14203.

Nissim L, Bar-Ziv RH. 2010. A tunable dual-promoter integrator for targeting of cancer cells. *Mol Syst Biol* **6:** 444.

Nomura Y, Kumar D, Yokobayashi Y. 2012. Synthetic mammalian riboswitches based on guanine aptazyme. *Chem Commun (Camb)* **48:** 7215–7217.

Nomura Y, Zhou L, Miu A, Yokobayashi Y. 2013. Controlling mammalian gene expression by allosteric hepatitis delta virus ribozymes. *ACS Synth Biol* **2:** 684–689.

Park JS, Rhau B, Hermann A, McNally KA, Zhou C, Gong D, Weiner OD, Conklin BR, Onuffer J, Lim WA. 2014. Synthetic control of mammalian-cell motility by engineering chemotaxis to an orthogonal bioinert chemical signal. *Proc Natl Acad Sci* **111:** 5896–5901.

Polstein LR, Gersbach CA. 2012. Light-inducible spatiotemporal control of gene activation by customizable zinc finger transcription factors. *J Am Chem Soc* **134:** 16480–16483.

Polstein LR, Gersbach CA. 2015. A light-inducible CRISPR-Cas9 system for control of endogenous gene activation. *Nat Chem Biol* **11:** 198–200.

Prochazka L, Angelici B, Haefliger B, Benenson Y. 2014. Highly modular bow-tie gene circuits with programmable dynamic behaviour. *Nat Commun* **5:** 4729.

Ramos JL, Martinez-Bueno M, Molina-Henares AJ, Teran W, Watanabe K, Zhang X, Gallegos MT, Brennan R, Tobes R. 2005. The TetR family of transcriptional repressors. *Microbiol Mol Biol Rev* **69:** 326–356.

Rinaudo K, Bleris L, Maddamsetti R, Subramanian S, Weiss R, Benenson Y. 2007. A universal RNAi-based logic evaluator that operates in mammalian cells. *Nat Biotechnol* **25:** 795–801.

Rössger K, Charpin-El Hamri G, Fussenegger M. 2013a. Reward-based hypertension control by a synthetic brain-dopamine interface. *Proc Natl Acad Sci* **110:** 18150–18155.

Rössger K, Charpin-El-Hamri G, Fussenegger M. 2013b. A closed-loop synthetic gene circuit for the treatment of diet-induced obesity in mice. *Nat Commun* **4:** 2825.

Rovner AJ, Haimovich AD, Katz SR, Li Z, Grome MW, Gassaway BM, Amiram M, Patel JR, Gallagher RR, Rinehart J, et al. 2015. Recoded organisms engineered to depend on synthetic amino acids. *Nature* **518:** 89–93.

Saito H, Inoue T. 2009. Synthetic biology with RNA motifs. *Int J Biochem Cell Biol* **41:** 398–404.

Saito H, Kobayashi T, Hara T, Fujita Y, Hayashi K, Furushima R, Inoue T. 2010. Synthetic translational regulation by an L7Ae–kink-turn RNP switch. *Nat Chem Biol* **6:** 71–78.

Saito H, Fujita Y, Kashida S, Hayashi K, Inoue T. 2011. Synthetic human cell fate regulation by protein-driven RNA switches. *Nat Commun* **2:** 160.

Saxena P, Charpin-El Hamri G, Folcher M, Zulewski H, Fussenegger M. 2016. Synthetic gene network restoring endogenous pituitary-thyroid feedback control in experimental Graves' disease. *Proc Natl Acad Sci* **113:** 1244–1249.

Schukur L, Geering B, Charpin-El Hamri G, Fussenegger M. 2015. Implantable synthetic cytokine converter cells with AND-gate logic treat experimental psoriasis. *Sci Transl Med* **7:** 318ra201.

Shechner DM, Hacisuleyman E, Younger ST, Rinn JL. 2015. Multiplexable, locus-specific targeting of long RNAs with CRISPR-Display. *Nat Methods* **12:** 664–670.

Stanley SA, Gagner JE, Damanpour S, Yoshida M, Dordick JS, Friedman JM. 2012. Radio-wave heating of iron oxide nanoparticles can regulate plasma glucose in mice. *Science* **336:** 604–608.

Stapleton JA, Endo K, Fujita Y, Hayashi K, Takinoue M, Saito H, Inoue T. 2012. Feedback control of protein expression in mammalian cells by tunable synthetic translational inhibition. *ACS Synth Biol* **1:** 83–88.

Tigges M, Marquez-Lago TT, Stelling J, Fussenegger M. 2009. A tunable synthetic mammalian oscillator. *Nature* **457:** 309–312.

Urban DJ, Roth BL. 2015. DREADDs (designer receptors exclusively activated by designer drugs): Chemogenetic tools with therapeutic utility. *Annu Rev Pharmacol Toxicol* **55:** 399–417.

Urlinger S, Baron U, Thellmann M, Hasan MT, Bujard H, Hillen W. 2000. Exploring the sequence space for tetracycline-dependent transcriptional activators: Novel mutations yield expanded range and sensitivity. *Proc Natl Acad Sci* **97:** 7963–7968.

Weber W, Fussenegger M. 2012. Emerging biomedical applications of synthetic biology. *Nat Rev Genet* **13:** 21–35.

Weber W, Fux C, Daoud-el Baba M, Keller B, Weber CC, Kramer BP, Heinzen C, Aubel D, Bailey JE, Fussenegger M. 2002. Macrolide-based transgene control in mammalian cells and mice. *Nat Biotechnol* **20:** 901–907.

Weber W, Schoenmakers R, Spielmann M, El-Baba MD, Folcher M, Keller B, Weber CC, Link N, van de Wetering P, Heinzen C, et al. 2003. Streptomyces-derived quorum-sensing systems engineered for adjustable transgene expression in mammalian cells and mice. *Nucleic Acids Res* **31:** e71.

Weber W, Link N, Fussenegger M. 2006. A genetic redox sensor for mammalian cells. *Metab Eng* **8:** 273–280.

Weber W, Kramer BP, Fussenegger M. 2007. A genetic time-delay circuitry in mammalian cells. *Biotechnol Bioeng* **98:** 894–902.

Weber W, Lienhart C, Baba MD, Fussenegger M. 2009. A biotin-triggered genetic switch in mammalian cells and mice. *Metab Eng* **11:** 117–124.

Wess J, Nakajima K, Jain S. 2013. Novel designer receptors to probe GPCR signaling and physiology. *Trends Pharmacol Sci* **34:** 385–392.

Wieland M, Fussenegger M. 2010. Ligand-dependent regulatory RNA parts for synthetic biology in eukaryotes. *Curr Opin Biotechnol* **21:** 760–765.

Wieland M, Fussenegger M. 2012. Engineering molecular circuits using synthetic biology in mammalian cells. *Annu Rev Chem Biomol Eng* **3:** 209–234.

Wieland M, Ausländer D, Fussenegger M. 2012. Engineering of ribozyme-based riboswitches for mammalian cells. *Methods* **56:** 351–357.

Win MN, Smolke CD. 2007. A modular and extensible RNA-based gene-regulatory platform for engineering cellular function. *Proc Natl Acad Sci* **104:** 14283–14288.

Win MN, Smolke CD. 2008. Higher-order cellular information processing with synthetic RNA devices. *Science* **322:** 456–460.

Wunnicke D, Strohbach D, Weigand JE, Appel B, Feresin E, Suess B, Müller S, Steinhoff HJ. 2011. Ligand-induced conformational capture of a synthetic tetracycline riboswitch revealed by pulse EPR. *RNA* **17:** 182–188.

Xie Z, Wroblewska L, Prochazka L, Weiss R, Benenson Y. 2011. Multi-input RNAi-based logic circuit for identification of specific cancer cells. *Science* **333:** 1307–1311.

Xie M, Ye H, Hamri GC, Fussenegger M. 2014. Antagonistic control of a dual-input mammalian gene switch by food additives. *Nucleic Acids Res* **42:** e116.

Ye H, Daoud-El Baba M, Peng RW, Fussenegger M. 2011. A synthetic optogenetic transcription device enhances blood-glucose homeostasis in mice. *Science* **332:** 1565–1568.

Ye H, Charpin-El Hamri G, Zwicky K, Christen M, Folcher M, Fussenegger M. 2013. Pharmaceutically controlled designer circuit for the treatment of the metabolic syndrome. *Proc Natl Acad Sci* **110:** 141–146.

Zalatan JG, Lee ME, Almeida R, Gilbert LA, Whitehead EH, La Russa M, Tsai JC, Weissman JS, Dueber JE, Qi LS, et al. 2015. Engineering complex synthetic transcriptional programs with CRISPR RNA scaffolds. *Cell* **160:** 339–350.

Zetsche B, Volz SE, Zhang F. 2015. A split-Cas9 architecture for inducible genome editing and transcription modulation. *Nat Biotechnol* **33:** 139–142.

Synthetic Biology in Cell and Organ Transplantation

Sean Stevens

Mammalian Synthetic Biology, Synthetic Genomics, Inc., La Jolla, California 92037

Correspondence: sstevens@syntheticgenomics.com

The transplantation of cells and organs has an extensive history, with blood transfusion and skin grafts described as some of the earliest medical interventions. The speed and efficiency of the human immune system evolved to rapidly recognize and remove pathogens; the human immune system also serves as a barrier against the transplant of cells and organs from even highly related donors. Although this shows the remarkable effectiveness of the immune system, the engineering of cells and organs that will survive in a host patient over the long term remains a steep challenge. Progress in the understanding of host immune responses to donor cells and organs, combined with the rapid advancement in synthetic biology applications, allows the rational engineering of more effective solutions for transplantation.

The development of synthetic biology as a discipline has greatly benefited the transplantation field. If one accepts the definition of synthetic biology as rational engineering of existing and novel biological pathways and materials, there are two broad areas of application within transplantation: the generation of organ structures from composite cells and matrices and the engineering of cellular genomes for improved survival and function in the host. The former is more relevant to organs; the latter applies to both cells and organs in transplantation.

Generation of organ and other biological structures in vitro, or tissue engineering, has been an active area of research for many years (Vacanti 2010). Application of more advanced technologies has greatly accelerated the generation of a variety of tissues and organ-like structures (Mou et al. 2015). The use of novel matrices and mechanical methods has allowed increasingly complex structures containing many different cell types (Murphy and Atala 2014). Hybrid approaches combining in vitro and in vivo organ generation are also active areas of investigation.

Modification of mammalian and other complex eukaryotic genomes has been a focus of research for decades; however, outside of mouse embryonic stem (ES) cells, most efforts have been limited to random integration of constructs in the genome using transfection or viral delivery. The application of more advanced synthetic biology techniques to mammalian genomes faces both biological and practical challenges.

First, mammalian genomes are much larger and encode substantially more genes, many of unknown function, compared with microbial genomes. Because the mammalian genome is responsible for the generation of a large variety

Cite this article as *Cold Spring Harb Perspect Biol* doi: 10.1101/cshperspect.a029561

of divergent cell types, complex regulatory circuits for gene expression are required. Therefore, predicting the effects of individual modifications on development and function of a multicellular organism is difficult.

Next, on a practical level, mammalian genomes are more difficult to modify than microbial genomes. Aside from larger genome sizes, mammalian DNA is more complex and repetitive compared with microbial genomes making synthesis and assembly more problematic. Mammalian genome engineering has accelerated as the efficiency and cost of DNA synthesis has decreased, and the size of constructs that can be produced has increased. More advanced assembly methods (Casini et al. 2015) and the use of intermediate cells such as yeast (Jakočiūnas et al. 2015) further facilitate large genome fragment construction. However, as DNA sizes increase, the isolation of intact constructs and methods for efficient delivery into cells remain limiting. The delivery of large constructs by fusion with yeast has been described, but they risk contamination from the donor cell genome (Li and Blankenstein 2013). Artificial mammalian chromosomes have been constructed (Katona 2015), but their deployment in animals exhibited inaccurate mitotic division and passage to daughter cells and poor inheritance (Tomizuka et al. 2000).

Synthesis and assembly of mammalian genomic DNA is only a part of the challenge. Once novel DNA is introduced into the cell, it must be inserted at the appropriate genomic site without introducing inappropriate sequences. Engineering of mouse ES cells takes advantage of endogenous homologous recombination activity; however, other mammalian cells have required creation of site-specific double-stranded breaks in the genome and enhancement of recombination. Technologies including zinc-finger nucleases (ZFNs), transcription activator-like effector nucleases (TALENs), and CRISPR-Cas9 have allowed more efficient double-stranded DNA breaks at virtually any place in the genome (Gaj et al. 2013). CRISPR-Cas9, in particular, has created a large amount of excitement because of its ease of use compared with ZFNs and TALENs; however, in all cases, the function is limited to the creation of double-stranded DNA breaks, which, although useful, does not address the low rate of homologous recombination. Ongoing efforts seek to improve homologous recombination or circumvent these limitations with sequence-specific recombination technologies, such as recombinases and integrases (Brown et al. 2011; Turan et al. 2013).

In summary, the application of synthetic biology approaches to cell and organ generation, modification, and function has enhanced the efficiency and complexity of tissue and genome engineering, rapidly advancing the field of transplantation.

FOCUS AREAS TO IMPROVE CELL AND ORGAN TRANSPLANTATION

Transplantation is an extremely well-researched area of medicine, with many rejection mechanisms defined at the cellular and molecular level. Although this provides a broad set of pathways and targets for intervention, the scale and complexity of the modifications required have made engineering of cells and organs for transplantation extremely difficult. Most efforts have focused on either ablation of immunogenic epitopes or random integration of constitutively expressed transgenes to control rejection responses. Recent improvements in synthetic biology toolsets now make large-scale mammalian genome modification feasible.

Although the immune mechanisms involved are similar, overall rejection responses to cells and organs differ. The ability to produce and transplant relatively uniform cell populations using less-invasive technologies results in less trauma and fewer immune stimuli compared with organ transplantation involving more damaging surgeries, which activates more complex rejection responses. Therefore, the engineering required to improve survival of cells is somewhat different than for organs in transplantation. There are several broad systems to be addressed for transplantation.

Reduction of Immunogenicity

Rejection responses in the absence of pharmacologic intervention can be extremely rapid, with complete destruction of donor tissue in just minutes. Even syngeneic cells can be immunogenic, as shown in experiments with induced pluripotent stem cells (iPSCs) (Sackett et al. 2016). A significant number of effective pharmacologic agents have been developed for the prevention of rejection responses (Allison 2016). Although drug treatment has made kidney, liver, and heart transplants almost routine procedures, the risk of rejection is ever-present, and long-term immunosuppression increases the chance of infection and cancer. The removal of immunogenic donor epitopes, or the prevention of their recognition by the immune system, reduces the risk of rejection of the donor tissue while retaining normal immune surveillance and function.

Elimination or Control of Rejection Mechanisms

Nearly all aspects of the innate and adaptive immune responses as well as related inflammation and coagulation pathways must be addressed in transplantation. Because these systems are interdependent and synergistic, the control of key regulatory points may have broader effects upon multiple responses. For example, inflammation is linked to innate immunity and contributes to complement and coagulation activation. The tissue damage incurred releases immune stimulators and antigens, further activating innate and adaptive immunity. Improved control early in rejection limits the damage of the donor cells and tissues, and thus reduces the risk of synergistic activation of pathways that lead to rejection.

Induction of Durable Tolerance

Immunogenicity of cells and organs can be altered depending on the form of antigen presentation, the microenvironment, and regulatory molecules to induce or suppress responses to a given antigen. In addition, specialized cell populations, such as regulatory T cells (Tregs), can reduce or eliminate inappropriate autoimmune reactions. Several approaches take advantage of existing regulatory mechanisms to create durable tolerance to donor tissues. The potential for tolerance is particularly attractive given the undesirable side effects of long-term immunosuppression and associated risks of infection and tumorigenesis.

Enhancement of Function

The ability to carry out large-scale genome engineering allows the possibility of improving donor cell and organ functions beyond simple replacement of defective host tissues. This could be as simple as providing optimal genetic isoforms or as complex as novel engineered signaling and response pathways. Desirable characteristics such as cell activation, drug resistance, and inducibility of gene expression are all possible.

CELL TRANSPLANTATION

Transplantation of cells is more straightforward than of organs, because of lower complexity and simpler delivery. Blood transfusion has been used for hundreds of years, and bone marrow transplantation is an almost routine clinical procedure. As protocols for production and modification of therapeutic cell populations have improved, the variety of potential applications has expanded and increased the need for mechanisms to overcome rejection.

Hematopoietic Cells

Bone marrow and hematopoietic stem cell transplantation has been extremely successful creating durable functional replacement of host immune system (Pagano and Lyon 2014). Culture and modification of hematopoietic stem cells (HSCs) in vitro allows rational engineering of the immune system. A recent example is the use of genome editing to eliminate the expression of CCR5, a receptor for HIV, which controls viral infection in animal models (Holt et al. 2010). Other approaches seek to introduce broadly neutralizing antibodies to HIV, or other pathogens for which vaccination is ineffective, into HSCs for expression by the immune system (Luo et al. 2009). Because it is possible to use patient-derived HSC for modification and reintroduction, the risk of rejection is significantly lower than with allogeneic HSCs.

β Islet Cells

The replacement of β islets in diabetes has been a high-priority application of cell transplantation. Many protocols using human- and animal-derived islets have been developed for transplant, albeit with limited success because of allo- and xenogeneic rejection responses. Encapsulation of β islets in matrices that allow the passage of smaller molecules such as insulin and glucagon, while limiting access to the immune system, have been partially effective (Ichii and Ricordi 2009). Recent data have shown that conversion of patient-derived iPSC to create syngeneic β islets may overcome rejection responses (Quiskamp et al. 2015).

Stem Cells

Nonhematopoietic cell transplantation has been more challenging to develop in part because many terminally differentiated cell types are not easily obtained or cultured. Transdifferentiation, direct conversion of one differentiated cell type into another, sought to overcome these limitations (Wang et al. 2015); however, not all cell types are amenable to this approach, because the regulatory genes necessary for conversion are not yet known.

The identification and development of human ES cells has enhanced the creation of a number of functional cell types and organoids (Girlovanu et al. 2015). Because the use of fetal cells carries significant ethical concerns, adult stem cells and, more recently, patient-derived iPSCs have become a greater focus. Differentiation of adult stem cells and iPSCs to heart, kidney, liver, and lung tissue has been very successful; however, the production of full-size human organs has not yet been realized, because of the lack of spatial cues and appropriate morphogenesis during embryonic development, and the complexity of multicellular tissues.

Despite being syngeneic with the host, iPSCs may still be immunogenic in vivo. Generation and culture of iPSCs leads to genomic mutations and epigenetic changes that may contribute to immunogenicity (Cao et al. 2014). Immunization data show that even completely conserved proteins can elicit a response; therefore, it may not be surprising that much more complex albeit syngeneic cells may also do so.

Immunotherapy

One of the most exciting applications of cell transplantation is in the area of cancer immunotherapy. Tumor studies show a significant intratumoral infiltration of immune cells, including T cells specific for tumor-derived peptides. Early studies isolated, expanded, activated, and reintroduced these tumor-infiltrating lymphocytes (TILs) into patients but met with limited success (Schober and Busch 2016). An improvement to this approach targets patient-derived cells with tumor-specific T-cell receptors (TCRs) with enhanced activity to improve antitumor activities (Kessels et al. 2001). More recently, chimeric antigen receptors (CARs), consisting of high-affinity binding domains to target tumor cell surface antigens linked to signaling domains that activate T-cell antitumor functions, have shown superior activity when introduced into patient T cells and used to treat tumors. The stunning successes of CAR-T cell approaches have led to further variations, including improved CAR-binding specificity and signaling, enhancement of T-cell effector mechanisms as well as expansion to alternate cell types, such as natural killer (NK) cells. The successes of CAR-T cells have mainly been limited to liquid tumors, such as leukemias and lymphomas, which possess highly specific surface targets. Solid tumors present a greater challenge because of the many immunosuppressive mechanisms used and greater target heterogeneity. Although synthetic biology approaches are being applied in many aspects of cancer immunotherapy, this topic has been extensively reviewed elsewhere (Chakravarti et al. 2016).

ORGAN TRANSPLANTATION

The transplantation of organs is more complex than the transplantation of cells, and even individual organs vary in their complexity. Heart, kidney, and liver transplantations are well developed clinically with extended survival times. Organs such as the lung are more challenging because of their greater sensitivity to trauma, shorter survival time, and more frequent complications. Although robustness is important to transplantation success, availability is the more critical unmet need. As clinical technologies have steadily improved in the past decades, the availability of donor organs has only slightly increased. Therefore, the gap between need and availability continues to widen.

Synthetic biology helps to address the critical lack of organs through both tissue-engineering strategies in vitro and in vivo to create organs and the use of complex genome engineering to address the host rejection mechanisms. Because of the rapid advancements in the synthetic biology tools available and better understanding of key rejection mechanisms, these approaches could not only meet the need for organs but eventually surpass human organs in function and survival.

Organ Differentiation

The ability to recreate embryonic development in the laboratory has been a focus of biological study for many years, leading to the possible generation of organs suitable for transplantation entirely in vitro (Rustad et al. 2010). Although a number of species can develop in culture, the complexity of mammalian development has been more difficult to reproduce. Some progress has been made toward production of functional artificial wombs, but application to humans is unlikely for some time and subject to significant ethical concerns (Lupton 1997). Relatively more progress has been made in the creation of organoids and organ-like structures in vitro, with the generation of increasingly complex structures and functions (Dye et al. 2016). The use of various natural and synthetic matrices allows the assembly of larger structures, which, combined with iPSC technologies, would be compatible with patients. However, without the structural support and cues provided by embryonic morphogenesis, the manufacture of functional adult organs entirely in vitro may be problematic.

Organ Printing

The use of 3D-printing technologies allows the deliberate printing of the final organ structure, overcoming the slower process of embryonic morphogenesis and thus creating organs tailored specifically for the patient (Patra and Young 2016). Combination of natural and synthetic scaffolds with the desired cell types in one process further improves the efficiency of production. This approach has been remarkably successful for more uniform and structural tissues such as bone and trachea (Fedorovich et al. 2010; Fishman et al. 2014). Attempts at creating more complex organ structures such as kidney or lung have proven more challenging, requiring accurate interactions between divergent cell types for function. For some organs, such as the lung, more than 40 cell types have been identified that require specific spatial positioning and cell–cell interactions to function properly (Nichols et al. 2014). As the sophistication and precision of 3D-printing techniques improves, organs of higher complexity will be possible.

Decellularization–Recellularization

An alternate approach to the generation of organs and 3D printing in vitro is to rebuild the organ structure from the genuine underlying noncellular scaffold, using cells compatible with the patient. This takes advantage of normal mechanisms of organ regeneration observed in recovery from injury, albeit at a larger scale and complexity. Starting with an existing organ, all of the cells are extracted using mild enzymatic and detergent treatments, leaving the decellularized structural elements (Song and Ott 2011). To this matrix are added new cells derived from, or compatible with, the patient, which are allowed to recellularize the structure, recreating the functional organ. Although this may allow the creation of compatible organs, there is still a requirement for the underlying matrix that needs to be addressed and presumably could be resolved with the use of porcine organs. The advent of iPSC technology provides an avenue to produce the cells necessary of recellularization. However, as with 3D printing, more complex organs require cells of many different types, not all of which are defined. In addition, embryonic morphogenesis may be required for some facets of organ structure and function that recellularization cannot effectively reproduce.

Nonhuman Blastocyst Complementation

A limitation for producing organs in vitro is the challenge of creating the appropriate structures and functions without following normal embryonic morphogenesis. One route to overcome this limitation is the use of human stem cells to complement defective organ production in a large animal species, such as pig (Nagashima and Fukuzawa 2016). In this approach, pigs are genetically modified to prevent production of a particular organ, with human stem cells introduced at the blastocyst stage. As the embryo develops, the lineages that cannot be produced from host animal cells are replaced by the human stem cells, with a goal of creating a fully functional human organ in the animal using the proper developmental pathways (Wu and Izpisua Belmonte 2016). A potential limitation of this approach is the difficulty in producing an organ entirely derived from human cells to avoid the xenogeneic rejection response to chimeric human–animal organs. Similarly, the use of allogeneic human cells may contribute to rejection responses. Although patient-derived iPSC may be used, the time it would take to generate the functional organ may be limiting. In addition, the risk of human cells contributing to other lineages within the animal, such as neurons, creates ethical concerns that need to be addressed.

Hybrid Methods

Although the above approaches are described separately, it is possible to combine technologies to arrive at more satisfactory solutions. For example, 3D-printed or decellularized–recellularized organs could be introduced into patients at a heterotopic site for a period of time to allow full regeneration, before orthotopic transplantation. Although theoretically advantageous, particularly if the organ was generated from patient-derived cells or iPSCs, the risk incurred by multiple procedures as well as the immediacy of the need of many patients makes the practical application challenging.

TRANSPLANTATION REJECTION MECHANISMS

Outside of transplantation between identical twins, all cell and organ transplants are at risk of immune rejection. The greater the number of genetic differences, the greater the chance of a host immune response, thus allogeneic organs create less severe reactions than xenogeneic organs.

The human immune response to porcine organs starts with hyperacute rejection, because of significant preexisting titers of antiporcine antibodies inducing complement-mediated destruction of the tissue. This is followed by acute cellular and vascular rejection, also unique to xenotransplantation (Auchincloss and Sachs 1998). Chronic rejection is common to both allo- and xenotransplant, and is more readily controlled by existing pharmacological treatments (Scherer et al. 2007).

The synergies among the various aspects of the immune response lead to rapid amplification; therefore, the treatment of rejection is extremely difficult once a response begins. The goals for genome-engineering strategies are to prevent the initiation of rejection while blunting any amplification once it occurs. Synthetic biology approaches allow the modification of multiple pathways simultaneously, including altering immunogenicity, immune regulation, and tolerance with rational design and multigenic changes. Examples of the systems that must be addressed are described below.

Histocompatibility

The immune system efficiently recognizes differences in protein sequence through the interaction of TCRs with peptides in complex with histocompatibility proteins (HLA). In transplantation, TCR–HLA interactions play a role in both direct and indirect antigen presentation. Direct antigen presentation is host T-cell recognition of donor HLA (or SLA in the case of porcine cells) complexed with donor peptide or donor HLA alone due to the lack of negative thymic selection of host T cells against the foreign haplotypes. Indirect presentation is due to the uptake of donor proteins by host antigen-presenting cells and display by host HLA to host T cells (Gould and Auchincloss 1999). The combination of direct and indirect responses results in cytolytic and helper T-cell activation and destruction of the transplanted cells and organs. Although the use of patient-derived cells or iPSCs overcomes many of the challenges of antigen presentation, use of allo- and xenogeneic cells provides increasingly large numbers of immunogenic epitopes that generate T-cell responses.

Modification of donor HLA to match host haplotypes is a potential route to eliminating peptide-independent direct responses of allo- and xenogeneic HLA. This would require rapid production of organs with patient-specific HLA haplotypes, which would not be practical for xenotransplantation but could be contemplated for in vitro organ production. However, this may have the unintended effect of enhancing the display of the immunogenic donor peptides by the modified donor cells, actually increasing host T-cell activity against the donor tissue. A complementary approach uses genome engineering to eliminate or reduce the expression of donor HLA and preventing antigen presentation (Hara et al. 2013). Although this approach reduces immunogenicity of donor cells, it increases the risk of infection and leads to "missing self" recognition by NK cells (van Bergen et al. 2009). Inhibitory receptors expressed by NK cells interact with cell surface HLA, preventing NK cells from targeting the cells. Therefore, ablation of donor HLA expression activates host NK cells causing donor cell destruction. Expression of ligands for NK inhibitory receptors, such as HLA-E or -G, will preclude NK cell activation (Crew 2007) and offset the elimination of antigen-presenting HLA from donor cells.

Eliminating antigen presentation also risks the inability to detect infection; therefore, a better option may be to edit immunogenic epitopes. Although CRISPR-Cas9 editing is useful for making small changes at multiple loci, the number of genetic changes required to target immunogenic loci in allotransplant is immense, with an even greater number for xenotransplant. The increased number of targets also increases risk of off-target mutations potentially creating unpredictable mutations.

Alteration of immunogenic sequences to match host sequences may be preferred over eliminating the expression of immunogenic loci, to avoid the loss of essential functional genes. However, changing coding sequences in simultaneous multiplexed CRISPR-Cas9 protocols is much less efficient than simply ablating expression, limiting the utility of this approach. Better prediction of immunogenicity through in silico modeling and in vitro assays could help prioritize required changes by focusing on the most highly reactive sequences. Unfortunately, the reliability of current prediction methods is insufficient for this task (Luo et al. 2015).

Donor-Specific Antibodies

The recognition of donor antigens by preexisting antibodies in host serum leads to hyperacute rejection via complement-mediated cell destruction of donor cells. One simple approach to reducing antibody-mediated mechanisms of rejection is to ablate the antigenic genes, or the pathways that create the antigens, in the case of glycan or other nonproteinaceous epitopes. Elimination of porcine xenoantigens has been successfully demonstrated with the "Gal" epitope (Gal α1–3Gal β1–(3)4GlcNAc-R), a porcine glycan residue that is one of the most reactive xenoantigens in humans (Galili 1999). Ablation of the transferase gene responsible for adding the Gal epitope to proteins significantly reduces the recognition of porcine cells by human serum (Sandrin et al. 1994). Several additional human-reactive porcine xenoantigens have been defined, such as Neu5Gc and B4GALNT2 (Byrne et al. 2015). It is likely that the elimination of additional xenoantigens may further reduce human antibody-mediated destruction of porcine cells; however, with each modification, there is a risk of eliminating key functions in porcine development or physiology.

Complement

Antibody-dependent cell-mediated cytotoxicity (ADCC) is due to the recognition of donor cell epitopes by natural and induced antibodies, resulting in the activation of the complement cascade and cell death (Sheen and Heeger 2015). The large diversity of antibody specificities can lead to inadvertent autoreactivity with cells in vivo; therefore, complement regulatory proteins (CRPs) are expressed that set a threshold for activation of ADCC to prevent accidental targeting (Mathern and Heeger 2015).

Human serum has significant titers of antibodies directed against porcine antigens and thus xenotransplanted cells are at high risk of antibody recognition and ADCC compared with allogeneic human cells (Schroeder et al. 2005). Furthermore, incompatibilities between porcine CRPs and their human counterparts in the complement cascade prevent appropriate regulation of complement activation leading to ADCC-derived damage to porcine tissues. The overexpression of human CRPs helps to regulate ADCC activity and has been a focus of transgenic animals for xenotransplantation (Miyagawa et al. 2010). The elimination of highly reactive antigens as described above, in combination with the expression of human CRPs, further reduces the risk of tissue destruction.

More precise methods of targeting in porcine cells allow the insertion of human CRPs under the control of the homologous porcine genes; however, normal levels of expression may be insufficient in xenotransplantation. Given the relatively greater number of antigens recognized by human antibodies, a higher level of CRP expression may be required compared with normal human tissue, and thus more dynamic methods of control of CRP expression may be required. One caution, which applies to manipulation of any porcine pathways, is that the endogenous porcine complement must be preserved to allow proper development and function of the porcine organ before removal for xenotransplantation; therefore, any engineering must take into account improvements transplantation in humans as well as maintaining proper function while in the pig.

Ischemia/Reperfusion Injury

Even before an organ is transplanted, it is subject to trauma during removal, preservation, and transport to the donor and subsequent surgery, leading to ischemia/reperfusion injury (I/RI). Initially, the loss of oxygenation causes interruption of metabolism and accumulation of metabolic intermediates. Later reintroduction of oxygenated blood can be even more damaging, overwhelming the antioxidant system and injuring the endothelium (Lutz et al. 2010). This induces expression of proinflammatory mediators and chemokines, followed by recruitment of neutrophils, macrophages, and other innate immune cells, rapidly causing tissue destruction and necrosis. The widespread inflammation also causes activation of the coagulation cascade and related thrombosis events.

Inflammation

In addition to I/RI mechanisms described above, the inflammatory process is reinforced by host innate immune responses. Damaged cells release innate immune activators, attracting cells to the site of the transplanted organ (Goldstein 2006). The influx of neutrophils, monocytes, macrophages, and other cells into the organ causes the release of additional proinflammatory mediators, resulting in further amplification of inflammation. Without resolution, the inflammation continues to damage tissue, continuing the cycle of activation and amplification, and inducing adaptive immune responses (Klose and Artis 2016). Unlike inflammation caused by infectious diseases, which can be resolved once the infectious agent is eliminated, the cells of the organ provide ongoing stimuli to the inflammatory response.

 Cite this article as *Cold Spring Harb Perspect Biol* doi: 10.1101/cshperspect.a029561

Inflammation is a major challenge in allo- and xenotransplantation regardless of source. Although transgenes have been introduced into the porcine genome to offset inflammatory responses, a greater number and variety of mechanisms need to be addressed (Ezzelarab and Cooper 2015a). Because a general reduction or elimination of inflammatory responses can create a risk of infection, a better solution is to tailor the engineering to limit the overreaction of rejection responses, while maintaining appropriate infectious disease responses. Creating genetic circuits to respond to inflammation in an inducible, self-limited manner will mimic the kinetics of a normal response, reducing pathogenic inflammation while maintaining infectious disease surveillance.

Coagulation

Activation of the coagulation cascade is a consequence of the inflammatory responses described above, leading to disseminated intravascular coagulation and thrombosis, rapid loss of platelet and clotting factors, decreased blood flow and ischemia/infarction (Schmelzle et al. 2010; Ezzelarab et al. 2015b). As with inflammation, coagulation is rapidly amplified after initiation and difficult to regulate once widespread. Similar to the complement cascade, there are a number of regulatory points in coagulation that, like CRPs, are not completely conserved between porcine and human systems and contribute to uncontrolled coagulation (Lin and Cooper 2009).

One route to improve coagulation regulation in xenotransplantation has been the overexpression of human coagulation regulators. Although there have been reports of improved coagulation regulation (Cowan et al. 2015), overexpression of the regulators can cause dysfunctions and death of modified pigs due to bleeding disorders. Improved expression constructs and integration of human regulator genes in place of the porcine genes may improve the defective phenotypes. Better regulation of expression through improved promoters and enhancers may be required. One challenge will be to implement appropriate post-xenotransplantation regulation while not disrupting endogenous porcine coagulation during pig development and growth.

Hematopoietic Chimerism

A particularly exciting approach to overcome transplant rejection is hematopoietic chimerism, the functional transplantation of donor hematopoietic cells, with or without secondary lymphoid tissues, to create durable immunological tolerance of the host to the donor antigens (Sykes 2007). This could significantly reduce, or even bypass, much of the rejection response and eliminate the requirement for long-term immunosuppression. Animal studies support the effectiveness of hematopoietic chimerism in allo- and xenotransplant tolerance (Vagefi et al. 2015). A key question is whether chimeric immune systems require central tolerance, peripheral tolerance, or both to be effective. In the former case, it may be essential to incorporate donor thymus tissue to allow reciprocal education of host and donor T cells (Barth et al. 2003). In the latter case, the ongoing production of potentially reactive T cells may require pharmacologic or other therapeutic treatments. A better understanding of the immunological mechanisms critical to establish and maintain tolerance will help establish protocols for clinical applications.

FUTURE CHALLENGES

The application of synthetic biology tools requires a better understanding of the genes affecting rejection to be fully effective. For example, recently published data regarding the engineering of a cell with a minimized genome shows that many of the genes required for life are of unknown function (Hutchison et al. 2016). The number of genes of unknown function will be much greater in mammalian cells and more challenging to define. More complex analytical tools are required for more precise definition of genetic mechanisms and cellular pathways. Similarly, understanding the regulation of the large number of interrelated genes must also improve to allow better prediction of effects of genome modification upon gene networks. Significant progress in genomics, transcriptomics, proteomics, metabolomics, immunomics, and other systems biology approaches is accelerating the development of essential data sets and predictive algorithms for more rational engineering of cells.

Thus far, much of the focus of genome engineering in transplantation has been primarily devoted to ablation and humanization of existing genes. The advancement of synthetic biology tools now allows design and implementation of completely novel genes and pathways to improve cell functions. Orthogonal signaling systems, unique metabolic pathways, engineered proteins, and many other novel solutions can now be applied in mammalian cells to enhance transplant survival and function.

Although mammalian genome engineering has accelerated greatly, new and better tools are still required to make further improvements. More efficient, sequence-specific DNA recombination, more uniform and higher editing rates, and reduced off-target editing are all highly desirable.

At the cell and tissue levels, more effective in vitro methods for functional in vitro organ production could overcome the need for in vivo methods such as blastocyst complementation or porcine organ engineering, and greatly speed production. The combination of patient-specific cells and in vitro organ production could avoid the need for sophisticated genome engineering entirely.

CONCLUSIONS

Synthetic biology is opening the door to approaches for cell and organ transplantation that would have been impossible even just a few years ago. Methods to produce compatible cells and organs in vitro and in vivo combined with increasingly rapid and precise methods for genome engineering have rapidly advanced (Mou et al. 2015). Synthetic biology technologies not only allow the large-scale changes necessary to address the many issues critical donor acceptance and tolerance but could lead to organs superior in function to human organs.

ACKNOWLEDGMENTS

Thanks to Guliang Xia, Bhaskar Thyagarajan, Adam Corper and the Mammalian Synthetic Biology team at Synthetic Genomics, Inc., Martine Rothblatt, David Ayares, Carol Phelps, Marc Lorber, and Jordan Shin of United Therapeutics, Craig Venter, Oliver Fetzer, Todd Peterson, Toby Richardson, Dan Gibson, Ariel Schwartz, Leah Soriaga at Synthetic Genomics, Megan Sykes, David Sachs, Richard N. Pierson III, and Agnes Azimzadeh.

REFERENCES

Allison TL. 2016. Immunosuppressive therapy in transplantation. *Nurs Clin North Am* **51:** 107–120.

Auchincloss H Jr, Sachs DH. 1998. Xenogeneic transplantation. *Annu Rev Immunol* **16:** 433–470.

Barth RN, Yamamoto S, LaMattina JC, Kumagai N, Kitamura H, Vagefi PA, Awwad M, Colvin RB, Cooper DK, Sykes M, et al. 2003. Xenogeneic thymokidney and thymic tissue transplantation in a pig-to-baboon model: I. Evidence for pig-specific T-cell unresponsiveness. *Transplantation.* **75:** 1615–1624.

Brown WR, Lee NC, Xu Z, Smith MC. 2011. Serine recombinases as tools for genome engineering. *Methods* **53:** 372–379.

Byrne GW, McGregor CG, Breimer ME. 2015. Recent investigations into pig antigen and anti-pig antibody expression. *Int J Surg* **23:** 223–228.

Cao J, Li X, Lu X, Zhang C, Yu H, Zhao T. 2014. Cells derived from iPSC can be immunogenic—Yes or no? *Protein Cell* **5:** 1–3.

Casini A, Storch M, Baldwin GS, Ellis T. 2015. Bricks and blueprints: Methods and standards for DNA assembly. *Nat Rev Mol Cell Biol* **16:** 568–576.

Chakravarti D, Cho JH, Weinberg BH, Wong NM, Wong WW. 2016. Synthetic biology approaches in cancer immunotherapy, genetic network engineering, and genome editing. *Integr Biol (Camb)* **8:** 504–517.

Cowan PJ, Robson SC. 2015. Progress towards overcoming coagulopathy and hemostatic dysfunction associated with xenotransplantation. *Int J Surg.* **23:** 296–300.

Crew MD. 2007. Play it in E or G: Utilization of HLA-E and -G in xenotransplantation. *Xenotransplantation* **14:** 198–207.

Dye BR, Miller AJ, Spence JR. 2016. How to grow a lung: Applying principles of developmental biology to generate lung lineages from human pluripotent stem cells. *Curr Pathobiol Rep* **4:** 47–57.

Ezzelarab MB, Cooper DK. 2015a. Systemic inflammation in xenograft recipients (SIXR): A new paradigm in pig-to-primate xenotransplantation? *Int J Surg* **23:** 301–305.

Ezzelarab MB, Ekser B, Azimzadeh A, Lin CC, Zhao Y, Rodriguez R, Echeverri GJ, Iwase H, Long C, Hara H, et al. 2015b. Systemic inflammation in xenograft recipients precedes activation of coagulation. *Xenotransplantation* **22:** 32–47.

Fedorovich NE, Alblas J, Hennink WE, Oner FC, Dhert WJ. 2011. Organ printing: The future of bone regeneration? *Trends Biotechnol* **29:** 601–606.

Fishman JM, Wiles K, Lowdell MW, De Coppi P, Elliott MJ, Atala A, Birchall MA. 2014. Airway tissue engineering: an update. *Expert Opin Biol Ther* **14:** 1477–1491.

Gaj T, Gersbach CA, Barbas CF III. 2013. ZFN, TALEN, and CRISPR/Cas-based methods for genome engineering. *Trends Biotechnol* **31:** 397–405.

Galili U. 1999. Significance of anti-Gal IgG in chronic xenograft rejection. *Transplant Proc* **31:** 940–941.

Girlovanu M, Susman S, Soritau O, Rus-Ciuca D, Melincovici C, Constantin AM, Mihu CM. 2015. Stem cells—Biological update and cell therapy progress. *Clujul Med* **88:** 265–271.

Goldstein DR. 2006. Toll like receptors and acute allograft rejection. *Transpl Immunol* **17:** 11–15.

Gould DS, Auchincloss H Jr. 1999. Direct and indirect recognition: The role of MHC antigens in graft rejection. *Immunol Today* **20:** 77–82.

Hara H, Witt W, Crossley T, Long C, Isse K, Fan L, Phelps CJ, Ayares D, Cooper DK, Dai Y, Starzl TE. 2013. Human dominant-negative class II transactivator transgenic pigs—Effect on the human anti-pig T-cell immune response and immune status. *Immunology* **140:** 39–46.

Holt N, Wang J, Kim K, Friedman G, Wang X, Taupin V, Crooks GM, Kohn DB, Gregory PD, Holmes MC, et al. 2010. Human hematopoietic stem/progenitor cells modified by zinc-finger nucleases targeted to CCR5 control HIV-1 in vivo. *Nat Biotechnol* **28:** 839–847.

Hutchison CA III, Chuang RY, Noskov VN, Assad-Garcia N, Deerinck TJ, Ellisman MH, Gill J, Kannan K, Karas BJ, Ma L, et al. 2016. Design and synthesis of a minimal bacterial genome. *Science* **351:** aad6253.

Ichii H, Ricordi C. 2009. Current status of islet cell transplantation. *J Hepatobiliary Pancreat Surg* **16:** 101–112.

Jakočiūnas T, Rajkumar AS, Zhang J, Arsovska D, Rodriguez A, Jendresen CB, Skjødt ML, Nielsen AT, Borodina I, Jensen MK, et al. 2015. CasEMBLR: Cas9-facilitated multiloci genomic integration of in vivo assembled DNA parts in *Saccharomyces cerevisiae*. *ACS Synth Biol* **4:** 1226–1234.

Katona RL. 2015. De novo formed satellite DNA-based mammalian artificial chromosomes and their possible applications. *Chromosome Res* **23:** 143–157.

Kessels HW, Wolkers MC, van den Boom MD, van der Valk MA, Schumacher TN. 2001. Immunotherapy through TCR gene transfer. *Nat Immunol* **2:** 957–961.

Klose CS, Artis D. 2016. Innate lymphoid cells as regulators of immunity, inflammation and tissue homeostasis. *Nat Immunol* **17:** 765–774.

Li L, Blankenstein T. 2013. Generation of transgenic mice with megabase-sized human yeast artificial chromosomes by yeast spheroplast-embryonic stem cell fusion. *Nat Protocol* **8:** 1567–1582.

Lin CC, Cooper DK, Dorling A. 2009. Coagulation dysregulation as a barrier to xenotransplantation in the primate. *Transpl Immunol* **21:** 75–80.

Luo H, Ye H, Ng HW, Shi L, Tong W, Mendrick DL, Hong H. 2015. Machine learning methods for predicting HLA-peptide binding activity. *Bioinform Biol Insights* **9:** 21–29.

Luo XM, Maarschalk E, O'Connell RM, Wang P, Yang L, Baltimore D. 2009. Engineering human hematopoietic stem/progenitor cells to produce a broadly neutralizing anti-HIV antibody after in vitro maturation to human B lymphocytes. *Blood* **113:** 1422–1431.

Lupton ML. 1997. Artificial wombs: Medical miracle, legal nightmare. *Med Law* **16:** 621–633.

Lutz J, Thürmel K, Heemann U. 2010. Anti-inflammatory treatment strategies for ischemia/reperfusion injury in transplantation. *J Inflamm (Lond)* **7:** 27.

Mathern DR, Heeger PS. 2015. Molecules great and small: The complement system. *Clin J Am Soc Nephrol* **10:** 1636–1650.

Miyagawa S, Yamamoto A, Matsunami K, Wang D, Takama Y, Ueno T, Okabe M, Nagashima H, Fukuzawa M. 2010. Complement regulation in the GalT KO era. *Xenotransplantation* **17:** 11–25.

Mou L, Chen F, Dai Y, Cai Z, Cooper DK. 2015. Potential alternative approaches to xenotransplantation. *Int J Surg* **23:** 322–326.

Murphy SV, Atala A. 3D bioprinting of tissues organs. *Nat Biotech* **32:** 773–785.

Pagano L, Lyon S. 2014. Celebrating 40 years of progress in bone marrow transplantation: A report from the 40th Annual Meeting of the European Society for Blood and Marrow Transplantation. *Future Microbiol* **9:** 1117–1121.

Nagashima H, Matsunari H. 2016. Growing human organs in pigs—A dream or reality? *Theriogenology.* **86:** 422–426.

Nichols JE, Niles JA, Vega SP, Argueta LB, Eastaway A, Cortiella J. 2014. Modeling the lung: Design and development of tissue engineered macro- and micro-physiologic lung models for research use. *Exp Biol Med (Maywood)* **239:** 1135–1169.

Patra S, Young V. 2016. A review of 3D printing techniques and the future in biofabrication of bioprinted tissue. *Cell Biochem Biophys* **74:** 93–98.

Quiskamp N, Bruin JE, Kieffer TJ. 2015. Differentiation of human pluripotent stem cells into β-cells: Potential and challenges. *Best Pract Res Clin Endocrinol Metab* **29:** 833–847.

Rustad KC, Sorkin M, Levi B, Longaker MT, Gurtner GC. 2010. Strategies for organ level tissue engineering. *Organogenesis* **6:** 151–157.

Sackett SD, Brown ME, Tremmel DM, Ellis T, Burlingham WJ, Odorico JS. 2016 Modulation of human allogeneic and syngeneic pluripotent stem cells and immunological implications for transplantation. *Transplant Rev (Orlando)* **30:** 61–70.

Sandrin MS, McKenzie IF. 1994. Gal α (1,3)Gal, the major xenoantigen(s) recognised in pigs by human natural antibodies. *Immunol Rev* **141:** 169–190.

Scherer MN, Banas B, Mantouvalou K, Schnitzbauer A, Obed A, Krämer BK, Schlitt HJ. 2007. Current concepts and perspectives of immunosuppression in organ transplantation. *Langenbecks Arch Surg* **392:** 511–523.

Schmelzle M, Schulte Esch J II, Robson SC. 2010. Coagulation, platelet activation and thrombosis in xenotransplantation. *Curr Opin Organ Transplant* **15:** 212–218.

Schober K, Busch DH. 2016. TIL 2.0: More effective and predictive T-cell products by enrichment for defined antigen specificities. *Eur J Immunol* **46:** 1335–1339.

Schroeder C, Allan JS, Nguyen BN, Wu G, Zhang T, Azimzadeh AM, Madsen JC, Schuurman HJ, Sachs DH, Pierson RN III. Hyperacute rejection is attenuated in GalT knockout swine lungs perfused ex vivo with human blood. *Transplant Proc* **37:** 512–513.

Sheen JH, Heeger PS. 2015. Effects of complement activation on allograft injury. *Curr Opin Organ Transplant* **20:** 468–475.

Song JJ, Ott HC. 2011. Organ engineering based on decellularized matrix scaffolds. *Trends Mol Med* **17:** 424–432.

Sykes M. 2007. Immune tolerance: mechanisms and application in clinical transplantation. *J Intern Med* **262:** 288–310.

Tomizuka K, Shinohara T, Yoshida H, Uejima H, Ohguma A, Tanaka S, Sato K, Oshimura M, Ishida I. 2000. Double trans-chromosomic mice: Maintenance of two individual human chromosome fragments containing Ig heavy and κ loci and expression of fully human antibodies. *Proc Natl Acad Sci* **97:** 722–727.

Turan S, Zehe C, Kuehle J, Qiao J, Bode J. 2013. Recombinase-mediated cassette exchange (RMCE)—A rapidly-expanding toolbox for targeted genomic modifications. *Gene* **515:** 1–27.

Vacanti J. 2010. Tissue engineering and regenerative medicine: From first principles to state of the art. *J Pediatr Surg* **45:** 291–294.

Vagefi PA, Shah JA, Sachs DH. 2015. Progress towards inducing tolerance of pig-to-primate xenografts. *Int J Surg* **23:** 291–295.

van Bergen J, Thompson A, Retière C, Trowsdale J, Koning F. 2009. Cutting edge: Killer Ig-like receptors mediate “missing self” recognition in vivo. *J Immunol* **182:** 2569–2572.

Wang H, Li X, Gao S, Sun X, Fang H. 2015. Transdifferentiation via transcription factors or microRNAs: Current status and perspective. *Differentiation* **90:** 69–76.

Wu J, Izpisua Belmonte JC. 2016. Interspecies chimeric complementation for the generation of functional human tissues and organs in large animal hosts. *Transgenic Res* **25:** 375–384.

 Cite this article as *Cold Spring Harb Perspect Biol* doi: 10.1101/cshperspect.a029561

Synthetic Morphogenesis

Brian P. Teague, Patrick Guye, and Ron Weiss

Synthetic Biology Center, Department of Biological Engineering, Massachusetts Institute of Technology, Cambridge, Massachusetts 02139

Correspondence: rweiss@mit.edu

Throughout biology, function is intimately linked with form. Across scales ranging from subcellular to multiorganismal, the identity and organization of a biological structure's subunits dictate its properties. The field of molecular morphogenesis has traditionally been concerned with describing these links, decoding the molecular mechanisms that give rise to the shape and structure of cells, tissues, organs, and organisms. Recent advances in synthetic biology promise unprecedented control over these molecular mechanisms; this opens the path to not just probing morphogenesis but directing it. This review explores several frontiers in the nascent field of synthetic morphogenesis, including programmable tissues and organs, synthetic biomaterials and programmable matter, and engineering complex morphogenic systems de novo. We will discuss each frontier's objectives, current approaches, constraints and challenges, and future potential.

What is the underlying basis of biological structure? Speculations were based on macroscopic observation until the 17th century, when Antonie van Leeuwenhoek and Robert Hooke discovered that organisms were composed of microscopic cells (Harris 1999), and that the size and shape of the cells affected the properties of the structures they formed. This hierarchical principle of organization remains a central tenet of the modern study of molecular morphogenesis: higher-level system properties emerge from, and are dependent on, lower-level properties of system components (Davies 2013). Many of these phenomena can be grouped into a small number of recurring motifs or morphogenic "modules," such as spatial information conveyed by a morphogen gradient or the role of programmed cell death in sculpting tissues and organs (Cachat et al. 2014). As the mechanistic underpinnings of these processes are elucidated, opportunities arise to use this knowledge to direct morphogenesis toward novel, useful ends. We call this emerging field of endeavor "synthetic morphogenesis." Inspired by and based on natural morphogenic systems, synthetic morphogenesis seeks to engineer, program, grow, and maintain biological systems with complex structures.

The field of synthetic biology is uniquely positioned to enable these efforts. The key to this synergy is the conception of synthetic biology as an engineering discipline (Arpino et al. 2013), combining traditional engineering concepts, such as reusable parts, modularity, and abstraction, with novel rules specifically suited to engineering biology (Slusarczyk et al. 2012). These concepts have served to propel the field's rapid development. In the last decade, synthetic

Cite this article as *Cold Spring Harb Perspect Biol* doi: 10.1101/cshperspect.a023929

biology has progressed from simple systems in model prokaryotes (Elowitz and Leibler 2000; Gardner et al. 2000) to sophisticated synthetic gene networks in almost every branch of life, including plants (Schaumberg et al. 2015) and mammalian systems both in vitro (Xie et al. 2011) and in vivo (Ye et al. 2011). As the discipline's tools become more powerful, they are beginning to enable the construction of systems that have dynamic behavior and nontrivial emergent properties similar to those of natural morphogenetic processes. Synthetic gene circuits can detect a cell's type (Miki et al. 2015), metabolic state (Callura et al. 2012), (bio-) chemical signals (Weber et al. 2007), and light (Müller et al. 2014; Schmidl et al. 2014); they can use these inputs, combinatorially or sequentially, to alter the cell's shape (Yeh et al. 2007), motility (Park et al. 2014), differentiation program (Wamaitha et al. 2015; Guye et al. 2016), or even kill the cell outright (Xie et al. 2011). Synthetic intercellular signaling allows cell populations to make decisions and coordinate behaviors both locally (Sprinzak et al. 2010; Matsuda et al. 2012) and globally (Tabor et al. 2009; Prindle et al. 2011; Chen et al. 2015b). These diverse sensors, actuators, and communication channels could implement complex morphogenic systems through a combination of top-down approaches in which cells are patterned by external signals, and bottom-up programs in which collective properties emerge through cells' local decision-making.

Although synthetic biology supplies tools to implement synthetic morphogenic systems, the design language is inspired by natural molecular morphogenesis. Metazoan development is a marvel of precise, flexible, autocorrecting self-assembly; as synthetic biology strives to improve synthetic gene circuits' robustness (Brophy and Voigt 2014) and complexity (Moon et al. 2012), natural complex systems inspire and inform best practices. Additionally, the forward design of synthetic morphogenetic systems requires reasoning across levels of scale and complexity. For example, tuning a synthetic tissue's mechanical properties might involve specifying the molecular components (signaling molecules, receptors, and cell–cell adhesion molecules) in individual cells, as well as the organization and interactions of the cells that compose such a tissue. The field of molecular morphogenesis has developed methods specifically for scaling this hierarchical organization, and many of the theoretical and computational tools used to describe natural systems will find ready use in designing synthetic ones.

Because the field of synthetic morphogenesis is in its infancy, we have structured this article as equal parts review and prospectus, exploring three frontiers that we believe will be particularly fruitful. The first section examines programmable tissues and organs for regenerative medicine, based on precise spatiotemporal control over cell-fate decisions enabled by synthetic gene networks. The second section explores the nascent field of "smart" programmable biomaterials and the impact they will have on the fabrication of the objects in our man-made world. The final section looks further afield, envisioning de novo minimal synthetic morphogenic systems built by rewiring natural morphogenic modules in ways outside the reach of evolution. Arranged in order from concrete to speculative, these three frontiers in synthetic morphogenesis promise an unprecedented opportunity to rationally engineer ourselves and the world around us.

FRONTIER ONE: PROGRAMMABLE TISSUES AND ORGANS

The field of regenerative medicine is based on the pursuit of therapies to repair or replace tissues and organs damaged by disease. Advances in the directed differentiation of stem cells, their development ex vivo into tissue buds or organoids, and advanced materials for tissue engineering all hold promise for future therapies, but each lacks the precise spatiotemporal control over cell fate required to engineer mature, functional tissues and organs for transplant. This section explores how synthetic gene networks synergize with these three emerging technologies, potentially resulting in more structured and mature tissues that bridge the gap between therapeutic promise and medical practice.

A key enabler of proposed regenerative therapies is pluripotent stem cells. Initially derived from human embryos (Thomson et al. 1998), stem cells retain the potential to differentiate into any cell type in the human body. This makes them a promising platform for regenerative medicine, especially because methods for returning terminally differentiated cells to a pluripotent state (Okita et al. 2007) have obviated both the immunogenic and ethical concerns surrounding human embryonic stem cells. Pure cultures of mature differentiated cells offer a promising therapeutic approach for diseases that affect a single cell type, such as Parkinson's disease (dopaminergic neurons) and heart disease (cardiomyocytes). In such cases, a cellular therapy could be derived from a patient's own stem cells, expanded and differentiated ex vivo, and then transplanted back into the patient. Although such approaches have met with some success (Kriks et al. 2011; Shiba et al. 2012), challenges surrounding the purity of differentiated cultures (Breitbach et al. 2007) and engraftment of mature cells into existing tissues make the approach difficult to generalize. Diseases causing widespread damages to tissues, such as liver cirrhosis or gross injuries to the skin, are likely best repaired with functional, structured, mature tissue that simple culture systems cannot recreate.

One development along the path toward mature, stem cell–derived tissues are recent methods for growing organoids. Cultured without pluripotency-maintaining factors, stem cells spontaneously form embryoid bodies (Itskovitz-Eldor et al. 2000), which are heterogeneous, disorganized structures that recapitulate some aspects of early embryogenesis. Subsequent culture with appropriate differentiation factors results in organoids, organized prototissues that recapitulate many aspects of their in vivo counterparts (Lancaster and Knoblich 2014). Organoid systems have been developed for many tissues, including cerebrum (Lancaster et al. 2013), liver (Huch et al. 2013), pancreas (Boj et al. 2014), and lung (Dye et al. 2015), and their experimental tractability makes them exciting disease- and drug-screening models (Ranga et al. 2014). For example, hepatic organoids from Alagille syndrome patients recapitulate the in vivo pathology of the disease (Huch et al. 2014), and pancreas organoids have been used to discover new drivers of ductal pancreatic cancer (Boj et al. 2014). Unfortunately, the use of organoids more broadly in regenerative medicine is hampered by their immaturity and heterogeneity. Organoid systems reported thus far only recapitulate early stages of tissue development, they only contain a subset of the cell types present in mature tissue, and many of their fine structures remain disorganized. For example, brain organoids display functional regions similar to those found in naturally developing embryos, but they are not reliably organized relative to one another (Lancaster et al. 2013). This is because organoid structures are developing outside of their natural context: In the developing embryo, cell fate depends on information from neighboring cells, longer-range diffusible signals, and extracellular matrix interactions. Without these signals, organoids develop solely based on local cell–cell interactions, which result in local organization but not large-scale mature tissue.

Tissue engineering approaches offer a promising route toward building larger-scale structures based on recent advances in materials science. Cells are seeded into/onto a functionalized biocompatible scaffold, which promotes their (re)organization into a functional tissue (Dvir et al. 2011). Importantly, the synthetic extracellular matrix is not just a mechanical substrate, but can provide physical and chemical cues to the cells that drive fate decisions and large-scale organization (Lutolf and Hubbell 2005). For example, seeding cardiomyocytes on an anisotropic matrix results in their lengthwise alignment and organization into contractile muscle fibers (Engelmayr et al. 2008), whereas tethering interferon-γ to chitosan hydrogels affects neural progenitor cells' lineage commitment (Leipzig et al. 2011). Such approaches leverage cells' innate ability to self-organize and differentiate, but they rely on static environments; in contrast, natural development occurs in a rich milieu of signals that change dynamically in time and space. These signals drive cell-fate decisions in natural development, and their dynamicity is difficult to replicate in engineered environments.

As the field of regenerative medicine matures, synthetic biology is poised to synergize with all three approaches in the development of complex tissues. Synthetic gene networks promise orthogonal control of differentiation and cell-fate decisions by offering programmable control of endogenous gene networks using signals that can be engineered. For example, recent work in our laboratory (Guye et al. 2016) described a simple gene circuit that drives ectopic expression of the master cell-fate regulator Gata6. Because the circuit was transduced with a lentivirus, cells showed a wide range of protein expression levels; a brief pulse of Gata6 led to symmetry breaking and segregation of the isogenic cell line into ectodermal, endodermal, and mesodermal subpopulations. These subpopulations developed further into a structured tissue, resulting in regions of fetal liver-like tissue surrounding populations of neuronal cells. The liver tissue evinces not only hepatocyte-like cells but also stromal cells, tubular structures, and hematopoetic processes, remarkable complexity and organization given the simplicity of the genetic construct used to create it.

Importantly, synthetic gene circuits can couple these cell-fate actuators to intracellular sensors of cell state. One example application might be to implement an automatic multistep stem cell differentiation protocol without laborious sequential application of chemical inducers (Schiesser et al. 2014). For example, a gene circuit could recapitulate the ontogeny of pancreatic β cells by sequentially expressing Gata6, Pdx1, Ngn3, and Pax4 to drive differentiation of stem cells through definitive endoderm (Wamaitha et al. 2015), pancreatic endoderm (Kubo et al. 2011), and β cells, respectively (Fig. 1, left) (Collombat et al. 2009). Each step of the process could be controlled by the activity of an endogenous transcription factor. For example, Pdx1 and Ngn3 expression could be driven by a transcription factor specific to definitive endoderm such as Fox2A (D'Amour et al. 2005). Such a synthetic multistep "sense-and-differentiate" gene circuit could guide cells down a desired differentiation path more efficiently, and with greater precision, than current methods.

A synthetic gene circuit for producing pure populations of terminally differentiated cells is useful in some contexts, as discussed above, but

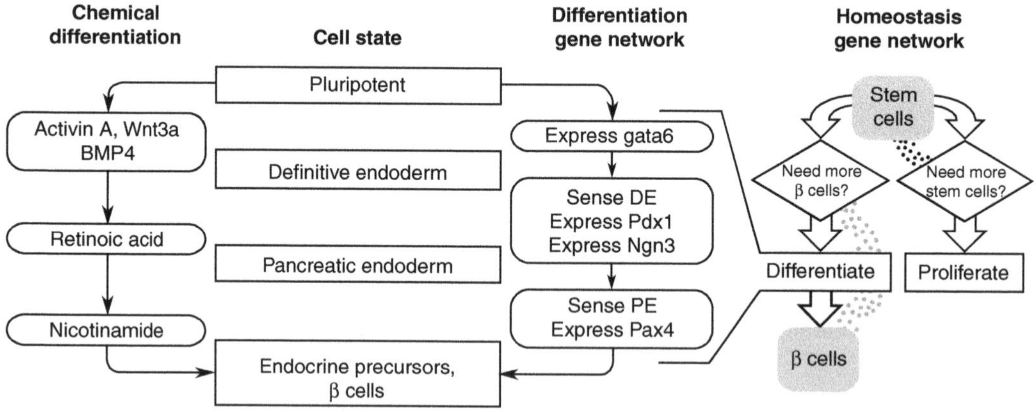

Figure 1. Synthetic gene networks drive cell-fate decisions using cell state sensors and extracellular signals. Pluripotent stem cells are directed to differentiate into β cells by sequential application of growth factors (Schiesser et al. 2014); the same differentiation program might be performed by a gene network that uses sequential expression of master cell-fate regulators such as Gata6 and Pdx1. Progression through the program could be controlled by endogenous sensors of cell state. This multistep sense-and-differentiate program could be part of a larger gene network, such as one that maintains tissue homeostasis (Miller et al. 2012). A gene network could sense the number of programmed stem cells, and terminally differentiated cells using diffusible signals then proliferate or differentiate to maintain a continuous supply of both.

it becomes particularly powerful when integrated into a larger synthetic gene network. For example, one could imagine a sense-and-differentiate gene circuit as part of a larger network devoted to maintaining tissue homeostasis in the face of autoimmune assault, as is frequently the case with type 1 diabetes (Zhang and Eisenbarth 2011). The loss of β cells in such patients is frequently a months-to-years-long process (Eisenbarth 2010), and transplanting β cells is ineffective because of the continuing autoimmune attack. A synthetic gene network might maintain β-cell levels by controlling the proliferation of engineered stem cells and their differentiation into β cells (Fig. 1, right). Such a network would sense the population levels of both stem cells and β cells based on the concentrations of diffusible intercellular signaling molecules, inducing proliferation of stem cells and differentiation into β cells when either population needed to be replenished (Miller et al. 2012).

The artificial tissue homeostasis example in Figure 1 relies on engineered intercellular signaling to coordinate cellular responses across a population. Engineered communication channels orthogonal to endogenous signaling will be particularly important in building gene networks for synthetic tissues because they allow engineered networks to recapitulate the larger-scale organization of natural development that many current synthetic tissues lack. For example, engineered diffusible signals can drive spatial patterning of gene expression in prokaryotes and eukaryotes (Basu et al. 2005; Chen and Weiss 2005; Bacchus et al. 2012), whereas synthetic juxtacrine signaling based on the Notch receptor can generate propagating, mutually exclusive states in neighboring cells (Fig. 2) (Sprinzak et al. 2010; Matsuda et al. 2012). The spatial patterns shown in these studies have many similarities to those created by organizing centers in metazoan development, such as the ones that specify the imaginal discs that later form the limbs in *Drosophila* (Brook et al. 1996). Similar patterns could be used to drive large-scale patterns in cell-fate decisions in developing synthetic tissues.

The opportunities to build tissues with synthetic biology present some exceptional challenges as well. Synthetic gene networks promise the ability to precisely direct cell fate by interacting with endogenous cell-fate pathways, but these pathways are frequently complex and sen-

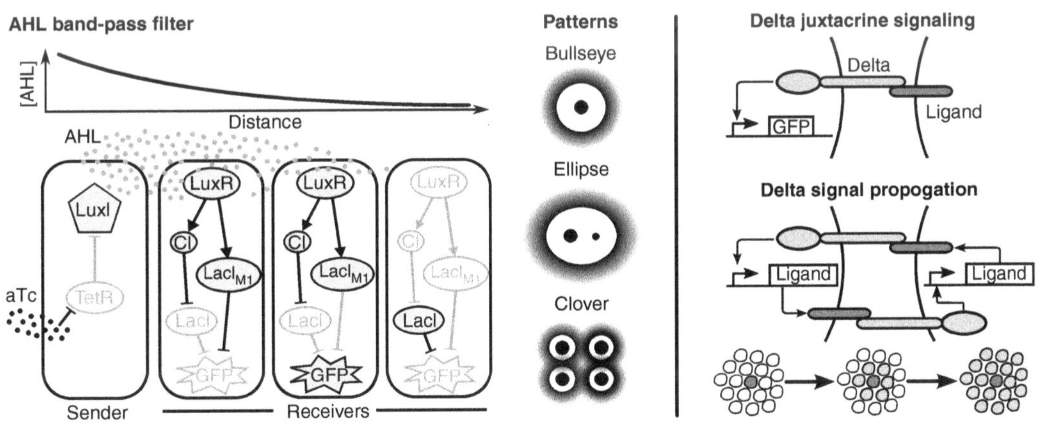

Figure 2. Engineered intercellular communication can coordinate spatial changes in gene expression. Communication channels are frequently based on diffusible signals such as acyl-homoserine lactone (AHL) (*left*) (Basu et al. 2005). The receiver gene network responds to intermediate concentrations of AHL, mimicking the "French flag" model of morphogen patterning (Wolpert 1969). Arranging senders in various starting configurations produces patterns similar to those found in natural development. Communication can also be based on juxtacrine signaling (*right*), in which a cell presenting a ligand on its surface activates signal transduction in adjacent cells (Sprinzak et al. 2010). A feedback loop between signal activation and ligand production leads to signal propagation (Matsuda et al. 2012). GFP, Green fluorescent protein.

sitive to context. What is worse, the effect of genetic, epigenetic, and environmental context on these endogenous networks is not fully known. Several design strategies will likely prove useful. For example, using orthogonal inter- and intracellular signaling components in synthetic gene networks will help insulate them from the rest of the cellular context (Brophy and Voigt 2014). Additionally, intracellular sensors can allow synthetic gene networks to account for cell state (Xie et al. 2011), whereas extracellular controls via small molecules (Weber et al. 2004), light (Müller et al. 2014), and even electric fields (Levin 2014) can be used to tune gene circuit behavior. These genetic controls nicely complement the macroscale layout imposed by scaffolds of biocompatible materials (Ma 2008) and micropatterned substrates that present extracellular cues at the length scale of single cells (Kinney and McDevitt 2013). Taken together, these strategies will allow synthetic gene networks to operate robustly in the face of uncertain or unknown cellular context.

The ready availability of patient-derived induced pluripotent stem cells has provided us access to the building blocks of our own development. These cells already contain the gene networks necessary to develop into tissues and organs, and the study of molecular development and morphogenesis continues to elucidate their composition and dynamic function. Current developments in synthetic biology promise the ability to precisely control the spatiotemporal activity of these gene networks, harnessing the function of these natural networks toward diagnostic and therapeutic ends. The in vitro, ex vivo, and (eventually) in vivo construction of complex human tissues promises to revolutionize both research into, and treatment of, many common human diseases.

FRONTIER TWO: SYNTHETIC BIOMATERIALS AND PROGRAMMABLE MATTER

Synthetic control of biological morphogenesis has broad applicability outside of biomedicine as well. Biological materials have traits that have proven impossible to duplicate with synthetic materials: the specific strength of wood, the toughness of dragline spider silk, or the selective permeability of endothelium. Efforts in material science to recapitulate these properties synthetically are often "biologically inspired," but they lack the sophistication and precision of the biological systems that produce natural biomaterials. Because these biological systems are genetically encoded, synthetic biology offers the possibility to interface with and supplement these natural gene networks with synthetic ones to modulate these materials' properties. This section explores the promises and challenges on the path to "programmable" biological materials.

Biological materials are composed of some combination of cells, biopolymers (e.g., cellulose, collagen, and keratin), and minerals (hydroxyapatite, calcium carbonate, and silica). Living systems assemble these components using a process of guided, feedback-controlled self-assembly (Davies 2013) at scales ranging from molecular to cellular to macroscale tissue arrangement. These multiple scales of feedback mean that biomaterial structure is universally hierarchical (Fratzl and Weinkamer 2007). For example, wood is anisotropic at multiple scales, from the arrangement of cellulose microfibrils in the cell wall, to the cells' parallel orientation along the shoot, to radial changes in cell morphology that give rise to a tree's annual rings. This anisotropy is controlled by genetic and molecular mechanisms. For example, cellulose microfibrils are deposited in a continuous, transverse spiral around the cells' long axis (McFarlane et al. 2014). Their (extracellular) orientation is controlled by the orientation of the (intracellular) microtubule network, which responds, in turn, to growth hormones and changes in turgor pressure. The transverse orientation gives rise to preferential expansion along the longitudinal axis of the stem. Plant hormones also act systemically to promote seasonal changes in the morphology of proliferating cells, leading to nested rings of early- and latewood (Ursache et al. 2013). This hierarchical structure, with its anisotropy at multiple scales, gives wood remarkable specific strength and fracture toughness (Committee on Synthetic

Hierarchical Structures 1994); similar hierarchical architectures underlie many other common biomaterials.

Early efforts to "program" biomaterials were based on genetically tractable biopolymers, particularly recombinant expression of proteins that could self-assemble into polymers. For example, elastin is an extracellular matrix protein that provides elasticity to connective tissue (Muiznieks et al. 2010). It readily self-assembles, and synthetic hydrogels made from recombinant elastin proteins undergo a temperature-sensitive phase change that makes them useful for drug delivery (Wright et al. 2002). Spider silk, on the other hand, is an extraordinarily tough material (Lazaris 2002); spider silk proteins expressed in *Escherichia coli* and extruded through an artificial spinneret have mechanical properties similar to the natural fiber (Xia et al. 2010). Their recombinant production allows for these protein-based materials to be readily manipulated, tuned, and functionalized beyond the capabilities of natural systems (Schacht and Scheibel 2014), for example, imbuing synthetic spider silk with antimicrobial properties by fusing recombinant silk proteins with antimicrobial peptides (Gomes et al. 2011). However, the materials form exclusively via entropic molecular self-assembly, which allows for none of the advantages of hierarchical, adaptive assembly seen in natural systems.

An alternative approach to smart biomaterials is to genetically modify the organism that produces a biomaterial to tune or functionalize it in situ. Diatoms, for example, have a glassy cell wall ("frustule") made of amorphous silica associated with a matrix of proteins (Gordon et al. 2009); fusing enzymes to these proteins immobilizes them on the diatom's cell wall, functionalizing this promising biomaterial (Sheppard et al. 2012). Bacterial biofilms have also been functionalized in this way. Recent work has shown functionalization of CsgA, the protein that forms *E. coli* biofilms, with peptide tags conferring activities such as substrate-specific adhesion and electrical conductivity (Fig. 3) (Chen et al. 2014; Nguyen et al. 2014).

Neither approach modifies the actual morphology of a natural biomaterial, however. Why is it so hard to change the shape of a biological structure in a tuned, specific way? For example, diatom cell walls have an array of astonishingly uniform pores, with promising nanotechnology applications ranging from molecular sieves to photonics (Gordon et al. 2009). Many of the molecular details about diatom cell wall biogenesis are known, including the protein scaffolds involved in silicate condensation and large-scale

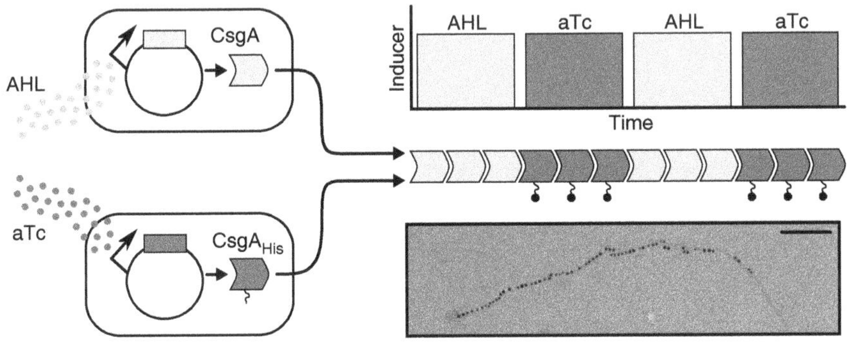

Figure 3. Chen et al. (2014) created a tunable biomaterial using curli fibers, a constituent of *Escherichia coli* biofilms. Curli fibers are made up of CsgA subunits, which self-assemble into higher-order structures. The investigators modified CsgA with a His tag that confers metal-binding capability; when they alternated expression of modified CsgA with unmodified CsgA (based on two different small-molecule inducers), the resulting curli fibers had regions that bound gold nanoparticles alternating with regions that did not, as shown with scanning electron microscopy. Scale bar, 100 nm. AHL, Acyl-homoserine lactone. (Image courtesy of Allen Chen and Tim Lu, both at the Massachusetts Institute of Technology.)

patterning (Kröger and Poulsen 2008; Scheffel et al. 2011), and yet we are still unable to alter pore size and arrangement in any directed fashion. In fact, none of the reported genetic modifications to the cell wall machinery resulted in any morphological changes at all!

The key insight is that the assembly of biomaterials, like most morphogenesis, is not entropic but, instead, guided by interlocking molecular feedback loops at multiple length scales (Davies 2013). A diatom's frustule, for example, is assembled via a multistep, multiscale process that involves silaic acid import, protein-catalyzed condensation into nanometer-scale particles of amorphous silica, nucleation of the new cell wall inside the silica deposition vesicle, and patterned silica deposition guided by both templating proteins and the diatom's cytoskeleton (Gordon et al. 2009; Hildebrand and Lerch 2015). In fact, for many biomaterials, we understand the nature of these feedback loops at the molecular level (Canty and Kadler 2002; McFarlane et al. 2014); but the relationships between physical and biochemical processes, molecular mechanisms, genetic control, and environmental factors are so complex as to defy intuition. In particular, specifying a desired "effect" (e.g., smaller pores in a diatom frustule) and then looking for a single molecular change that will "cause" it is challenging because of the difficulty in causal reasoning about feedback-based behavior (Åström and Murray 2008).

One way to program specific morphologies might be to modulate these natural feedback loops with external control instead of allowing structures to emerge from the detailed configuration of biophysical, molecular biological, and genetic processes. Natural biological systems are capable of sensing environmental conditions and responding to them on multiple scales, from single cells that move along chemotactic gradients to trees that grow around boulders. One way to take advantage of these natural morphogenic systems is by directing them using external factors, such as a silken canopy "grown" by silkworms crawling along a scaffolding (Oxman et al. 2014) or a functional large intestine created by seeding a collagen scaffold with intestinal progenitor cells (Shaffiey et al. 2016). External control has been used to control synthetic morphogenic systems as well. A priori arrangement of "sender" and "receiver" cells can create patterns caused by spatial variation in the concentration of a small molecule intercellular signal (Basu et al. 2005; Chen and Weiss 2005), as can spatial activation of synthetic gene networks using light (Gautier et al. 2014). More recent reports have used gene circuits to perform complex spatial computations such as Boolean logic (Tamsir et al. 2011) and edge detection (Tabor et al. 2009). Interfacing these engineered control mechanisms with endogenous gene networks will allow for top-down control of the molecular mechanisms that drive biomaterial morphogenesis.

Ultimately, the ability to synthetically modulate a biomaterial's structure at the microscale will depend on a detailed systems-level understanding of the natural system's biogenesis. Because synthetic gene networks act at cellular and subcellular levels, careful modeling and experiments will be needed to elucidate the connection between molecular behavior and material properties. Eventually, these approaches will enable the hybrid design of programmable materials in which microscale control of a material's properties is provided by a synthetic gene network whose spatiotemporal activity is controlled by external signals. The programmable matter enabled by these synthetic gene networks promises to marry the remarkable functionality of natural biomaterials with the engineerability required by advanced manufacturing and fabrication.

FRONTIER THREE: SYNTHETIC MORPHOGENESIS AND SYNTHETIC LIFE

Synthetic tissues and tunable biomaterials are special cases of a more general goal: the ability to predictably engineer novel biological shape- and pattern-forming systems. This echoes the goal of synthetic biology more broadly, to predictably engineer cells' behaviors by programming them with synthetic gene networks. Such predictable engineering is difficult in the complex, uncertain environs of naturally evolved cells, spurring recent efforts to engineer a "minimal" life form (Forster and Church 2006) by either minimizing an existing organism's genome (Glass et al. 2006; Pósfai et al. 2006; Gibson et al. 2010; Hutchison et al. 2016) or by chemically synthesizing autonomously replicating cells de novo (Blain and Szostak 2014). Both approaches aim to better understand the requirements for life in a "simple" minimal system (de Lorenzo and Danchin 2008), and to develop a predictable chassis on which to base further engineering efforts (Keasling 2008; Vickers et al. 2010).

Might a similar approach be applied to engineering multicellular structures? What might such an approach look like? In the same way that minimized cells can help biologists understand the basic functions required for self-sufficient life, minimal morphogenic systems could help us understand the rules underpinning multicellular morphogenesis. In the same way, minimized cells could provide a well-controlled environment for rational engineering; synthetic morphogenic systems could allow us to explore configurations and biological functions not witnessed in evolution (Dichtel-Danjoy and Félix 2004), from prosaic tunable materials for industrial and manufacturing applications to fantastical, functional forms such as trees that grow into tree houses (Ginsberg 2013). These morphogenic systems could be based on engineered or synthetic minimized cells, or they might be built on top of more complex cells with more "built-in" functionality.

Novel biological shapes and patterns have been difficult to engineer due in large part to the complexity of natural morphogenic systems and the way they are organized. As the previous sections discussed, biological structures are hierarchical, emerging from interactions at the molecular, macromolecular, cellular, multicellular, tissue, and organ levels. This hierarchical arrangement is the result of evolutionary processes in which biological complexity evolved stepwise, fixing core cellular processes first and then using them to build higher-order structures (Kirschner and Gerhart 1998). For example, the epithelium is one of the key evolutionary innovations of multicellular organisms (Leys and Riesgo 2012) because it allows organisms to spatially and biochemically isolate an "inside" cavity from the organism's surroundings. Epithelial cells depend on cadherins and catenins for cell adhesion and on the cytoskeleton to develop cell polarity and directional transport (Fath et al. 1993). Both cadherins (Abedin and King 2008) and catenins (Dickinson et al. 2011) are present in premetazoan organisms, and bacteria have cytoskeleton homologs of both actin and tubulin (Busiek and Margolin 2015), but only their coordinate regulation allows for the formation of higher-order structures. In turn, epithelia serve as the building blocks for larger structures, such as tissues and organs (Jamora et al. 2003), as spatial cues induce their folding, branching, and interactions with other cell types. This kind of hierarchical structure enables rapid evolution and exploration of phenotypic space (Kirschner and Gerhart 1998) by allowing for easy reuse of underlying morphogenic modules. However, it makes engineering lower levels of the hierarchy difficult because changing the behavior of one gene or gene network alters the behavior or structure of every context in which that gene or network is used.

The structure and evolutionary history of morphogenic processes have important implications for building novel morphogenic systems. Instead of attempting to modify or co-opt underlying cellular processes, perhaps it will be easier to follow evolution's lead and layer synthetic, engineered control mechanisms on top of preexisting morphogenic machinery. One might imagine beginning with a very simple chassis, such as a minimized *E. coli* or *Mycoplasma* species, then elaborating multicellular morphogenic modules on top of the existing molecular mechanisms, coordinating multicellular behavior with signals and networks orthogonal to native machinery. For example, the molecular basis of *E. coli* biofilm formation is relatively well understood (Beloin et al. 2008); placing these mechanisms under the control of both local, short-range intercellular signaling (for bottom-up pattern formation) and global chemical or light-based control could serve to "domesticate" them (Fig. 4, left) (Chen et al. 2015a), giving biological engineers control

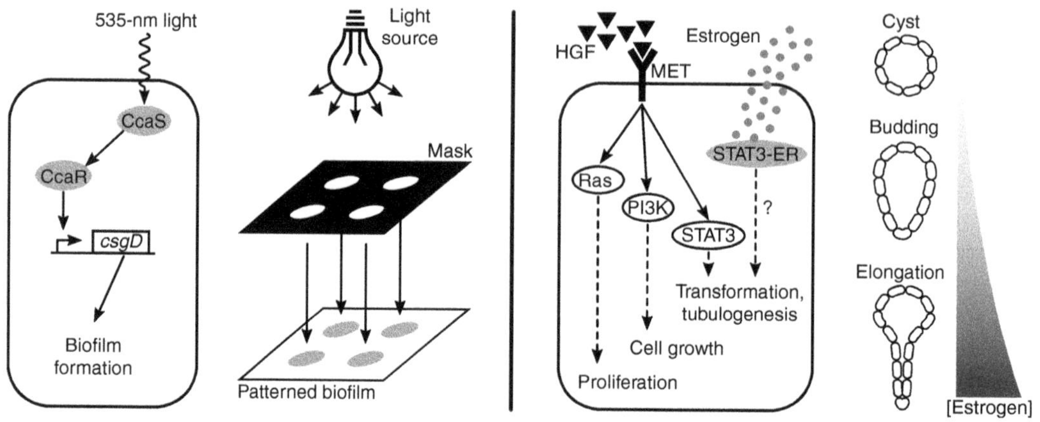

Figure 4. Proposed synthetic morphogenic systems. Patterned biofilm formation (*left*) may be achieved by connecting a synthetic optogenetic sensor, such as CcaS/CcaR (Schmidl et al. 2014), to a master regulator of biofilm production, such as *csgD* (Brombacher et al. 2003). A model mammalian morphogenic system (*right*) may offer a richer palette of morphogenic behaviors. For example, hepatocyte growth factor (HGF) induces tubulogenesis in Madin–Darby canine kidney (MDCK) cells grown in three-dimensional culture (Zegers 2014). The HGF receptor, MET, activates a number of downstream signaling pathways; controlling one via synthetic receptor, such as STAT3 fused to an estrogen receptor (ER) (STAT3-ER) (Matsuda et al. 1999), may allow tunable control of some of the morphogenic modules that lead to tubulogenesis.

over both when and where biofilms form and their fine-scale properties.

Layering synthetic control mechanisms on top of preexisting morphogenic modules in a "simple" chassis such as *E. coli* partially addresses issues of pleiotropy that plague current efforts, because engineers can change the behavior of the morphogenic modules (such as functionalizing the CsgA protein; Nguyen et al. 2014), confident that they understand those changes' effects in the context of a broader system. However, it may be that the palette of morphogenic behaviors available in a bacterial system is not broad enough. We might be better served by a more complex chassis in which a richer array of multicellular morphogenic modules are present but not yet dynamically regulated. For example, mammalian Madin–Darby canine kidney (MDCK) cells (Gaush et al. 1966) are a common in vitro model for epithelia; in three-dimensional culture, they spontaneously form hollow cysts that mimic in vivo epithelial morphology, and can be then induced to form branching tubes by the addition of hepatocyte growth factor (HGF) (Zegers 2014). Three-dimensional MDCK cultures could be a model system for studying external control of tube formation and extension (Fig. 4, right). Directing tubulogenesis in a system that is already "primed" to do so may be a more attractive engineering prospect than attempting to recapitulate it via the underlying morphogenic building blocks, such as cell–cell adhesion, cell division and apoptosis, and the establishment of apicobasal polarity. This is both because reusing mechanisms that are already present reduces the amount of engineering required to form higher-level structures, and because modulating these processes from the top of the complexity hierarchy is more in line with how these structures evolved naturally. This promises fewer pleiotropic effects, more modularity, greater robustness, and a broader space of phenotypes to explore.

The strategy of overlaying synthetic control on top of high-level morphogenic mechanisms is also poised to take advantage of our increasing control over the differentiation of pluripotent stem cells (Kinney and McDevitt 2013). From this perspective, different cell types have available to them different palettes of morphogenic modules that could be composed synthetically. The kinds of structures we could expect to synthesize from epithelial cells (sheets, domes, tubes) are very different than those we could imagine deriving from fibroblasts (connective and structural tissue components) or myocytes (contractile structures). Not only do the basic morphogenic modules differ between these various cell types, but their coordinate regulation does as well. Unfortunately, for many structures of interest, we do not understand the molecular basis of morphogenesis well enough to co-opt it; but by continuing to build in vitro models of various natural morphogenic systems (Foty and Steinberg 2005; Zegers 2014), we expect to learn about not only the molecular networks that control them but also means by which we can control them synthetically.

The phrase "de novo synthetic morphogenesis" conjures synthetic cells created in the laboratory, living systems entirely orthogonal to natural ones, designed rationally, and freed from evolution's series of "bad hacks." Although promising advances are being made in the field of artificial life, it seems wasteful and presumptuous to throw away four billion years of evolution and the dynamic, robust, flexible set of core molecular mechanisms, gene networks, and morphogenic modules that have resulted. By choosing carefully the level of the complexity hierarchy with which our synthetic gene circuits interact, we can minimize the issues of pleiotropy and context sensitivity that hamper current efforts.

CONCLUSION

Complex synthetic multicellular structures could lead to advances in medicine, manufacturing, and many other human endeavors. To deliver on these promises, however, the field needs a set of design rules akin to the ones the broader field of synthetic biology is developing (Brophy and Voigt 2014). We close with a look at some of the challenges that are common not only to the three frontiers discussed in this review but indeed the entire field of synthetic morphogenesis. We also consider lessons learned from the study of natural morphogenic systems, as well as subjects further afield.

First, why are natural morphogenic systems difficult to alter rationally? One reason lies in their structure. Such systems are modular and hierarchical, just like many complex engineered systems (Csete and Doyle 2002). This structure is the result of biological systems' evolutionary history, which evolved complexity by fixing core molecular processes and then overlaying dynamic regulation on top of them (Kirschner and Gerhart 1998). This resulted in the evolution of multicellularity, followed by intercellular signaling, the fixation of phylotypic body plans, and then elaborations on those body plans: appendages, the neural crest, and reproductive specializations (Kuratani 2009). Each level of complexity was selected for robustness and stability, which supported phenotypic exploration of additional complexity. This modular hierarchy is a key evolutionary driver of morphogenesis because rewiring modules' connectivity (via the signals that regulate them) requires fewer genetic changes and is less likely maladaptive than altering the molecular species that compose them (Pigliucci 2008).

Unfortunately, this evolutionary history also makes rational changes to natural morphogenesis difficult because these morphogenic modules are not isolable from each other. Morphogenic motifs are reused in many different contexts, as are the signals that activate them; this reuse leads to pleiotropy and muddies the boundaries between morphogenic modules. Instead, these modules assume much of their identity from the system in which they are embedded (Csete and Doyle 2002). The formation of a structure is not controlled by a single gene but the time-varying interactions of many molecular species. A prospective biological engineer is forced to contend with complexity and feedback control between modules and across levels of hierarchy; the global behavior of the system emerges from the interactions of its constituent parts.

Why are morphogenic systems so complex? One might intuit that such complexity would result in systems that are fragile, but of course natural morphogenesis is marvelously robust. The study of biology through the lens of complex systems seems to indicate several reasons for this reliability. First, morphogenic and developmental systems make wide use of feedback (Davies 2013). Negative feedback makes systems robust to perturbations, and positive feedback can amplify stimuli to move systems from one state to another. Second, natural biological systems are composed of overlapping and redundant parts, and there are frequently several mechanisms that give rise to the same phenotypic output (Kitano 2004). These increases in complexity actually serve to make morphogenic systems more robust, not less.

In fact, it may be that the tradeoff between robustness, performance, and complexity is an underlying law not only of biological systems but of complex systems more generally (Csete and Doyle 2002). This may be the reason, for example, that development is slow; complex programs, executed slowly, lead to a more robust outcome than simple processes proceeding quickly. Conversely, simulations of minimal metabolic networks suggest that they are more fragile than naturally evolved ones (Gabaldón et al. 2007). As engineered systems begin to achieve complexity comparable to biological ones, explorations of the tradeoff between complexity and robustness in natural and engineered biological systems may have lessons to teach engineers of complex mechanical and electronic systems too.

Another constellation of open questions concerns modularity and context sensitivity. This review has used the term "module" without giving it a precise definition; comparisons of biological organisms at both the gene (Stuart et al. 2003) and protein network (Wuchty et al. 2003) levels show evolutionarily conserved networks of interacting genes and, of course, cells coordinately regulate suites of genes in response to both intracellular and extracellular conditions. In spite of this, we do not understand the organization of biological systems well enough to harness the abstraction that is implied by a modular organization. How can molecular and morphogenic modules be rearranged, rewired, or replaced to modify existing systems or build new ones? How can one best account for pleiotropy and context sensitivity, or isolate engineered systems from it? What is the right balance between leveraging natural systems and building synthetic, orthogonal systems alongside or on top of them?

Biological systems manipulate matter, perform chemistry, and process information precisely, robustly, and efficiently; the promise of synthetic biology, writ broad, is to harness this sophistication to engineer biological systems reliably and predictably. Great strides have been made in engineering useful, predictable functions into single cells, but life in nature is never unicellular. Life is structured, and cells are organized to maximize productivity, protect against environmental assault, and thus improve the transmission of genetic information. Synthetic biologists are just beginning to explore the design and control of biological structures, leading to a new field of synthetic morphogenesis. The field is still embryonic, but we look forward excitedly to its development.

ACKNOWLEDGMENTS

We thank Samira Kiani for stimulating discussions as this manuscript was being developed. This work is supported by National Science Foundation grants CBET-0939511, EEC-054 0879, CNS-1446607, and EFMA-1137269.

REFERENCES

Abedin M, King N. 2008. The premetazoan ancestry of cadherins. *Science* **319:** 946–948.

Arpino JA, Hancock EJ, Anderson J, Barahona M, Stan GB, Papachristodoulou A, Polizzi K. 2013. Tuning the dials of Synthetic Biology. *Microbiology* **159:** 1236–1253.

Åström KJ, Murray RM. 2008. *Feedback systems: An introduction for scientists and engineers*. Princeton University Press, Princeton, NJ.

Bacchus W, Lang M, El-Baba MD, Weber W, Stelling J, Fussenegger M. 2012. Synthetic two-way communication between mammalian cells. *Nat Biotechnol* **30:** 991–996.

Basu S, Gerchman Y, Collins CH, Arnold FH, Weiss R. 2005. A synthetic multicellular system for programmed pattern formation. *Nature* **434:** 1130–1134.

Beloin C, Roux A, Ghigo JM. 2008. *Escherichia coli* biofilms. *Curr Top Microbiol Immunol* **322:** 249–289.

Blain JC, Szostak JW. 2014. Progress toward synthetic cells. *Annu Rev Biochem* **83:** 615–640.

Boj SF, Hwang C-I, Baker LA, Chio IIC, Engle DD, Corbo V, Jager M, Ponz-Sarvise M, Tiriac H, Spector MS, et al. 2014. Organoid models of human and mouse ductal pancreatic cancer. *Cell* **160:** 324–338.

Breitbach M, Bostani T, Roell W, Xia Y, Dewald O, Nygren JM, Fries JWU, Tiemann K, Bohlen H, Hescheler J, et al.

2007. Potential risks of bone marrow cell transplantation into infarcted hearts. *Blood* **110:** 1362–1369.

Brombacher E, Dorel C, Zehnder AJB, Landini P. 2003. The curli biosynthesis regulator CsgD co-ordinates the expression of both positive and negative determinants for biofilm formation in *Escherichia coli*. *Microbiology* **149:** 2847–2857.

Brook WJ, Diaz-Benjumea FJ, Cohen SM. 1996. Organizing spatial pattern in limb development. *Annu Rev Cell Dev Biol* **12:** 161–180.

Brophy JAN, Voigt CA. 2014. Principles of genetic circuit design. *Nat Methods* **11:** 508–520.

Busiek KK, Margolin W. 2015. Bacterial actin and tubulin homologs in cell growth and division. *Curr Biol* **25:** R243–R254.

Cachat E, Liu W, Hohenstein P, Davies JA. 2014. A library of mammalian effector modules for synthetic morphology. *J Biol Eng* **8:** 26.

Callura JM, Cantor CR, Collins JJ. 2012. Genetic switchboard for synthetic biology applications. *Proc Natl Acad Sci* **109:** 5850–5855.

Canty E, Kadler K. 2002. Collagen fibril biosynthesis in tendon: A review and recent insights. *Comp Biochem Physiol A Mol Integr Physiol* **133:** 979–985.

Chen M-T, Weiss R. 2005. Artificial cell–cell communication in yeast *Saccharomyces cerevisiae* using signaling elements from *Arabidopsis thaliana*. *Nat Biotechnol* **23:** 1551–1555.

Chen AY, Deng Z, Billings AN, Seker UOS, Lu MY, Citorik RJ, Zakeri B, Lu TK. 2014. Synthesis and patterning of tunable multiscale materials with engineered cells. *Nat Mater* **13:** 515–523.

Chen AY, Zhong C, Lu TK. 2015a. Engineering living functional materials. *ACS Synth Biol* **4:** 8–11.

Chen Y, Kim JK, Hirning AJ, Josić K, Bennett MR. 2015b. Emergent genetic oscillations in a synthetic microbial consortium. *Science* **349:** 986–989.

Collombat P, Xu X, Ravassard P, Sosa-Pineda B, Dussaud S, Billestrup N, Madsen OD, Serup P, Heimberg H, Mansouri A. 2009. The ectopic expression of Pax4 in the mouse pancreas converts progenitor cells into α and subsequently β cells. *Cell* **138:** 449–462.

Committee on Synthetic Hierarchical Structures. 1994. *Hierarchical structures in biology as a guide for new materials technology*, National Academy of Sciences, Washington, DC.

Csete ME, Doyle JC. 2002. Reverse engineering of biological complexity. *Science* **295:** 1664–1669.

D'Amour KA, Agulnick AD, Eliazer S, Kelly OG, Kroon E, Baetge EE. 2005. Efficient differentiation of human embryonic stem cells to definitive endoderm. *Nat Biotechnol* **23:** 1534–1541.

Davies JA. 2013. *Mechanisms of morphogenesis*, 2nd ed. Academic, Cambridge, MA.

de Lorenzo V, Danchin A. 2008. Synthetic biology: Discovering new worlds and new words. *EMBO Rep* **9:** 822–827.

Dichtel-Danjoy M-L, Félix M-A. 2004. Phenotypic neighborhood and micro-evolvability. *Trends Genet* **20:** 268–276.

Dickinson DJ, Nelson WJ, Weis WI. 2011. A polarized epithelium organized by β- and α-catenin predates cadherin and metazoan origins. *Science* **331:** 1336–1339.

Dvir T, Timko BP, Kohane DS, Langer R. 2011. Nanotechnological strategies for engineering complex tissues. *Nat Nanotechnol* **6:** 13–22.

Dye BR, Hill DR, Ferguson MAH, Tsai Y-H, Nagy MS, Dyal R, Wells JM, Mayhew CN, Nattiv R, Klein OD, et al. 2015. In vitro generation of human pluripotent stem cell derived lung organoids. *eLife* **4:** e05098.

Eisenbarth GS. 2010. Banting Lecture 2009: An unfinished journey: Molecular pathogenesis to prevention of type 1A diabetes. *Diabetes* **59:** 759–774.

Elowitz MB, Leibler S. 2000. A synthetic oscillatory network of transcriptional regulators. *Nature* **403:** 335–338.

Engelmayr GC, Cheng M, Bettinger CJ, Borenstein JT, Langer R, Freed LE. 2008. Accordion-like honeycombs for tissue engineering of cardiac anisotropy. *Nat Mater* **7:** 1003–1010.

Fath KR, Mamajiwalla SN, Burgess DR. 1993. The cytoskeleton in development of epithelial cell polarity. *J Cell Sci Suppl* **17:** 65–73.

Forster AC, Church GM. 2006. Towards synthesis of a minimal cell. *Mol Syst Biol* **2:** 45.

Foty RA, Steinberg MS. 2005. The differential adhesion hypothesis: A direct evaluation. *Dev Biol* **278:** 255–263.

Fratzl P, Weinkamer R. 2007. Nature's hierarchical materials. *Prog Mater Sci* **52:** 1263–1334.

Gabaldón T, Peretó J, Montero F, Gil R, Latorre A, Moya A. 2007. Structural analyses of a hypothetical minimal metabolism. *Philos Trans R Soc Lond B Biol Sci* **362:** 1751–1762.

Gardner TS, Cantor CR, Collins JJ. 2000. Construction of a genetic toggle switch in *Escherichia coli*. *Nature* **403:** 339–342.

Gaush CR, Hard WL, Smith TF. 1966. Characterization of an established line of canine kidney cells (MDCK). *Proc Soc Exp Biol Med* **122:** 931–935.

Gautier A, Gauron C, Volovitch M, Bensimon D, Jullien L, Vriz S. 2014. How to control proteins with light in living systems. *Nat Chem Biol* **10:** 533–541.

Gibson DG, Glass JI, Lartigue C, Noskov VN, Chuang R-Y, Algire MA, Benders GA, Montague MG, Ma L, Moodie MM, et al. 2010. Creation of a bacterial cell controlled by a chemically synthesized genome. *Science* **329:** 52–56.

Ginsberg AD. 2013. The pre future of synthetic biology, www.nextnature.net/2013/05/the-prefuture-of-synthetic-biology.

Glass JI, Assad-Garcia N, Alperovich N, Yooseph S, Lewis MR, Maruf M, Hutchison CA, Smith HO, Venter JC. 2006. Essential genes of a minimal bacterium. *Proc Natl Acad Sci* **103:** 425–430.

Gomes SC, Leonor IB, Mano JF, Reis RL, Kaplan DL. 2011. Antimicrobial functionalized genetically engineered spider silk. *Biomaterials* **32:** 4255–4266.

Gordon R, Losic D, Tiffany MA, Nagy SS, Sterrenburg FAS. 2009. The glass menagerie: Diatoms for novel applications in nanotechnology. *Trends Biotechnol* **27:** 116–127.

Guye P, Ebrahimkhani MR, Kipniss N, Velazquez JJ, Schoenfeld E, Kiani S, Griffith LG, Weiss R. 2016. Genetically engineering self-organization of human pluripotent stem cells into a liver bud-like tissue using Gata6. *Nat Commun* **7:** 10243.

Harris H. 1999. *The birth of the cell.* Yale University Press, New Haven, CT.

Hildebrand M, Lerch SJL. 2015. Diatom silica biomineralization: Parallel development of approaches and understanding. *Semin Cell Dev Biol* **46:** 27–35.

Huch M, Dorrell C, Boj SF, van Es JH, Li VSW, van de Wetering M, Sato T, Hamer K, Sasaki N, Finegold MJ, et al. 2013. In vitro expansion of single $Lgr5^+$ liver stem cells induced by Wnt-driven regeneration. *Nature* **494:** 247–250.

Huch M, Gehart H, van Boxtel R, Hamer K, Blokzijl F, Verstegen MMA, Ellis E, van Wenum M, Fuchs SA, de Ligt J, et al. 2014. Long-term culture of genome-stable bipotent stem cells from adult human liver. *Cell* **160:** 1–14.

Hutchison CA, Chuang R-Y, Noskov VN, Assad-Garcia N, Deerinck TJ, Ellisman MH, Gill J, Kannan K, Karas BJ, Ma L, et al. 2016. Design and synthesis of a minimal bacterial genome. *Science* doi: 10.1126/science.aad6253.

Itskovitz-Eldor J, Schuldiner M, Karsenti D, Eden A, Yanuka O, Amit M, Soreq H, Benvenisty N. 2000. Differentiation of human embryonic stem cells into embryoid bodies compromising the three embryonic germ layers. *Mol Med* **6:** 88–95.

Jamora C, DasGupta R, Kocieniewski P, Fuchs E. 2003. Links between signal transduction, transcription and adhesion in epithelial bud development. **422:** 317–322.

Keasling JD. 2008. Synthetic biology for synthetic chemistry. *ACS Chem Biol* **3:** 64–76.

Kinney MA, McDevitt TC. 2013. Emerging strategies for spatiotemporal control of stem cell fate and morphogenesis. *Trends Biotechnol* **31:** 78–84.

Kirschner M, Gerhart J. 1998. Evolvability. *Proc Natl Acad Sci* **95:** 8420–8427.

Kitano H. 2004. Biological robustness. *Nat Rev Genet* **5:** 826–837.

Kriks S, Shim J-W, Piao J, Ganat YM, Wakeman DR, Xie Z, Carrillo-Reid L, Auyeung G, Antonacci C, Buch A, et al. 2011. Dopamine neurons derived from human ES cells efficiently engraft in animal models of Parkinson's disease. *Nature* **480:** 547–551.

Kröger N, Poulsen N. 2008. Diatoms—From cell wall biogenesis to nanotechnology. *Annu Rev Genet* **42:** 83–107.

Kubo A, Stull R, Takeuchi M, Bonham K, Gouon-Evans V, Sho M, Iwano M, Saito Y, Keller G, Snodgrass R. 2011. Pdx1 and Ngn3 overexpression enhances pancreatic differentiation of mouse ES cell-derived endoderm population. *PLoS ONE* **6:** e24058.

Kuratani S. 2009. Modularity, comparative embryology and evo–devo: Developmental dissection of evolving body plans. *Dev Biol* **332:** 61–69.

Lancaster MA, Knoblich JA. 2014. Organogenesis in a dish: Modeling development and disease using organoid technologies. *Science* **345:** 1247125.

Lancaster MA, Renner M, Martin C-A, Wenzel D, Bicknell LS, Hurles ME, Homfray T, Penninger JM, Jackson AP, Knoblich JA. 2013. Cerebral organoids model human brain development and microcephaly. *Nature* **501:** 373–379.

Lazaris A. 2002. Spider silk fibers spun from soluble recombinant silk produced in mammalian cells. *Science* **295:** 472–476.

Leipzig ND, Wylie RG, Kim H, Shoichet MS. 2011. Differentiation of neural stem cells in three-dimensional growth factor–immobilized chitosan hydrogel scaffolds. *Biomaterials* **32:** 57–64.

Levin M. 2014. Molecular bioelectricity: How endogenous voltage potentials control cell behavior and instruct pattern regulation in vivo. *Mol Biol Cell* **25:** 3835–3850.

Leys SP, Riesgo A. 2012. Epithelia, an evolutionary novelty of metazoans. *J Exp Zool B Mol Dev Evol* **318:** 438–447.

Lutolf MP, Hubbell JA. 2005. Synthetic biomaterials as instructive extracellular microenvironments for morphogenesis in tissue engineering. *Nat Biotechnol* **23:** 47–55.

Ma PX. 2008. Biomimetic materials for tissue engineering. *Adv Drug Deliv Rev* **60:** 184–198.

Matsuda T, Nakamura T, Nakao K, Arai T, Katsuki M, Heike T, Yokota T. 1999. STAT3 activation is sufficient to maintain an undifferentiated state of mouse embryonic stem cells. *EMBO J* **18:** 4261–4269.

Matsuda M, Koga M, Nishida E, Ebisuya M. 2012. Synthetic signal propagation through direct cell–cell interaction. *Sci Signal* **5:** ra31.

McFarlane HE, Döring A, Persson S. 2014. The cell biology of cellulose synthesis. *Annu Rev Plant Biol* **65:** 69–94.

Miki K, Endo K, Takahashi S, Funakoshi S, Takei I, Katayama S, Toyoda T, Kotaka M, Takaki T, Umeda M, et al. 2015. Efficient detection and purification of cell populations using synthetic MicroRNA switches. *Cell Stem Cell* **16:** 699–711.

Miller M, Hafner M, Sontag E, Davidsohn N, Subramanian S, Purnick PEM, Lauffenburger D, Weiss R. 2012. Modular design of artificial tissue homeostasis: Robust control through synthetic cellular heterogeneity. *PLoS Comput Biol* **8:** e1002579.

Moon TS, Lou C, Tamsir A, Stanton BC, Voigt CA. 2012. Genetic programs constructed from layered logic gates in single cells. *Nature* **491:** 249–253.

Muiznieks LD, Weiss AS, Keeley FW. 2010. Structural disorder and dynamics of elastin. *Biochem Cell Biol* **88:** 239–250.

Müller K, Engesser R, Timmer J, Zurbriggen MD, Weber W. 2014. Orthogonal optogenetic triple-gene control in mammalian cells. *ACS Synth Biol* **3:** 796–801.

Nguyen PQ, Botyanszki Z, Tay PKR, Joshi NS. 2014. Programmable biofilm-based materials from engineered curli nanofibres. *Nat Commun* doi: 10.1038/ncomms 5945.

Okita K, Ichisaka T, Yamanaka S. 2007. Generation of germline-competent induced pluripotent stem cells. *Nature* **448:** 313–317.

Oxman N, Laucks J, Kayser M, Nuro-Royo J, Gonzales-Uribe C. 2014. Silk pavilion: A case study in fiber-based digital fabrication. In *Proceedings of the Fabricate Conference*, pp. 248–255. Zurich, Switzerland.

Park JS, Rhau B, Hermann A, McNally KA, Zhou C, Gong D, Weiner OD, Conklin BR, Onuffer J, Lim WA. 2014. Synthetic control of mammalian-cell motility by engineering chemotaxis to an orthogonal bioinert chemical signal. *Proc Natl Acad Sci* **111:** 5896–5901.

Pigliucci M. 2008. Is evolvability evolvable? *Nat Rev Genet* **9:** 75–82.

Pósfai G, Plunkett G, Fehér T, Frisch D, Keil GM, Umenhoffer K, Kolisnychenko V, Stahl B, Sharma SS, de Arruda M, et al. 2006. Emergent properties of reduced-genome *Escherichia coli*. *Science* **312:** 1044–1046.

Prindle A, Samayoa P, Razinkov I, Danino T, Tsimring LS, Hasty J. 2011. A sensing array of radically coupled genetic "biopixels." *Nature* **481:** 39–44.

Ranga A, Gjorevski N, Lutolf MP. 2014. Drug discovery through stem cell-based organoid models. *Adv Drug Deliv Rev* **69–70:** 19–28.

Schacht K, Scheibel T. 2014. Processing of recombinant spider silk proteins into tailor-made materials for biomaterials applications. *Curr Opin Biotechnol* **29:** 62–69.

Schaumberg KA, Antunes MS, Kassaw TK, Xu W, Zalewski CS, Medford JI, Prasad A. 2015. Quantitative characterization of genetic parts and circuits for plant synthetic biology. *Nat Methods* **13:** 94–100.

Scheffel A, Poulsen N, Shian S, Kröger N. 2011. Nanopatterned protein microrings from a diatom that direct silica morphogenesis. *Proc Natl Acad Sci* **108:** 3175–3180.

Schiesser JV, Micallef SJ, Hawes S, Elefanty AG, Stanley EG. 2014. Derivation of insulin-producing β-cells from human pluripotent stem cells. *Rev Diabet Stud* **11:** 6–18.

Schmidl SR, Sheth RU, Wu A, Tabor JJ. 2014. Refactoring and optimization of light-switchable *Escherichia coli* two-component systems. *ACS Synth Biol* **3:** 820–831.

Shaffiey SA, Jia H, Keane T, Costello C, Wasserman D, Quidgley M, Dziki J, Badylak S, Sodhi CP, March JC, et al. 2016. Intestinal stem cell growth and differentiation on a tubular scaffold with evaluation in small and large animals. *Regen Med* **11:** 45–61.

Sheppard VC, Scheffel A, Poulsen N, Kröger N. 2012. Live diatom silica immobilization of multimeric and redox-active enzymes. *Appl Environ Microbiol* **78:** 211–218.

Shiba Y, Fernandes S, Zhu W-Z, Filice D, Muskheli V, Kim J, Palpant NJ, Gantz J, Moyes KW, Reinecke H, et al. 2012. Human ES-cell-derived cardiomyocytes electrically couple and suppress arrhythmias in injured hearts. *Nature* **489:** 322–325.

Slusarczyk AL, Lin A, Weiss R. 2012. Foundations for the design and implementation of synthetic genetic circuits. *Nat Rev Genet* **13:** 406–420.

Sprinzak D, Lakhanpal A, Lebon L, Santat LA, Fontes ME, Anderson GA, Garcia-Ojalvo J, Elowitz MB. 2010. *Cis*-interactions between Notch and Delta generate mutually exclusive signalling states. *Nature* **465:** 86–90.

Stuart JM, Segal E, Koller D, Kim SK. 2003. A gene-coexpression network for global discovery of conserved genetic modules. *Science* **302:** 249–255.

Tabor JJ, Salis HM, Simpson ZB, Chevalier AA, Levskaya A, Marcotte EM, Voigt CA, Ellington AD. 2009. A synthetic genetic edge detection program. *Cell* **137:** 1272–1281.

Tamsir A, Tabor JJ, Voigt CA. 2011. Robust multicellular computing using genetically encoded NOR gates and chemical "wires." *Nature* **469:** 212–215.

Thomson JA, Shapiro SS, Waknitz MA, Marshall VS. 1998. Embryonic stem cell lines derived from human blastocysts. *Science* **282:** 1145–1147.

Ursache R, Nieminen K, Helariutta Y. 2013. Genetic and hormonal regulation of cambial development. *Physiol Plant* **147:** 36–45.

Vickers CE, Blank LM, Krömer JO. 2010. Grand challenge commentary: Chassis cells for industrial biochemical production. *Nat Chem Biol* **6:** 875–877.

Wamaitha SE, del Valle I, Cho LTY, Wei Y, Fogarty NME, Blakeley P, Sherwood RI, Ji H, Niakan KK. 2015. Gata6 potently initiates reprograming of pluripotent and differentiated cells to extraembryonic endoderm stem cells. *Genes Dev* **29:** 1239–1255.

Weber W, Rimann M, Spielmann M, Keller B, Daoud-El Baba M, Aubel D, Weber CC, Fussenegger M. 2004. Gas-inducible transgene expression in mammalian cells and mice. *Nat Biotechnol* **22:** 1440–1444.

Weber W, Daoud-El Baba M, Fussenegger M. 2007. Synthetic ecosystems based on airborne inter- and intrakingdom communication. *Proc Natl Acad Sci* **104:** 10435–10440.

Wolpert L. 1969. Positional information and the spatial pattern of cellular differentiation. *J Theor Biol* **25:** 1–47.

Wright ER, McMillan RA, Cooper A, Apkarian RP, Conticello VP. 2002. Thermoplastic elastomer hydrogels via self-assembly of an elastin-mimetic triblock polypeptide. *Adv Funct Mater* **12:** 149–154.

Wuchty S, Oltvai ZN, Barabási A-L. 2003. Evolutionary conservation of motif constituents in the yeast protein interaction network. *Nat Genet* **35:** 176–179.

Xia X-X, Qian Z-G, Ki CS, Park YH, Kaplan DL, Lee SY. 2010. Native-sized recombinant spider silk protein produced in metabolically engineered *Escherichia coli* results in a strong fiber. *Proc Natl Acad Sci* **107:** 14059–14063.

Xie Z, Wroblewska L, Prochazka L, Weiss R, Benenson Y. 2011. Multi-input RNAi-based logic circuit for identification of specific cancer cells. *Science* **333:** 1307–1311.

Ye H, Daoud-El Baba M, Peng R-W, Fussenegger M. 2011. A synthetic optogenetic transcription device enhances blood-glucose homeostasis in mice. *Science* **332:** 1565–1568.

Yeh BJ, Rutigliano RJ, Deb A, Bar-Sagi D, Lim WA. 2007. Rewiring cellular morphology pathways with synthetic guanine nucleotide exchange factors. *Nature* **447:** 596–600.

Zegers MM. 2014. 3D in vitro cell culture models of tube formation. *Semin Cell Dev Biol* **31:** 132–140.

Zhang L, Eisenbarth GS. 2011. Prediction and prevention of Type 1 diabetes mellitus. *J Diabetes* **3:** 48–57.

Synthetic Botany

Christian R. Boehm,[1,4] Bernardo Pollak,[1,4] Nuri Purswani,[2] Nicola Patron,[3] and Jim Haseloff[1]

[1]Department of Plant Sciences, University of Cambridge, Cambridge CB2 3EA, United Kingdom

[2]The IBM Place I, Singapore, 486072, Singapore

[3]Earlham Institute, Norwich NR4 7UG, United Kingdom

Correspondence: jh295@cam.ac.uk

Plants are attractive platforms for synthetic biology and metabolic engineering. Plants' modular and plastic body plans, capacity for photosynthesis, extensive secondary metabolism, and agronomic systems for large-scale production make them ideal targets for genetic reprogramming. However, efforts in this area have been constrained by slow growth, long life cycles, the requirement for specialized facilities, a paucity of efficient tools for genetic manipulation, and the complexity of multicellularity. There is a need for better experimental and theoretical frameworks to understand the way genetic networks, cellular populations, and tissue-wide physical processes interact at different scales. We highlight new approaches to the DNA-based manipulation of plants and the use of advanced quantitative imaging techniques in simple plant models such as *Marchantia polymorpha*. These offer the prospects of improved understanding of plant dynamics and new approaches to rational engineering of plant traits.

The development of new technologies for the production of larger and improved quantities of goods from less feedstock has defined human innovation for thousands of years, especially in food production. To increase agricultural productivity, plant breeders have been selecting for advantageous traits in crops since 9000–10,000 BC (Zohary et al. 2012). One notable success is the selective breeding of modern maize from teosinte (Beadle 1939; Doebley 2004), in which a handful of genetic differences caused substantial changes in ear morphology, kernel number, and crop yield.

Technologies such as mutagenesis and introgressive hybridization, developed in the early to mid-20th century, are commonly used to increase genetic diversity in breeding populations of food crops, which show lower allelic variation compared with wild populations. Selective breeding still plays a major role in the production of new varieties but the emergence of modern plant biotechnology has led to a more targeted approach to increasing crop yields. Traits of agricultural importance successfully introduced to plants using recombinant DNA technology include herbicide resistance (Comai et al. 1985), drought resistance (Kumar et al. 2014), pest resistance (Bates et al. 2005), pathogen resistance (Brunner et al. 2011; Horvath et al. 2012; Jones et al. 2014), abiotic stress resistance (Jaglo-Ottosen et al. 1998), enhanced photosynthetic capacity (Ku et al. 2001), im-

[4]These authors contributed equally to this work.

Cite this article as *Cold Spring Harb Perspect Biol* doi: 10.1101/cshperspect.a023887

proved nitrogen use efficiency (Yanagisawa et al. 2004), and added nutritional value (Ye et al. 2000). However, although plants hold a unique promise for bioproduction at the gigatonne scale, efforts in genetic engineering of plants are lagging compared with microbial systems (Antunes et al. 2009, 2011; Liu et al. 2011, 2013; Koschmann et al. 2012; Wend et al. 2013; Zurcher et al. 2013; Fethe et al. 2014; Müller et al. 2014).

To exploit the unused potential of plants in expressing complex traits, we require (1) efficient tools and methods for genetic engineering, (2) simpler multicellular plant chassis that are amenable to rapid, high-throughput analysis, and (3) control over the biosynthesis, transport, and storage of metabolites in specialized cells within complex plant tissues. In the following, we highlight progress made in these three key areas of concern for the future development of plant synthetic biology.

NUCLEAR TRANSFORMATION IN PLANTS

Implementation of synthetic biology in plants calls for efficient methods for genetic manipulation. The most widely used technique for transformation of most plant species relies on the native capacity of virulent strains of *Agrobacterium tumefaciens* to infect plant tissue and to transfer a segment of its DNA (T-DNA) to the host cell (reviewed in Gelvin 2003). Because this method was adapted to allow delivery of transgenes, it has been used extensively: thousands of transformants can be obtained in a single experiment (Meyerowitz 1989; Ishizaki et al. 2008), providing the throughput required for forward genetic screens. The random nature of this transformation method has been used to mine the underlying biology of plants by mutational T-DNA insertion as well as gene trap methods (reviewed in Springer 2000). Enhancer traps have been used to indicate the presence of nearby endogenous enhancer elements driving an orthogonal transcriptional activator such as GAL4-VP16. T-DNA insertion in the vicinity of an enhancer allows temporal and/or tissue-specific expression of the activator, leading to restricted reporter gene expression according to the endogenous enhancer (Johnson et al. 2005; Laplaze et al. 2005; Gardner et al. 2009).

In addition to stable nuclear transformation mediated by *Agrobacterium*, transient methods have been developed for quantification of gene expression (Kapila et al. 1997) and bioproduction. For instance, *Nicotiana benthamiana* leaves can be infiltrated with *Agrobacterium* at multiple regions of a single leaf and screened for gene expression a few days after infection. This allows relatively fast characterization of libraries of genetic components in a common genetic background (Sparkes et al. 2006; Engler et al. 2014; Brückner et al. 2015). Transient leaf agroinfiltration in *Nicotiana* has been adopted as an efficient method for the optimization of metabolic pathways, such as artemisinin (van Herpen et al. 2010) and triterpene biosynthesis pathways (reviewed in Thimmappa et al. 2014), and the bioproduction of vaccines (D'Aoust et al. 2008, Mardanova et al. 2015). Large investments have been made in both the private and the public sectors to scale up production for vaccines against viruses like Ebola (ZMapp, Mapp Biopharmaceuticals) and influenza (Medicago). Protoplasts, single cells derived from tissues by digestion of cell walls, can also be transformed transiently. Protoplast transformation has been performed by electroporation (Fromm et al. 1985; Ou-Lee et al. 1986; Hauptmann et al. 1987; Negrutiu et al. 1987; Nishiguchi et al. 1987; Jones et al. 1989) or incubation in a PEG solution (Krens et al. 1982; Potrykus et al. 1985), and is established as one of the preferred methods for studying signaling pathways (reviewed in Sheen 2001). Furthermore, high-throughput protoplast transformation has been used to perform quantitative characterization of large libraries of genetic elements in *Arabidopsis thaliana* (*Arabidopsis*) and *Sorghum bicolor* (sorghum) (Schaumberg et al. 2016).

Precise methods for genomic inspection and reverse genetics screens have also been implemented in plants: for instance, CRISPR-Cas9 genome editing has been applied in a number of plant model systems including *Arabidopsis*, *Nicotiana spp.*, *Solanum lycopersicum* (tomato), *Oryza sativa* (rice), *Triticum aestivum* (wheat), *Zea mays* (maize), *Citrus sinensis* (orange), and

Marchantia polymorpha (*Marchantia*), at varying efficiencies (reviewed in Bortesi and Fischer 2015). Notably, CRISPR-Cas9 genome editing has been successfully used to promote homologous recombination and to simultaneously target multiple loci in *Arabidopsis* and *Nicotiana* (Li et al. 2014), as well as to induce targeted deletions in *Nicotiana* and rice (Lowder et al. 2015). Such methods are enabling unprecedented efficiency of construct delivery, genome refactoring, and chassis engineering in plants.

A COMMON SYNTAX FOR SYNTHETIC PLANT GENES

Increasingly, synthetic biologists rely on the use of modular DNA components to implement genetic circuits, along with a facile chassis for prototyping and troubleshooting, and tools for predicting behavior from mechanistic models. The implementation of assembly standards that allow parts, even those from multiple manufacturers, to be assembled together has underpinned invention in engineering disciplines and the generation of libraries of well-characterized standardized components is at the core of the synthetic biology paradigm.

The BioBrick assembly standard was the first widely adopted biological standard (Knight 2003; Shetty et al. 2008) and large number of parts, primarily for engineering prokaryotes, have been submitted to the Registry of Standard Biological Parts (Endy 2005). Eukaryotic organisms such as plants, however, require a different genetic syntax for the assembly of genes and pathways. Additionally, plasmids used for plant transformation often require specific features, such as the ability to replicate in a shuttle chassis such as *A. tumefaciens*. Recently, several alternative DNA assembly technologies have gained prominence because they offer the advantage of allowing the assembly of multiple parts in a single reaction (Engler et al. 2008; Gibson et al. 2009; Quan and Tian 2009; Li and Elledge 2012; De Kok et al. 2014). Although overlap-dependent methods such as Gibson assembly are powerful, they require custom oligonucleotides and amplification of even well characterized standard parts for each new assembly (Ellis et al. 2009; Liu et al. 2013; Patron 2014) unless standardized overlaps are used (Torella et al. 2013; Casini et al. 2014). The application of Type IIS restriction enzymes for assembling standard parts, known widely as "Golden Gate Cloning," has become widely used as an alternative approach because parts can be exchanged and assembled cheaply, easily, and in an automatable way without proprietary tools and reagents (Engler et al. 2009; Sarrion-Perdigones et al. 2011; Werner et al. 2012). Many commonly used sequences have been adapted for Type IIS assembly by various plant research laboratories (Sarrion-Perdigones et al. 2011; Weber et al. 2011; Emami et al. 2013; Lampropoulos et al. 2013; Binder et al. 2014; Engler et al. 2014; Vafaee et al. 2014), and a common syntax to enable the exchange of interoperable DNA parts for plants has recently been agreed by a large section of the plant research community (see Fig. 1; Patron et al. 2015).

This common syntax describes 12 fusion sites to enable the facile assembly of eukaryotic transcriptional units and establishes standards that allow standard parts to be exchanged and reused without adaptation. This agreement will aid in establishing registries of genetic components for plants and thus facilitate the design of genetic circuits. The availability of well-characterized parts is the first stage for establishing a higher-order abstraction to implement more complex functions in plants. Combining extensive collections of genetic elements together with high-throughput quantification methods will provide robust measurements to inform mathematical models that describe the statistical normalization methods required for developing parts and devices that will work as expected in complex multicellular organisms.

CONTROL OF TRANSGENE EXPRESSION IN PLANTS

Control of transgene expression can be implemented at multiple levels in plants and other eukaryotes. Regulation can be exerted at the transcriptional, posttranscriptional, and trans-

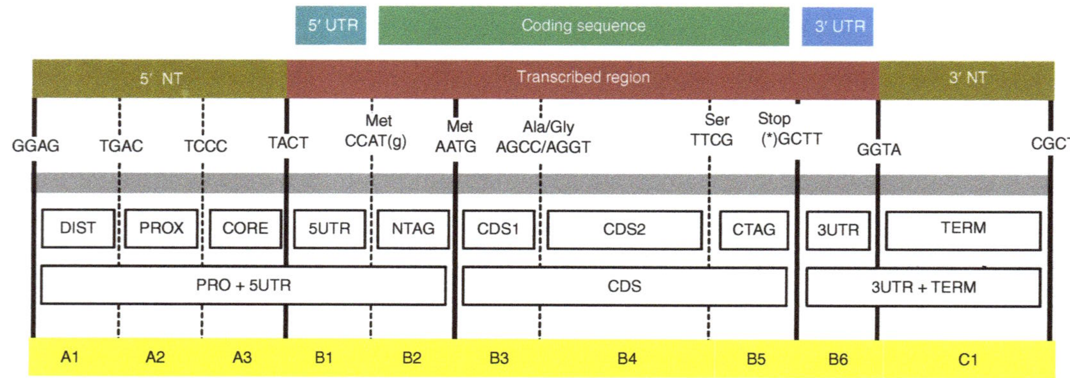

Position	Name	Function	5′ overhang	3′ overhang
A1	DIST	Distal promoter region, *cis* regulator, or transcriptional enhancer	GGAG	TGAC
A2	PROX	Proximal promoter region or transcriptional enhancer	TGAC	TCCC
A3	CORE	Minimal promoter region, including transcription start site	TCCC	TACT
B1	5UTR	5′ untranslated region	TACT	CCAT
B2	NTAG	Amino-terminal coding region	CCAT	AATG
B3	CDS1	Coding region—optional amino-terminal coding region	AATG	AGCC/AGGT
B4	CDS2	Coding region—no start or stop codon	AGCC/AGGT	TTCG
B5	CTAG	Carboxy-terminal coding region	TTCG	GCTT
B6	3UTR	3′ untranslated region	GCTT	GGTA
C1	TERM	Transcription terminator, including polyadenylation signal	GGTA	CGCT

Figure 1. Summary of the Phytobrick syntax for standardized plant DNA part composition and assembly. The Phytobrick syntax is a consolidation of Golden Gate, MoClo, and Golden Braid standards (Patron et al. 2015) and defines 12 fusion sites that divide eukaryotic genes into 10 basic functional units (A1–C1). The domains are listed with a brief description of their encoded function. Phytobrick parts can comprise the region between an adjacent pair of fusion sites or span many sites, and consist of portion(s) of a gene cloned into a plasmid flanked by a convergent pair of *BsaI* Type IIS restriction endonuclease recognition sequences. Parts can be assembled into complete transcriptional units in a one-pot, one-step digestion–ligation reaction provided compatible overhangs are produced on digestion and the acceptor plasmid has divergent *BsaI* recognition sites as well as a unique bacterial selection cassette.

Cite this article as *Cold Spring Harb Perspect Biol* doi: 10.1101/cshperspect.a023887

lational levels by both endogenous and synthetic orthogonal systems. Among proven orthogonal systems for modulation of gene expression are transactivation systems like the aforementioned GAL4 activator, a tTA tetracycline inducible system based on the *Escherichia coli* tetR repressor (Weinmann et al. 1994), and an IPTG-inducible pOp/LhG4 system based on the *E. coli lac* operon (Moore et al. 1998). The various inducible expression systems implemented in plants to date include the dexamethasone responsive rat glucocorticoid (GR) ligand-binding domain (Aoyama and Chua 1997; Craft et al. 2005; Samalova et al. 2005), ethanol/acetaldehyde-inducible ALCR transcription factor and *alcA* promoter (Caddick et al. 1998; Salter et al. 1998; Roslan et al. 2001), DNA-binding domain of the *lexA* oestrogen receptor (Bruce et al. 2000; Zuo et al. 2000), copper-inducible *ace1* promoter (Mett et al. 1993), and ecdysone receptor (EcR) ligand binding-domain inducible by insecticide methoxyfenozide (Martinez et al. 1999; Padidam et al. 2003; Koo et al. 2004). Other useful circuits comprise the Cre-Lox recombinase system that can be used for induction of transcription (Hoff et al. 2001) or transgene excision (Chakraborti et al. 2008), and a catalytically inactive version of Cas9 fused to regulatory domains. The latter has been shown to perform effective transcriptional activation and repression of an endogenous gene in *Nicotiana* (Piatek et al. 2015) and *Arabidopsis* (Lowder et al. 2015).

Posttranscriptional control over gene expression in plants has been enabled by a variety of RNA-based technologies. Artificial microRNAs (amiRNAs), for example, exploit the endogenous silencing machinery to repress gene expression. The extent of amiRNA-mediated repression can be tuned by altering their expression level (Yu and Pilot 2014). Systems for optimizing the specificity and the efficiency of the amiRNAs (Li et al. 2013) as well as for obtaining inducible multigene silencing (Goh et al. 2012) have also been implemented. Another approach to posttranscriptional control of gene expression in plants is based on riboswitches. Riboswitches are ligand-binding nucleic acid aptamers that are able to modulate gene expression by sequestering or exposing regulatory regions in transcripts via alternative nucleotide base-pairing (Anthony et al. 2012). The thiamine pyrophosphate riboswitch is an endogenous example in plants: It was identified in the 3′ end of the untranslated region of the *Arabidopsis thiC* gene, and exerts control by affecting pre-mRNA splicing of the transcript (Sudarsan et al. 2003; Wachter et al. 2007). Methods have been developed to tune the sensitivities of riboswitches for their cognate ligands (Beisel and Smolke 2009), and to expand their chemical diversity (reviewed in Link and Breaker 2009).

Technologies for nuclear transformation and control over gene expression have been combined to implement synthetic genetic logic systems in plants. Examples include a light-inducible gene expression system activated by red light and switched off by far-red light (Müller et al. 2014) as well as a range of in vivo biosensors to monitor levels of cytokinin (Müller and Sheen 2008; Zurcher et al. 2013), auxin (Wend et al. 2013), plant pathogens (Liu et al. 2011; Fethe et al. 2014), and elicitors (Koschmann et al. 2012). These are examples of synthetic circuits that have been constructed in plants using methods for DNA manipulation, transgene regulation, transformation, and analysis to yield basic systems for chassis engineering and genetic network design in plant synthetic biology.

METABOLIC ENGINEERING IN CELLULAR COMPARTMENTS

Although plant biotechnology has traditionally relied on the cumulative introduction of genes followed by selection (Bates et al. 2005), new technologies promise to streamline this process: methods such as genome-editing (reviewed in Raitskin and Patron 2016; and in Schiml and Puchta 2016), construction of synthetic eukaryotic chromosomes (Annaluru et al. 2014), or the application of operon-like gene clusters (Field et al. 2011; Nützmann and Osbourn 2014) enable unprecedented control over metabolic pathways.

Plants are ideally suited for the implementation of cooperative biosynthetic processes and dedicated storage compartments. One example of a highly beneficial metabolic interaction between different plant cell types is the cooperation of bundle sheath and mesophyll, which markedly enhances the efficiency of photosynthesis in C4 plants (Hibberd and Covshoff 2010). Another prominent instance of multicellular compartmentalization strategy is naturally observed in glandular trichomes of the wormwood *Artemisia annua*, which are specialized for production and storage of the antimalarial compound artemisinin (Graham et al. 2010). The potential of glandular tissues in various plant species for metabolic engineering of alkaloid, terpenoid, and fatty acid metabolism for applications in biotechnology has been reviewed in more detail elsewhere (McCaskill and Croteau 1999; Schilmiller et al. 2008).

Marchantia AS A BASAL MODEL CHASSIS FOR PLANT SYNTHETIC BIOLOGY

One of the major challenges in harnessing the potential of plant systems lies in our limited understanding of complex processes in cellular development. Morphogenetic processes in plants are driven by genetic programs, which shape the collective behavior of cohorts of cells. Vice versa, groups of cells display self-organizing properties, which affect the expression of individual genes. Plant metabolic processes take place in dedicated tissue compartments and are subject to multiscale feedback regulation. Thus, there is a need to develop suitable platforms for elucidating and engineering patterning processes defining the anatomy of plant tissues.

Ideally, such a platform requires a model system that serves as a link between microbes and higher plants. Genetic engineering of crop plants still remains a cumbersome undertaking owing to slow life cycles: Obtaining mature plants from seeds takes ~4 months for tomato (Kimura and Sinha 2008), 5 months for maize (Green and Phillips 1975), and 7 months for wheat (McHughen 1983). Furthermore, polyploidy (Vaughan et al. 2007) and high levels of genetic redundancy in higher plants (Dean et al. 1999) require mutagenesis of several loci to dissect gene function. *Arabidopsis* became popular as a model organism for molecular studies in the 1980s mostly because of its relative simplicity among angiosperm plants (Somerville and Koornneef 2002). Features of *Arabidopsis* include its phylogenetic relevance, a comparatively rapid life cycle of 2–3 months, small stature, abundant seed production, and relative ease of transformation (Flavell 2009). Although easier to work with than many crop species, *Arabidopsis* still shows notable levels of genetic redundancy (Briggs et al. 2006), and crosses to obtain plants homozygous for integrated transgenes (generally integrated hemizygously) are laborious (Wijnker et al. 2012).

Promising alternative candidate species of lower complexity can be found in the group of bryophytes, descendants of the earliest terrestrial plants (Goffinet and Shaw 2009). The best studied of these are the moss *Physcomitrella patens* and the thalloid liverwort *M. polymorpha*. Although both species show ample promise for bioengineering, *Marchantia* offers several unique advantages including a shorter life cycle, the production of *gemmae* as accessible and robust vegetative propagules (Shimamura 2012), a more streamlined genome (Sasaki et al. 2007; Zobell et al. 2010), and a simple spore-based transformation method for genetic manipulation.

In particular, *Marchantia* represents an excellent model for studying aspects of plant development and morphology. *Marchantia* forms simple, sheet-like tissues that can be characterized in terms of its distinct surfaces (Heberlein 1929): The lower surface shows root-like cells, called rhizoids, which are responsible primarily for the uptake of water and organic nutrients as well as the anchoring of the plant body to the substrate. The body of the thallus features scattered differentiated cells called oil bodies, which have been shown to play a role in isoprenoid metabolism (Suire et al. 2000). Finally, the upper surface is composed of primitive modular complexes for photosynthesis, each of which contains a permanently open pore for gas exchange. The most striking features of the upper surface of the thallus arguably are conical splash cups. Within them, *gemmae*

 Cite this article as *Cold Spring Harb Perspect Biol* doi: 10.1101/cshperspect.a023887

originate from single cells and are attractive specimens for engineering morphology, given their robustness, plasticity, and accessibility (Vötching 1885; Fitting 1935). In contrast to most developing organs in higher plants, *gemmae* are not buried within maternal or support tissues. Furthermore, *gemmae* retain their viability for approximately one year if stored at 4°C in agar, and are capable of vigorous germination on soil as well as artificial and sterile media (Miller 1964).

The morphological simplicity of *Marchantia* is matched by highly streamlined genetics. The sequences of the Y-chromosome, mitochondrial, and plastid genomes are available (Ohyama et al. 1986; Oda et al. 1992; Yamato et al. 2007), whereas an assembled nuclear genome sequence is expected to be published shortly, and can be found in draft form online (marchantia.info). The size of the nuclear genome has been estimated at 280 Mb by flow cytometry, and ~20,000 protein-coding genes have been predicted. In comparison to the 32,670 genes present in *Arabidopsis* and the 35,938 genes in *P. patens* (Yamato and Kohchi 2012), lower levels of gene redundancy appear to be present in *Marchantia*. For instance, a study of receptor-like kinases (RLK) concluded that the *Arabidopsis* genome contains 52 different RLK-families comprised of >600 *RLK* genes, whereas in *Marchantia* 26 RLK-families comprised of only 29 *RLK* genes were found (Sasaki et al. 2007). Another example is illustrated by MIKC* MADS box transcription factors: *P. patens* has been shown to include 11 members of this gene family (Rensing et al. 2008) whereas *Marchantia* only encodes one of these.

Marchantia possesses features that qualify it as a plant model chassis for quick and easy genetic engineering. Like moss, the life cycle of *Marchantia* is characterized by a dominant haploid gametophyte phase (O'Hanlon 1926). This allows the immediate expression of mutant genotypes in the phenotype. Through propagation of *gemmae*, haploid isogenic plants are obtained. Because *gemma* development originates from a single cell, transgenic *gemmae* do not display chimeric gene expression patterns. Further, during the gametophytic phase, it is possible to determine the sex of haploid plants by means of a PCR-based assay (Okada et al. 2000), allowing easy handling and management of female and male lines. To prepare crossing, the sexual phase of the life cycle can be induced by exposure to far-red light (Nakazato et al. 1999), and fertilization is performed by transferring sperm from mature antheridiophores to archegonia. Following fertilization and zygote formation, the life cycle continues in its diploid phase, leading to formation of spores packaged within yellow sporangia. Each cross can produce approximately 7 million spores that can be stored at −80°C for several years. Spore development takes place as soon the spores find a suitable substrate for germination. Unlike seeds of higher plants, spores are unicellular and do not possess a testa, allowing direct visualization of all early processes in development. During spore germination, spores undergo asymmetric cell division, initiate photosynthesis after differentiation of plastids, and begin to develop differentiated tissues within only a few days (see Fig. 2). Following germination, new plants grow vigorously on solid (Takenaka et al. 2000) or liquid media (Ishizaki et al. 2008) in sterile culture, and robustly regenerate from vegetative *gemmae* (Miller 1964), protoplasts (Ono et al. 1979), or tissue cuttings (Kubota et al. 2013). Rapid photoautotrophic growth of *Marchantia* cultures has been reported with a 1.76-day doubling time for culture mass, a faster rate than *Arabidopsis* (Katoh et al. 1979). The high regenerative capacity of liverworts is independent of externally supplied hormones (Vötching 1885), and well exceeds the reported regeneration efficiency of *Arabidopsis* (Valvekens et al. 1992) or wheat (McHughen 1983). The life cycle of *Marchantia* takes ~2 months to complete (Shimamura 2012), which is similar to *Arabidopsis* and faster than in *P. patens* (~3 months [Cove 2005]).

Marchantia is one of few plant species for which chloroplast transformation protocols have been confirmed by at least two independent studies (Bock 2015). Stable transplastomic lines of *Marchantia* encoding dual spectinomycin and streptomycin resistance were generated through particle bombardment of sporelings

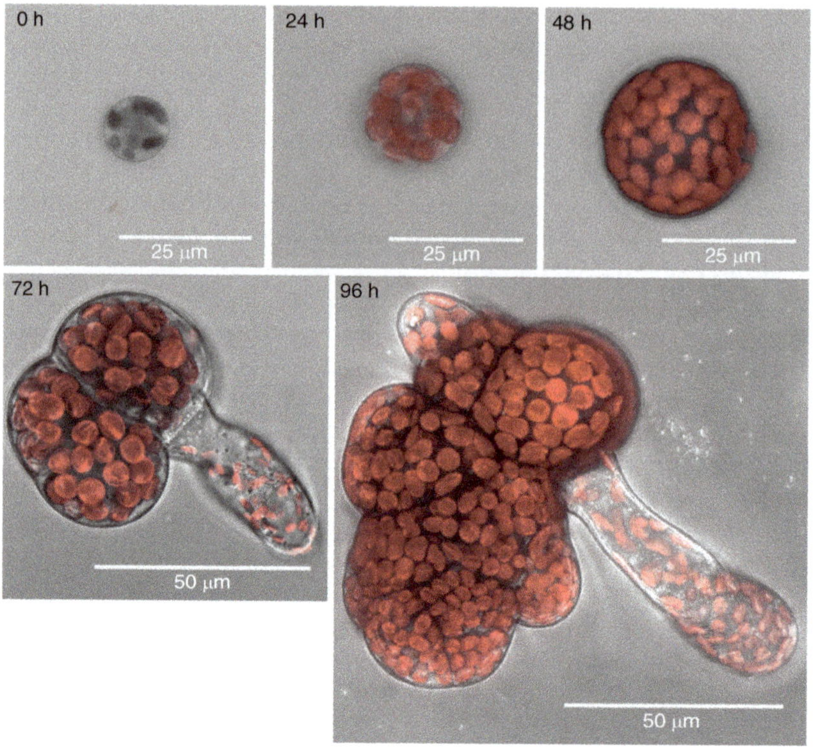

Figure 2. Spore germination. *Marchantia* spores were germinated on a nutrient agar surface. The spores were examined under a 63×, NA 1.2 objective at 0, 24, 48, 72, and 96 h after germination using a Leica SP5 confocal laser scanning microscope. A 488-nm laser was used to collect transmission images (grayscale channel), and these were overlaid with images of chlorophyll fluorescence (488 nm excitation, 680–700 nm emission, red channel). Z-series of images (2 µm apart) were collected and merged to provide views of the developing sporelings at different stages of growth. Scale bars are indicated in each image.

(Chiyoda et al. 2007). This approach was recently extended to the expression of a fluorescent reporter gene from the *Marchantia* plastome (Boehm et al. 2016), which allows characterization of plastid promoters in this model system (see Fig. 3). A notable feature of the *Marchantia* plastome is the apparent absence of mechanisms for RNA editing (Ohyama et al. 2009), which may substantially simplify rational engineering of this organellar genome. Although simpler organisms may lack some of the complexity of angiosperms, the basal plant model *Marchantia* can enhance our still fragmented understanding of fundamental aspects of gene regulation, expression, and function in plant metabolism and development as well as provide a more rapid and tractable testbed for synthetic biology.

GENETIC MANIPULATION OF CHLOROPLASTS

In bridging the gap between microbes and plants as platforms for synthetic biology, a particularly attractive chassis is the chloroplast. Sharing its evolutionary heritage with cyanobacteria, this organelle possesses prokaryote-like regulation of gene expression (Mayfield et al. 1995), and is the major biosynthetic compartment in plants. The high biosynthetic capacity of chloroplasts has been used for production of a number of biofuel enzymes (Verma et al. 2010) as well as resistance proteins, antibodies, biopharmaceutical proteins, and vaccine antigens (Wani et al. 2010). Accumulation of large amounts of target protein has been achieved in chloroplasts, exceeding 70% of the

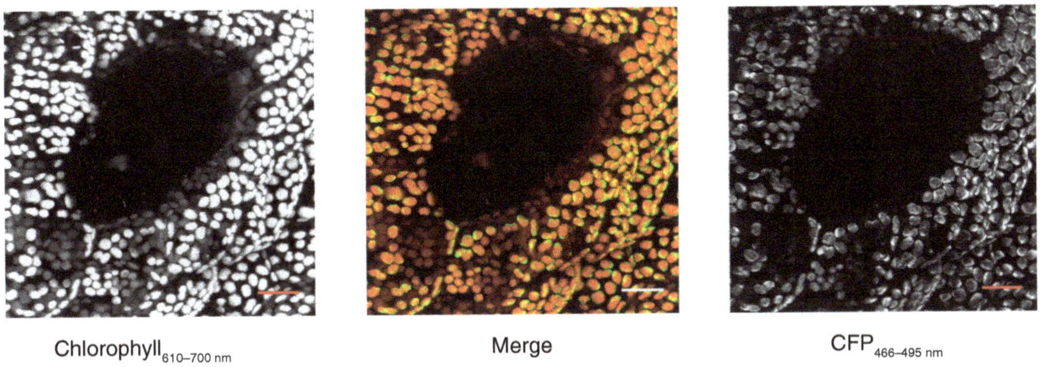

Figure 3. Transgenic chloroplasts of *Marchantia polymorpha* expressing the cyan fluorescent protein mTurquoise2 under control of the tobacco *psbA* promoter. Mature thallus of *Marchantia* was mounted in water under a coverslip, and examined under a 63×, NA 1.2 objective using a Leica SP5 confocal laser-scanning microscope. Cells surrounding an air pore were imaged using 458-nm laser excitation. Emission wavelengths were collected for chloroplast autofluorescence (610–700 nm, *left*) and cyan fluorescent protein (CFP) fluorescence (466–495 nm, *right*), with merged images in red and green channels, respectively, shown *center*. Scale bar, 20 μm.

plant's total soluble protein (Oey et al. 2009). No gene silencing has been observed in chloroplasts despite such high accumulation of foreign transcripts (169 times higher than in nuclear transgenic plants, Lee et al. 2003) or foreign protein (46% of total leaf protein [De Cosa et al. 2001]). Metabolic engineering for the production of bioplastic monomers (Bohmert-Tatarev et al. 2011) and compounds of nutritional relevance (Craig et al. 2008; Hasunuma et al. 2008; Apel and Bock 2009) has also been applied to chloroplasts. The high biosynthetic capacity of the chloroplast is closely linked to the polyploid nature of the system: At 10–100 chloroplasts per cell, and 10–1000 genomes per chloroplast (Bendich 1987), stable integration of a transgene into the chloroplast genome enables a substantial amplification in transgene copy number. The high ploidy of the plastome can entail the need for several rounds of selective regeneration following transgene introduction to segregate transplastomic organelles and establish homoplasmy (Maliga 2004). However, homoplasmy has also been achieved immediately after primary selection (Chiyoda et al. 2014).

The process of transgene integration into the plastome following delivery into the chloroplast by means of particle bombardment (Svab et al. 1990) or PEG-mediated transformation (Golds et al. 1993) is mediated by homologous recombination. This mechanism has enabled the introduction of point mutations in a site-specific manner (Przibilla et al. 1991). As a consequence, the risk of inconsistent gene expression by the positional effect in T-DNA-mediated nuclear transformation is alleviated in the chloroplast environment (Daniell et al. 2002). The high efficiency of homologous recombination in the chloroplast (Blowers et al. 1989) also allows the simultaneous introduction of modifications to several sites of the chloroplast genome by means of cotransformation (Kindle et al. 1991), encouraging the implementation of phenotypic traits based on multiple foreign genes. Multiple genes may be conveniently organized in operon-like polycistronic units (Hasunuma et al. 2008), which can be processed into more efficiently translated monocistronic transcripts by the incorporation of intercistronic expression elements (Lu et al. 2013). Previous studies have also highlighted the activity of chloroplast promoters in bacteria (Brixey et al. 1997) and of bacterial promoters in the chloroplast (Newell et al. 2003). Another bacterial gene expression element, the widely used lac repressor from *E. coli*, has been adapted for IPTG-inducible chloroplast transgene expression (Mühlbauer and Koop 2005). Alternatively, chloroplast-based translational regulation of transgene expression can be implemented via riboswitches under control of the externally applied ligand theopylline (Verhounig et al. 2010; Emadpour et al. 2015).

Large DNA fragments exceeding 50 kb have been successfully incorporated into the plastome (Adachi et al. 2007). In fact, on the basis of recent advances in DNA synthesis and assembly technology, synthesis of entire tailor-made plastid genomes has become feasible (Scharff and Bock 2014). Today, metabolic engineering in microbial systems is predominantly driven by generation of large numbers of circuit variants followed by selection of a desired phenotype. Translation of this approach into plants has not been widely adopted primarily because of the requirement for generation and analysis of a sufficient number of transgenic events in the context of long generation times and polyploidy of established model species. Compared with random integration in the nuclear genome by transgenesis or interbreeding (Karunanandaa et al. 2005), integration in the chloroplast genome is achieved by homologous recombination and thus DNA is inserted at a chosen location. This allows a far smaller number of events to be compared as they are free from position effects as well as from epigenetic gene silencing, which is only encountered in the nuclear genome (Lu et al. 2013). Another notable advantage of transplastomic technology over nuclear transformation lies in maternal inheritance of plastid DNA, which greatly reduces the risk of unwanted transgene transmission via pollen and markedly increases the level of containment for field release of transgenic plants.

MODELS FOR MULTICELLULAR GROWTH AND INTEGRATED TECHNIQUES FOR QUANTITATIVE MICROSCOPY AND IMAGE ANALYSIS

Specialized structures in plants carry out a number of highly efficient processes for photosynthesis, secondary metabolite production, and compound storage. These structures are generated during morphogenesis through the concerted behavior of cell populations during development. Our ability to modify plant form is currently limited by our understanding and control of multicellular processes during morphogenesis. Increasing evidence from mutagenesis studies in angiosperms has suggested that morphogenesis is shaped by a combination of processes embodied in classical organismal (Green 1980) and cellular (Beemster et al. 2003) theories: the "neo"-cell theory proposes that feedback and interplay between cellular and organism-wide processes cooperate in the patterning of tissues (Tsukaya 2003). Evidence indicates that the polarity of cell division and elongation is regulated by the plant cell cytoskeleton and interaction with local cell wall determinants, wall strain, or geometry (Fig. 4). Subsequent genetic interactions mediated by protein or phytohormone traffic between cells can provide positional signals to trigger gene expression, cell proliferation, differentiation, and organ formation (Bohn-Courseau 2010). Consequent cellular growth generates physical strains that are instantaneously transmitted across tissues and constrain growth. The physical constraints on cell size and shape regulate timing and orientation of individual cell divisions and expansion, and therefore guide morphogenesis. This results in hierarchical, highly parallel, and feedback-regulated systems in which the DNA program and phenotype are not directly related, but part of an emergent self-organizing system. DNA-based reprogramming of plants can predictably alter individual cell behaviors, but the grand challenge of plant synthetic biology is to understand and predict how altered cell logic can propagate through large cell populations, modify tissue physics and local patterns of cellular growth, and produce defined morphologies. Effective multiscale modeling of cellular growth will be a prerequisite for predictable engineering of plant morphogenesis.

The past decade has seen the emergence of increasingly effective models for plant growth in which the software describes the genetic, cellular, and biophysical properties of growing tissues (Jönsson and Krupinski 2010; De Vos et al. 2012; Prusinkiewicz and Runions 2012). Although these models have been developed primarily to provide insight into plant develop-

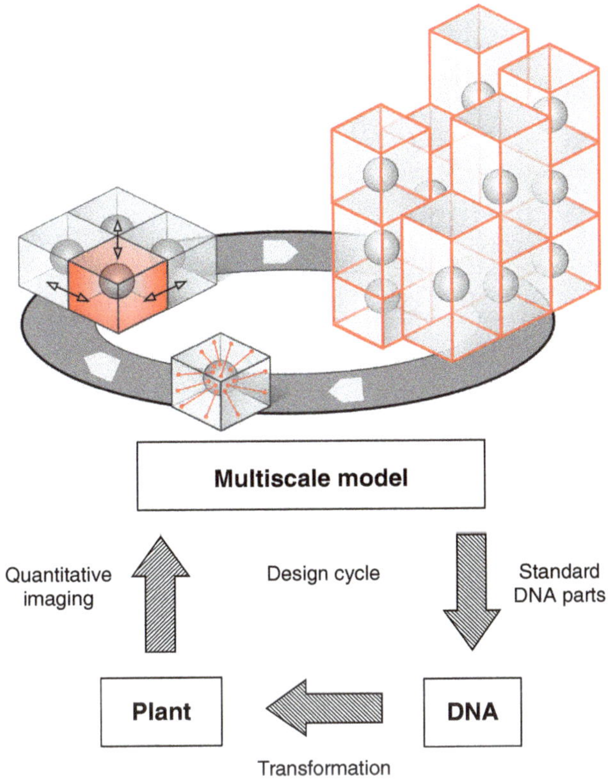

Figure 4. Multiscale model of plant growth for engineering synthetic botanical forms. To model and predict the form of reprogrammed plants, integrated, multiscale models for plant growth are required. These software models need to capture (1) the interaction between cytoskeletal elements and local cell wall determinants, strain or geometry regulating the polarity of cell division and elongation, (2) genetic interactions between neighboring cells that can trigger gene expression and cell proliferation and differentiation, and (3) cellular growth that results in physical strains that are transmitted across tissues and constrain cell growth, because physical constraints on cell size and shape regulate timing and orientation of individual cell divisions and guide morphogenesis. Multiscale models provide an essential tool for engineering multicellular systems. Standardized DNA parts facilitate assembly of DNA circuits that may be introduced into plant systems by transformation, and the performance of DNA-based circuits can be measured using quantitative imaging techniques. Although a genetic circuit may regulate or alter the behavior of an individual cell in an easily predictable fashion, the consequences of altered cell interactions, propagation of changes across large cell populations, changes in tissue-wide physical and chemical interactions, and feedback on the properties of individual cells are difficult to predict. However, this type of system, characterized by cross talk and emergent properties, can be captured accurately by multiscale models. The models form an essential part of any design–build–test cycle for DNA-based engineering of plants.

mental processes, they allow the integration of genetic, cellular, and whole-tissue properties, in a way that facilitates DNA-based reprogramming of large-scale plant growth. To complement these software tools, synthetic biologists require an experimental platform that allows simultaneous measurements of gene expression, growth, and division of individual cells at the scale of the entire organism. Sporelings and *gemmae*, the reproductive propagules of *Marchantia*, are well suited for such purpose. Their prostrate morphology and exposed mode of development facilitate the application of quantitative imaging techniques (Fig. 5): all

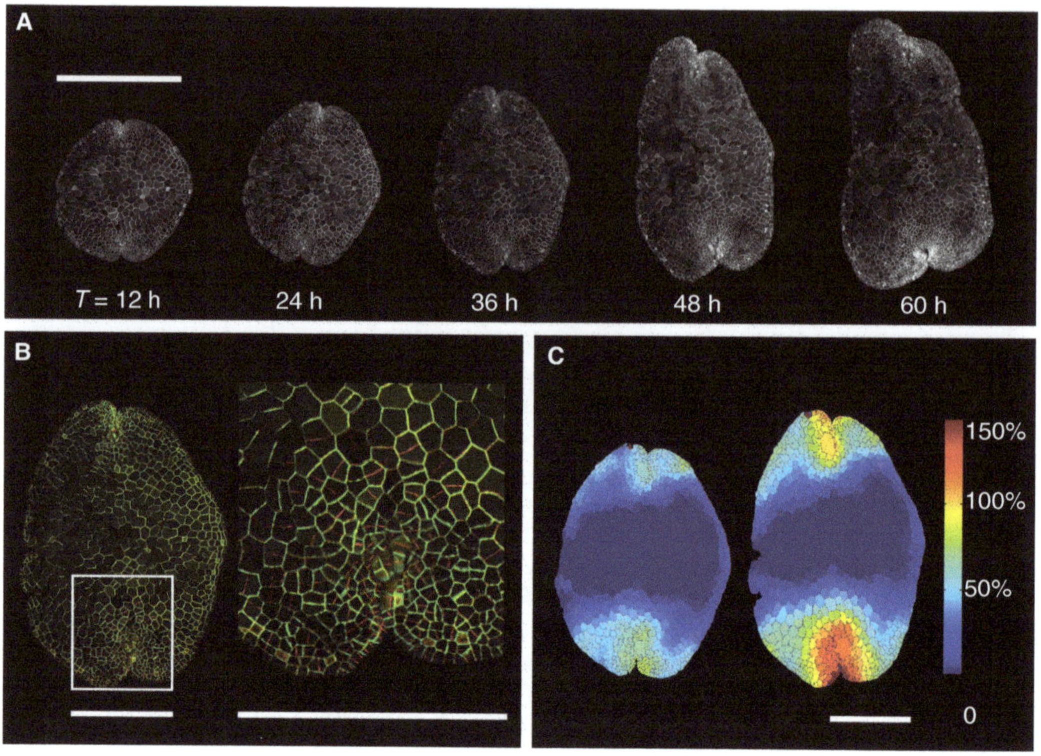

Figure 5. *Marchantia gemmae* as testbeds for quantitative parameterization of plant growth. (*A*) A transgenic line of *Marchantia* was generated that expressed a green fluorescent protein localized to the plasma membrane. A *gemma* from this transgenic line was transferred to a nutrient agar surface and examined after 12, 24, 36, 48, and 60 h of growth using a Leica SP5 confocal laser scanning microscope. Z-series of optical sections were collected for the same *gemma*, and maximum intensity projections are shown for each point during growth. Scale bar, 500 μm. (*B*) The 24-h (green channel) and 36-h (red channel) images from the time course were matched using warp-registration image-processing techniques and overlaid. A white box is positioned over one of the apical notches, and this corresponds to the enlarged view shown in the *inset*. The frequency and orientation of apex-localized cell divisions can be directly visualized in a single *gemma*. Scale bar, 200 μm. (*C*) The plasma membrane-localized marker allows accurate segmentation of cell geometry during growth of living plants. Quantitative parameters such as cell expansion rate can be mapped across a single *gemma*. Measurements of percent clonal sector expansion per 12 h are shown as a color map. Scale bar, 200 μm.

cells in the surface layer are accessible for imaging to quantify gene expression based on *in planta* cytometry (Federici et al. 2012). Robust protocols to map the dynamics of cell expansion, division, and differentiation on the surface of the *gemma* during the early stages of *gemma* development have been developed (N Purswani, unpubl.).

This combination of advanced imaging techniques and in silico models has proven useful for examining and predicting laws for cell division and tissue growth in multicellular algae (Dupuy et al. 2010; Besson and Dumais 2011). In an engineering context, these methods can form an essential part of a design–build–test cycle for reprogramming plant growth (see Fig. 4). Prototype designs can be assembled from modular DNA parts and transformed into plants, in which the properties of a synthetic gene circuit and impact on plant growth can be measured by quantitative microscopy. Multiscale software models can allow evaluation and even prediction of emergent processes. The simplicity, experimental accessibility, and

 Cite this article as *Cold Spring Harb Perspect Biol* doi: 10.1101/cshperspect.a023887

morphological plasticity of *Marchantia* sporelings and *gemmae* (Vötching 1885; Fitting 1935), ease of quantitative measurements, and growing sophistication of software models for cellular growth provide a pathway to the rational design of morphogenetic programs in plants.

CONCLUDING REMARKS

Marchantia provides an experimental model to help bridge the gap between the relative simplicity of microbes and the complexity of higher plants. In particular, *Marchantia* is well-suited for prototyping genetic circuits before their application in a higher plant species of interest because of its amenability to high-throughput transformation and screening of large numbers of circuit variants: As an integral part of its rapid life cycle, *Marchantia* produces millions of spores, which can be transformed within a week using available protocols (Ishizaki et al. 2008; Tsuboyama and Kodama 2014). Owing to the haploid-dominant nature of *Marchantia*, phenotypic effects in transgenic plants can be immediately screened without the need for repeated selective regeneration to obtain homozygous lines. Screening can be performed by means of existing quantitative microscopy methods (Federici et al. 2012), and image processing and parameters can be extrapolated for physico-genetic modeling of multicellular plant tissues (Dupuy et al. 2010) to refine the next iteration of the design cycle. In the future, this process will benefit not only from the improvement of current techniques for DNA assembly, plant transformation, quantitative microscopy, and computational modeling, but especially from the availability of the *Marchantia* nuclear genome. This sequence information will guide the discovery of cell-type- and developmental stage–specific marker genes. The identification of tissue and cell-type-specific promoters in *Marchantia* has particular merit not only for developmental studies but also for metabolic engineering. Notably, compounds derived from plant secondary metabolism still play a dominant role in the discovery of new biopharmaceuticals (Newman and Cragg 2016), despite the advances of modern chemistry. In this light, the ultimate challenge for biological engineers will not be manipulating individual genes, but exercising control over the collective behavior of metabolic pathways and cohorts of cells and refactoring a multicellular organism's body plan to generate specialized structures and organs for manufacturing and storage of compounds of interest. For example, applications of spatially controlled gene expression in *Marchantia* may include targeted manipulation of carbon fixation in assimilatory filaments (Goffinet and Shaw 2009) or production of biofuel components in oil cells (Suire et al. 2000). Control circuits and synthetic pathways proven in this context may serve as valuable tools for some of the largest contemporary challenges in plant metabolic engineering, such as the refactoring of nitrogen fixation pathways in plants (Rogers and Oldroyd 2014) or the introduction of C4 photosynthesis into C3 crops (Leegood 2013).

ACKNOWLEDGMENTS

The authors acknowledge support by the Gates Cambridge Trust (to C.R.B.), Becas Chile and the Cambridge Trust (to B.P.), and the Biotechnology and Biological Sciences Research Council and Engineering and Physical Sciences Research Council (OpenPlant Grant No. BB/L014130/1) to N.P. and J.H.

REFERENCES

Adachi T, Takase H, Tomizawa K. 2007. Introduction of a 50 kbp DNA fragment into the plastid genome. *Biosci Biotechnol Biochem* **71:** 2266–2273.

Annaluru N, Muller H, Mitchell LA, Ramalingam S, Stracquadanio G, Richardson SM, Dymond JS, Kuang Z, Scheifele LZ, Cooper EM, et al. 2014. Total synthesis of a functional designer eukaryotic chromosome. *Science* **344:** 55–58.

Anthony PC, Perez CF, Garcia-Garcia C, Block SM. 2012. Folding energy landscape of the thiamine pyrophosphate riboswitch aptamer. *Proc Natl Acad Sci* **109:** 1485–1489.

Antunes MS, Morey KJ, Tewari-Singh N, Bowen TA, Smith JJ, Webb CT, Hellinga HW, Medford JI. 2009. Engineering key components in a synthetic eukaryotic signal transduction pathway. *Mol Syst Biol* **5:** 270.

Antunes MS, Morey KJ, Smith JJ, Albrecht KD, Bowen TA, Zdunek JK, Troupe JF, Cuneo MJ, Webb CT, Hellinga HW, et al. 2011. Programmable ligand detection system in plants through a synthetic signal transduction pathway. *PLoS ONE* **6:** e16292.

Aoyama T, Chua N-H. 1997. A glucocorticoid-mediated transcriptional induction system in transgenic plants. *Plant J* **11:** 605–612.

Apel W, Bock R. 2009. Enhancement of carotenoid biosynthesis in transplastomic tomatoes by induced lycopene-to-provitamin A conversion. *Plant Physiol* **151:** 59–66.

Bates SL, Zhao JZ, Roush RT, Shelton AM. 2005. Insect resistance management in GM crops: Past, present and future. *Nat Biotechnol* **23:** 57–62.

Beadle G. 1939. Teosinte and the origin of maize. *J Hered* **30:** 245–247.

Beemster GTS, Fiorani F, Inzé D. 2003. Cell cycle: The key to plant growth control? *Trends Plant Sci* **8:** 154–158.

Beisel CL, Smolke CD. 2009. Design Principles for riboswitch function. *PLoS Comput Biol* **5:** e1000363.

Bendich AJ. 1987. Why do chloroplasts and mitochondria contain so many copies of their genome? *BioEssays* **6:** 279–282.

Besson S, Dumais J. 2011. Universal rule for the symmetric division of plant cells. *Proc Natl Acad Sci* **108:** 6294–6299.

Binder A, Lambert J, Morbitzer R, Popp C, Ott T, Lahaye T, Parniske M. 2014. A modular plasmid assembly kit for multigene expression, gene silencing and silencing rescue in plants. *PLoS ONE* **9:** e88218.

Blowers AD, Bogorad L, Shark KB, Sanford JC. 1989. Studies on *Chlamydomonas* chloroplast transformation: Foreign DNA can be stably maintained in the chromosome. *Plant Cell* **1:** 123–132.

Bock R. 2015. Engineering plastid genomes: Methods, tools, and applications in basic research and biotechnology. *Annu Rev Plant Biol* **66:** 211–241.

Boehm CR, Ueda M, Nishimura Y, Shikanai T, Haseloff J. 2016. A cyan fluorescent reporter expressed from the chloroplast genome of *Marchantia polymorpha*. *Plant Cell Physiol* **57:** 291–299.

Bohmert-Tatarev K, McAvoy S, Daughtry S, Peoples OP, Snell KD. 2011. High levels of bioplastic are produced in fertile transplastomic tobacco plants engineered with a synthetic operon for the production of polyhydroxybutyrate. *Plant Physiol* **155:** 1690–1708.

Bohn-Courseau I. 2010. Auxin: A major regulator of organogenesis. *C R Biol* **333:** 290–296.

Bortesi L, Fischer R. 2015. The CRISPR/Cas9 system for plant genome editing and beyond. *Biotechnol Adv* **33:** 41–52.

Briggs G, Osmont K, Shindo C, Sibout R, Hardtke C. 2006. Unequal genetic redundancies in *Arabidopsis*—A neglected phenomenon? *Trends Plant Sci* **11:** 492–498.

Brixey PJ, Guda C, Daniell H. 1997. The chloroplast psbA promoter is more efficient in *Escherichia coli* than the T7 promoter for hyperexpression of a foreign protein. *Biotechnol Lett* **19:** 395–400.

Bruce W, Folkerts O, Garnaat C, Crasta O, Roth B, Bowen B. 2000. Expression profiling of the maize flavonoid pathway genes controlled by estradiol-inducible transcription factors CRC and P. *Plant Cell* **12:** 65–80.

Brückner K, Schäfer P, Weber E, Grützner R, Marillonnet S, Tissier A. 2015. A library of synthetic transcription activator-like effector-activated promoters for coordinated orthogonal gene expression in plants. *Plant J* **82:** 707–716.

Brunner S, Hurni S, Herren G, Kalinina O, von Burg S, Zeller SL, Schmid B, Winzeler M, Keller B. 2011. Transgenic *Pm3b* wheat lines show resistance to powdery mildew in the field. *Plant Biotechnol J* **9:** 897–910.

Caddick MX, Greenland AJ, Jepson L, Krause KP, Qu N, Riddell KV, Salter MG, Schuch W, Sonnewald U, Tomsett AB. 1998. An ethanol inducible gene switch for plants used to manipulate carbon metabolism. *Nat Biotechnol* **16:** 177–180.

Casini A, MacDonald JT, de Jonghe J, Christodoulou G, Freemont PS, Baldwin GS, Ellis T. 2014. One-pot DNA construction for synthetic biology: The modular overlap-directed assembly with linkers (MODAL) strategy. *Nucleic Acids Res* **42:** e7.

Chakraborti D, Sarkar A, Mondal HA, Schuermann D, Hohn B, Sarmah BK, Das S. 2008. Cre/*lox* system to develop selectable marker free transgenic tobacco plants conferring resistance against sap sucking homopteran insect. *Plant Cell Rep* **27:** 1623–1633.

Chiyoda S, Linley PJ, Yamato KT, Fukuzawa H, Yokota A, Kohchi T. 2007. Simple and efficient plastid transformation system for the liverwort *Marchantia polymorpha* L. suspension-culture cells. *Transgenic Res* **16:** 41–49.

Chiyoda S, Yamato KT, Kohchi T. 2014. Plastid transformation of sporelings and suspension-cultured cells from the liverwort *Marchantia polymorpha* L. *Methods Mol Biol* **1132:** 439–447.

Comai L, Facciotti D, Hiatt WR, Thompson G, Rose RE, Stalker DM. 1985. Expression in plants of a mutant *aroA* gene from *Salmonella typhimurium* confers tolerance to glyphosate. *Nature* **317:** 741–744.

Cove D. 2005. The moss *Physcomitrella patens*. *Annu Rev Genet* **39:** 339–358.

Craft J, Samalova M, Baroux C, Townley H, Martinez A, Jepson I, Tsiantis M, Moore I. 2005. New pOp/LhG4 vectors for stringent glucocorticoid-dependent transgene expression in *Arabidopsis*. *Plant J* **41:** 899–918.

Craig W, Lenzi P, Scotti N, De Palma M, Saggese P, Carbone V, McGrath Curran N, Magee AM, Medgyesy P, Kavanagh TA, et al. 2008. Transplastomic tobacco plants expressing a fatty acid desaturase gene exhibit altered fatty acid profiles and improved cold tolerance. *Transgenic Res* **17:** 769–782.

Daniell H, Khan MS, Allison L. 2002. Milestones in chloroplast genetic engineering: An environmentally friendly era in biotechnology. *Trends Plant Sci* **7:** 84–91.

D'Aoust MA, Lavoie PO, Couture MMJ, Trépanier S, Guay JM, Dargis M, Mongrand S, Landry N, Ward BJ, Vézina LP. 2008. Influenza virus-like particles produced by transient expression in *Nicotiana benthamiana* induce a protective immune response against a lethal viral challenge in mice. *Plant Biotechnol J* **6:** 930–940.

Dean RE, Dahlberg JA, Hopkins MS, Mitchell SE, Kresovich S. 1999. Genetic redundancy and diversity among "Orange" accessions in the U.S. National Sorghum Collection as assessed with simple sequence repeat (SSR) markers. *Crop Sci* **39:** 1215–1221.

De Cosa B, Moar W, Lee SB, Miller M, Daniell H. 2001. Overexpression of the *Bt cry2Aa2* operon in chloroplasts leads to formation of insecticidal crystals. *Nat Biotechnol* **19:** 71–74.

De Kok S, Stanton LH, Slaby T, Durot M, Holmes VF, Patel KG, Platt D, Shapland EB, Serber Z, Dean J, et al. 2014. Rapid and reliable DNA assembly via ligase cycling reaction. *ACS Synth Biol* **3:** 97–106.

De Vos D, Dzhurakhalov A, Draelants D, Bogaerts I, Kalve S, Prinsen E, Vissenberg K, Vanroose W, Broeckhove J, Beemster GTS. 2012. Towards mechanistic models of plant organ growth. *J Exp Bot* **63:** 3325–3337.

Doebley J. 2004. The genetics of maize evolution. *Annu Rev Genet* **38:** 37–59.

Dupuy L, Mackenzie J, Haseloff J. 2010. Coordination of plant cell division and expansion in a simple morphogenetic system. *Proc Natl Acad Sci* **107:** 2711–2716.

Ellis T, Wang X, Collins JJ. 2009. Diversity-based, model-guided construction of synthetic gene networks with predicted functions. *Nat Biotechnol* **27:** 465–471.

Emadpour M, Karcher D, Bock R. 2015. Boosting riboswitch efficiency by RNA amplification. *Nucleic Acids Res* **43:** e66.

Emami S, Yee MC, Dinneny JR. 2013. A robust family of Golden Gate *Agrobacterium* vectors for plant synthetic biology. *Front Plant Sci* **4:** 339.

Endy D. 2005. Foundations for engineering biology. *Nature* **438:** 449–453.

Engler C, Kandzia R, Marillonnet S. 2008. A one pot, one step, precision cloning method with high throughput capability. *PLoS ONE* **3:** e3647.

Engler C, Gruetzner R, Kandzia R, Marillonnet S. 2009. Golden gate shuffling: A one-pot DNA shuffling method based on type IIs restriction enzymes. *PLoS ONE* **4:** e5553.

Engler C, Youles M, Gruetzner R, Ehnert TM, Werner S, Jones JDG, Patron NJ, Marillonnet S. 2014. A golden gate modular cloning toolbox for plants. *ACS Synth Biol* **3:** 839–843.

Federici F, Dupuy L, Laplaze L, Heisler M, Haseloff J. 2012. Integrated genetic and computation methods for in planta cytometry. *Nat Methods* **9:** 483–485.

Fethe MH, Liu W, Burris JN, Millwood RJ, Mazarei M, Rudis MR, Yeaman DG, Dubosquielle M, Stewart CN. 2014. The performance of pathogenic bacterial phytosensing transgenic tobacco in the field. *Plant Biotechnol J* **12:** 755–764.

Field B, Fiston-Lavier AS, Kemen A, Geisler K, Quesneville H, Osbourn AE. 2011. Formation of plant metabolic gene clusters within dynamic chromosomal regions. *Proc Natl Acad Sci* **108:** 16116–16121.

Fitting H. 1935. Untersuchungen über die induktion der dorsiventralität bei den keimenden brutkörpern *Marchantia* und *Lunularia*. *Jahrb Wiss Bot* **82:** 333.

Flavell R. 2009. Role of model plant species. *Methods Mol Biol* **513:** 1–18.

Fromm M, Taylor LP, Walbot V. 1985. Expression of genes transferred into monocot and dicot plant cells by electroporation. *Proc Natl Acad Sci* **82:** 5824–5828.

Gardner MJ, Baker AJ, Assie JM, Poethig RS, Haseloff JP, Webb AA. 2009. *GAL4 GFP* enhancer trap lines for analysis of stomatal guard cell development and gene expression. *J Exp Bot* **60:** 213–226.

Gelvin SB. 2003. *Agrobacterium*-mediated plant transformation: The biology behind the "Gene-Jockeying" Tool. *Microbiol Mol Biol Rev* **67:** 16–37.

Gibson DG, Young L, Chuang RY, Venter JC, Hutchison CA, Smith HO. 2009. Enzymatic assembly of DNA molecules up to several hundred kilobases. *Nat Methods* **6:** 343–345.

Goffinet B, Shaw AJ. 2009. *Bryophyte biology*, 2nd ed. Cambridge University Press, Cambridge.

Goh HH, Sloan J, Dorca-Fornell C, Fleming A. 2012. Inducible repression of multiple expansin genes leads to growth suppression during leaf development. *Plant Physiol* **159:** 1759–1770.

Golds T, Maliga P, Koop HU. 1993. Stable plastid transformation in PEG-treated protoplasts of *Nicotiana tabacum*. *Nat Biotechnol* **11:** 95–97.

Graham IA, Besser K, Blumer S, Branigan CA, Czechowski T, Elias L, Guterman I, Harvey D, Isaac PG, Khan AM, et al. 2010. The genetic map of *Artemisia annua* L. identifies loci affecting yield of the antimalarial drug artemisinin. *Science* **327:** 328–331.

Green PB. 1980. Organogenesis—A biophysical view. *Annu Rev Plant Physiol* **31:** 51–82.

Green CE, Phillips RL. 1975. PPlant regeneration from tissue cultures of maize. *Crop Sci* **15:** 417–421.

Hasunuma T, Miyazawa SI, Yoshimura S, Shinzaki Y, Tomizawa KI, Shindo K, Choi SK, Misawa N, Miyake C. 2008. Biosynthesis of astaxanthin in tobacco leaves by transplastomic engineering. *Plant J* **55:** 857–868.

Hauptmann RM, Ozias-Akins P, Vasil V, Tabaeizadeh Z, Rogers SG, Horsch RB, Vasil IK, Fraley RT. 1987. Transient expression of electroporated DNA in monocotyledonous and dicotyledonous species. *Plant Cell Rep* **6:** 265–270.

Heberlein EA. 1929. Morphological studies on a new species of *Marchantia*. *Bot Gaz* **88:** 417–429.

Hibberd JM, Covshoff S. 2010. The regulation of gene expression required for C_4 photosynthesis. *Annu Rev Plant Biol* **61:** 181–207.

Hoff T, Schnorr KM, Mundy J. 2001. A recombinase-mediated transcriptional induction system in transgenic plants. *Plant Mol Biol* **45:** 41–49.

Horvath DM, Stall RE, Jones JB, Pauly MH, Vallad GE, Dahlbeck D, Staskawicz BJ, Scott JW. 2012. Transgenic resistance confers effective field level control of bacterial spot disease in tomato. *PLoS ONE* **7:** e42036.

Ishizaki K, Chiyoda S, Yamato KT, Kohchi T. 2008. *Agrobacterium*-mediated transformation of the haploid liverwort *Marchantia polymorpha* L., an emerging model for plant biology. *Plant Cell Physiol* **49:** 1084–1091.

Jaglo-Ottosen KR, Gilmour SJ, Zarka DG, Schabenberger O, Thomashow MF. 1998. *Arabidopsis CBF1* overexpression induces *COR* genes and enhances freezing tolerance. *Science* **280:** 104–106.

Johnson AAT, Hibberd JM, Gay C, Essah PA, Haseloff J, Tester M, Guiderdoni E. 2005. Spatial control of transgene expression in rice (*Oryza sativa* L.) using the GAL4 enhancer trapping system. *Plant J* **41:** 779–789.

Jones H, Ooms G, Jones MGK. 1989. Transient gene expression in electroporated *Solanum* protoplasts. *Plant Mol Biol* **13:** 503–511.

Jones JDG, Witek K, Verweij W, Jupe F, Cooke D, Dorling S, Tomlinson L, Smoker M, Perkins S, Foster S. 2014. Elevating crop disease resistance with cloned genes. *Philos Trans R Soc B Biol Sci* **369:** 20130087.

Jönsson H, Krupinski P. 2010. Modeling plant growth and pattern formation. *Curr Opin Plant Biol* **13:** 5–11.

Kapila J, De Rycke R, Van Montagu M, Angenon G. 1997. An *Agrobacterium*-mediated transient gene expression system for intact leaves. *Plant Sci* **122:** 101–108.

Karunanandaa B, Qi Q, Hao M, Baszis SR, Jensen PK, Wong YHH, Jiang J, Venkatramesh M, Gruys KJ, Moshiri F, et al. 2005. Metabolically engineered oilseed crops with enhanced seed tocopherol. *Metab Eng* **7:** 384–400.

Katoh K, Ohta Y, Hirose Y, Iwamura T. 1979. Photoautotrophic growth of *Marchantia polymorpha* L. cells in suspension culture. *Planta* **144:** 509–510.

Kimura S, Sinha N. 2008. How to grow tomatoes. *CSH Protoc* **2008:** pdb.prot5081.

Kindle KL, Richards KL, Stern DB. 1991. Engineering the chloroplast genome: Techniques and capabilities for chloroplast transformation in *Chlamydomonas reinhardtii*. *Proc Natl Acad Sci* **88:** 1721–1725.

Knight T. 2003. *Idempotent vector design for standard assembly of biobricks.* MIT Artificial Intelligence Laboratory; MIT Synthetic Biology Working Group.

Koo JC, Asurmendi S, Bick J, Woodford-Thomas T, Beachy RN. 2004. Ecdysone agonist-inducible expression of a coat protein gene from tobacco mosaic virus confers viral resistance in transgenic *Arabidopsis*. *Plant J* **37:** 439–448.

Koschmann J, Machens F, Becker M, Niemeyer J, Schulze J, Bulow L, Stahl DJ, Hehl R. 2012. Integration of bioinformatics and synthetic promoters leads to the discovery of novel elicitor-responsive *cis*-regulatory sequences in *Arabidopsis*. *Plant Physiol* **160:** 178–191.

Krens FA, Molendijk L, Wullems GJ, Schilperoort RA. 1982. In vitro transformation of plant protoplasts with Ti-plasmid DNA. *Nature* **296:** 72–74.

Ku MS, Cho D, Li X, Jiao DM, Pinto M, Miyao M, Matsuoka M. 2001. Introduction of genes encoding C4 photosynthesis enzymes into rice plants: Physiological consequences. *Novartis Found Symp* **236:** 100–116.

Kubota A, Ishizaki K, Hosaka M, Kohchi T. 2013. Efficient *Agrobacterium*-mediated transformation of the liverwort *Marchantia polymorpha* using regenerating Thalli. *Biosci Biotechnol Biochem* **77:** 167–172.

Kumar A, Dixit S, Ram T, Yadaw RB, Mishra KK, Mandal NP. 2014. Breeding high-yielding drought-tolerant rice: Genetic variations and conventional and molecular approaches. *J Exp Bot* **65:** 6265–6278.

Lampropoulos A, Sutikovic Z, Wenzl C, Maegele I, Lohmann JU, Forner J. 2013. GreenGate−−A novel, versatile, and efficient cloning system for plant transgenesis. *PLoS ONE* **8:** e83043.

Laplaze L, Parizot B, Baker A, Ricaud L, Martinière A, Auguy F, Franche C, Nussaume L, Bogusz D, Haseloff J. 2005. GAL4-GFP enhancer trap lines for genetic manipulation of lateral root development in *Arabidopsis thaliana*. *J Exp Bot* **56:** 2433–2442.

Lee SB, Kwon HB, Kwon SJ, Park SC, Jeong MJ, Han SE, Byun MO, Daniell H. 2003. Accumulation of trehalose within transgenic chloroplasts confers drought tolerance. *Mol Breed* **11:** 1–13.

Leegood RC. 2013. Strategies for engineering C_4 photosynthesis. *J Plant Physiol* **170:** 378–388.

Li MZ, Elledge SJ. 2012. SLIC: A method for sequence- and ligation-independent cloning. *Methods Mol Biol* **852:** 51–59.

Li JF, Chung HS, Niu Y, Bush J, McCormack M, Sheen J. 2013. Comprehensive protein-based artificial microRNA screens for effective gene silencing in plants. *Plant Cell* **25:** 1507–1522.

Li JF, Zhang D, Sheen J. 2014. Cas9-based genome editing in *Arabidopsis* and tobacco. *Methods Enzymol* **546:** 459–472.

Link KH, Breaker RR. 2009. Engineering ligand-responsive gene-control elements: Lessons learned from natural riboswitches. *Gene Ther* **16:** 1189–1201.

Liu W, Mazarei M, Rudis MR, Fethe MH, Stewart CN. 2011. Rapid in vivo analysis of synthetic promoters for plant pathogen phytosensing. *BMC Biotechnol* **11:** 108.

Liu W, Yuan JS, Stewart CN Jr. 2013. Advanced genetic tools for plant biotechnology. *Nat Rev Genet* **14:** 781–793.

Lowder LG, Zhang D, Baltes NJ, Paul JW, Tang X, Zheng X, Voytas DF, Hsieh TF, Zhang Y, Qi Y. 2015. A CRISPR/Cas9 toolbox for multiplexed plant genome editing and transcriptional regulation. *Plant Physiol* **169:** 971–985.

Lu Y, Rijzaani H, Karcher D, Ruf S, Bock R. 2013. Efficient metabolic pathway engineering in transgenic tobacco and tomato plastids with synthetic multigene operons. *Proc Natl Acad Sci* **110:** E623.

Maliga P. 2004. Plastid transformation in higher plants. *Annu Rev Plant Biol* **55:** 289–313.

Mardanova ES, Kotlyarov RY, Kuprianov VV, Stepanova LA, Tsybalova LM, Lomonosoff GP, Ravin NV. 2015. Rapid high-yield expression of a candidate influenza vaccine based on the ectodomain of M2 protein linked to flagellin in plants using viral vectors. *BMC Biotechnol* **15:** 42.

Martinez A, Sparks C, Hart CA, Thompson J, Jepson I. 1999. Ecdysone agonist inducible transcription in transgenic tobacco plants. *Plant J* **19:** 97–106.

Mayfield SP, Yohn CB, Cohen A, Danon A. 1995. Regulation of chloroplast gene expression. *Annu Rev Plant Physiol Plant Mol Biol* **46:** 147–166.

McCaskill D, Croteau R. 1999. Strategies for bioengineering the development and metabolism of glandular tissues in plants. *Nat Biotechnol* **17:** 31–36.

McHughen A. 1983. Rapid regeneration of wheat in vitro. *Ann Bot* **51:** 851–853.

Mett VL, Lochhead LP, Reynolds PH. 1993. Copper-controllable gene expression system for whole plants. *Proc Natl Acad Sci* **90:** 4567–4571.

Meyerowitz EM. 1989. *Arabidopsis*, a useful weed. *Cell* **56:** 263–269.

Miller MW. 1964. A technique for isolating and culturing gemmae of *Marchantia polymorpha* L. under aseptic conditions. *Bryologist* **67:** 317–320.

Moore I, Galweiler L, Grosskopf D, Schell J, Palme K. 1998. A transcription activation system for regulated gene expression in transgenic plants. *Proc Natl Acad Sci* **95:** 376–381.

Mühlbauer SK, Koop H-U. 2005. External control of transgene expression in tobacco plastids using the bacterial lac repressor. *Plant J* **43:** 941–946.

Müller B, Sheen J. 2008. Cytokinin and auxin interaction in root stem-cell specification during early embryogenesis. *Nature* **453:** 1094–1097.

Müller K, Siegel D, Rodriguez Jahnke F, Gerrer K, Wend S, Decker EL, Reski R, Weber W, Zurbriggen MD. 2014. A red light-controlled synthetic gene expression switch for plant systems. *Mol BioSyst* **10:** 1679–1688.

Nakazato T, Kadota A, Wada M. 1999. Photoinduction of spore germination in *Marchantia polymorpha* L. is mediated by photosynthesis. *Plant Cell Physiol* **40:** 1014–1020.

Negrutiu I, Shillito R, Potrykus I, Biasini G, Sala F. 1987. Hybrid genes in the analysis of transformation conditions. *Plant Mol Biol* **8:** 363–373.

Newell CA, Birch Machin I, Hibberd JM, Gray JC. 2003. Expression of green fluorescent protein from bacterial and plastid promoters in tobacco chloroplasts. *Transgenic Res* **12:** 631–634.

Newman DJ, Cragg GM. 2016. Natural products as sources of new drugs from 1981 to 2014. *J Nat Prod* **79:** 629–661.

Nishiguchi M, Sato T, Motoyoshi F. 1987. An improved method for electroporation in plant protoplasts: Infection of tobacco protoplasts by tobacco mosaic virus particles. *Plant Cell Rep* **6:** 90–93.

Nützmann HW, Osbourn A. 2014. Gene clustering in plant specialized metabolism. *Curr Opin Biotechnol* **26:** 91–99.

Oda K, Yamato K, Ohta E, Nakamura Y, Takemura M, Nozato N, Akashi K, Kanegae T, Ogura Y, Kohchi T, et al. 1992. Complete nucleotide sequence of the mitochondrial DNA from a liverwort, *Marchantia polymorpha. Plant Mol Biol Report* **10:** 105–163.

Oey M, Lohse M, Kreikemeyer B, Bock R. 2009. Exhaustion of the chloroplast protein synthesis capacity by massive expression of a highly stable protein antibiotic. *Plant J* **57:** 436–445.

O'Hanlon ME. 1926. Germination of spores and early stages in development of gametophyte of *Marchantia polymorpha. Bot Gaz* **82:** 215–222.

Ohyama K, Fukuzawa H, Kohchi T, Shirai H, Sano T, Sano S, Umesono K, Shiki Y, Takeuchi M, Chang Z, et al. 1986. Chloroplast gene organization deduced from complete sequence of liverwort *Marchantia polymorpha* chloroplast DNA. *Nature* **322:** 572–574.

Ohyama K, Takemura M, Oda K, Fukuzawa H, Kohchi T, Nakayama S, Ishizaki K, Fujisawa M, Yamato K. 2009. Gene content, organization and molecular evolution of plant organellar genomes and sex chromosomes: Insights from the case of the liverwort *Marchantia polymorpha. Proc Jpn Acad Ser B Phys Biol Sci* **85:** 108–124.

Okada S, Fujisawa M, Sone T, Nakayama S, Nishiyama R, Takenaka M, Yamaoka S, Sakaida M, Kono K, Takahama M, et al. 2000. Construction of male and female PAC genomic libraries suitable for identification of Y-chromosome-specific clones from the liverwort, *Marchantia polymorpha. Plant J* **24:** 421–428.

Ono K, Ohyama K, Gamborg OL. 1979. Regeneration of the liverwort *Marchantia polymorpha* L. from protoplasts isolated from cell suspension culture. *Plant Sci Lett* **14:** 225–229.

Ou-Lee TM, Turgeon R, Wu R. 1986. Expression of a foreign gene linked to either a plant-virus or a *Drosophila* promoter, after electroporation of protoplasts of rice, wheat, and sorghum. *Proc Natl Acad Sci* **83:** 6815–6819.

Padidam M, Gore M, Lu DL, Smirnova O. 2003. Chemical-inducible, ecdysone receptor-based gene expression system for plants. *Transgenic Res* **12:** 101–109.

Patron NJ. 2014. DNA assembly for plant biology: Techniques and tools. *Curr Opin Plant Biol* **19:** 14–19.

Patron NJ, Orzaez D, Marillonnet S, Warzecha H, Matthewman C, Youles M, Raitskin O, Leveau A, Farré G, Rogers C, et al. 2015. Standards for plant synthetic biology: A common syntax for exchange of DNA parts. *New Phytol* **208:** 13–19.

Piatek A, Ali Z, Baazim H, Li L, Abulfaraj A, Al-Shareef S, Aouida M, Mahfouz MM. 2015. RNA-guided transcriptional regulation in planta via synthetic dCas9-based transcription factors. *Plant Biotechnol J* **13:** 578–589.

Potrykus I, Paszkowski J, Saul MW, Petruska J, Shillito RD. 1985. Molecular and general genetics of a hybrid foreign gene introduced into tobacco by direct gene transfer. *MGG Mol Gen Genet* **199:** 169–177.

Prusinkiewicz P, Runions A. 2012. Computational models of plant development and form. *New Phytol* **193:** 549–569.

Przibilla E, Heiss S, Johanningmeier U, Trebst A. 1991. Site-specific mutagenesis of the D1 subunit of photosystem II in wild-type *Chlamydomonas. Plant Cell* **3:** 169–174.

Quan J, Tian J. 2009. Circular polymerase extension cloning of complex gene libraries and pathways. *PLoS ONE* **4:** e6441.

Raitskin O, Patron NJ. 2016. Multi-gene engineering in plants with RNA-guided Cas9 nuclease. *Curr Opin Biotechnol* **37:** 69–75

Rensing SA, Lang D, Zimmer AD, Terry A, Salamov A, Shapiro H, Nishiyama T, Perroud P-F, Lindquist EA, Kamisugi Y, et al. 2008. The physcomitrella genome reveals evolutionary insights into the conquest of land by plants. *Science* **319:** 64–69.

Rogers C, Oldroyd GED. 2014. Synthetic biology approaches to engineering the nitrogen symbiosis in cereals. *J Exp Bot* **65:** 1939–1946.

Roslan HA, Salter MG, Wood CD, White MRH, Croft KP, Robson F, Coupland G, Doonan J, Laufs P, Tomsett AB, et al. 2001. Characterization of the ethanol-inducible alc gene-expression system in *Arabidopsis thaliana. Plant J* **28:** 225–235.

Salter MG, Paine JA, Riddell K V, Jepson I, Greenland AJ, Caddick MX, Tomsett AB. 1998. Characterisation of the ethanol-inducible *alc* gene expression system for transgenic plants. *Plant J* **16:** 127–132.

Samalova M, Brzobohaty B, Moore I. 2005. pOp6/LhGR: A stringently regulated and highly responsive dexamethasone-inducible gene expression system for tobacco. *Plant J* **41:** 919–935.

Sarrion-Perdigones A, Falconi EE, Zandalinas SI, Juárez P, Fernández-del-Carmen A, Granell A, Orzaez D. 2011. GoldenBraid: An iterative cloning system for standardized assembly of reusable genetic modules. *PLoS ONE* **6:** e21622.

Sasaki G, Katoh K, Hirose N, Suga H, Kuma K, Miyata T, Su ZH. 2007. Multiple receptor-like kinase cDNAs from liverwort *Marchantia polymorpha* and two charophycean green algae, *Closterium ehrenbergii* and *Nitella axillaris*: Extensive gene duplications and gene shufflings in the early evolution of streptophytes. *Gene* **401:** 135–144.

Scharff LB, Bock R. 2014. Synthetic biology in plastids. *Plant J* **78:** 783–798.

Schaumberg KA, Antunes MS, Kassaw TK, Xu W, Zalewski CS, Medford JI, Prasad A. 2016. Quantitative characterization of genetic parts and circuits for plant synthetic biology. *Nat Methods* **13:** 94–100.

Schilmiller AL, Last RL, Pichersky E. 2008. Harnessing plant trichome biochemistry for the production of useful compounds. *Plant J* **54:** 702–711.

Schiml S, Puchta H. 2016. Revolutionizing plant biology: Multiple ways of genome engineering by CRISPR/Cas. *Plant Methods* **12:**8

Sheen J. 2001. Signal transduction in maize and *Arabidopsis* mesophyll protoplasts. *Plant Physiol* **127:** 1466–1475.

Shetty RP, Endy D, Knight TF. 2008. Engineering BioBrick vectors from BioBrick parts. *J Biol Eng* **2:** 5.

Shimamura M. 2012. An introduction to the taxonomy and morphology of *Marchantia polymorpha*. *BSJ-Review* **3:** 84–113.

Somerville C, Koornneef M. 2002. A fortunate choice: The history of *Arabidopsis* as a model plant. *Nat Rev Genet* **3:** 883–889.

Sparkes IA, Runions J, Kearns A, Hawes C. 2006. Rapid, transient expression of fluorescent fusion proteins in tobacco plants and generation of stably transformed plants. *Nat Protoc* **1:** 2019–2025.

Springer PS. 2000. Gene traps: Tools for plant development and genomics. *Plant Cell* **12:** 1007–1020.

Sudarsan N, Barrick JE, Breaker RR. 2003. Metabolite-binding RNA domains are present in the genes of eukaryotes. *RNA* **9:** 644–647.

Suire C, Bouvier F, Backhaus RA, Bégu D, Bonneu M, Camara B. 2000. Cellular localization of isoprenoid biosynthetic enzymes in *Marchantia polymorpha*. Uncovering a new role of oil bodies. *Plant Physiol* **124:** 971–978.

Svab Z, Hajdukiewicz P, Maliga P. 1990. Stable transformation of plastids in higher plants. *Proc Natl Acad Sci* **87:** 8526–8530.

Takenaka M, Yamaoka S, Hanajiri T, Shimizu-Ueda Y, Yamato KT, Fukuzawa H, Ohyama K. 2000. Direct transformation and plant regeneration of the haploid liverwort *Marchantia polymorpha* L. *Transgenic Res* **9:** 179–185.

Thimmappa R, Geisler K, Louveau T, O'Maille P, Osbourn A. 2014. Triterpene biosynthesis in plants. *Annu Rev Plant Biol* **65:** 225–257.

Torella JP, Boehm CR, Lienert F, Chen JH, Way JC, Silver PA. 2013. Rapid construction of insulated genetic circuits via synthetic sequence-guided isothermal assembly. *Nucleic Acids Res* **42:** 1–9.

Tsuboyama S, Kodama Y. 2014. AgarTrap: A simplified *Agrobacterium*-mediated transformation method for sporelings of the liverwort *Marchantia polymorpha* L. *Plant Cell Physiol* **55:** 229–236.

Tsukaya H. 2003. Organ shape and size: A lesson from studies of leaf morphogenesis. *Curr Opin Plant Biol* **6:** 57–62.

Vafaee Y, Staniek A, Mancheno-Solano M, Warzecha H. 2014. A modular cloning toolbox for the generation of chloroplast transformation vectors. *PLoS ONE* **9:** e110222.

Valvekens D, Van Lijsebettens M, Van Montagu M. 1992. *Arabidopsis* regeneration and transformation (Root Explant System). In *Plant tissue culture manual*, pp. 1–17. Springer, Boston.

van Herpen TWJM, Cankar K, Nogueira M, Bosch D, Bouwmeester HJ, Beekwilder J. 2010. *Nicotiana benthamiana* as a production platform for artemisinin precursors. *PLoS ONE* **5:** e14222.

Vaughan DA, Balazs E, Heslop-Harrison JS. 2007. From crop domestication to super-domestication. *Ann Bot* **100:** 893–901.

Verhounig A, Karcher D, Bock R. 2010. Inducible gene expression from the plastid genome by a synthetic riboswitch. *Proc Natl Acad Sci* **107:** 6204–6209.

Verma D, Kanagaraj A, Jin S, Singh ND, Kolattukudy PE, Daniell H. 2010. Chloroplast-derived enzyme cocktails hydrolyse lignocellulosic biomass and release fermentable sugars. *Plant Biotechnol J* **8:** 332–350.

Vötching H. 1885. Über die regeneration der Marchantiaceen. *Jahrbuecher fur Wissenschaftliche Bot* **16:** 367–414.

Wachter A, Tunc-Ozdemir M, Grove BC, Green PJ, Shintani DK, Breaker RR. 2007. Riboswitch control of gene expression in plants by splicing and alternative 3′ end processing of mRNAs. *Plant Cell* **19:** 3437–3450.

Wani S, Haider N, Kumar HB, Singh N. 2010. Plant plastid engineering. *Curr Genomics* **11:** 500–512.

Weber E, Engler C, Gruetzner R, Werner S, Marillonnet S. 2011. a modular cloning system for standardized assembly of multigene constructs. *PLoS ONE* **6:** e16765.

Weinmann P, Gossen M, Hillen W, Bujard H, Gatz C. 1994. A chimeric transactivator allows tetracycline-responsive gene expression in whole plants. *Plant J* **5:** 559–569.

Wend S, Bosco CD, Kämpf MM, Ren F, Palme K, Weber W, Dovzhenko A, Zurbriggen MD. 2013. A quantitative ratiometric sensor for time-resolved analysis of auxin dynamics. *Sci Rep* **3:** 769–782.

Werner S, Engler C, Weber E, Gruetzner R, Marillonnet S. 2012. Fast track assembly of multigene constructs using Golden Gate cloning and the MoClo system. *Bioeng Bugs* **3:** 38–43.

Wijnker E, van Dun K, de Snoo CB, Lelivelt CLC, Keurentjes JJB, Naharudin NS, Ravi M, Chan SWL, de Jong H, Dirks R. 2012. Reverse breeding in *Arabidopsis thaliana* generates homozygous parental lines from a heterozygous plant. *Nat Genet* **44:** 467–470.

Yamato KT, Kohchi T. 2012. The "sneak" preview of the *Marchantia polymorpha* genome. *BSJ Rev* **3:** 71–83.

Yamato KT, Ishizaki K, Fujisawa M, Okada S, Nakayama S, Fujishita M, Bando H, Yodoya K, Hayashi K, Bando T, et al. 2007. Gene organization of the liverwort Y chromosome reveals distinct sex chromosome evolution in a haploid system. *Proc Natl Acad Sci* **104:** 6472–6477.

Yanagisawa S, Akiyama A, Kisaka H, Uchimiya H, Miwa T. 2004. Metabolic engineering with Dof1 transcription factor in plants: Improved nitrogen assimilation and growth under low-nitrogen conditions. *Proc Natl Acad Sci* **101:** 7833–7838.

Ye X, Al-Babili S, Klöti A, Zhang J, Lucca P, Beyer P, Potrykus I. 2000. Engineering the provitamin A (β-carotene) biosynthetic pathway into (carotenoid-free) rice endosperm. *Science* **287:** 303–305.

Yu S, Pilot G. 2014. Testing the efficiency of plant artificial microRNAs by transient expression in *Nicotiana benthamiana* reveals additional action at the translational level. *Front Plant Sci* **5:** 622.

Zobell O, Faigl W, Saedler H, Munster T. 2010. MIKC* MADS-Box Proteins: Conserved regulators of the gametophytic generation of land plants. *Mol Biol Evol* **27:** 1201–1211.

Zohary D, Hopf M, Weiss E. 2012. *Domestication of plants in the old world: The origin and spread of domesticated plants in Southwest Asia, Europe, and the Mediterranean basin.* Oxford University Press, New York.

Zuo J, Niu QW, Chua NH. 2000. An estrogen receptor-based transactivator XVE mediates highly inducible gene expression in transgenic plants. *Plant J* **24:** 265–273.

Zurcher E, Tavor-Deslex D, Lituiev D, Enkerli K, Tarr PT, Muller B. 2013. A robust and sensitive synthetic sensor to monitor the transcriptional output of the cytokinin signaling network in planta. *Plant Physiol* **161:** 1066–1075.

Phage Therapy in the Era of Synthetic Biology

E. Magda Barbu, Kyle C. Cady, and Bolyn Hubby

Synthetic Genomics, La Jolla, California 92037

Correspondence: bhubby@syntheticgenomics.com

For more than a century, bacteriophage (or phage) research has enabled some of the most important discoveries in biological sciences and has equipped scientists with many of the molecular biology tools that have advanced our understanding of replication, maintenance, and expression of genetic material. Phages have also been recognized and exploited as natural antimicrobial agents and nanovectors for gene therapy, but their potential as therapeutics has not been fully exploited in Western medicine because of challenges such as narrow host range, bacterial resistance, and unique pharmacokinetics. However, increasing concern related to the emergence of bacteria resistant to multiple antibiotics has heightened interest in phage therapy and the development of strategies to overcome hurdles associated with bacteriophage therapeutics. Recent progress in sequencing technologies, DNA manipulation, and synthetic biology allowed scientists to refactor the entire bacterial genome of *Mycoplasma mycoides*, thereby creating the first synthetic cell. These new strategies for engineering genomes may have the potential to accelerate the construction of designer phage genomes with superior therapeutic potential. Here, we discuss the use of phage as therapeutics, as well as how synthetic biology can create bacteriophage with desirable attributes.

Lysogenic (temperate) and lytic bacteriophages are viruses that specifically infect bacteria, and have long been considered as natural antimicrobial agents for treatment of bacterial infections (d'Herelle 1931; Twort 1936). On binding to a bacterial surface moiety, phages inject their genetic material into the host cell. In the case of temperate phage, the lysogenic cycle is initiated by viral genome integration into the bacterial chromosome (prophage) where it remains until specific lytic genes are activated and phage virions are produced, eventually lysing the host cell. Before the activation of the lytic cycle, the phage genome remains repressed and can be transferred to daughter cells during bacterial replication. Temperate phages are largely not considered for therapeutic use, as many have been shown to encode bacterial virulence factors or horizontally transfer virulence genes in a process known as transduction (Boyd and Brussow 2002; Ptashne 2004; Tinsley et al. 2006). In contrast, virulent phages immediately begin replication and quickly subvert host-cell metabolism, which results in the destruction of the host cell within minutes to hours (Fig. 1) (Sulakvelidze et al. 2001; Hanlon 2007). Virulent phages do not enter a prophage state and have not shown the ability to transfer virulence factors between bacterial cells, making them suitable for phage therapy (Hendrix 2003; Hatfull 2008).

The clinical potential of virulent bacteriophages as antibacterial agents was first recog-

Cite this article as *Cold Spring Harb Perspect Biol* doi: 10.1101/cshperspect.a023879

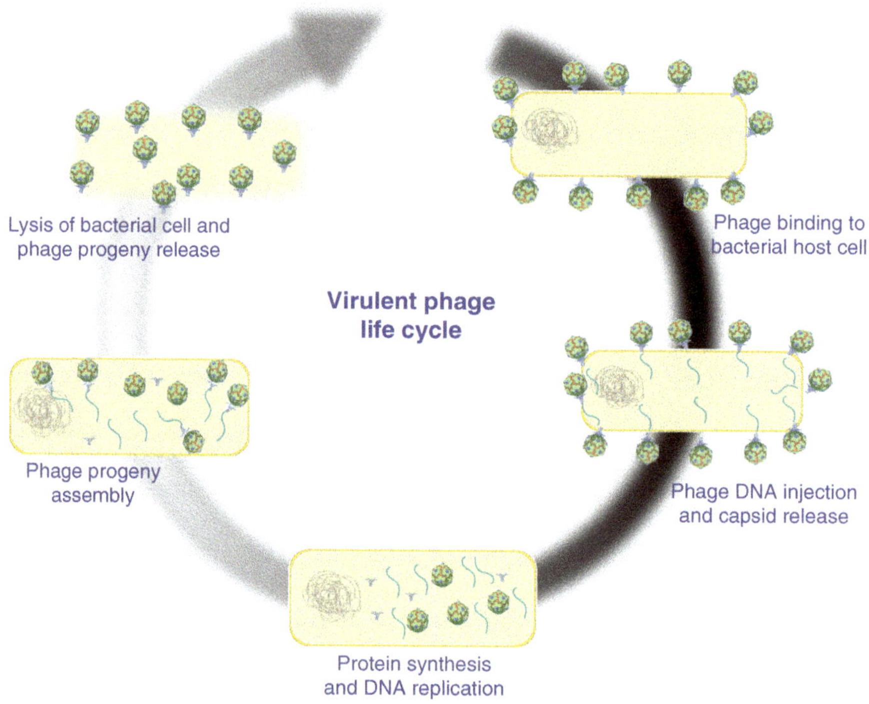

Figure 1. Lytic cycle of bacteriophage. Schematic representation of virulent phage infection, replication, and lysis of bacterial host cells.

nized by Felix d'Herelle, one of the pioneers of the phage therapy field (d'Herelle et al. 1931). Although early successes of personalized clinical applications with tailored formulations were reported, the use of phage therapy against bacterial pathogens in large patient populations was never achieved in part because of its displacement by antibiotics (Merril et al. 2003, 2006). The growing threat posed by highly antibiotic-resistant bacteria combined with the decreasing numbers of newly licensed drugs has recently triggered interest in the development of phage antimicrobials. However, several challenges associated with phage therapy limit their therapeutic potential.

OVERCOMING PHAGE THERAPY CHALLENGES IN THE ERA OF SYNTHETIC BIOLOGY

Host Specificity

Viral attachment and genome injection are regimented and stepwise processes that are tightly controlled by specific receptor–ligand interactions (Furukawa et al. 1983). Phage–host selectivity permits phage therapy to be directed at specific bacterial populations, preventing secondary infections by leaving nontarget bacteria unaffected (Sulakvelidze et al. 2001). Host discernment prevents phage therapy from enriching for resistance, a major problem associated with antibiotic use. Historically, the diversity and malleability of bacterial surface ligands, even at the species level, makes it difficult to identify phage capable of treating a clinical infection without prescreening against phage libraries (Keen 2012). The specificity of viral–host interactions has made phage therapy attractive, but has also required either the targeting of pathogens that are largely phylogenetically constrained (clonal) or the use of large phage cocktails capable of covering the diversity of a bacterial population (O'Flaherty et al. 2005; McVay et al. 2007).

The plummeting cost of DNA sequencing and new molecular biology techniques may allow scientists to leverage phenotypic screening, comparative genomics, and viral engineering to overcome the challenges associated with host specificity. As bacterial phylogeny correlates with phage host range, the ability to rapidly sequence clinical isolates will increase both the accuracy and the speed at which lytic phage can be paired with potentially susceptible bacterial infections (DeLappe et al. 2003). Last, newly developed viral engineering and DNA manipulation methods could be used to expand the host range of bacteriophages through rational design, potentially decreasing the number of phages needed to cover bacterial diversity. For example, ϕX174 genome (5.4 kb) was assembled from a pool of synthetic nucleotides and successfully launched in bacteria (Smith et al. 2003). Furthermore, the entire bacterial genome of *Mycoplasma mycoides* was chemically synthesized from DNA fragments (Gibson et al. 2009). Other technologies, such as bacteriophage recombineering of electroporated DNA (BRED) and genome editing via targetrons and recombinases (GETR), have been developed to edit phage genomes (Marinelli et al. 2008; Enyeart et al. 2013). Assembly of genomes in yeast artificial chromosome and plasmids and clustered regularly interspaced short palindromic repeats (CRISPR)-Cas-editing systems in bacteria have also been recently used to manipulate phage DNA (Lu and Koeris 2011; Jaschke et al. 2012; Kiro et al. 2014; Martel and Moineau 2014; Nobrega et al. 2015). Together, these methods can be integrated to construct synthetic phages with desired characteristics and precise genomic content (Fig. 2).

Development of Resistance

The ability of bacteria to rapidly acquire resistance to a single selective pressure hinders the success of both standard small molecule and nonstandard antimicrobials, such as phages (Neu 1992; Labrie et al. 2010). Phage resistance most commonly arises through the down-regulation, mutation, or shielding of the viral receptor (Labrie et al. 2010). This simple fact has encouraged the use of phages that bind to receptors that are highly conserved or required for in vivo virulence (O'Flaherty et al. 2005; Ojala et al. 2013). However, bacteria can also acquire resistance to phages through a diverse array of restriction modification, toxin–antitoxin, and CRISPR-Cas systems (Labrie et al. 2010). Unlike purified small molecules, phages respond to selective pressure through evolution, which has supplied scientists with a plethora of genetically encoded counter-resistance systems capable of overcoming bacterial resistance mechanisms (Samson et al. 2013). In a clinical setting, it is important to remember that phage and antibiotic resistance mechanisms do not overlap; therefore, loss of sensitivity to phage is unlikely to affect antibiotic susceptibility (Labrie et al. 2010; Blair et al. 2015). In fact, CRISPR-Cas-mediated resistance to phage is associated with decreased antibiotic resistance in enterococci (Palmer and Gilmore 2010).

The success of chemical cocktails, phage cocktails, and multivalent vaccines has shown that targeting multiple conserved but independent ligands drastically reduces the incidence of pathogen evasion. Thus, endowing single bacteriophage with an independent bactericidal payload could prevent bacterial resistance to phages (Fig. 3).

Unique Pharmacological Properties of Viral Therapeutics

Biologics are large molecules with desirable therapeutic and prophylactic uses, which are very complex and are often 200 to more than 1000 times the size of small molecule drugs (Kingham et al. 2014). Similar to biologics, the large size of phage poses distinctive pharmacological challenges when compared with chemical products. Additionally, therapeutic viruses, such as oncolytic virions or bacteriophage, are capable of replicating in vivo, further differentiating them from static biologics or chemicals (Chiocca 2002).

The large size of biologics prevents them from diffusing into cells or biofilms, crossing the blood–brain barrier, and makes administration of large doses challenging (Baumann 2006). Fortunately, bacteriophages are capable of density-dependent productive amplification

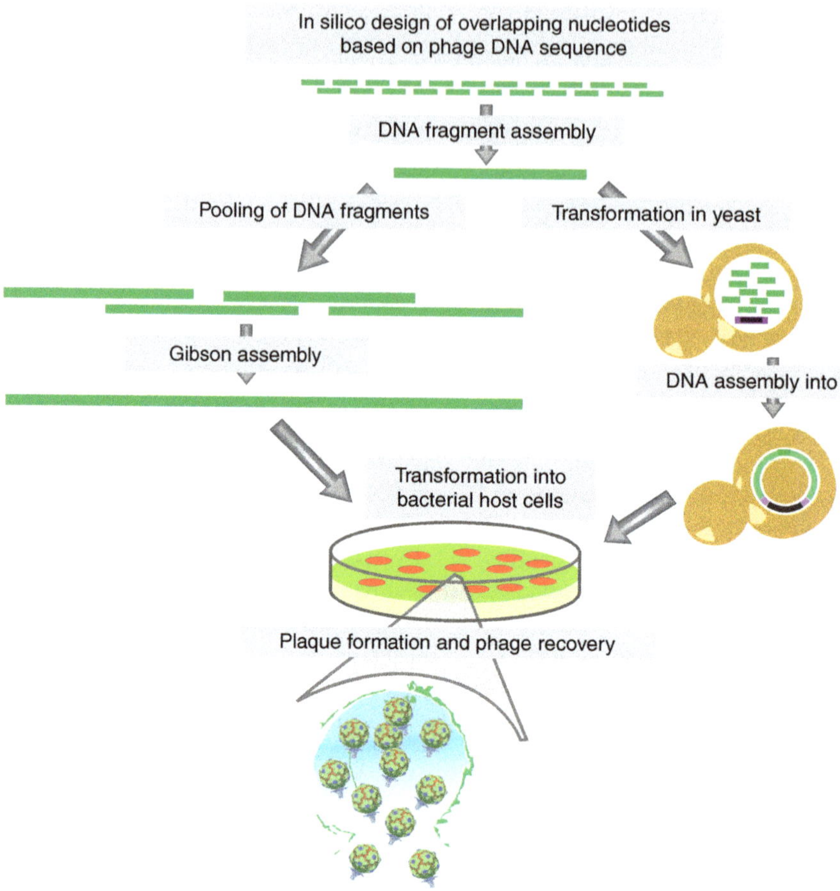

Figure 2. Synthetic assembly of bacteriophage genomes. Step-by-step construction of phage genomes starting from overlapping nucleotides and recovery of functional phage particles from host cells following transformation. (*Left*) Chemical assembly of DNA fragments synthesized from overlapping nucleotides using Gibson assembly. (*Right*) Assembly of DNA fragments synthesized from overlapping nucleotides in yeast artificial chromosome (YAC, in purple).

within a target pathogen at the site of infection. Whereas autoamplification may require a certain concentration of phage particles to accumulate at the site of infection, it is believed that amplification can substantially lower the dose required to treat an infection (Abedon 2011). As replication absolutely requires the presence of the target bacteria, phages are self-limiting and quickly cleared following pathogen eradication (Abedon 2011). Phage replication results in the lysis of target cells and potential release of bacterial toxins, an important concern especially for endotoxin encoding Gram-negative pathogens (Hagens et al. 2004). However, no toxic effects have been reported in bacteremic patients successfully treated with phages (Kutter et al. 2014).

Using molecular biology, scientists have already created bacteriophages coated in cell-permeable peptides, allowing them to diffuse into human cells or through the blood–brain barrier (Staquicini et al. 2011; Rangel et al. 2012). Additionally, phages are known to encode and can be engineered to produce enzymes, allowing them to degrade bacterial biofilms (Lu and Collins 2007). Researchers have also created nonreplicating or lysis-deficient phages to deliver bacterial static toxins, enabling inhibition of

 Cite this article as *Cold Spring Harb Perspect Biol* doi: 10.1101/cshperspect.a023879

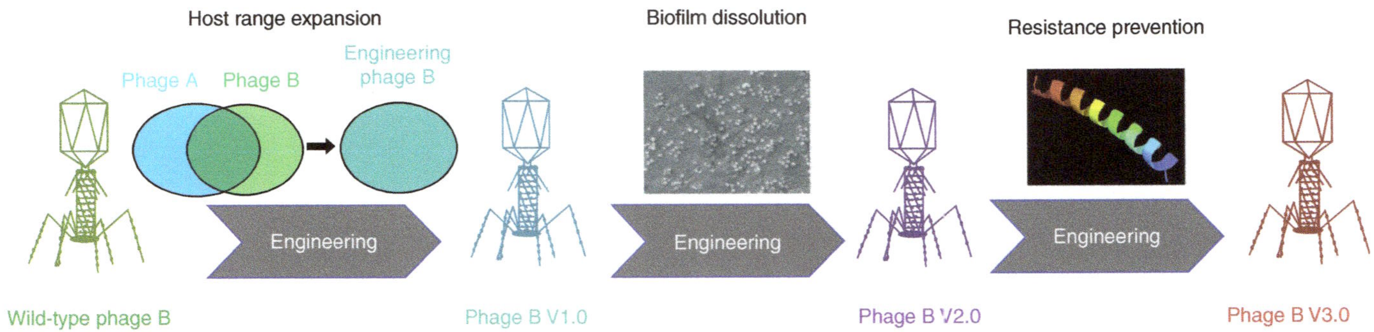

Figure 3. Iterative engineering of phage genomes to create phages with superior therapeutic potential. Phage B V1.0 represents a phage with enhanced host range (includes the host range specific to phage A). Phage B V2.0 delineates iterative engineering of phage B V1.0 to include moieties aiding in biofilm disruption. Similarly, phage B V3.0 captures the properties of phage B V2.0 and is engineered to express antimicrobial payloads that may aid in elimination of bacterial subpopulation that lost susceptibility to phage.

bacterial growth without release of toxins through lysis (Hagens et al. 2004; Matsuda et al. 2005; Fairhead 2009). Whereas viral size is unlikely to be meaningfully altered, genetically programing phages with the ability to produce small peptides with useful properties could offset many of the size constraints inherent to classic biologics (Fig. 3).

Human Immune Response and Phage Clearance

The mammalian innate immune system recognizes and clears viral particles, creating a major hurdle for any viral-based therapeutic (Geier et al. 1973). Phagocytic cells located in the reticular connective tissue play a key role in rapidly clearing circulating phage particles through the innate immune mononuclear phagocyte system (MPS) (Merril et al. 1996). Whereas the MPS may form the first major barrier to phage circulation, the adaptive immune response provides the second. The adaptive response mediated by B/T cells takes 5–7 d to produce robust antibody titers against phage in naïve humans, providing an ample window for phage therapeutic activity (Ochs et al. 1971; Wedgwood et al. 1975). However, it is hypothesized that the adaptive memory response may prevent the reuse of phage therapy vectors within the same patient, as protective antibodies will already have been raised (Górski et al. 2012).

Long-circulating phage harboring single amino acid mutations in the major capsid protein have already been selected in vivo and shown to provide an increased evasion of MPS by more than 10,000-fold, drastically improving phage therapy efficacy (Merril et al. 1996; Vitiello et al. 2005). Thus, engineering of mutations that enhance pharmacokinetics properties of phage capsids may improve the therapeutic potential of multiple phage vectors for antimicrobial and phage-based therapeutic nanovector applications.

Regulatory and Production Challenges

Bacteriophages are large and complex therapeutics more similar to other genetically engineered biologics than small molecule antibiotics. Like other biologics, phage are difficult to fully characterize, and are sensitive to manufacturing and handling conditions, making their production much more complex than chemical drugs (Kingham et al. 2014). However, a handful of phage products have been approved by American regulatory agencies, including the U.S. Food and Drug Administration (FDA), granting multiple bacteriophages generally recognized as safe (GRAS) status when used as antimicrobial agents in human food (Soni and Nannapaneni 2010; Carter et al. 2012; Woolston et al. 2013). Further, phage cocktails for human therapeutic use have been produced for decades within the former Soviet Republics of Eastern Europe, indicating some level of safety and efficacy (Sulakvelidze et al. 2001). Scalable oncolytic virus current good manufacturing practice (cGMP)-compliant processes provide significant insight into the methods and regulatory oversight required for viral therapeutic production. Interestingly, the substantially higher cost of producing biologics than chemical therapeutics has not prevented their rapid ascension into the ranks of the best selling drugs in the United States. Unfortunately, phage therapeutic production costs relative to multidrug-resistant (MDR) infection rates may still provide the largest single barrier to phage therapy success. Many phage therapy products currently being tested in clinical trials or used in Eastern Europe constitute a cocktail of multiple independent types of phages (Wright et al. 2009; Abedon et al. 2011). Current Western medical standards would require each independent phage to be produced and purified under cGMP-compliant processes, multiplying production costs to prohibitive levels (Kingham et al. 2014).

The ability to generate whole bacteriophage genomes directly from synthetic oligonucleotides will be attractive to regulators, as it decreases contamination with adventitious agents. Additionally, launching and amplifying bacteriophage in cell-free reactions could significantly lower production costs and contamination (Kerr and Sadowski 1974; Gunther et al. 1993; Shin et al. 2012). These synthetic biology advances coupled with our growing ability to engineer bacteriophage with expanded host range, antiresistance mechanisms, and immune evasion capabilities may lower the number of phages and cost of production to a level attractive to regulators and commercial entities (Merril et al. 1996; Tetart et al. 1996, 1998; Samson et al. 2013). Last, the decreasing cost of sequencing will greatly aid the speed and accuracy of matching a particular therapeutic phage with a susceptible strain, further limiting the known challenge of phage host range.

PHAGES AS ANTI-INFECTIVES

Numerous strategies have been used to show that virulent phages are realistic alternatives to standard antibiotics. However, insight into the potential clinical and therapeutic relevance of preclinical efficacy models of phage treatment is often obscured by a lack of standardized techniques and protocols and, as a consequence, vast differences in therapeutic outcomes. Whereas some studies report superior therapeutic value (McVay et al. 2007; Oliveira et al. 2010; Gu et al. 2012; Hall et al. 2012; Jaiswal et al. 2013), others note marginal reduction in bacterial loads (Rozema et al. 2009). Nonetheless, results stemming from recent attempts to show efficacy of phage therapy exemplify how delivery routes, single or multiple treatments, monophage versus polyphage therapy, and dosage/timing of administration contribute to therapeutic success in preclinical models of infection.

Lessons Learned from Preclinical Phage Therapy

Oral Administration of Phages

Per os administration of phage has been found suitable for treatment of gastrointestinal infections. For example, a cocktail of three bacteriophages administered orally in a single dose (10^8 PFU/mL) resulted in significant reduction of *Escherichia coli* O157:H7 in the feces of mice (99%). Further examination revealed that the most successful administration, as determined by reduction in *E. coli* load, was daily oral ingestion of high amounts of phage (10^{10} PFU/mL) (Tanji et al. 2005). Similar studies found that a threshold of an orally administered phage cocktail was required to reduce the colonization of mice infected with pediatric diarrhea *E. coli* clinical isolates (Chibani-Chennoufi et al. 2004). The effect of phage treatment on fecal counts was negligible, which led the investigators to hypothesize that either phages lose infectivity because of exposure to acid during gastric passage or that once in the gut *E. coli* becomes resistant to phage infection. Although subsequent investigations showed that intestinal transit has no impact on phage infectivity, protection from gastric acidity by microencapsulation or coadministration with antacid may enhance their efficacy (Stanford et al. 2010). Thus, acid sensitivity must be characterized for individual phage before oral administration.

The preventative and therapeutic efficacy of oral administration of phage was also examined in a chicken model of *Campylobacter jejuni* decolonization (Wagenaar et al. 2005). To show therapeutic effect, poultry received an oral dose of 10^5 CFU *C. jejuni*, followed by inoculation with a single phage (10^{10} PFU) for six successive days starting 5 d after the bacterial challenge. Initially, a 30-fold decrease in *C. jejuni* viable counts was observed in the therapeutic group. However, after 5 d, phage and bacteria reached equilibrium most likely because of the presence of a resistant subpopulation of microbes. Hence, using a cocktail or an engineered phage with a wide host range rather than a single phage may overcome bacterial resistance.

The timing of treatment also affects the success of phage therapy. Prophylaxis with phage KPP1 in a model of *Pseudomonas aeruginosa*–induced gut-derived septicemia had little to no effect on mice survival, whereas concomitant administration of phage and bacteria induced significant protection against infection.

Parenteral Administration of Phages

Intramuscular injection (i.m.) of different concentrations of phage R in chickens and calves suffering from *E. coli*–derived septicemia showed that significant protection is conferred by a high concentration of phages (Barrow et al.

1998). Similarly, intraperitoneal (i.p.) injection of ENB6 phage in a vancomycin-resistant *Enterococcus faecium* mouse bacteremia model within an hour of bacterial challenge rescued all experimental animals (Biswas et al. 2002). Several other reports highlighted the rapid distribution and bioavailability of phage on i.m. or i.p. injection. For instance, phage ΦMR11 appeared in the blood shortly after i.p. injection into a *Staphylococcus aureus* mouse bacteremia model, and high levels of circulating phage could be detected until bacteria were eliminated (Matsuzaki et al. 2003). Most importantly, McVay et al. (2007) showed that, in the case of a *P. aeruginosa*–specific phage cocktail, i.p. injection of phage has enhanced their therapeutic potential. In the absence of phage administration, almost all wounded mice (94%) succumbed to the infection within 3 d. In contrast, the rate of mortality was reduced to 12% when the phage cocktail was delivered by i.p. injection. A low percentage of animals (28% or 22%, respectively) were rescued by i.m. or subcutaneous phage delivery. In summary, these results suggest that the route of parenteral administration must be carefully chosen based on empirical phage pharmacokinetics data and tailored to the type of infection requiring treatment. At the same time, the dosage levels appear to directly correlate with the success of phage therapy.

Local Application of Phage

Successful topical administration of phage has been widely reported. Although early studies focused almost solely on wound healing, more recent investigations also include otic and ocular applications (Kumari et al. 2009; Hawkins et al. 2010; Fukuda et al. 2012). For certain microbes, skin and mucosal colonization represents a prerequisite for infection (Casadevall and Pirofski 2000). Therefore, effective decontamination with an antimicrobial that has no effect on normal microbiota significantly decreases the risk of disease. Early attempts to use phage for decolonization of chicken skin indicated that *Salmonella enterica*, *C. jejuni*, and *Campylobacter* spp. could be effectively eliminated when the multiplicity of infection (MOI) of viral particles to bacteria was above 100. However, the minimum levels of phage particles necessary to significantly reduce colonization were specific for each organism within a species (Goode et al. 2003).

Recent reports highlighted the successful treatment of *Klebsiella pneumoniae*–induced burn wound infection in mice with the *Klebsiella*-specific phage Kpn5 incorporated in a 3% hydroxypropylmethylcellulose hydrogel (Kumari et al. 2010). On bacterial challenge, mice were treated once with low and high doses of phage hydrogel (MOI of 1 or 100, respectively). Significant levels of protection against invasive *K. pneumonia* infection were achieved only by high titers of phage. These results indicate that the dosage of phage is important for effective phage therapy. However, whether application of high titers of phage triggers bacterial death because of a productive infection or lysis from without remains to be determined. It is possible that the need for excessive MOIs may be overcome by multiple applications of phage following an empirically determined treatment schedule.

Otic administration of a phage cocktail in dogs suffering from *P. aeruginosa* otitis has shown great promise. Two days after instillation of a single dose of phage into the auditory canal of one ear, bacterial counts decreased significantly. Concomitantly, phage titers increased and no adverse effects were registered. Although the cocktail formulation was not disclosed, otic phage therapy may be a suitable alternative to antibiotics. Moreover, amplification of phage at the site of infection indicates that the infectious agent was susceptible to at least one phage administered and the dose applied was amenable to a productive viral infection (Hawkins et al. 2010).

Inhalation of Phage

In vivo efficacy of phage therapy against *Burkholderia cepacia* (isolated from the sputum of patients with cystic fibrosis) respiratory tract infections was evaluated in a mouse model of acute lung infection (Carmody et al. 2010).

Twenty-four hours postintratracheal bacterial challenge, mice were treated by i.p. injection or intranasal inhalation with a single phage at an MOI of 100. No significant differences in lung bacterial load or inflammatory cytokine levels were found between untreated mice and rodents treated intranasally with phage. In contrast to inhalation, systemic administration of phages was moderately effective, suggesting that the systemic application of phages may facilitate accessibility to bacteria present in the lungs.

Debarbieux et al. (2010) reported multiple studies of successful intranasal phage treatment of *P. aeruginosa* pulmonary infection in a mouse model. In a first study, phage were administered either 24 h before or 2 h after bacterial challenge with a bioluminescent strain of *P. aeruginosa*. No significant difference in survival of untreated animals and mice treated with the lowest dose of phage (MOI of 0.1) was observed. A substantially higher dose of phage (MOI of 10) facilitated complete recovery of infected mice. Further analysis revealed that phage propagated within lungs, underscoring the importance of dosage and timing in phage therapy. Given that treatment was delivered shortly after bacterial infection, it is difficult to predict whether inhalation of phage would be successful for treating lower pulmonary tract or chronic infections. These data indicate that inhalation of phage may confer therapeutic and prophylactic protection against pulmonary infection.

Prevention of Biofilm Formation and Medical Device–Associated Infections

Indwelling medical devices are often used to restore function or regenerate tissue. However, the propensity to become colonized by bacteria and serve as abiotic support for biofilm formation leading to medical device–associated infections undermines their use in medical practice. Over the past couple of decades, therapeutic management of implant-associated infections has been further complicated by the emergence of antibiotic-resistant bacteria (Arias and Murray 2009; Boucher et al. 2013).

Mitigation of *Staphylococcus epidermidis* biofilm formation on hydrogel-coated catheters has been achieved by pretreatment with bacteriophage (Curtin and Donlan 2006). In similar studies, Fu et al. (2010) investigated the effect of pre-, post-, and recurring treatment of Foley catheters with a cocktail of phage on *P. aeruginosa* biofilm formation. Significant reduction of bacterial density was observed regardless of the timing of treatment (Fu et al. 2010). However, regrowth of biofilm occurred between 24 and 48 h. Importantly, sequential treatments had triggered mature biofilm dissolution. Another study reported prevention of biofilm formation on Foley catheters following impregnation of neutral hydrogel (Lubri-Sil, Bard, Covington, GA) loaded with lytic bacteriophages (Carson et al. 2010). The densities of *E. coli* and *Proteus mirabilis* on the silicone tubing were reduced by 90% after 24 h of exposure to the gel. These results highlight the potential use of phage cocktails for mitigation of biofilm formation by clinically relevant bacteria on the surfaces of indwelling medical devices. Moreover, wide host range phage engineered to express biofilm disruptive enzymes may become the next-generation coating materials for indwelling devices.

Anti-Infective Phage Therapy in Humans

Before the late 1930s, phage therapy was used worldwide to treat infections. In the United States, phage therapy was of interest mainly between the 1920s and the 1930s. However, the advent of antibiotics and a couple of controversial opinions regarding the preparation and efficacy of phage lysates led to decreased interest in phage therapy (Eaton and Bayne-Jones 1934). Nonetheless, seven phage products for human use were produced by Eli Lilly (Indianapolis, IN). Another product, Staph Phage Lysate, was developed by Delmont Laboratories (Swarthmore, PA) and licensed for human use in 1959 and veterinary use in 1987 for treatment of canine pyoderma. Phage therapy continued to be used in France until the 1990s when phage preparations were discontinued. Ever since then, phage therapists have continued their practice with products obtained from Poland and the United States, where phage therapy is included in the current health care standard (Abedon et al. 2011).

Many studies have been conducted to evaluate the efficacy and safety of phage therapy and much clinical experience has been accumulated, especially in Eastern Europe (reviewed in Alisky et al. 1998; Summers 2001; Sulakvelidze et al. 2001; Górski et al. 2009; Kutter et al. 2010; Abedon et al. 2011; Burrowes et al. 2011; Harper et al. 2011; Chanishvili 2012). However, these trials did not follow current Western rigorous standards. The first double-blind, randomized, placebo-controlled phase I trial to show the safety of phage treatment was performed by Nestle Research Center (Lausanne, Switzerland) (Bruttin and Brussow 2005). Investigators showed no significant side effects following administration of phage and also showed that oral administration of T4 did not disturb the natural gut *E. coli* population. In subsequent studies, the group performed a detailed metagenomics analysis of their entire T4-related antidiarrheagenic phage collection and assessed the clinical risk of a subset of phages following oral administration in healthy adults (Sarker et al. 2012).

The first reported randomized controlled trial evaluated both safety and efficacy of phage therapy against chronic otitis caused by antibiotic-resistant *P. aeruginosa* (Wright et al. 2009). No treatment-related adverse events were reported, and the results showed that topical administration of a therapeutic cocktail (Biophage-PA, Biocontrol, UK) in the ear results in an improvement of clinical manifestation of the infection. However, the patients were selected based on being infected with a *P. aeruginosa* strain susceptible to one or more of the phage included in the cocktail, which shows that, similar to antibiotics, a phage antibiogram is necessary before treatment.

The first FDA-approved phase I clinical trial of phage therapy was performed in 2007 at the Southwest Regional Wound Care Center in Lubbock, Texas to evaluate local administration of a small set of well-characterized phage in patients with chronic venous leg ulcers (Rhoads et al. 2009). The study showed that topical phage administration neither had any adverse effects nor affects wound healing. Although outside of the scope of the clinical trial, efficacy could not be established. Currently, several other clinical studies are being conducted, which highlights the interest of the scientific community in developing bacteriophage as antimicrobials (Kutter et al. 2014). However, the development of phage for contemporary clinical use will likely rely on the synthesis of phage with built-in characteristics (such as wide host range, biofilm-degrading enzymes, and secondary antimicrobial payloads) that can be evaluated in rigorous clinical trials (Fig. 4).

PHAGE IN THE TREATMENT OF CANCER AND DEGENERATIVE DISEASES

Natural phages have been long considered poor gene delivery vehicles because they only infect bacteria (Ivanenkov et al. 1999). Recognizing the potential of phages for gene therapy, Hajitou et al. (2006) engineered a new generation of hybrid prokaryotic–eukaryotic nanovectors as a chimera between eukaryotic adeno-associated virus (AAV) and the filamentous M13 bacteriophage, referred to as adeno-associated virus/phage (AAVP). This phage particle expresses the cyclic peptide RGD4C (CDCRGDCFC) ligand on the phage pIII minor coat protein, allowing systemic and specific targeting to the $\alpha v\beta 3$-integrin receptor, which is present primarily on tumor vasculature and tumor cells, and is expressed at barely detectable levels in normal endothelium and tissues (Folkman 1997). The hybrid nanovector genome was also modified by inserting an engineered AAV (recombinant adeno-associated virus [rAAV]) transgene cassette into an intergenomic region of the phage genome, under the regulation of the cytomegalovirus (CMV) promoter and flanked by full-length inverted terminal repeats (ITRs) from AAV serotype 2. The ITRs improve transduction efficiency and enhance transgene expression by maintaining and forming concatemers of the eukaryotic transgene cassette (Hajitou et al. 2006; Trepel et al. 2009). Further, this vector has been engineered to carry the herpes simplex thymidine kinase gene cassette suitable for tumor treatment with ganciclovir and positron

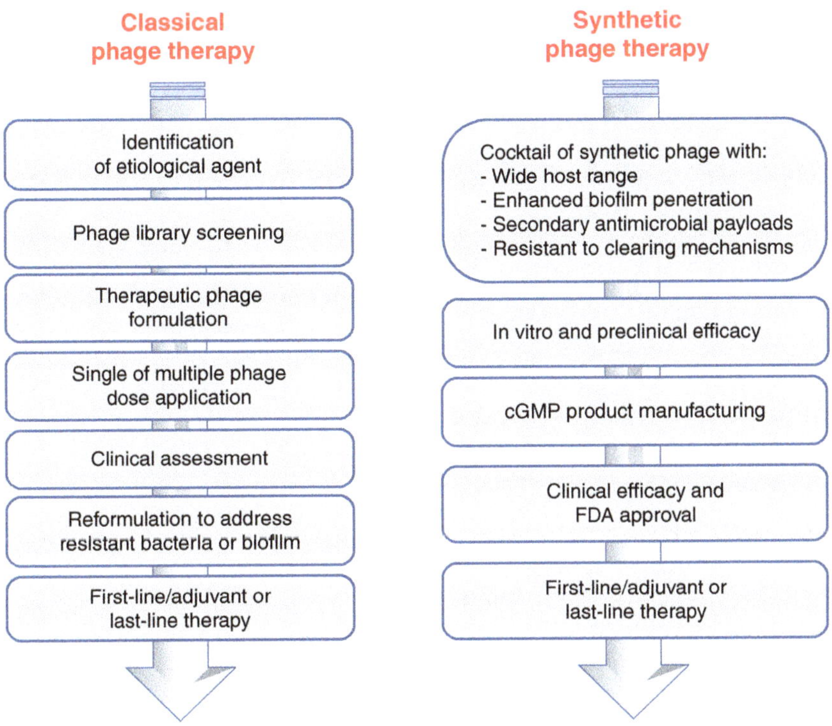

Figure 4. Phage therapy approaches. Classical (or Eastern European approach) relies on extensive banks of phages that can be used for treatment on a case-by-case basis by a physician specializing in phage therapy and adjusted based on clinical assessment. The synthetic phage therapy approach aims to develop phage with drug-like properties that can be used, much like antibiotics, in standardized Western treatment approaches. cGMP, Current good manufacturing practice.

emission tomography imaging for theranostic approaches (Soghomonyan et al. 2007; Hajitou et al. 2007, 2008). The RGD4C-AAVP has been used to deliver tumor necrosis factor α (TNF-α) to the angiogenic vasculature of human melanoma xenografts in nude mice (Tandle et al. 2009). Following systemic administration of phage, the TNF-α expression was shown to be specifically localized in tumors, leading to apoptosis in tumor blood vessels and significant inhibition of tumor growth, while remaining virtually undetectable in all other tissues, notably the liver and spleen. The efficacy of targeted RGD4C-AAVP expressing the TNF-α has been also shown in dogs with soft tissue sarcoma (Paoloni et al. 2009). This vector was modified even further to replace the CMV promoter with the tumor-specific promoter Grp78, which drives expression only in the targeted tumor vasculature (Kia et al. 2012).

Recent studies have shown that the linear structure of M13 bacteriophage permits binding to β-amyloid and α-synuclein proteins, leading to plaque disaggregation in models of Alzheimer's and Parkinson's diseases. Convection-enhanced delivery of M13 phage to the brain of nonhuman primates confirmed distribution into both white and gray matter, which makes filamentous phage very attractive theranostics for direct plaque dissolution, as well as delivery of therapeutic and imaging agents into the brain (Frenkel and Solomon 2002; Ksendzovsky et al. 2012).

PHAGE-BASED VACCINOLOGY

Recombinant DNA technology allows synthesis of vaccine candidates that are subunits of pathogens. However, such vaccines have limited immunogenic features and require adjuvants or effective delivery systems for proper activation of the immune system (Petrovsky and Aguilar 2004; Barrett and Stanberry 2008). Thus, phages are now being evaluated as vaccine delivery agents because of their inherited ability to stimulate humoral and cell-mediated immunity (Burdin et al. 2004). Two strategies are often combined to produce phage vaccines: (1) phage display, when virions are decorated with peptides selected for their ability to bind antigen-presenting cells; and (2) phage DNA vaccines, when viral DNA is engineered to carry a foreign antigen gene under the control of a strong eukaryotic promoter and has the ability to deliver this element to mammalian cells (Clark and March 2004a; Zanghi et al. 2007).

Filamentous phage M13 has been the first virus manipulated to express a melanoma-specific tumor antigen fragment and has been successfully used to raise an immune response capable of reducing tumor growth in animal models (Benhar 2001; Fang et al. 2005). Currently, several vaccines for infectious diseases are prepared by using the T4 phage display system, which has shown promising results in animal models (Jiang et al. 1997). Similarly, phage T7 has been engineered to display vascular endothelial growth factor (VEGF) and has been successfully used to break immunologic tolerance and produce a strong immune response against Lewis lung cell carcinoma (Li et al. 2006).

Alternatively, phage can be exploited to transfer genes into mammalian cells. In these vectors, the antigens, under the control of a eukaryotic promoter, are cloned inside of a nonessential region of a phage genome. When injected in a mammalian system, these phage particles, acting as a DNA vaccine, can induce potent immune response by expressing foreign antigen inside of anaphase-promoting complexes (APCs) or other cells (Clark and March 2006). Several λ-based DNA vaccines for infectious diseases have been prepared that have shown promising results in animal models (Clark and March 2004b; March et al. 2004). It should be noted that these viruses are manipulated to penetrate mammalian cells and, in the absence of engineering, phages are only capable of infecting bacteria.

CONCLUDING REMARKS

The global escalation of bacterial resistance to antibiotics raises the possibility of returning to the clinical equivalent of the preantibiotic era and requires the development of new classes of antimicrobials that could strengthen the effectiveness of current use drugs. Phage therapy is a promising, yet challenging, strategy for the development of new antimicrobial therapies that can enable more strategic treatment approaches that do not affect beneficial normal microbiota. Another major benefit of phage therapy is that the mode of action of bacteriophage is distinct from those of antibiotics; thus, bacterial strains that are resistant to antibiotics are nevertheless susceptible to phage infection. However, limitations such as narrow host range, bacterial resistance to phage, cost of manufacturing, and challenging dose finding/delivery methods need to be addressed. Recent advances in synthetic genome assembly and viral genome engineering can be used to create phage with superior therapeutic properties. Designer synthetic phage may enable new intellectual property and attract interest from large pharmaceutical companies in these products. Moreover, these techniques can also be applied to engineering novel theranostic nanovectors with improved tissue penetration and payload carrying potential for treatment of cancer and other diseases.

REFERENCES

Abedon S. 2011. Phage therapy pharmacology: Calculating phage dosing. *Adv Appl Microbiol* **77:** 1–40.

Alisky J, Iczkowski K, Rapoport A, Troitsky N. 1998. Bacteriophages show promise as antimicrobial agents. *J Infect* **36:** 5–15.

Arias CA, Murray BE. 2009. Antibiotic-resistant bugs in the 21st century—A clinical super-challenge. *N Engl J Med* **360:** 439–443.

Barrett ADT, Stanberry LR. 2008. *Vaccines for biodefence and emerging and neglected diseases.* Academic, Waltham, MA.

Barrow P, Lovell M, Berchieri A Jr. 1998. Use of lytic bacteriophage for control of experimental *Escherichia coli* septicemia and meningitis in chickens and calves. *Clin Diag Lab Immun* **5:** 294–298.

Baumann A. 2006. Early development of therapeutic biologics—Pharmacokinetics. *Curr Drug Metab* **7:** 15–21.

Benhar I. 2001. Biotechnological applications of phage and cell display. *Biotech Adv* **19:** 1–33.

Biswas B, Adhya S, Washart P, Paul B, Trostel AN, Powell B, Carlton R, Merril CR. 2002. Bacteriophage therapy rescues mice bacteremic from a clinical isolate of vancomycin-resistant *Enterococcus faecium*. *Infect Immun* **70:** 204–210.

Boucher HW, Talbot GH, Benjamin DK Jr, Bradley J, Guidos RJ, Jones RN, Murray BE, Bonomo RA, Gilbert D, Infectious Diseases Society of America. 2013. 10 × '20 Progress—Development of new drugs active against Gram-negative bacilli: An update from the Infectious Diseases Society of America. *Clin Infect Dis* **56:** 1685–1694.

Boyd EF, Brussow H. 2002. Common themes among bacteriophage-encoded virulence factors and diversity among the bacteriophages involved. *Trends Microbiol* **10:** 521–529.

Bruttin A, Brussow H. 2005. Human volunteers receiving *Escherichia coli* phage T4 orally: A safety test of phage therapy. *Antimicrob Agents Chemother* **49:** 2874–2878.

Burdin N, Guy B, Moingeon P. 2004. Immunological foundations to the quest for new vaccine adjuvants. *BioDrugs* **18:** 79–93.

Burrowes B, Harper DR, Anderson J, McConville M, Enright MC. 2011. Bacteriophage therapy: Potential uses in the control of antibiotic-resistant pathogens. *Expert Rev Anti Infect Ther* **9:** 775–785.

Carmody LA, Gill JJ, Summer EJ, Sajjan US, Gonzalez CF, Young RF, LiPuma JJ. 2010. Efficacy of bacteriophage therapy in a model of *Burkholderia cenocepacia* pulmonary infection. *J Infect Dis* **201:** 264–271.

Carson L, Gorman SP, Gilmore BF. 2010. The use of lytic bacteriophages in the prevention and eradication of biofilms of *Proteus mirabilis* and *Escherichia coli*. *FEMS Immunol Med Microbiol* **59:** 447–455.

Carter CD, Parks A, Abuladze T, Li M, Woolston J, Magnone J, Senecal A, Kropinski AM, Sulakvelidze A. 2012. Bacteriophage cocktail significantly reduces *Escherichia coli* O157: H7 contamination of lettuce and beef, but does not protect against recontamination. *Bacteriophage* **2:** 178–185.

Casadevall A, Pirofski LA. 2000. Host–pathogen interactions: Basic concepts of microbial commensalism, colonization, infection, and disease. *Infect Immun* **68:** 6511–6518.

Chanishvili N. 2012. Phage therapy—History from Twort and d'Herelle through Soviet experience to current approaches. *Adv Vir Res* **83:** 3–40.

Chibani-Chennoufi S, Sidoti J, Bruttin A, Kutter E, Sarker S, Brussow H. 2004. In vitro and in vivo bacteriolytic activities of *Escherichia coli* phages: Implications for phage therapy. *Antimicrob Agents Chemother* **48:** 2558–2569.

Chiocca EA. 2002. Oncolytic viruses. *Nat Rev Cancer* **2:** 938–950.

Clark JR, March JB. 2004a. Bacterial viruses as human vaccines? *Exp Rev Vaccines* **3:** 463–476.

Clark JR, March JB. 2004b. Bacteriophage-mediated nucleic acid immunization. *FEMS Immunol Med Microbiol* **40:** 21–26.

Clark JR, March JB. 2006. Bacteriophages and biotechnology: Vaccines, gene therapy and antibacterials. *Trends Biotechnol* **24:** 212–218.

Curtin JJ, Donlan RM. 2006. Using bacteriophages to reduce formation of catheter-associated biofilms by *Staphylococcus epidermidis*. *Antimicrob Agents Chemother* **50:** 1268–1275.

Debarbieux L, Leduc D, Maura D, Morello E, Criscuolo A, Grossi O, Balloy V, Touqui L. 2010. Bacteriophages can treat and prevent *Pseudomonas aeruginosa* lung infections. *J Infect Dis* **201:** 1096–1104.

DeLappe N, O'Halloran F, Fanning S, Corbett-Feeney G, Cheasty T, Cormican M. 2003. Antimicrobial resistance and genetic diversity of *Shigella sonnei* isolates from western Ireland, an area of low incidence of infection. *J Clin Microbiol* **41:** 1919–1924.

d'Herelle F. 1931. An address on bacteriophagy and recovery from infectious diseases. *Can Med Assoc J* **24:** 619–628.

Eaton MD, Bayne-Jones S. 1934. Bacteriophage therapy: Review of the principles and results of the use of bacteriophage in the treatment of infections. *JAMA* **103:** 1769–1776.

Enyeart PJ, Chirieleison SM, Dao MN, Perutka J, Quandt EM, Yao J, Whitt JT, Keatinge-Clay AT, Lambowitz AM, Ellington AD. 2013. Generalized bacterial genome editing using mobile group II introns and Cre-lox. *Mol Syst Biol* **9:** 685.

Fairhead H. 2009. SASP gene delivery: A novel antibacterial approach. *Drug News Perspect* **22:** 197–203.

Fang J, Wang G, Yang Q, Song J, Wang Y, Wang L. 2005. The potential of phage display virions expressing malignant tumor specific antigen MAGE-A1 epitope in murine model. *Vaccine* **23:** 4860–4866.

Folkman J. 1997. Addressing tumor blood vessels. *Nat Biotech* **15:** 510.

Frenkel D, Solomon B. 2002. Filamentous phage as vector-mediated antibody delivery to the brain. *Proc Natl Acad Sci* **99:** 5675–5679.

Fu W, Forster T, Mayer O, Curtin JJ, Lehman SM, Donlan RM. 2010. Bacteriophage cocktail for the prevention of biofilm formation by *Pseudomonas aeruginosa* on catheters in an in vitro model system. *Antimicrob Agents Chemother* **54:** 397–404.

Fukuda K, Ishida W, Uchiyama J, Rashel M, Kato S, Morita T, Muraoka A, Sumi T, Matsuzaki S, Daibata M, et al. 2012. *Pseudomonas aeruginosa* keratitis in mice: Effects of topical bacteriophage KPP12 administration. *PloS ONE* **7:** e47742.

Furukawa H, Kuroiwa T, Mizushima S. 1983. DNA injection during bacteriophage T4 infection of *Escherichia coli*. *J Bacteriol* **154:** 938–945.

Geier MR, Trigg ME, Merril CR. 1973. Fate of bacteriophage λ in non-immune germ-free mice. *Nature* **246:** 221–223.

Gibson DG, Young L, Chuang RY, Venter JC, Hutchison CA III, Smith HO. 2009. Enzymatic assembly of DNA molecules up to several hundred kilobases. *Nat Methods* **6:** 343–345.

Goode D, Allen VM, Barrow PA. 2003. Reduction of experimental *Salmonella* and *Campylobacter* contamination of chicken skin by application of lytic bacteriophages. *Appl Env Microbiol* **69:** 5032–5036.

Górski A, Miedzybrodzki R, Borysowski J, Weber-Dabrowska B, Lobocka M, Fortuna W, Letkiewicz S, Zimecki M,

Filby G. 2009. Bacteriophage therapy for the treatment of infections. *Curr Opin Investig Drugs* **10:** 766–774.

Górski A, Miedzybrodzki R, Borysowski J, Dabrowska K, Wierzbicki P, Ohams M, Korczak-Kowalska G, Olszowska-Zaremba N, Lusiak-Szelachowska M, Klak M, et al. 2012. Phage as a modulator of immune responses: Practical implications for phage therapy. *Adv Virus Res* **83:** 41–71.

Gu J, Liu X, Li Y, Han W, Lei L, Yang Y, Zhao H, Gao Y, Song J, Lu R, et al. 2012. A method for generation phage cocktail with great therapeutic potential. *PloS ONE* **7:** e31698.

Gunther EJ, Murray NE, Glazer PM. 1993. High efficiency, restriction-deficient in vitro packaging extracts for bacteriophage λ DNA using a new *E. coli* lysogen. *Nucleic Acids Res* **21:** 3903–3904.

Hagens S, Habel A, von Ahsen U, von Gabain A, Blasi U. 2004. Therapy of experimental pseudomonas infections with a nonreplicating genetically modified phage. *Antimicrob Agents Chemother* **48:** 3817–3822.

Hajitou A, Trepel M, Lilley CE, Soghomonyan S, Alauddin MM, Marini FC 3rd, Restel BH, Ozawa MG, Moya CA, Rangel R, et al. 2006. A hybrid vector for ligand-directed tumor targeting and molecular imaging. *Cell* **125:** 385–398.

Hajitou A, Rangel R, Trepel M, Soghomonyan S, Gelovani JG, Alauddin MM, Pasqualini R, Arap W. 2007. Design and construction of targeted AAVP vectors for mammalian cell transduction. *Nat Protocols* **2:** 523–531.

Hajitou A, Lev DC, Hannay JA, Korchin B, Staquicini FI, Soghomonyan S, Alauddin MM, Benjamin RS, Pollock RE, Gelovani JG, et al. 2008. A preclinical model for predicting drug response in soft-tissue sarcoma with targeted AAVP molecular imaging. *Proc Natl Acad Sci* **105:** 4471–4476.

Hall AR, De Vos D, Friman VP, Pirnay JP, Buckling A. 2012. Effects of sequential and simultaneous applications of bacteriophages on populations of *Pseudomonas aeruginosa* in vitro and in wax moth larvae. *Appl Env Microbiol* **78:** 5646–5652.

Hanlon GW. 2007. Bacteriophages: An appraisal of their role in the treatment of bacterial infections. *Int J Antimicrob Agents* **30:** 118–128.

Harper DR, Anderson J, Enright MC. 2011. Phage therapy: Delivering on the promise. *Ther Deliv* **2:** 935–947.

Hatfull GF. 2008. Bacteriophage genomics. *Curr Opin Microbiol* **11:** 447–453.

Hawkins C, Harper D, Burch D, Anggard E, Soothill J. 2010. Topical treatment of *Pseudomonas aeruginosa* otitis of dogs with a bacteriophage mixture: A before/after clinical trial. *Vet Microbiol* **146:** 309–313.

Hendrix RW. 2003. Bacteriophage genomics. *Curr Opin Microbiol* **6:** 506–511.

Ivanenkov V, Felici F, Menon AG. 1999. Uptake and intracellular fate of phage display vectors in mammalian cells. *Biochem Biophys Acta* **1448:** 450–462.

Jaiswal A, Koley H, Ghosh A, Palit A, Sarkar B. 2013. Efficacy of cocktail phage therapy in treating *Vibrio cholerae* infection in rabbit model. *Microbes Infect* **15:** 152–156.

Jaschke PR, Lieberman EK, Rodriguez J, Sierra A, Endy D. 2012. A fully decompressed synthetic bacteriophage ϕX174 genome assembled and archived in yeast. *Virology* **434:** 278–284.

Jiang J, Abu-Shilbayeh L, Rao VB. 1997. Display of a PorA peptide from *Neisseria meningitidis* on the bacteriophage T4 capsid surface. *Infect Immun* **65:** 4770–4777.

Keen EC. 2012. Phage therapy: Concept to cure. *Frontiers Microbiol* **3:** 238.

Kerr C, Sadowski PD. 1974. Packaging and maturation of DNA of bacteriophage T7 in vitro. *Proc Natl Acad Sci* **71:** 3545–3549.

Kia A, Przystal JM, Nianiaris N, Mazarakis ND, Mintz PJ, Hajitou A. 2012. Dual systemic tumor targeting with ligand-directed phage and Grp78 promoter induces tumor regression. *Mol Cancer Ther* **11:** 2566–2577.

Kingham R, Klasa G, Carver KH. 2014. Key regulatory guidelines for the development of biologics in the United States and Europe. In *Biological drug products: Development and strategies*, 1st ed. (ed. Wang W and Singh M), pp. 75–109. Wiley, New York.

Kiro R, Shitrit D, Qimron U. 2014. Efficient engineering of a bacteriophage genome using the type I-E CRISPR-Cas system. *RNA Biol* **11:** 42–44.

Ksendzovsky A, Walbridge S, Saunders RC, Asthagiri AR, Heiss JD, Lonser RR. 2012. Convection-enhanced delivery of M13 bacteriophage to the brain. *J Neurosurgery* **117:** 197–203.

Kumari S, Harjai K, Chhibber S. 2009. Efficacy of bacteriophage treatment in murine burn wound infection induced by *Klebsiella pneumoniae. J Microbiol Biotechnol* **19:** 622–628.

Kumari S, Harjai K, Chhibber S. 2010. Topical treatment of *Klebsiella pneumoniae* B5055 induced burn wound infection in mice using natural products. *J Infect Dev Ctries* **4:** 367–377.

Kutter E, De Vos D, Gvasalia G, Alavidze Z, Gogokhia L, Kuhl S, Abedon ST. 2010. Phage therapy in clinical practice: Treatment of human infections. *Curr Pharmacol Biotechnol* **11:** 69–86.

Kutter E, Borysowski J, Międzybrodzki R, Górski A, Weber-Dąbrowska B, Kutateladze M, Alavidze Z, Goderdzishvili M, Adamia R. 2014. Clinical phage therapy. In *Phage therapy: Current research and applications* (ed. Borysowski J, et al.), pp. 257–288. Caister Academic, Norfolk, UK.

Labrie SJ, Samson JE, Moineau S. 2010. Bacteriophage resistance mechanisms. *Nat Rev Microbiol* **8:** 317–327.

Li XH, Tang L, Liu D, Sun HM, Zhou CC, Tan LS, Wang LP, Zhang PD, Zhang SQ. 2006. Antitumor effect of recombinant T7 phage vaccine expressing xenogenic vascular endothelial growth factor on Lewis lung cancer in mice. *Chinese J Cancer* **25:** 1221–1226.

Lu TK, Collins JJ. 2007. Dispersing biofilms with engineered enzymatic bacteriophage. *Proc Natl Acad Sci* **104:** 11197–11202.

Lu TK, Koeris MS. 2011. The next generation of bacteriophage therapy. *Curr Opin Microbiol* **14:** 524–531.

March JB, Clark JR, Jepson CD. 2004. Genetic immunisation against hepatitis B using whole bacteriophage λ particles. *Vaccine* **22:** 1666–1671.

Marinelli LJ, Piuri M, Swigonova Z, Balachandran A, Oldfield LM, van Kessel JC, Hatfull GF. 2008. BRED: A simple and powerful tool for constructing mutant and recombinant bacteriophage genomes. *PloS ONE* **3:** e3957.

Martel B, Moineau S. 2014. CRISPR-Cas: An efficient tool for genome engineering of virulent bacteriophages. *Nucleic Acids Res* **42:** 9504–9513.

Matsuda T, Freeman TA, Hilbert DW, Duff M, Fuortes M, Stapleton PP, Daly JM. 2005. Lysis-deficient bacteriophage therapy decreases endotoxin and inflammatory mediator release and improves survival in a murine peritonitis model. *Surgery* **137:** 639–646.

Matsuzaki S, Yasuda M, Nishikawa H, Kuroda M, Ujihara T, Shuin T, Shen Y, Jin Z, Fujimoto S, Nasimuzzaman MD, et al. 2003. Experimental protection of mice against lethal *Staphylococcus aureus* infection by novel bacteriophage phi MR11. *J Infect Dis* **187:** 613–624.

McVay CS, Velasquez M, Fralick JA. 2007. Phage therapy of *Pseudomonas aeruginosa* infection in a mouse burn wound model. *Antimicrob Agents Chemother* **51:** 1934–1938.

Merril CR, Biswas B, Carlton R, Jensen NC, Creed GJ, Zullo S, Adhya S. 1996. Long-circulating bacteriophage as antibacterial agents. *Proc Natl Acad Sci* **93:** 3188–3192.

Merril CR, Scholl D, Adhya SL. 2003. The prospect for bacteriophage therapy in Western medicine. *Nat Rev Drug Discov* **2:** 489–497.

Merril CR, Scholl D, Adhya S. 2006. Phage therapy. In *The bacteriophages*, 2nd ed. (ed. Calendar RL, Abedon S), pp. 725–741. Oxford University Press, London.

Neu HC. 1992. The crisis in antibiotic resistance. *Science* **257:** 1064–1073.

Nobrega FL, Costa AR, Kluskens LD, Azeredo J. 2015. Revisiting phage therapy: New applications for old resources. *Trends Microbiol* **23:** 185–191.

Ochs HD, Davis SD, Wedgwood RJ. 1971. Immunologic responses to bacteriophage ϕX 174 in immunodeficiency diseases. *J Clin Invest* **50:** 2559–2568.

O'Flaherty S, Ross RP, Meaney W, Fitzgerald GF, Elbreki MF, Coffey A. 2005. Potential of the polyvalent anti-*Staphylococcus* bacteriophage K for control of antibiotic-resistant staphylococci from hospitals. *Appl Env Mocrobiol* **71:** 1836–1842.

Ojala V, Laitalainen J, Jalasvuori M. 2013. Fight evolution with evolution: Plasmid-dependent phages with a wide host range prevent the spread of antibiotic resistance. *Evol Appl* **6:** 925–932.

Oliveira A, Sereno R, Azeredo J. 2010. In vivo efficiency evaluation of a phage cocktail in controlling severe colibacillosis in confined conditions and experimental poultry houses. *Vet Microbiol* **146:** 303–308.

Palmer KL, Gilmore MS. 2010. Multidrug-resistant enterococci lack CRISPR-*cas*. *mBio* **1:** e00227.

Paoloni MC, Tandle A, Mazcko C, Hanna E, Kachala S, Leblanc A, Newman S, Vail D, Henry C, Thamm D, et al. 2009. Launching a novel preclinical infrastructure: Comparative oncology trials consortium directed therapeutic targeting of TNF α to cancer vasculature. *PloS ONE* **4:** e4972.

Petrovsky N, Aguilar JC. 2004. Vaccine adjuvants: Current state and future trends. *Immunol Cell Biol* **82:** 488–496.

Ptashne M. 2004. *A genetic switch: Phage λ revisited.* Cold Spring Harbor Laboratory Press, Cold Spring Harbor, NY.

Rangel R, Guzman-Rojas L, le Roux LG, Staquicini FI, Hosoya H, Barbu EM, Ozawa MG, Nie KD Jr, Langley RR, et al. 2012. Combinatorial targeting and discovery of ligand-receptors in organelles of mammalian cells. *Nat Commun* **3:** 788.

Rhoads DD, Wolcott RD, Kuskowski MA, Wolcott BM, Ward LS, Sulakvelidze A. 2009. Bacteriophage therapy of venous leg ulcers in humans: Results of a phase I safety trial. *J Wound Care* **18:** 237–238, 240–233.

Rozema EA, Stephens TP, Bach SJ, Okine EK, Johnson RP, Stanford K, McAllister TA. 2009. Oral and rectal administration of bacteriophages for control of *Escherichia coli* O157:H7 in feedlot cattle. *J Food Prot* **72:** 241–250.

Samson JE, Magadan AH, Sabri M, Moineau S. 2013. Revenge of the phages: Defeating bacterial defences. *Nat Rev Microbiol* **11:** 675–687.

Sarker SA, McCallin S, Barretto C, Berger B, Pittet AC, Sultana S, Krause L, Huq S, Bibiloni R, Bruttin A, et al. 2012. Oral T4-like phage cocktail application to healthy adult volunteers from Bangladesh. *Virology* **434:** 222–232.

Shin J, Jardine P, Noireaux V. 2012. Genome replication, synthesis, and assembly of the bacteriophage T7 in a single cell-free reaction. *ACS Synth Biol* **1:** 408–413.

Smith HO, Hutchison CA 3rd, Pfannkoch C, Venter JC. 2003. Generating a synthetic genome by whole genome assembly: phiX174 bacteriophage from synthetic oligonucleotides. *Proc Natl Acad Sci* **100:** 15440–15445.

Soghomonyan S, Hajitou A, Rangel R, Trepel M, Pasqualini R, Arap W, Gelovani JG, Alauddin MM. 2007. Molecular PET imaging of *HSV1-tk* reporter gene expression using [^{18}F]FEAU. *Nat Protoc* **2:** 416–423.

Soni KA, Nannapaneni R. 2010. Bacteriophage significantly reduces *Listeria monocytogenes* on raw salmon fillet tissue. *J Food Prot* **73:** 32–38.

Srinivasan S, Alexander JF, Driessen WH, Leonard F, Ye H, Liu X, Arap W, Pasqualini R, Ferrari M, Godin B. 2013. Bacteriophage associated silicon particles: Design and characterization of a novel theranostic vector with improved payload carrying potential. *J Mater Chem B Mater Biol Med* **1:** 5218–5229.

Stanford K, McAllister TA, Niu YD, Stephens TP, Mazzocco A, Waddell TE, Johnson RP. 2010. Oral delivery systems for encapsulated bacteriophages targeted at *Escherichia coli* O157:H7 in feedlot cattle. *J Food Prot* **73:** 1304–1312.

Staquicini FI, Ozawa MG, Moya CA, Driessen WH, Barbu EM, Nishimori H, Soghomonyan S, Flores LG II, Liang X, Paolillo V, et al. 2011. Systemic combinatorial peptide selection yields a non-canonical iron-mimicry mechanism for targeting tumors in a mouse model of human glioblastoma. *J Clin Invest* **121:** 161–173.

Sulakvelidze A, Alavidze Z, Morris JG Jr. 2001. Bacteriophage therapy. *Antimicrob Agents Chemother* **45:** 649–659.

Summers WC. 2001. Bacteriophage therapy. *Ann Rev Microbiol* **55:** 437–451.

Tandle A, Hanna E, Lorang D, Hajitou A, Moya CA, Pasqualini R, Arap W, Adem A, Starker E, Hewitt S, et al. 2009. Tumor vasculature-targeted delivery of tumor necrosis factor-α. *Cancer* **115:** 128–139.

Tanji Y, Shimada T, Fukudomi H, Miyanaga K, Nakai Y, Unno H. 2005. Therapeutic use of phage cocktail for controlling *Escherichia coli* O157:H7 in gastrointestinal tract of mice. *J Biosci Bioeng* **100:** 280–287.

Tetart F, Repoila F, Monod C, Krisch HM. 1996. Bacteriophage T4 host range is expanded by duplications of a small domain of the tail fiber adhesin. *J Mol Biol* **258:** 726–731.

Tetart F, Desplats C, Krisch HM. 1998. Genome plasticity in the distal tail fiber locus of the T-even bacteriophage: Recombination between conserved motifs swaps adhesin specificity. *J Mol Biol* **282:** 543–556.

Tinsley CR, Bille E, Nassif X. 2006. Bacteriophages and pathogenicity: More than just providing a toxin? *Microbes Infect* **8:** 1365–1371.

Trepel M, Stoneham CA, Eleftherohorinou H, Mazarakis ND, Pasqualini R, Arap W, Hajitou A. 2009. A heterotypic bystander effect for tumor cell killing after adeno-associated virus/phage-mediated, vascular-targeted suicide gene transfer. *Mol Cancer Ther* **8:** 2383–2391.

Twort FW. 1936. Further investigations on the nature of ultra-microscopic viruses and their cultivation. *J Hygiene* **36:** 204–235.

Vitiello CL, Merril CR, Adhya S. 2005. An amino acid substitution in a capsid protein enhances phage survival in mouse circulatory system more than a 1000-fold. *Virus Res* **114:** 101–103.

Wagenaar JA, Van Bergen MA, Mueller MA, Wassenaar TM, Carlton RM. 2005. Phage therapy reduces *Campylobacter jejuni* colonization in broilers. *Vet Microbiol* **109:** 275–283.

Wedgwood RJ, Ochs HD, Davis SD. 1975. The recognition and classification of immunodeficiency diseases with bacteriophage ϕX 174. *Birth Defects Orig Artic Ser* **11:** 331–338.

Woolston J, Parks AR, Abuladze T, Anderson B, Li M, Carter C, Hanna LF, Heyse S, Charbonneau D, Sulakvelidze A. 2013. Bacteriophages lytic for rapidly reduce contamination on glass and stainless steel surfaces. *Bacteriophage* **3:** e25697.

Wright A, Hawkins CH, Anggard EE, Harper DR. 2009. A controlled clinical trial of a therapeutic bacteriophage preparation in chronic otitis due to antibiotic-resistant *Pseudomonas aeruginosa*; a preliminary report of efficacy. *Clin Otolaryngol* **34:** 349–357.

Zanghi CN, Sapinoro R, Bradel-Tretheway B, Dewhurst S. 2007. A tractable method for simultaneous modifications to the head and tail of bacteriophage λ and its application to enhancing phage-mediated gene delivery. *Nucleic Acids Res* **35:** e59.

Index

N

O

P

www.ingramcontent.com/pod-product-compliance
Lightning Source LLC
Chambersburg PA
CBHW042013200926
51615CB00029B/6
* 9 7 8 1 6 2 1 8 2 1 1 8 2 *